Advances in Intelligent and Soft Computing 91

Editor-in-Chief: J. Kacprzyk

Advances in Intelligent and Soft Computing

Further volumes of this series can be found on our homepage: springer.com

Vol. 79. A.P. de Leon F. de Carvalho, S. Rodríguez-González, J.F. De Paz Santana, and J.M. Corchado Rodríguez (Eds.)
Distributed Computing and Artificial Intelligence, 2010
ISBN 978-3-642-14882-8

Vol. 80. N.T. Nguyen, A. Zgrzywa, and A. Czyzewski (Eds.)
Advances in Multimedia and Network Information System Technologies, 2010
ISBN 978-3-642-14988-7

Vol. 81. J. Düh, H. Hufnagl, E. Juritsch, R. Pfliegl, H.-K. Schimany, and Hans Schönegger (Eds.)
Data and Mobility, 2010
ISBN 978-3-642-15502-4

Vol. 82. B.-Y. Cao, G.-J. Wang, S.-L. Chen, and S.-Z. Guo (Eds.)
Quantitative Logic and Soft Computing 2010
ISBN 978-3-642-15659-5

Vol. 83. J. Angeles, B. Boulet, J.J. Clark, J. Kovecses, and K. Siddiqi (Eds.)
Brain, Body and Machine, 2010
ISBN 978-3-642-16258-9

Vol. 84. Ryszard S. Choraś (Ed.)
Image Processing and Communications Challenges 2, 2010
ISBN 978-3-642-16294-7

Vol. 85. Á. Herrero, E. Corchado, C. Redondo, and Á. Alonso (Eds.)
Computational Intelligence in Security for Information Systems 2010
ISBN 978-3-642-16625-9

Vol. 86. E. Mugellini, P.S. Szczepaniak, M.C. Pettenati, and M. Sokhn (Eds.)
Advances in Intelligent Web Mastering – 3, 2011
ISBN 978-3-642-18028-6

Vol. 87. E. Corchado, V. Snášel, J. Sedano, A.E. Hassanien, J.L. Calvo, and D. Ślęzak (Eds.)
Soft Computing Models in Industrial and Environmental Applications, 6th International Workshop SOCO 2011
ISBN 978-3-642-19643-0

Vol. 88. Y. Demazeau, M. Pěchouček, J.M. Corchado, and J.B. Pérez (Eds.)
Advances on Practical Applications of Agents and Multiagent Systems, 2011
ISBN 978-3-642-19874-8

Vol. 89. J.B. Pérez, J.M. Corchado, M.N. Moreno, V. Julián, P. Mathieu, J. Canada-Bago, A. Ortega, and A.F. Caballero (Eds.)
Highlights in Practical Applications of Agents and Multiagent Systems, 2011
ISBN 978-3-642-19916-5

Vol. 90. J.M. Corchado, J.B. Pérez, K. Hallenborg, P. Golinska, and R. Corchuelo (Eds.)
Trends in Practical Applications of Agents and Multiagent Systems, 2011
ISBN 978-3-642-19930-1

Vol. 91. A. Abraham, J.M. Corchado, S.R. González, J.F. De Paz Santana (Eds.)
International Symposium on Distributed Computing and Artificial Intelligence, 2011
ISBN 978-3-642-19933-2

Ajith Abraham, Juan M. Corchado,
Sara Rodríguez González, and
Juan F. De Paz Santana (Eds.)

International Symposium on Distributed Computing and Artificial Intelligence

Springer

Editors

Prof. Ajith Abraham
Machine Intelligence Research Labs
(MIR Labs)
Scientific Network for Innovation and
Research Excellence (SNIRE)
P.O. Box 2259
Auburn, WA 98071-2259
USA

Prof. Juan M. Corchado Rodríguez
University of Salamanca
Department of Computing Science and
Control, Faculty of Science
Plaza de la Merced S/N
37008 Salamanca
Spain
E-mail: corchado@usal.es

Prof. Sara Rodríguez González
University of Salamanca
Department of Computing Science
Faculty of Science
Plaza de la Merced S/N
37008 Salamanca
Spain

Prof. Juan F. De Paz Santana
University of Salamanca
Department of Computing Science
Faculty of Science
Plaza de la Merced S/N
37008 Salamanca
Spain

ISBN 978-3-642-19933-2   e-ISBN 978-3-642-19934-9

DOI 10.1007/978-3-642-19934-9

Advances in Intelligent and Soft Computing   ISSN 1867-5662

Library of Congress Control Number: 2011923214

Typeset & Cover Design: Scientific Publishing Services Pvt. Ltd., Chennai, India.

Printed on acid-free paper
5 4 3 2 1 0
springer.com

Preface

The International Symposium on Distributed Computing and Artificial Intelligence 2011 (DCAI 2011) is a stimulating and productive forum where the scientific community can work towards future cooperation on Distributed Computing and Artificial Intelligence areas. This conference is the forum to present application of innovative techniques to complex problems. Artificial intelligence is changing our society. Its application in distributed environments, such as internet, electronic commerce, environment monitoring, mobile communications, wireless devices, distributed computing, to cite some, is continuously increasing, becoming an element of high added value with social and economic potential, both industry, life quality and research. These technologies are changing constantly as a result of the large research and technical effort being undertaken in universities, companies. The exchange of ideas between scientists and technicians from both academic and industry is essential to facilitate the development of systems that meet the demands of today's society.

This edition of DCAI brings together past experience, current work and promising future trends associated with distributed computing, artificial intelligence and their application to provide efficient solutions to real problems. This symposium is organized by the Bioinformatics, Intelligent System and Educational Technology Research Group (http://bisite.usal.es/) of the University of Salamanca. The present edition has been held in Salamanca, Spain, from 6 to 8 April 2011.

This symposium continues to grow and prosper in its role as one of the premier conferences devoted to the quickly changing landscape of distributed computing, artificial intelligence and the application of AI to distributed systems. This year's technical program presented both high quality and diversity, with contribution in well established and evolving areas of research. This year, 72 papers were submitted from over 13 different countries (Spain, Japan, Germany, India, Brazil, France, USA, Russia, Thailand, Poland, Qatar, Portugal, Egypt), representing a truly "wide area network" of research activities. The DCAI'11 technical program has 55 selected papers (50 long papers, 5 short papers).

We thank the Local Organization members and the Program Committee members for their hard work, which was essential for the success of DCAI´11 and the support obtained by Junta de Castilla y León (Spain).

Salamanca
April 2011

Ajith Abraham
Juan M. Corchado
Sara Rodríguez
Juan F. De Paz

Organization

General Chairs

Sigeru Omatu	Osaka Institute of Technology (Japan)
José M. Molina	Universidad Carlos III de Madrid (Spain)
James Llinas	State University of New York (USA)
Andre Ponce de Leon F. de Carvalho	University of Sao Paulo (Brazil)

Scientific Chair

Ajith Abraham	Machine Intelligence Research Labs (MIR Labs), USA

Organizing Committee

Juan M. Corchado (Chairman)	University of Salamanca (Spain)
Sara Rodríguez (Co-Chairman)	University of Salamanca (Spain)
Juan F. De Paz	University of Salamanca (Spain)
Javier Bajo	Pontifical University of Salamanca (Spain)
Dante I. Tapia	University of Salamanca (Spain)
Fernando de la Prieta Pintado	University of Salamanca (Spain)
Davinia Carolina Zato Domínguez	University of Salamanca (Spain)
Cristian I. Pinzón	University of Salamanca (Spain)
Rosa Cano	University of Salamanca (Spain)
Emilio S. Corchado	University of Salamanca (Spain)
Manuel P. Rubio	University of Salamanca (Spain)
Belén Pérez Lancho	University of Salamanca (Spain)
Angélica González Arrieta	University of Salamanca (Spain)
Vivian F. López	University of Salamanca (Spain)
Ana de Luís	University of Salamanca (Spain)
Ana B. Gil	University of Salamanca (Spain)
Mª Dolores Muñoz Vicente	University of Salamanca (Spain)
Jesús García Herrero	University Carlos III of Madrid (Spain)

Scientific Committee

Adriana Giret	Politechnich University of Valencia (Spain)
Agapito Ledezma	University Carlos III of Madrid (Spain)
Alberto Fernández	University Rey Juan Carlos (Spain)
Alicia Troncoso Lora	University Pablo de Olavide (Spain)
Álvaro Herrero	University of Burgos (Spain)
Ana Carolina Lorena	Federal University of ABC (Brazil)
Andre Coelho	University of Fortaliza (Brazil)
Ángel Alonso	University of León (Spain)
Ângelo Costa	University of Minho (Portugal)
Antonio Moreno	University Rovira y Virgili (Spain)
Araceli Sanchís	University Carlos III of Madrid (Spain)
B. Cristina Pelayo García-Bustelo	University of Oviedo (Spain)
Beatriz López	University of Girona (Spain)
Bianca Innocenti	University of Girona (Spain)
Bogdan Gabrys	Bournemouth University (UK)
Bruno Baruque	University of Burgos (Spain)
Carina González	University of La Laguna (Spain)
Carlos Carrascosa	Politechnich University of Valencia (Spain)
Carlos Soares	University of Porto (Portugal)
Carmen Benavides	University of Leon (Spain)
Daniel Gayo Avello	University of Oviedo (Spain)
Daniel Glez-Peña	University of Vigo (Spain)
Darryl Charles	University of Ulster (North Irland)
David de Francisco	Telefónica I+D (Spain)
David Griol Barres	University Carlos III of Madrid (Spain)
Davide Carneiro	University of Minho (Portugal)
Dídac Busquets	University of Girona (Spain)
Eduardo Hruschka	University of Sao Paulo (Brazil)
Eladio Sanz	Unversity of Salamanca (Spain)
Eleni Mangina	University College Dublin (Ireland)
Emilio Corchado	University of Burgos (Spain)
Eugénio Oliveira	University of Porto (Portugal)
Evelio J. González	University of La Laguna (Spain)
Faraón Llorens Largo	University of Alicante (Spain)
Fernando Díaz	Univesity of Valladolid (Spain)
Fidel Aznar Gregori	University of Alicante (Spain)
Florentino Fdez-Riverola	University of de Vigo (Spain)
Francisco Pujol López	Polytechnic University of Alicante (Spain)
Fumiaki Takeda	Kochi University of Technology (Japan)
Gary Grewal	University of of Guelph (Canada)
Germán Gutiérrez	University Rey Juan Carlos (Spain)
Helder Coelho	University of Lisbon (Portugal)

Hideki Tode	Osaka Prefecture University (Japan)
Ivan López Arévalo	Lab. of Information Technology Cinvestavpas (Mexico)
Javier Carbó	University Carlos III of Madrid (Spain)
Javier Martínez Elicegui	Telefónica I+D (Spain)
Jesús García Herrero	University Carlos III of Madrid (Spain)
Joao Gama	University of Porto (Portugal)
José M. Molina	University Carlos III of Madrid (Spain)
José R. Méndez	University of Vigo (Spain)
José R. Villar	University of Oviedo (Spain)
José V. Álvarez-Bravo	University of Valladolid (Spain)
Juan A. Botia	University of Murcia (Spain)
Juan Manuel Cueva Lovelle	University of Oviedo (Spain)
Juan Gómez Romero	University Carlos III of Madrid (Spain)
Juan Pavón	Complutense University of Madrid (Spain)
Kazutoshi Fujikawa	Nara Institute of Science and Technology (Japan)
Lourdes Borrajo	University of Vigo (Spain)
Luis Alonso	University of Salamanca (Spain)
Luis Correia	University of Libon (Portugal)
Luis F. Castillo	Autonomous University of Manizales (Colombia)
Luís Lima	Polytechnic of Porto, (Portugal)
Manuel González-Bedia	University of Zaragoza (Spain)
Manuel Resinas	University of Sevilla (Spain)
Marcilio Souto	Federal University of Rio Grande do Norte (Brazil)
Margarida Cardoso	ISCTE (Portugal)
Maria del Mar Pujol López	University of Alicante (Spain)
Michifumi Yoshioka	Osaka Prefecture University (Japan)
Miguel Ángel Patricio	University Carlos III of Madrid (Spain)
Miguel Rebollo	University of Vigo (Spain)
Naoki Mori	Osaka Prefecture University (Japan)
Nora Muda	National University of Malaysia (Malaysia)
Norihiko Ono	University of Tokushima (Japan)
Oscar Sanjuan Martínez	University of Oviedo (Spain)
Paulo Novais	Polytechnic University of Minho (Portugal)
Pawel Pawlewski	Poznan University of Technology (Poland)
Rafael Corchuelo	Catholic University of Sevilla (Spain)
Ramón Rizo	University of Alicante (Spain)
Ricardo Campello	University of Sao Paulo (Brazil)
Ricardo Costa	Polytechnic of Porto (Portugal)

Rodrigo Mello	University of Sao Paulo (Brazil)
Rubén Fuentes	Complutense University of Madrid (Spain)
Rui Camacho	University of Porto (Portugal)
Shanmugasundaram Hariharan	B.S. Abdur Rahman University (India)
Silvana Aciar	University of Girona (Spain)
Teresa Ludermir	Universidade Federal de Pernambuco (Brazil)
Vicente Botti	Politechnich University of Valencia (Spain)
Vicente Julián	Politechnich University of Valencia (Spain)
Victor J. Sosa-Sosa	Laboratory of Information Technology (LTI) (México)
Zbigniew Pasek	IMSE/University of Windsor (Canada)

Contents

Bioinformatics, Biomedical Systems

Multiagent Systems

New Algorithms

Data Mining, Information Extraction, Semantic

Mobile Systems, Locating Systems

Tracking, Robotic Systems, Control Systems

Intelligent Electronic Nose System Independent on Odor Concentration

S. Omatu and M. Yano

Abstract. Conventional odor classification methods have been considered under steady state conditions of temperature, humidity, density, etc. In real applications, those conditions may not occur and they will be variable from time to time. Therefore, it is necessary to find some features which are independent on various environmental conditions. In this paper, using a derivative of odor density and odor sensing data in log scale, we will find a feature of the odor which is independent on the density. Using this feature, we construct an electronic nose which is independent on odor density based on a neural network. The neural network used here is a competitive neural network by the learning vector quantization (LVQ). Various odors are measured with an array of many metal oxide gas sensors. After removing noises from the odor data which are measured under the various concentrations, we take the maximum values among the time series data of odors. to reduce the effect of concentration, we use a normalization method to reduce the fluctuation of the data due to the concentration levels. Those data are used to classify the various odors of tees and coffees. The classification results show the effectiveness of the proposed method.

Keywords: Odor robust features, neural networks, learning vector quantization, odor classification.

1 Introduction

Recently, electronic nose (EN) systems have been studied and much progress has been developed from viewpoints of technology and commerce. The ex-

S. Omatu · M. Yano
Osaka Institute of Technology,
5-16-1 Omiya, Asahi-ku, Osaka, 535-8585, Japan
e-mail: omatu@rsh.oit.ac.jp,yano@elc.oit.ac.jp
http://www.oit.ac.jp

A. Abraham et al. (Eds.): International Symposium on DCAI, AISC 91, pp. 1–9.
springerlink.com

pression of the EN refers to the capability of reproducing human sense of smell using sensor arrays and pattern recognition systems.

James A. Milke [1] has proved that two kinds of MOGS have the ability to classify several sources of fire more precisely than with conventional smoke detector. However, his results achieve only 85% of correct classification.

Despite the growing number of applications, much development work is still required. There are a number of limitations to current commercial instruments, including the high cost, large size and weight, humidity and temperature dependence, poor reproducibility and repeatability, high power consumption, relatively low sensibility, long response times and recovery periods. For example, some applications to be used in the field usually require short response time and robust against environmental changes, including humidity and temperature changes, such as fire detector, gas leakage detector.

This paper proposes a new type of an EN system to classify the various odors under the various densities based on a competitive neural network based on the learning vector quantization (LVQ) method. The odor data are measured by an odor sensor array made of semi-conductor metal oxide gas sensors (MOGSs). We have used fourteen MOGSs of FIGARO Technology Ltd in Japan. We consider two types of data for classification in the experiment. The first type is four kinds of tees and the second one is five kinds of coffees of similar properties. The classification results of tees and coffees are about 96% and about 89%, respectively, which is much better than the results in [1]–[5].

2 Princple of MOGS

MOGS used in this paper is the most widely used sensor for making an array of artificial olfactory receptors in the EN system. These sensors are commercially available as the chemical sensor for detecting some specific odors. Generally, an MOGS is applied in many kinds of electrical appliances such as a microwave oven to detect the food burning, an alcohol breath checker to check the drunkenness, an air purifier to check the air quality, and so on.

Various kinds of metal oxide, such as SnO_2, ZnO_2, WO_2,TiO_2 are coated on the surface of semi-conductor, but the most widely applied metal oxide is SnO_2. These metal oxides have a chemical reaction with the oxygen in the air and the chemical reaction changes when the adsorbing gas is detected. The scheme of chemical reaction of an MOGS when adsorbing with the CO gas is shown as follows:

$$\frac{1}{2}O_2 + (SnO_{2-x}) \rightarrow O^{-}ad(SnO_{2-x}) \tag{1}$$

$$CO + O^{-}ad(SnO_{2-x}) \rightarrow CO_2 + (SnO_{2-x}) \tag{2}$$

The relationship between sensor resistance and the concentration of deoxidizing gas can be expressed by the following equation over a certain range of gas concentration [3]:

$$R_s = A[C]^{-\alpha} \tag{3}$$

where R_s =electrical resistance of the sensor, A = constant, C =gas concentration, and α =slope of R_s curve.

Generally, it is designed to detect some specific odor in electrical appliances such as an air purifier, a breath alcohol checker, and so on. Each type of MOGS has its own characteristics in the response to different gases. When combining many MOGS together, the ability to detect the odor is increased. An EN system shown in Fig. 1 has been developed, based on the concept of human olfactory system. The combination of MOGS, listed in Table 1, are used as the olfactory receptors in the human nose.

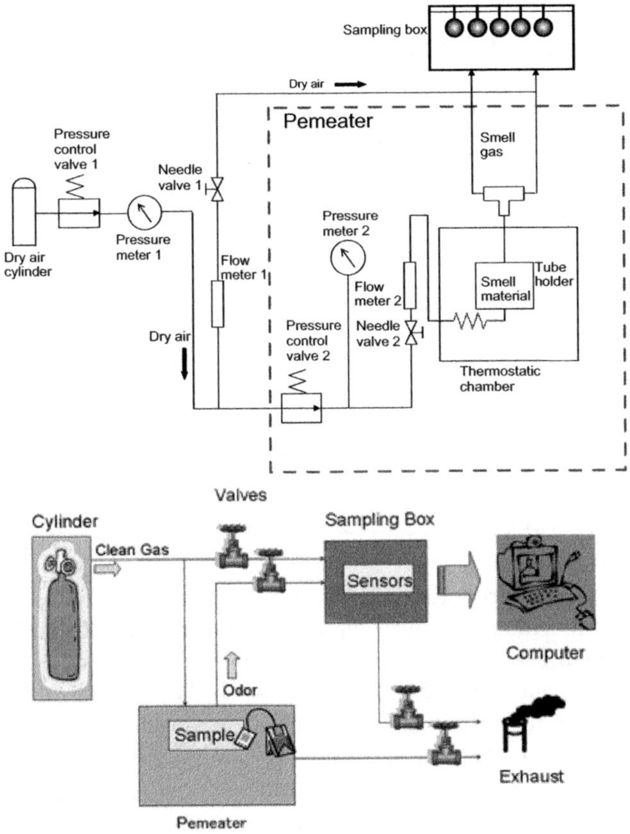

Fig. 1 Structure of the electronic nose system

Table 1 List of MOGS used in the Experiment from FIGARO Technology Inc.

Sensor No.	Sensor model	Main detecting gas
1	TGS2600	tobacco, cooking odor
2	TGS2602	hydrogen sulfide, VOC, ammonia
3	TGS2610	LP gas, butane, propane
4	TGS2611	methane
5	TGS2620	alcohol, organic solvent
6,7	TGS826	ammonia, amine compounds
8,9	TGS816	methane, propane, butane(flammable gas)
10	TGS821	hydrogen gas
11	TGS832	chlorofluorocarbon gas
12	TGS825	hydrogen sulfide
13	TGS830	chlorofluorocarbon gas
14	TGS822	alcohol, organic solvent

3 Experimental Data Collection

The odor data used here are shown Table 2, which is measured by the EN system explained in the previous section and used in the later classification.

Table 2 Kinds and number of samples used the classification.

Tea			Coffee		
Label	Materials	No. of samples	Label	Materials	No. of samples
A	English tea	20	A	Mocha coffee1	35
B	green tea	20	B	Mocha coffee2	35
C	barley tea	20	C	Mocha coffee3	35
D	oolong tea	20	D	Kilimanjaro coffee	35
			E	char-grilled coffee	35

Note that in Table 2 Mocha coffees of the labels A,B,and C are selected from different companies.

The sample of the raw data of Experiment1 is shown in Fig. 2. From the beginning of experiment, we use clean gas and during 2,200s and 2,600s. Then we use the clean gas to erase the odor during 2,600s and 3,800s. After that the same process has been continued. For each data set, the sampling period is 1s and the voltage signal in the clean gas is measured at the beginning of the repetition of experiment. To reduce the noise, we use smoothing filter to the measurement odor data $v'_s(t)$ such that

$$\bar{v}_s(t) = \frac{1}{3}\sum_{i=0}^{3} v'_s(t-i). \tag{4}$$

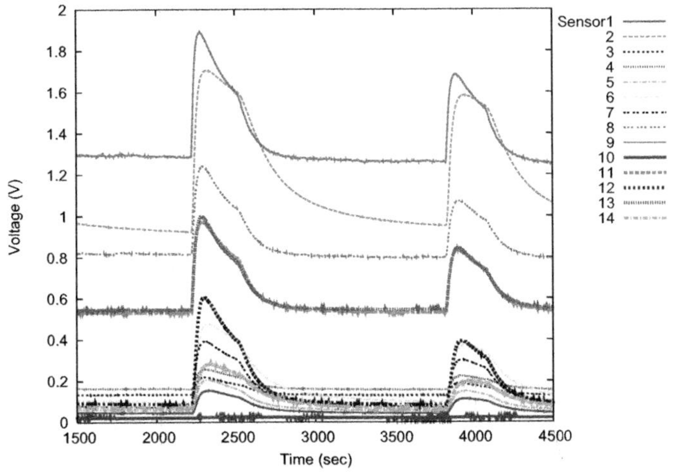

Fig. 2 Full time series data $v'_s(t)$ from a coffee odor in Experiment 1

Then we take the difference $\Delta\bar{v}_s(t) = \bar{v}_s(t) - \bar{v}_s(t-1)$ and if this $\Delta\bar{v}_s(t) \geq \theta$ for more than N sensors where N is a predetermined number, we assume that the odor has been transported to the odor sensors. Here, we take $\theta = 0.001$ and $N = 7$. We assume that the standard value $\bar{v}_s^{std}$ to determine the deference of the odor voltage and the clean gas voltage is an average of the clean gas voltages for five seconds before the odor data begins, that is,

$$\bar{v}_s^{std} = \frac{1}{5} \sum_{i=0}^{5} \bar{v}_s(t). \tag{5}$$

Then the effective voltage of the odor $v_s(t)$ is given by

$$v_s(t) = \bar{v}_s(t) - \bar{v}_s^{std} \tag{6}$$

After testing one odor the MOGS need to be cleaned by removing the tested odor and supplying only the clean gas until the output of the MOGS returns to a stable point. Then a new sample can be tested, repeatedly. This process is just like the human nose which needs to breathe the fresh air before being able to recognize a new odor accurately. Some time series data from the experiment show that all odors approach the saturation stages within the measuring periods.

The levels of odor data are different according to the concentration values of the odor. Generally, we take the maximum value of the time series data as the representative characteristic of the odor for the sensor, which means that the maximum value reflects the steady state of the time series due to the mechanism of the sensing devices of MOGSs. Figure 3 shows the relation of the concentration levels and measurement data.

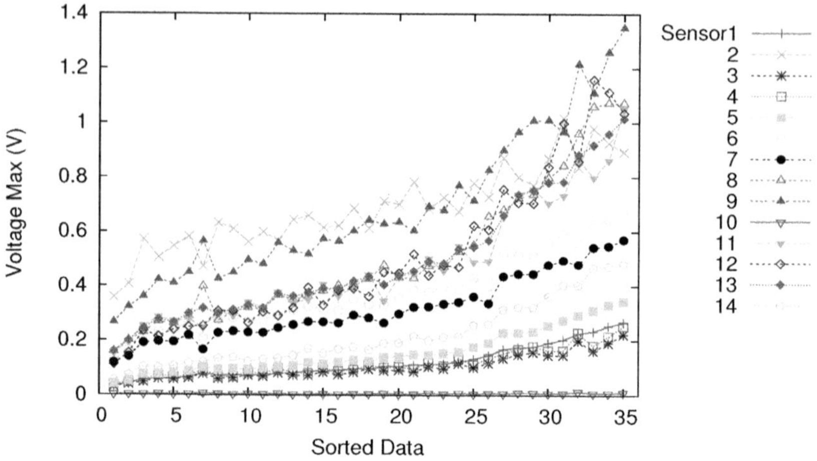

Fig. 3 Sorted data according to the concentration levels from a coffee odor in Experiment II

In order to delete the dependence on the concentration of the odors, we arrange those data such that the horizontal axis is the maximum values of a typical sensor (TGS2600) for experimental trials and the vertical axis show the ratio of the measurement values of the different sensors to those of the typical sensor (TGS2600) selected above. To classify the linear regression lines into some groups, we will define some notations. Let the maximal value of the odor c data for the measurement trial p denote $v_{p,s}^c$. we denote the average of $v_{p,s}^c$ with respect to s by μ_p^c, that is,

$$\mu_p^c = \frac{1}{S}\sum_{s=1}^{S} v_{p,s}^c, c = 1, 2, ..., C, p = 1, 2, ..., P \quad (7)$$

where C is the total number of odor kinds and P is the total number of experimental trials.

For a fixed odor c, we plot the data $\mu_p^c, v_{p,s}^c$ in the logarithmic scale in the plane. Then we can get a group of several kinds of lines given by the following regression lines as shown in Fig. 4:

$$f_s^c(\mu_p^c) = \alpha_s^c \mu_p^c + b_s^c, c = 1, 2, ..., C, p = 1, 2, ..., P. \quad (8)$$

Therefore, we can regard $f_s^c(\mu_p^c)$ or their regression coefficients α_s^c, b_s^c as the representative parameters of the odor data $v_{p,s}^c$ for the odor $c, c = 1, 2, ..., C$, the sensor $s, s = 1, 2, ..., S$,and the experimental trial $p, p = 1, 2, ..., P$.

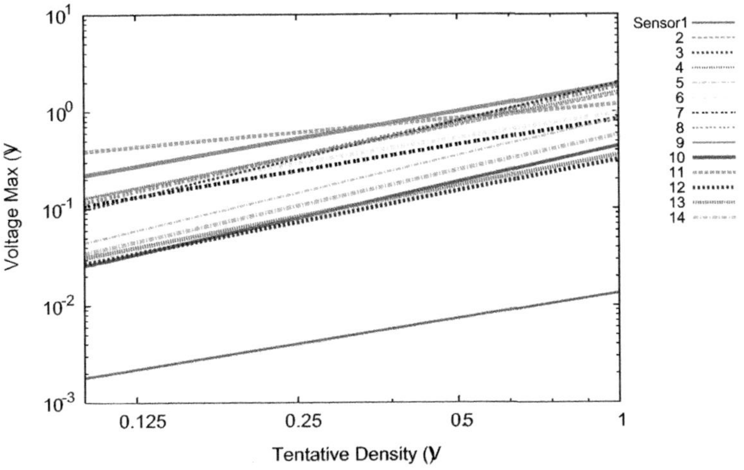

Fig. 4 Linear regression lines in Experiment II

4 Classification Algorithm

We must train the regression coefficients α_s^c, b_s^c for training samples. We assume that some part P' of total experimental trials P is the training data and the remaining one is the test data. For training phase, we adopt the competitive neural network based on LVQ, which has been used to train the regression coefficients α_s^c, b_s^c with respect to p.

After training, we calculate the value μ_p by the following equation:

$$\mu_p = \frac{1}{S}\sum_{s=1}^{S} v_{p,s} forallp \tag{9}$$

where $v_{p,s}, s = 1, 2, ..., S$ are the maximal values of $v_s(t), s = 1, 2, ..., S$ for the measurement trial p. Using μ_p, we get the estimated value $\hat{v}_{p,s}^c$ by the following equation:

$$\hat{v}_{p,s}^c = \hat{f}_s^c(\mu_p^c) = \alpha_s^c \mu_p + b_s^c, s = 1, 2, ..., S, c = 1, 2, ..., C. \tag{10}$$

Furthermore, we find c^o such that

$$c^o = arg \min_c d(v_p, \hat{v}_p^c) \tag{11}$$

$$d(v_{p,s}, \hat{v}_{p,s}^c) = \sqrt{\sum_{s=1}^{S} (v_{p,s} - \hat{v}_{p,s}^c)^2} \tag{12}$$

Then the measurement data $v_{p,s}, s = 1, 2, ..., S$ have been classified into the odor c^o which satisfies (11).

5 Classification Results

We have examined two examples, Experiment 1 and Experiment II stated in Section 3. In Experiment I, the training sample number $P' = 15$ and test sample number is five. By changing the training data set for 100 times and checked the classification accuracy for the test data samples. Thus, the total number of classification of 500 test samples is checked. The results are summarized in Table 3. Average of the classification is 96.15% and the classification is sufficient for real applications.

Table 3 Classification results for Experiment I

	Classification results(96.15%)					
odor data	A	B	C	D	Total	Correct %
A	500	0	0	0	500	100.0%
B	0	494	6	0	500	98.80%
C	0	71	429	0	500	85.80%
D	0	0	0	500	500	100.0%

Experiment II is to classify the different kinds of coffee. As mentioned in Section 3, odor data A, B, and C are the coffees of Mocha made from different companies. If we regard those cluster as the same, the classification rates become more. Roughly speaking, the classification of coffee is also very good for real application although those data look similar.

Table 4 Classification results for Experiment II

	Classification results(88.80%)						
odor data	A	B	C	D	E	Total	Correct%
A	1190	253	29	27	1	1500	79.33%
B	225	1237	9	10	19	1500	82.47%
C	142	7	1325	26	0	1500	88.33%
D	9	14	3	1437	37	1500	95.80%
E	0	18	0	11	1471	1500	98.07%

6 Conclusions

We have presented the reliability of a new EN system designed from various kinds of MOGS. The EN has the ability to identify various sources with more than 90% of accuracy. It can be concluded that the EN is suitable for various applications in real world to classify the difficult odors.

Acknowledgement. This research has been supported by Grant-in-Aid for Challenging Exploratory Research No. 22656089 of Japan Society for the Promotion of Science and we wish to thank JSPS for their support.

References

1. Milke, J.A.: Application of Neural Networks for discriminating Fire Detectors. In: 10th International Conference on Automatic Fire Detection, AUBE 1995, Duisburg, Germany, pp. 213–222 (1995)
2. Charumporn, B., Yoshioka, M., Fujinaka, T., Omatu, S.: An Electronic Nose System Using Back Propagation Neural Networks with a Centroid Training Data Set. In: Proc. Eighth International Symposium on Artificial Life and Robotics, Japan, pp. 605–608 (2003)
3. General Information for TGS sensors, Figaro Engineering, `http://www.figarosensor.com/products/general.pdf`
4. Carlson, W.L., Thorne, B.: Applied Statistical Methods. Prentice-Hall International, Englewood Cliffs (1997)
5. Fujinaka, T., Yoshioka, M., Omatu, S., Kosaka, T.: Intelligent Electronic Nose Systems for Fiore Detection Systems Based on Neural Networks. In: The Second International Conference on Advanced Engineering Computing and Applications in Sciences, Valencia, Spain, pp. 73–76 (2008)

Building Biomedical Text Classifiers under Sample Selection Bias

R. Romero, E.L. Iglesias, and L. Borrajo

Abstract. Scientific papers are a primary source of information for investigators to know the current status in a topic or compare their results with other colleagues. However, mining biomedical texts remains to be a great challenge by the huge volume of scientific databases stored in the public databases and their imbalanced nature, with only a very small number of relevant papers to each user query. Classifying in the presence of data imbalances presents a great challenge to machine learning. Techniques such as support-vector machines (SVMs) have excellent performance for balanced data, but may fail when applied to imbalanced datasets. In this paper, we study the effects of undersampling, resampling and subsampling balancing strategies on four different biomedical text classifiers (with lineal, sigmoid, exponential and polynomial SVM kernels, respectively). Best results were obtained by normalized lineal and sigmoid kernels using the subsampling balancing technique. These results have been compared with those obtained by other authors using the TREC Genomics 2005 public corpus.

Keywords: Biomedical text mining, classification techniques, SVMs, imbalanced data.

1 Introduction

Currently there are many resources that store on-line biomedical publications. The most widely used is the *PubMed* database of the *National Center of Biotechnology Information (NCBI)*. It contains more than 16 millions of abstracts of the medical publications database Medline, and is requested by millions of users everyday.

In the last decade several text mining methods have been proposed to automate the process of searching and classifying information in the on-line biomedical

R. Romero · E.L. Iglesias · L. Borrajo
Univ. of Vigo, Campus As Lagoas s/n 32004 Ourense Spain
e-mail: `{rrgonzalez, eva, lborrajo}@uvigo.es`

A. Abraham et al. (Eds.): International Symposium on DCAI, AISC 91, pp. 11–18.
springerlink.com

publications. However, to the best of our knowledge, results are not enough good mainly due to the imbalanced nature of the biomedical papers.

The data imbalance problem exists in a broad range of experimental data, but only recently has it attracted close attention for researchers [4, 17]. Data imbalance occurs when the majority class is represented by a large portion of all the examples, while the other, the minority class, has only a small percentage [14]. When a text classifier encounters an imbalanced document corpus, the performance of machine learning algorithms often decreases [11].

Sampling strategies such as over and subsampling are popular in tackling the problem of class imbalance [1, 7, 15, 19]. In this work, we study their effects on four different SVM kernels (lineal, polynomial, exponential and sigmoid) when used to classify biomedical texts. The *subsampling* algorithm decreases artificially the number of samples that belongs to majority class, while the *oversampling* algorithm redistributes the number of samples that belongs to minority class regarding the majority one. Finally, we apply both concepts at the same time increasing or decreasing the number of instances that belongs to each class to obtain a balanced dataset. In this case we are talking about *resampling*.

2 Methods

2.1 Text Classification Process

The biomedical text classification process proposed here is divided into four tasks:

1. **Annotation:** This task processes the documents extracting the most relevant keywords. The annotation process can be quite complex depending on the techniques to apply. In this research we have used a tool called GATE [9] with an annotation plugin called Abner-Tagger [12]. The entity recognizer (Abner) uses a dictionary in order to preprocess the documents. Our proofs are based on the NLPBA dictionary [8]. The annotation task applies these tools in order to create a dataset (the *sparse matrix*) compounded by vectors. In this matrix each row is a mathematical representation of a document. On the other hand, in order to increase the data usability, we applied a *tf-idf* normalizer. As a result of this task, we create test and train sparse matrices.
2. **Operation sets:** During the classification process, train and test matrices must have the same number of attributes in a particular order. When the applied dictionaries are too large, as in our case, may occur that a lot of relevant attributes belonging to train matrix do not belong to the second one. This situation generates a test matrix with meaningless data.
   We have solved this problem applying an intersection over these matrices in order to reduce their dimensionalities and make them computable.
3. **Instance filtering and attribute selection:** These tasks permit to balance the number of instances that belong to each class and apply algorithms in order to

decrease the number of attributes. In our case, subsampling, oversampling and resampling techniques, before mentioned, have been used.

4. **Classification:** In this task different reasoning models to classify texts in relevant or not relevant are applied. Specifically, we have used an implementation of a SVM with four kernels (lineal, polynomial, radial and sigmoid), as detailed below.

2.2 Dataset

To perform tests and to compare results with those obtained by other authors we have used the *Text Retrieval Conferences (TREC)* [16] public corpus. In 2005, TREC provided a set of evaluation tasks to know the state of the art of applying information extraction techniques to problems in biology. Specifically, the goal of the TREC Genomics Track was to create test collections for evaluation of information retrieval and related tasks in the genomics domain [10].

2.3 Model Evaluation

In this section we are going to explain the measures used in order to represent the results in the whole process. In this way, Precision (1) represents the percentage of relevant documents correctly classified over all documents. Recall (2) represents the percent of relevant documents which were correctly classified. F-Measure (3) stablishes a relation between Recall and Precision that represents the weighted harmonic mean and shows the correlation between them.

Latest measure, Utility (4), is often applied in text categorization. This measure contains coefficients for the utility of retrieving a relevant and a nonrelevant document. It is composed by the best possible score U_{max} and the raw score U_{raw}(5), where U_r (6) is the relative utility of relevant document and U_{nr} is the relative utility of nonrelevant document. For our purposes, we assume that U_{nr} is -1.

$$Precision = \frac{|\{relevant\ documents\} \cap \{retrieved\ documents\}|}{|\{retrieved\ documents\}|} \quad (1)$$

$$Recall = \frac{|\{relevant\ documents\} \cap \{retrieved\ documents\}|}{|\{relevant\ documents\}|} \quad (2)$$

$$F-measure = 2 \cdot \frac{Precision \cdot Recall}{Precision + Recall} \quad (3)$$

$$U_{norm} = \frac{U_{raw}}{U_{max}} \quad (4)$$

$$U_{raw} = (U_r \cdot relevant\ docs\ retrieved) + (U_{nr} \cdot nonrelevant\ docs\ retrieved) \quad (5)$$

$$U_r = \frac{all\ possible\ negatives}{all\ possible\ positives} \quad (6)$$

3 Experimental Results

The tests were based on the process explained in the previous section.

For the subsampling we used different distribution values, $\{10, 5, 4, 3, 2, 1\}$. These values represent the distribution factor between classes; i.e., a distribution of 10 means that the majority class will be reduced until it contains a number of instances 10 times more than the minority class. Thus, a distribution with a factor 1 represents an uniform distribution. The random selection algorithm was chosen in order to remove the number of instances.

For the resampling process we used a distribution bias between $\{$ 1.0, 0.75, 0.5, 0,25, 0.0 $\}$. A distribution factor of 1.0 means an uniform distribution, and a distribution of 0.25 means that the number of relevant document would be increased in a factor of 0.25 compared with the majority class.

At last, in the classification process we used an implementation of a SVM with four kernels (lineal, polynomial, radial and sigmoid). In order to support the software we employed a library called LibSVM [6] that implements a kernel method based on costs (C-SVM). Furthermore, we used some parameters like *probability estimates* to generate probabilities instead of $[-1, +1]$ values for SVM output classification, or *normalize* to scale attribute values between $[-1, +1]$.

The kernels used for testing part are represented by the following equations.

$$Lineal : U \cdot V \tag{7}$$

$$Polynomial : (Gamma \cdot u \cdot v + Coef)^{Degree} \tag{8}$$

$$Exponential : exp(-Gamma \cdot |u - v|^2) \tag{9}$$

$$Sigmoid : tanh(Gamma \cdot u' \cdot v + Coef) \tag{10}$$

In the polynomial kernel (8) we tested the *Degree* and *Gamma* values between 1 to 7. With exponential (9) and sigmoid kernels (10) we made tests using values for the *Gamma* parameter between 1 to 15. Other parameters that appear in kernel equations like *Coef*, which means a single coefficient, was set to the default value 0.0. Finally, the cost parameter associated to this SVM, namely C-SVM, was set to 1.0.

In the figures we used the following acronyms: *N* means Normalize, *P* means Probabilistic, *NP* is equal to both Normalize and Probabilistic, and $G[X]$ is the Gamma parameter with X corresponding to the different values of the parameter. In order to represent each plot, boundaries for all measures *(Utility, Precision, Recall and F-Measure)* are denoted between 0 to 1, where a value close to 1 is much better than a value close to 0.

In Fig. 1 we use the utility measure to compare the effect of balanced and non-balanced techniques. As shown, the poor results are obtained when any class of balance is used.

Regarding resampling and subsampling, we got the best results with an uniform distribution between classes, i.e., subsampling factor equal to 1 and resampling equal to 1.0. To calculate the utility measure we used these parameters.

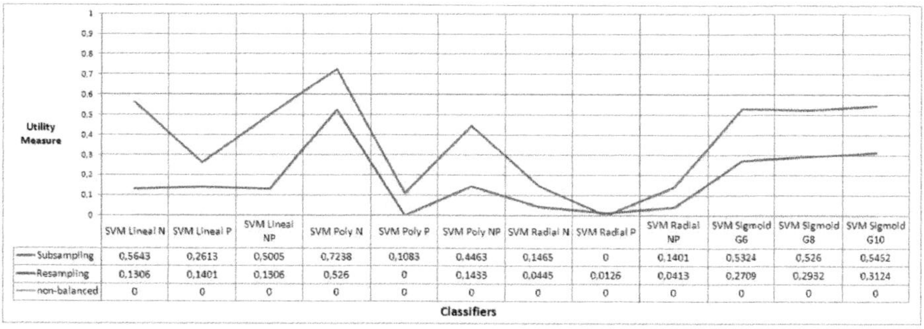

Fig. 1 Comparative between classificators using balanced and non-balanced sampling techniques

In Fig. 1 we also can see a global peak using a normalized polynomial kernel. Normalized kernels got good results in general, between 0.5 and 0.724, except in case of radial. It is interesting to note that the use of normalization in sigmoid kernel did not improve the results. Therefore, we do not include it in plots.

Regarding the balance techniques we can observe that subsampling got better results than resampling. In this situation we can conclude that if we have an overtrained class with samples too similar, it is very difficult to make a good classification.

Comparing statistical measures, in Fig. 2 and Fig. 3 we can see some plots based on *Utility, Recall, Precision and F-Measure*. Each plot belongs to subsampling and resampling, respectively. As can be seen, the best results were obtained by polynomial kernel for the utility measure. Regarding to this measure is only based on relevant documents retrieved, is highly recommended to take also a look to recall and precision, or directly F-Measure. If we analyse the results based on these latest measures, we can conclude that best results were obtained by the normalized lineal and sigmoid kernels, not by the normalized polynomial, even using different balancing techniques.

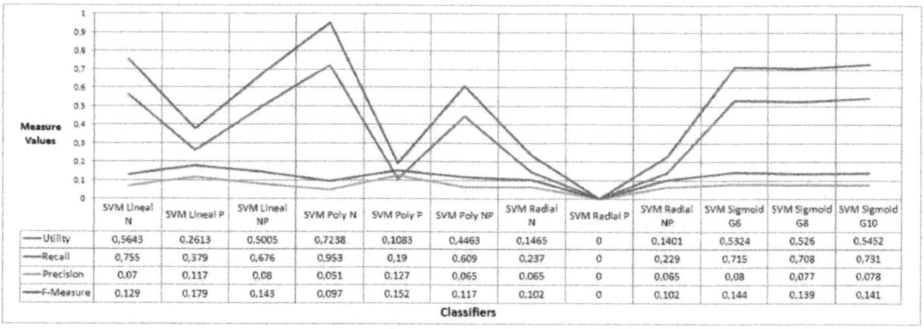

Fig. 2 Results for classification task using the subsampling balance technique

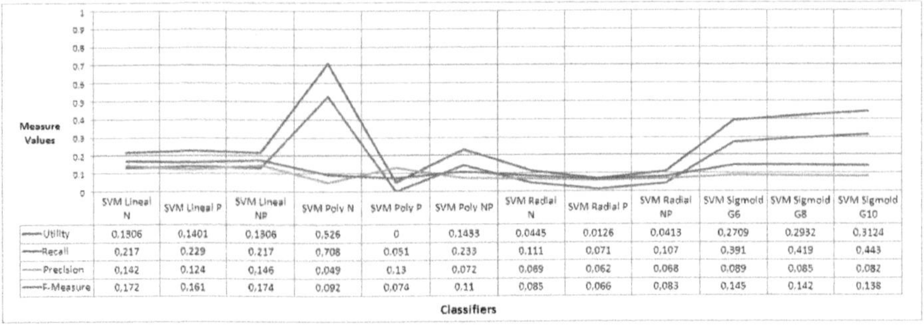

Fig. 3 Results for classification task using the resampling balance technique

Respect to the sigmoid kernel, this kernel got good results in almost all tests with small variations.

4 Discussion

In order to appraise our results, we have researched on similar works published in the TREC Genomics 2005 communications. In Table 1 we show the run results by run name, group name and performance measures. The first two entries in the table correspond to the best values obtained in the competition.

As we can see, our results are similar to [5]. They used a SVM based on a RBF kernel adapted to each situation, and a kernelized Naive Bayes classifier.

Table 1 Comparative based on TREC Genomics 2005 results, sorted by utility measure

Tag	Group	Precision	Recall	F-Measure	Utility
aibamadz05[2]	ibm.zhang	0.4669	0.9337	0.6225	0.8710
ABBR003SThr[13]	ibm.kanungo	0.4062	0.9458	0.5683	0.8645
ASVMN03[13]	ibm.kanungo	0.4019	0.9127	0.5580	0.8327
aNLMB[3]	nlm-umd.aronson	0.3391	0.9398	0.4984	0.8320
aQUT14[20]	queensu.shatkay	0.3582	0.8675	0.5070	0.7760
aMUSCUIUC3[18]	uiuc.zhai	0.4281	0.8072	0.5595	0.7438
SVM Poly N	uvigo.es	0.0510	0.9530	0.0970	0.7238
aUCHSCnb1En3[5]	ucolorado.cohen	0.5080	0.7651	0.6106	0.7215
SVM Lineal N	uvigo.es	0.0700	0.7550	0.1290	0.5643
SVM Sigmoid G10	uvigo.es	0.0780	0.7310	0.1410	0.5452
SVM Lineal NP	uvigo.es	0.0800	0.6760	0.1430	0.5005
SVM Poly NP	uvigo.es	0.0650	0.6090	0.1170	0.4463
aUCHSCsvm[5]	ucolorado.cohen	0.7957	0.4458	0.5714	0.4391
aNLMF[3]	nlm-umd.aronson	0.2219	0.5301	0.3129	0.4208
	Minimum	*0.2191*	*0.2500*	*0.2387*	*0.2009*
	Median	*0.3572*	*0.8931*	*0.5065*	*0.7773*
	Maximum	*0.7957*	*0.9578*	*0.6667*	*0.8710*

On the other hand, it is interesting to note that these studies have used preprocessing techniques like chi-square [20] to reduce features dimensionality or another kind of thresholds like *Rcut, Pcut or Scut* [13, 3], but in any case balance techniques have been used.

Finally, we have made a comparative between classification techniques and we have demonstrated that our balancing techniques are effective, because in some cases we have achieved better results than similar works.

5 Conclusions

In this research we apply four different SVM kernels to build a biomedical text classification system. Due to the unbalanced nature of this information, various data balancing techniques are also applied.

The experimental results show that, regarding to the utility measure, the best classifier is based on the polynomial kernel. On the other hand, if we take in account other measures (precision, recall and f-measure) the best balanced result were achieved by linear and sigmoid kernels.

These results demonstrate the advantages of the polynomial SVM techniques to classify biomedical texts handling the imbalance problem with subsampling.

Acknowledgements. This work has been partially funded by the Spanish Ministry of Science and Innovation, the Plan E from the Spanish Government and the European Union from the ERDF (TIN2009-14057-C03-02).

References

1. Anand, A., Pugalenthi, G., Fogel, G.B., Suganthan, P.N.: An approach for classification of highly imbalanced data using weighting and undersampling. Amino Acids 39, 1385–1391 (2010)
2. Ando, R.K., Dredze, M., Zhang, T.: Trec 2005 genomics track experiments at ibm watson. In: Proceedings of TREC 2005. NIST Special Publication(2005)
3. Aronson, A.R.: Fusion of knowledge-intensive and statistical approaches for retrieving and annotating textual genomics documents. In: Proc TREC 2005, pp. 36–45 (2005)
4. Barandela, R., Sánchez, J.S., García, V., Rangel, E.: Strategies for learning in class imbalance problems. Pattern Recognition 36(3), 849–851 (2003)
5. Caporaso, J.G.: Concept recognition and the trec genomics tasks. the fourteenth text retrieval. In: Conference Proceedings (TREC 2005). National Institute for Standards and Technology, Gaithersburg (2005)
6. Chang, C.-C., Lin, C.-J.: LIBSVM: a library for support vector machines (2001), Software available at `http://www.csie.ntu.edu.tw/~cjlin/libsvm`
7. Chawla, N.V., Japkowicz, N., Kotcz, A.: Editorial: special issue on learning from imbalanced data sets. SIGKDD Explorations 6(1), 1–6 (2004)
8. Collier, N., Ruch, P., Nazarenko, A. (eds.): JNLPBA 2004: Proceedings of the International Joint Workshop on Natural Language Processing in Biomedicine and its Applications. ACL, Morristown (2004)

9. Cunningham, H., Wilks, Y., Gaizauskas, R.J.: Gate - a general architecture for text engineering (1996)
10. Hersh, W., Cohen, A., Yang, J., Bhupatiraju, R.T., Roberts, P., Hearst, M.: Trec 2005 genomics track overview. In: TREC 2005 Notebook, pp. 14–25 (2005)
11. Kang, P., Cho, S.: EUS SVMs: Ensemble of Under-Sampled SVMs for Data Imbalance Problems. In: King, I., Wang, J., Chan, L.-W., Wang, D. (eds.) ICONIP 2006. LNCS, vol. 4232, pp. 837–846. Springer, Heidelberg (2006)
12. Settles, B.: ABNER: An open source tool for automatically tagging genes, proteins, and other entity names in text. Bioinformatics 21(14), 3191–3192 (2005)
13. Si, L., Kanungo, T.: Thresholding strategies for text classifiers: Trec 2005 biomedical triage task experiments. the fourteenth text retrieval. In: Conference Proceedings (TREC 2005). National Institute for Standards and Technology, Gaithersburg (2005), http://trec.nist.gov/pubs/trec14/papers/carnegie-mu-kanungo.geo.pdf
14. Tan, S.: Neighbor-weighted k-nearest neighbor for unbalanced text corpus. Expert Systems with Applications 28(4), 667–671 (2005)
15. Tang, Y., Zhang, Y., Chawla, N.V.: Svms modeling for highly imbalanced classification. IEEE Transactions on Systems, Man, and Cybernetics, Part B: Cybernetics 39(1), 281–288 (2009)
16. Voorhees, E.M., Buckland, L.P. (eds.): Proceedings of the Fourteenth Text REtrieval Conference, TREC 2005, Gaithersburg, Maryland, November 15-18. National Institute of Standards and Technology (NIST), Special Publication 500-266 (2005)
17. Weiss, G.M.: Mining with rarity: a unifying framework. SIGKDD Explor. Newsl. 6, 7–19 (2004)
18. Zhaif, C.: Uiuc/musc at trec 2005 genomics track. the fourteenth text retrieval. In: Conference Proceedings (TREC 2005). National Institute for Standards and Technology, Gaithersburg (2005), http://trec.nist.gov/pubs/trec14/papers/uillinoisuc.geo.pdf
19. Zhang, J., Mani, I.: knn approach to unbalanced data distributions: A case study involving information extraction. In: Proceedings of the ICML 2003 Workshop on Learning from Imbalanced Datasets (2003)
20. Zheng, Z.H.: Applying probabilistic thematic clustering for classification in the trec 2005 genomics track. the fourteenth text retrieval. In: Conference Proceedings (TREC 2005). National Institute for Standards and Technology, Gaithersburg (2005), http://trec.nist.gov/pubs/trec14/papers/queensu.geo.pdf

A Multi-agent Model with Dynamic Leadership for Fault Diagnosis in Chemical Plants

Benito Mendoza, Peng Xu, and Limin Song

Abstract. Timely fault detection and diagnosis are critical matters for modern chemical plants and refineries. Traditional approaches to fault detection and diagnosis of those complex systems produce centralized models that are very difficult to maintain. In this article, we introduce a biologically inspired multi-agent model which exploits the concept of leadership; that is, when a fault is detected one agent emerges as leader and coordinates the fault classification process. The proposed model is flexible, modular, decentralized, and portable. Our experimental results show that even using simple detection and diagnosis methods, the model can achieve comparable results to those from sophisticated centralized approaches.

Keywords: Multi-agent systems modeling, distributed data fusion, fault diagnosis, collective consensus.

1 Introduction

A highly effective fault management system for modern complex plants plays an important role in avoiding unplanned capacity loss, increasing plant efficiency, and improving operator safety. Key features of a successful fault management system include short fault detection and diagnosis delay time, high accuracy, high reliability, good maintainability, and robustness. In addition, the system's architecture and its modules should be portable so that transferring a system developed for one plant to many other plants can be easily done. Conventional fault management follows a centralized approach that achieves a good balance between high accuracy and delay time. However, they are very rigid and hard to maintain to reflect plant changes. More important, having a single central point for decision making raises great concerns about the system's reliability and robustness.

In this paper, we present a biologically inspired multi-agent model that is able to overcome these problems. Specifically, our work is inspired by fish, birds, and

Benito Mendoza · Peng Xu · Limin Song
ExxonMobil Research and Engineering Company
Annandale, NJ 08801, USA
e-mail: {benito.g.mendoza,peng.xu,limin.song}@exxonmobil.com

A. Abraham et al. (Eds.): International Symposium on DCAI, AISC 91, pp. 19–26.
springerlink.com

other species in which a few informed or knowledgeable individuals influence the schooling behavior of the entire group [4]. Similarly, in our approach, each agent acquires measurements from one or a few sensors and detects faults based on the acquired data; the first agent that detects a fault becomes the leading agent and collects fault detection/diagnosis information from other agents. It performs fault diagnosis and abandons its leadership once the fault is mitigated. We believe that this mechanism for dynamic coordination and information fusion is more efficient and reliable than conventional centralized approaches.

This paper is organized as follows. Section 2 details the proposed approach. Section 3 shows how we implement our model for a specific plant model and presents experimental results. Finally, Section 4 concludes the paper and points out future research directions.

2 The Proposed Approach

Our work is inspired by the group behavior of some animal species. Specifically, we are motivated by the mechanism of fish schooling and bird flocking. It is well known that these complex behaviors are results of relatively simple and local interactions among group members [4]. Typically, a few informed or knowledgeable individuals emerge as "leaders" and influence the entire group by persuading their neighbors. Similarly, the initial followers induce other followers, which form the observed collective behavior. This kind of coordination has been exploited in some multi-agent models [10]. These models are very attractive because of their natural capabilities to adapt to changes in the environment, resulting in high reliability and robustness.

In our proposed model, each agent monitors a sensor or a group of sensors in a plant. Each agent is able to determine the plant's local state (normal or abnormal) and communicates it to other agents. The first agent that detects an abnormal event assumes the leader role. A leader agent is in charge of aggregating all other agents' state information. As a result of the communication process, a spatio-temporal pattern of abnormal events or disturbances emerges. The leader agent identifies the corresponding type of the fault by comparing the emerged pattern against existing fault templates that are stored in its local knowledge base.

Fig. 1 illustrates our proposed solution. The bottom layer shows that each agent is connected to one or a few sensors. At time 1, one agent (black filled) detects a disturbance and becomes the leader agent. As time progresses, more agents (gray filled) detect disturbances and they send their state information to the leader. The upper layer shows that at time *t+i*, the leader agent declares that the plant is back to its normal state and abandons its leadership; as a result, agents are not communicating with each other any more until a new fault emerges.

2.1 Communication and Coordination

We assume that the physical communication channels allow communication among all the agents. The messages among agents contain the following

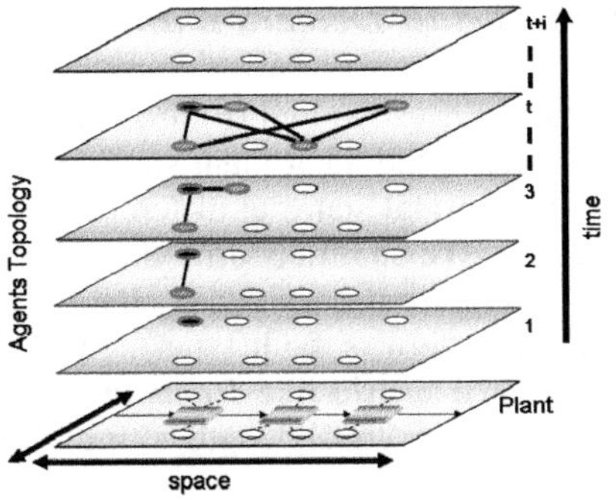

Fig. 1 Illustration of the proposed solution.

components: sender's ID, receivers' IDs, time stamp, and content. The content determines the purpose of the message as follows.

- Declaration of leadership. An agent declares leadership to all other agents if it detects a disturbance and there is no leader in the system.
- Acceptance or rejection of leadership declaration. An agent accepts the declaration if it has not detected the disturbance. Agents that simultaneously declare leadership roles will reject other agents' declarations. They will negotiate with each other until one leader is elected.
- Plant's state. Each non-leader agent sends information about the local plant state and the leader agent announces whether the entire plant is normal or not.

2.2 Agent's Knowledge Base

Each agent's knowledge base consists of two components. The first component is a *local model* for fault detection and diagnosis based on the local data it acquires from the plant. During the design phase, one has the flexibility to choose from various fault detection algorithms. They might be simple algorithms like control charts, Principal Component Analysis (PCA) [8] or any of its variations (e.g., dynamic PCA), or more sophisticated algorithms like support vector machines or artificial neural networks [1]. The second component is a *global fault diagnosis model*, which is used only when the agent assumes the leader role. Each agent stores a few templates for the faults that it- is very likely to detect earlier than other agents. In this way, the fault templates are stored in a distributed manner and none of the agents is overloaded with templates. As mentioned before, once an agent assumes the leadership role, it uses these templates for fault diagnosis. In this paper, we describe a specific type of spatio-temporal template (see the next section) for fault detection and diagnosis. The strength of agent based approaches, however, allows different type of templates to be incorporated into the model.

2.3 Spatio-Temporal Fault Detection and Diagnosis

Imagine that the patterns in Fig. 2 (a) and (b) are two known types of faults in a plant. Each row corresponds to the local plant states that an agent is monitoring. The filled cells represent abnormal states at different time and the black row corresponds to the leader agent. We can transform the rows in each pattern into a binary string, where 0 and 1 represent normal and abnormal states respectively. For example, row T3 from the pattern of fault type f_1 forms the string 0001101100. We can use a simple distance metric like *Hamming distance*, which counts the number of different bits of two equal length strings, to compare an observed pattern with a stored template.

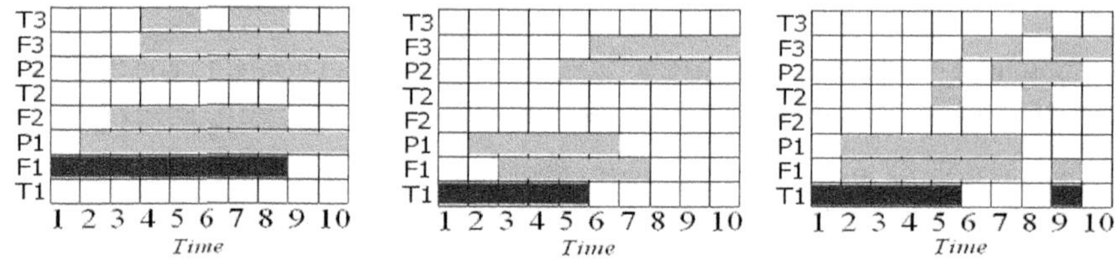

(a) Pattern of fault type f_1 (b) Pattern of fault type f_2 (c) Noisy instance of f_2

Fig. 2 Graphical representation of fault patterns. (a) and (b) are fault patterns, and (c) is a noisy instance of (b).

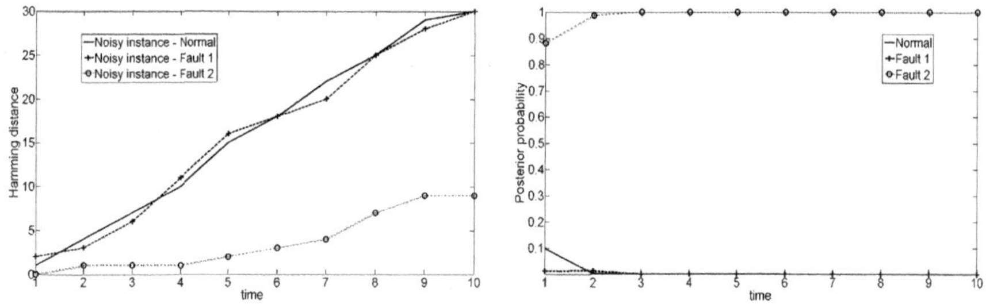

(a) (b)

Fig. 3 Fault diagnosis illustration. (a) Hamming distance for the observation given in Figure 2 (c) to normal and fault patterns as a function of time, (b) Posterior probabilities for the same observation using the Naïve Bayes method.

Fig. 3 (a) shows the Hamming distance between the pattern in Fig. 2 (c) and the normal state (i.e., a pattern with no colored cells) and the fault patterns in Fig. 2 (a) and (b) as a function of time. It shows clearly that as time progresses the diagnosis converges to the ground truth, which is fault 2. Note that Hamming distance is just one of the many methods our model can incorporate. Other candidates include Bayesian methods (see Fig. 3 (b) as an example), Dynamic Time Warping, and other popular information fusion methods.

2.4 Discussion

Our model is different from existing agent approaches [2, 5, 7, 9] in that each agent can be treated as a smart virtual sensor in a distributed sensor network and each agent's functionalities are essentially the same to other agents. These functionalities are: a) acquiring data from one or a few sensors, b) providing diagnostic information based on the acquired data, and c) accumulating diagnostic information from other agents if it is the first one detecting a fault, and making the final diagnostic decision. The advantages of our approach are as follows.

1. It is portable and easy to maintain. Since agents are fungible, they can be easily transferred to a new monitoring system. Each agent can be maintained (e.g., updated, added, removed) easily to reflect local changes.
2. It is flexible because it can incorporate a large variety of detection and diagnostic methods. Since each agent is working independently, one can choose the best method for that agent based on the characteristics of the data.
3. It requires less communication capacity compared to centralized methods and many other agent-based methods because agents share processed information instead of raw sensor data. Such a design ensures that the critical information will be transmitted in a timely manner. It also reinforces the maintainability of the system since adding new sensors will not significantly overload the communication network. The system may still function well during a partial communication interruption.

Our approach, however, does not take the full advantage of the bio-inspired method described in [4]; that is, no agent is trying to influence other agents. We have envisioned that an agent, once it has detected a fault, should be able to pass corresponding information to its neighbors (defined by the topology of the plant) so that the neighbor agents can be in a more "alert" mode (e.g., adjusting detection threshold, invoking more sophisticated detection algorithms) for fault detection and diagnosis. This will be investigated in our future work.

3 Implementation and Experiments

We implemented our multi-agent model using JADE[1] for fault detection and diagnosis of the Tennessee-Eastman Process (TEP) [3], a well known challenging problem for industrial process monitoring. The TEP simulates an industrial plant with 41 measured variables and 12 manipulated variables.

In this implementation, an agent monitors only one variable or sensor. The detection method is a simple six-sigma approach, which defines upper and lower normal operation limits by $\mu \pm 3\sigma$ (μ is the mean and σ is the standard deviation of the variable) and any value out of these limits is considered abnormal. Data of the normal state are used to calculate the corresponding mean and standard deviation; data corresponding to faults are used to create the templates in the agents'

[1] `http://jade.tilab.com/`

knowledge bases. In this implementation, we assume all types of possible faults are known a priori. The templates are distributed among the agents based on the training data, where for each training example a corresponding template is added to the database of the agent(s) that first detected the fault.

A non-leader agent only communicates with the leader agent if the monitored local state changes (e.g., from normal to abnormal). In this implementation each agent has a numeric ID related to sensor it is monitoring. To resolve conflicts when more than one agent claims the leadership, a simple rule is put in place: when more than one agent claim leadership simultaneously, the one with the lowest ID prevails as leader and the rest give up. Since other agents will follow the same strategy, they would only coordinate with the resultant leader of this tie break rule; there is no incentive for any agent to try to remain as leader. We have developed a simple similarity metric similar to Hamming distance in our implementation. This metric requires less computation and less storage space for templates. For details, refer to [6].

There are twenty-one types of fault in TEP. Faults 3, 9, 15, 19, and 21 are very subtle disturbances and nearly all methods in the literature fail to detect them reliably [1]. Thus, they are not included in our experiment. Our training data set consists of 10 simulation runs for each of the remaining types of faults and 10 runs of the normal operation. Each simulation run contains 1920 data points with sampling rate as 1 point every 3 minutes. Note that in TEP each data point is a vector containing 53 elements (41 measurement variables and 12 manipulated variables). For each simulation containing a fault, the fault is introduced right in the middle; that is, starting from data point 961, the plant is at the faulty state.

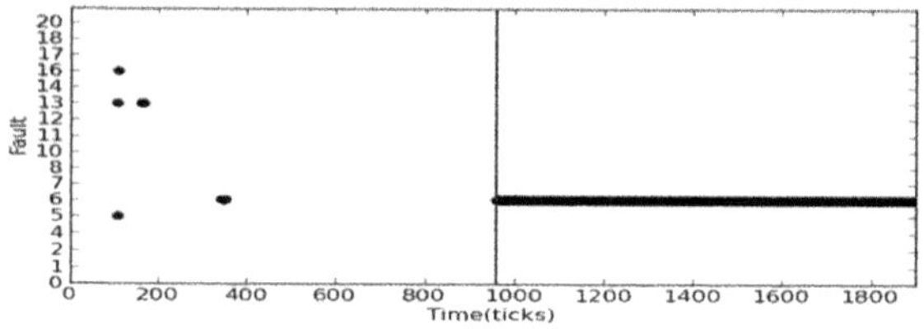

Fig. 4 Diagnostic result for fault 6.

Fig. 4 illustrates the fault diagnosis result for fault 6. The vertical line indicates the time the fault is introduced. Before that, most of the time our model determines that the system is at normal state. The system does produce a few false alarms; however, they are too short to be considered individually. The average false alarm rate is around 0.01%. Starting from point 961, the system consistently indicates that fault 6 occurs in the plant.

We evaluate our method's performance in terms of fault classification accuracy and detection delay time, which is the interval between the time when a fault is introduced and the time when it is detected. Some faults have longer delay diagnosis time as they are either subtle or they need longer time to propagate through the plant. Table 1 shows our experimental results compared against a holistic centralized method based on support vector machine (SVM). Faults 6, 7, 14 and 17 are

Table 1 Average results on ten test simulations

Fault	MAS		SVM	
	Accuracy	Delay	Accuracy	Delay
1	0.89	13	**0.98**	**3**
2	0.89	13	**0.98**	**11**
4	0.89	12	**0.95**	**2**
5	0.88	9	**0.97**	**2**
6	**1**	3	0.98	**1**
7	**1**	3	0.98	**1**
8	0.77	**27**	**0.90**	33
10	0.56	47	**0.81**	**28**
11	0.65	10	**0.86**	**8**
12	0.8	**14**	**0.90**	16
13	0.84	**64**	**0.90**	72
14	0.96	6	**0.97**	**3**
16	0.26	81	**0.77**	**19**
17	**0.95**	**25**	0.94	**25**
18	0.86	51	**0.92**	**46**
20	0.74	56	**0.91**	**40**

classified with high accuracy (95% to 100%) and faults 1, 2, 4, 5, 12, 13, and 18 are classified with accuracy between 80% and 90%.

In our experiment, we have purposely chosen a naïve method (i.e., six-sigma) for fault detection to test the limitation of our model's fault classification capability. Our analysis shows that by incorporating the temporal information, the results are comparable to more sophisticated pattern classification algorithms like SVM. In addition, our SVM uses a more sophisticated feature extraction method than the six-sigma approach. This partially explains why its performance is better. If we incorporate this feature extraction capability and advanced pattern classification methods into the agents, our agent based model will have much better performance.

4 Conclusions and Future Work

We present a multi-agent model for fault detection and diagnosis that is decentralized, robust, and flexible. In this model, agents having simple computational tasks are distributed across the plant and monitor local equipment. A leader agent, who collects information from other agents, progressively classifies the fault by matching the emerging spatio-temporal pattern with known patterns stored in its knowledge base. Since each agent has a potential to be the leader, our approach avoids having a fixed central point of failure or bottleneck, making it more robust to handle unexpected situations such as malfunctioning of a few agents.

Our experiments show that even using simple detection and classification methods in the agents, the performance of our model can be comparable to that of a centralized approach. In addition, agents can be easily modified to reflect local changes on the plant without affecting the overall architecture of the system.

One potential shortcoming of our approach is that an agent may assume a leadership role due to a false alarm. For most situations, it will give up its leadership quickly since the false alarm will not last long. However, if this false alarm is right before a true fault whose diagnosis should be led by other agents, the leader may fail to classify it since it does not have the right knowledge. In this case, it should give up its leadership and let the more appropriate agent be the leader. We will incorporate this mechanism into our model in the future.

In the future, our model will also incorporate a collective consensus protocol, similar to the one presented in [10], where a group of leaders jointly make the decision. In addition, we will investigate a mechanism for grouping the sensors and assigning them to an agent dynamically so that data correlation within a group of sensors can be exploited.

References

1. Chiang, L.H., Russell, E.L., Braatz, R.D.: Fault Detection and Diagnosis in Industrial Systems. Springer, London (2001)
2. Crowder, J.A.: Multiple information agents for real-time ISHM: Architectures for real-time warfighter support. In: Proc. of Int. Conf. on Artificial Intelligence (2010)
3. Downs, J.J., Vogel, E.F.: A plant-wide industrial process control problem. Computers and Chemical Engineering 17(3), 245–255 (1993)
4. King, A., Cowlishaw, G.: Leaders, followers and group decision-making. Communicative & Integrative Biology 2(2), 147–150 (2009)
5. Mangina, E.E., McArthur, S.D.J., McDonald, J.R.: COMMAS (COndition Monitoring Multi-Agent System). Autonomous Agents and Multi-Agent Systems 4(3), 279–282 (2001)
6. Mendoza, B., Xu, P., Song, L.: A multi-agent model for fault diagnosis in petrochemical plants. In: Proc. of 2011 IEEE Sensors Applications Symposium (2011)
7. Perk, S., Teymour, F., Cinar, A.: Statistical monitoring of complex chemical processes using agent-based systems. Industrial & Engineering Chemistry Research 49(11), 5080–5093 (2010)
8. Raich, A., Cinar, A.: Multivariate statistical methods for monitoring continuous processes: Assessment of discrimination power of disturbance models and diagnosis of multiple disturbances. Chemometrics and Intelligent Laboratory Systems 30, 37–48 (1995)
9. Seng, N.Y., Srinivasana, R.: Multi-agent based collaborative fault detection and identification in chemical processes. Eng. Applications of Artificial Intelligence 23(6), 934–949 (2010)
10. Yu, C.H., Werfel, J., Nagpal, R.: Collective decision-making in multi-agent systems by implicit leadership. In: Proc. of 9th Int. Conf. on Autonomous Agents and Multiagent Systems (2010)

Matching and Retrieval of Medical Images

Amir Rajaei and Lalitha Rangarajan

Abstract. Digital imaging has revolutionized the field of medical imaging and has led to the development of sophisticated computer hardware technologies and specialized software that empower physicians to better distinguish abnormalities, characterize findings, supervise interventions and predict prognosis. In fact, CAD is one of the major research subjects in medical imaging and diagnostic radiology. These benefits have motivated researchers to develop dedicated systems to specific medical domains from clinical decision making to medical education and research. The medical imaging field has generated additional interest in methods and tools for the management and analysis of these images. It is important to extend such applications by supporting the retrieval of medical images by content.

Keywords: Computer Aided Diagnosis, Health Database Management, Medical Image Retrieval, Semantic, Relevance Feedback.

1 Introduction

Content Based Medical Image Retrieval (CBMIR) is quite different from Content Based Image Retrieval (CBIR) as the retrieval similarity must consider the medical context, recognizing the specific organs with their relative locations as well as the user's individualized subjectivity. Consequently, general CBIR systems cannot guarantee a meaningful query completion when used within the medical context. The results are rather poor when CBIR systems are used to retrieve medical image [12]. CBMIR ranges from clinical decision support to medical education and research. In medical research, researchers can use CBMIR to find images with similar pathological areas and investigate their association. Medical education can lead lectures to easily fin images with particular pathological attributes that can imply particular diseases [31]. Medical images are multimodal, heterogeneous and higher dimensional with temporal properties which distinguish them from images in the other domains. In Computer Aided Diagnosis (CAD), the basic objective is to provide diagnostic support to the physicians or radiologist by displaying relevant

Amir Rajaei · Lalitha Rangarajan
Department fo Studies in Computer Science, University of Mysore
570006, Mysore, India
e-mail: amir_rajaei@hotmail.com, lali85arun@yahoo.co.in

A. Abraham et al. (Eds.): International Symposium on DCAI, AISC 91, pp. 27–33.
springerlink.com

past cases. Different types medical images like X-ray, Ct, MRI, SPECT, PET, Ultrasound are playing an important role in detecting anatomical and functional information of the body part for diagnosis. The physicians and radiologists are provided by medical images for diagnosis and therapy [4]. However, handling a large number of images in the health Database Management is to integrate imaging modalities and interfaces with hospitals and department information systems in order to mange the storage and distribution of images to radiologists, physicians, specialist, clinics and imaging centers [29]. Most traditional medical image retrieval systems exploit adding metadata such as captioning, keywords or descriptions to the images. So, retrieval can be performed over the annotated words. Annotating process of images is not only time-consuming, laborious and expensive but also a subjective task due to the experts and keywords are inadequate to represent the image content. CBMIR systems can be used for searching and retrieving different kinds of medical images from large databases on the bases of the visual content of the images. In this context, the design of efficient matching and retrieval techniques in medical image database becomes an important issue. Quick accessing and referencing to the specific organ of the body in real time require perceiving some spatial topologies and understanding the high-level semantic information of the images. In addition, images should be categorized according to the imaging modalities and be classified based on the body organs and orientation for a better kind of matching and retrieval.

The rest of the paper is organized as follows: In section2, we have a brief review on segmentation of medical images. Various methods for feature extraction are discussed in section 3. Feature selection and representation are explained in section 4. In section 5, we present the methods which have been used for classification of medical images. Matching and Retrieval which are the main issue of this doctoral paper are covered in section 6. Finally, overall conclusions are presented in section 7.

2 Segmentation

Segmentation is a basic operation in medical image processing and it involves the recognition and localization of sub patterns within an image. Low resolution and strong noises are two common characteristics which medical images cannot be precisely segmented and extracted for the visual content of their features. Many attempts have been done in this area but still segmentation is a challenging task. Dynamic Edge Tracing with Recovery and Classification (DTRAC) was introducing by Withey et al. [33].In [28], the employed Minimum Spanning Tree (MST) for determining the set of nodes representing the connected components. They represented silent objects of an image by proposing an efficient procedure for extracting the boundary of visual objects from connected components. Li et al. [34] used watershed, gradient based segmentation technique. 2D discrete cosine transforms which has high computational efficiency and accuracy in segmentation was exploited by Pun and Zhu [21].

3 Feature Extraction

In literature, different modalities of medical images could be discriminated using basic low-level characteristics such as particular colors, textures or shapes and they are at the base of most image analysis methods. In[24], they used the concept of color conversion to add color to gray scale images. Since the majority of medical images are in gray level, the general retrieval based on color feature is useless. The retrieval based on texture and shape features are adapted wide spread. Texture is a primary feature of medical images which supplies us with much clinical texture information and most kinds of diseases can be automatically distinguished by distribution of texture feature. Spatial Gray LEVEL Dependence (SGLD) method was employed by Shshadri et al. [26]. However, this technique is the worst in terms speed and its computational complexity depends on the size of co-occurrence matrix. In [32], they recommended a Gabor filtering method,. Gray Level histogram moment statistical texture analysis method was used by Pharwaha and Singh [19]. In their method, they applied Non-Shannon measure to compute entropy. In[22],exploited texture feature (energy and entropy) based on Ridgelet transform. Traditional Gray- level Histogram texture descriptor obtained from the Harlick descriptors to perform homogeneity, energy, contras and correlation was proposed by Bugatti et al. [3]. The shape of the objects plays an essential role among the different aspects of visual information. It is a very powerful feature when used in similarity, search and retrieval. Liu and Tong [14] proposed a novel salient detector based on spectrum energy variation. In [36], they optimized the feature extraction based on geometrical shape, Edge Chain Code (ECC) and Geometric Moment Invariant (GMI). Vijay et al. [30] used the Generic Fourier Descriptor (GFD) with brightness as additional parameter to have good retrieval accuracy.

The combination of different feature extractors of an image can improve the performance of medical systems. In [1], they combined rotation invariant Contourlet transform and Fourier descriptors to extract shape feature. Youssif et al. [37] approach was based on color (HSV using quadratic distance equation), texture (pyramid structure wavelet) and shape (Fourier descriptor). A medical image may implicitly require various kinds of visual reasoning about the meaning or the purpose of different objects in an image. The discrepancy between the actual information and its representation using the computed feature values is known as semantic gap. One approach to solving this problem is to associate high-level semantic information with low-level visual data. The local Fuzzy Fractal Dimension (LFFD) was proposed to extract local fractal feature of medical images [38]. Fractal feature was applied in medical images retrieval by Wu et al. [34]. In [8], they applied kernel density estimation statistical model to describe the complicated medical image data. Finally, the hill climbing strategy of artificial intelligent was proposed to extract the semantic features.

4 Feature Selection and Representation

Improving an image retrieval technique requires modifying the image representations.

It is well known that the choice of an adequate feature vector develops the accuracy of image retrieval. Feature selection is achieved by removing irrelevant, redundant and noisy features, selects the subset features which can achieve the best performance in terms of accuracy and computation time. The methodology used in [20,24] was comprised of continuous and probabilistic image representation scheme using Gaussian Mixture Modeling (GMM) along with information-theoretic image matching measure (KL). Texture based symbolic feature for medical image representation was proposed by Florea et al. [9] to assess the performance of a new image symbolic descriptor for image modality, anatomic, region and view angle image categorization. Their approach achieved high recognition rate. In [27], they employed a wrapper strategy that searches for best-reduced feature set. The quality of feature subset was evaluated in term of measuring the ranking quality which was evolved by genetic algorithm.

5 Classification

The medical image classification is an important issue for computer-aided diagnosis. Traditionally, medical images have been classified by experts. Problem of medical image classification is a new and great challenge, because of great imbalance between classes, visual similarities between some classes, variety in one class and difficulty to define discriminative of visual features. Mueen et al.[17] were focused on Support Vector Machine(SVM). In [13], the presented a novel method for medical image classification using Fuzzy Support Vector Machine (FSVM). In[10, 23], they provided a technique for the determination of optimal cluster number form the supplied set of initial cluster centers in K-mean clustering algorithm. Their proposed technique had low complexity, high transparency and accuracy. Sharma et al. [25] analyzed medical images based on hybridization of syntactic and statistical approaches using Artificial Neural Network (ANN).

6 Matching and Retrieval

Knowledge base is created through construction of multilevel classification. The process of object recognition consists of matching features extracted from a given input image with those models. The accuracy of the system can be improved by an iterative refinement process of retrieved images guided by user's interaction known as the relevance feedback mechanism. Web MIRS system proposed by Long R et al. [15] retrieved X-ray image based on automated image segmentation, image feature extraction and organization along with associated textual data. Image Retrieval in Medical Application (IRMA) system could handle retrieval from a large set of radiological image obtained from hospitals based on various textural features [12]. Med GIFT retrieved divers medical images. High- dimensional feature space of various low-level feature was used as visual terms analogous to the use of keywords in a text-based retrieval approach [18]. Some other existed medical images system can be named as ASSERT, CasImage, NHANES II, FSSEM, COBRA, I2C. In [16], they presented an intermediate level image representation based on category membership scores, feature-level fusion by probabilistic classifier combinations and an adaptive similarity fusion scheme. Cheng et al. [5] extended the concepts of SIFT technique for extracting and matching discriminative feature. Jing and Yang [11] used the Discriminate Adaptive Nearest Neighbors (DANN) metric to identify the similar cases from a library for a given query. Xu et al. [35] proposed an innovative Partial Shape Matching (PSM) technique using Dynamic Programming (DP) for the retrieval of X-ray images. In [6,7], they exploited Latent Semantic Indexing (LSI) technology to implement image retrieval. It was based on its semantic information. Bueno et al. [2] presented the Fractal Scaled Product Metric(FPM) as a similarity assessment.

7 Discussion and Conclusion

The goals of medical systems have often been defined to deliver the needed information at the right time, right place and too the right person for improving the quality and efficiency of care processes. We are going to develop a system to assist in healthcare and clinical decision making for diagnosis of diseases. It is understood through a detailed survey that many contributions are reported on segmentation of medical images and low-level visual feature extraction on color, texture and shape. However, in specific medical domain applications, the semantic content is more desirable since it facilitates the physicians to have better and more accurate classification and retrieval system through retrieving the relevant cases of particular diseases. Therefore, one key issue to be faced is the identification and extraction of semantic information from the visual data through very little work is reported on high-level semantic feature extraction in medical images.

References

1. Arun, K.S., Menon, H.P.: Content Based Medical Image Retrieval by Combining Rotation Invariant Contourlet Features and Fourier Descriptor. International Journal of Recent Trends in Engineering 2(2), 35–39 (2009)
2. Bueno, R., Kaster, D.S., Paterlini, A.A., Traina, A.J.M.: Unsupervised Scaling of Multi-descriptor Similarity Functions for Medical Images Datasets. In: 22nd IEEE International Symposium on Computer-Based Medical Systems, pp. 1–8 (2009)
3. Bugatti, P.H., Ribeiro, M.X., Traina, J.M., Traina Jr., C.: Content-based retrieval of medical images by continuous feature selection. In: IEEE International Symposium on Computer-Based Medical Systems, pp. 272–277 (2008)
4. Chatzishristofis, S.A., Boutalis, Y.S.: Content Based Radiology Image Retrieval using a Fuzzy Rule Based Scalable Composite Descriptor 46, 493–519 (2010)
5. Cheng, W., Hamarneh, G.: N-SIFT: N-Dimensional Scale Invariant Feature Transform for Matching Medical Images. In: 4th IEEE International Symposium on Biomedical Imaging: From Nano to Macro, pp. 720–723 (2007)
6. Chen, L., Zeng, J., Pei, J.: Classifying Noisy and Incomplete Medical Data by a Differential Latent Semantic Indexing Approach (2007)

7. Chen, Q., Tai, X., Jiang, B., Li, G., Zhao, J.: Medical Image Retrieval Based on Latent Semantic Indexing. In: IEEE International Conference on Computer Science and Software Engineering, pp. 561–564 (2008)
8. Conghua, X., Yuqing, S., Jinyi, C.: A New Method of Semantic Feature Extraction for Medical Image Data. Wuhan University Journal of Natural Science 1 11(5), 1152–1156 (2006)
9. Florea, F., Barbu, E., Rogozan, A., Benshair, A.: Using Texture-Base Symbolic Features for Medical Image Representation. In: IEEE 18th International Conference on Pattern Recognition, pp. 946–949 (2006)
10. Ganguly, D., Mukherjee, S., Naskar, S., Murherjee, P.: A Novel Approach for Determination of Optimal Number of Cluster. In: IEEE International Conference on Computer and Automation Engineering, pp. 113–117 (2009)
11. Jing, H., Yang, Y.: Image Retrieval for Computer-Aided Diagnosis of Breast Cancer. In: IEEE Southwest Symposium on Image Analysis & Interpretation, pp. 9–12 (2010)
12. Lehman, T.M., Wein, B.B., Dahmen, J., Vogelsang, F., Kohnen, M.: Content-Based Image Retrieval in Medical Application. In: Proceeding of SPIE, pp. 312–320 (2000)
13. Li, B., Xu, Q.: Medical Image Classification Based 0n Fuzzy Support Vector Machines. In: IEEE International Conference on Intelligent Computation Technology and Automation, pp. 145–149 (2008)
14. Liu, W., Tong, Q.Y.: Medical Image Retrieval Using Salient Point Detector. In: IEEE Annual Conference Engineering In Medicine and Biology, pp. 6352–6355 (2005)
15. Long, R.T., Thoma, G.R.: Land Marking and Feature Localization in Spine X-ray. Journal of Electrical Imaging 10(4), 936–956 (2001)
16. Mahmudar Rahman, M., Desai, B.C., Bhattacharya, P.: Medical image retrieval with probabilistic multi-class support vector machine classifiers and adaptive similarity fusion. Computerized Medical Imaging and Graphics 32, 95–108 (2008)
17. Mueen, A., Sapiyan Baba, M., Zainuddin, R.: Multilevel Feature Extraction and X-Ray Image Classification. Journal of Applied Sciences 7(8), 1224–1229 (2007)
18. Muller, H., Rosset, A., Vallee, J., Geissbuhler, A.: Integrating Content-Based Visual Access Methods Into a Medical Case Database. Proc. Med. Infom. Europe., 480–485 (2003)
19. Pharwaha, A.P.S., Singh, B.: Shannon and Non-Shannon Measures of Entropy for Statistical Texture Feature Extraction in Digitized Mammograms. In: World Congress on Engineering and Computer Science, USA (2009)
20. Pinha, A., Greenspan, H.: A Continuous and Probabilistic Framework for Medical Image Representation and Categorization. In: Proceeding of SPIE, pp. 230–238 (2004)
21. Pum, C., Zhu, H.: Image Segmentation Using Discrete Cosine Texture Feature. International Journal of Computers 4, 19–26 (2010)
22. Rantnaparkhe, V.R., Manthalkar, R.R., Joshi, Y.V.: Texture Characterization of CT Images Based on Ridgelet Transform. ICGST-GVIP Journal 8(V), 43–50 (2009)
23. Ray, C., Sasmal, K.: A New Approach for Clustering of x-ray Images. International Journal of Computer Science Issue 7(4), 8, 22–26 (2010)
24. Sathik, M.: Feature Extraction on Colored X-Ray Images by Bit-plane Slicing Technologies. International Journal of Engineering Science and Technology 2(7), 2820–2824 (2010)
25. Sharma, N., Ray, A.K., Sharma, S., Shukla, K.K., Prandhan, S., Aggarwal, L.M.: Segmentation and Classification of Medical Images Using Texture-Primitive Features: Application of BAM-type Artificial Neural Network. Journal of Medical Physics 33(3), 119–126 (2008)

26. Sheshadri, H.S., Kandaswamy, A.: Experimental Investigation on Breast Tissue Classification Based On Statistical Feature Extraction of Mammograms. Computerized Medical Imaging and Graphic 13, 46–48 (2007)
27. Silva, S.F., Traina, J.M.: Ranking Evolution Functions to Improve Genetic Feature Selection in Content-Based Image Retrieval of Mammograms. In: IEEE 22nd International Symposium on Computer-Based Medical Systems, pp. 1–8 (2009)
28. Stanescu, L., Burdescu, D.D.: Medical Image Segmentation- A Comparison of Two Algorithms. In: IEEE International Workshop on Medical Measurements and Applications, pp. 165–170 (2010)
29. Trojacanecc, K., Dimitrovski, I., Loskovska, S.: Content Based Image Retrieval in Medical Applications: An Improvement of the Two- Level Architecture. In: IEEE EUROCON, pp. 118–121 (2009)
30. Vijay, A., Bhattacharya, M.: Content-Based Medical Image Retrieval Using the Generic Fourier Descriptor with Brightness. In: 2nd International Conference on Machine Vision, pp. 330–332 (2010)
31. Wei, C., Li, C., Wilson, R.: A Content-Based Approach to Medical Image Database Retrieval. In: Ma, Z. (ed.) Database Modeling for Industrial Data Management: Emerging Technologies and Applications, pp. 258–290. Idea Group Publishing, USA (2005)
32. Wei, C.H., Li, Y., Li, C.T.: Effective Extraction of Gabor Features for Adaptive Mammogram Retrieval. In: IEEE International Conference on Multimedia and Expo., pp. 1503–1506 (2007)
33. Withey, D.J., Pedrycz, W., Koles, Z.J.: Computer Vision and Image Understanding 113, 1039–1052 (2009)
34. Wu, J., Jiang, C., Yao, L.: Medical Image Retrieval Based on Fractal Dimension. In: 9th International Conference for Young Scientists, pp. 2959–2961 (2008)
35. Xu, X., Lee, D., Antani, S.: A Spine X-ray Image Retrieval System Using Partial Shape Matching. IEEE Transactions on Information Technology in Biomedicine 12(1), 100–108 (2008)
36. Yin, Y., Tian, G.Y.: Feature Extraction and Optimization for X-Ray Weld Image Classification. In: 17th World Conference on Nondestructive Testin, China (2008)
37. Youssif, A.A., Darwish, A.A., Mohamed, R.A.: Content based medical image retrieval based on Pyramid Structure Wavelet. International Journal of Computer Science and Network Security 10(3), 157–164 (2010)
38. Zhang, X., Meng, Q.: Local Fuzzy Fractal Dimension and its Application in Medical Image Processing. Artificial Intelligence in Medicine 32, 29–36 (2004)

Advanced System for Management and Recognition of Minutiae in Fingerprints

Angélica González, José Gómez, Miguel Ramón, and Luis García

Abstract. This article briefly describes the advanced computer system designed for the recognition of minutiae in fingerprints digital images. The system provides both automatic and manual extraction of relevant data from the fingerprints images, storing that information in a database. Provides statistical calculations, including calculations for cumulative frequency analysis; this is an important parameter for calculating distinction rates. The system is enabled to differentiate by sex, finger, fingerprint type and sector that has been divided the dactylogram.

Keywords: Minutiae recognition, Automatic minutiae extraction, Fingerprints statistical calculation, AFIS (Automated Fingerprint Identication System), Dactylogram.

1 Introduction

The main objective is developing a comprehensive management system for fingerprints digitalized images and minutiae; extracting the minutiae automatically and providing functionality to system users to make manual corrections at any time. All these operations will be recorded in the system database, which will compile the information of the minutiae and fingerprints for subsequent statistical analysis.

The second objective is to seek a complementary method for fingerprint identification by the automatic calculation of frequencies of the characteristic points, the minutiae are located by determining the number of ridges that are crossed by an imaginary line that connects each minutia and fingerprint center and also by a Cartesian grid that can generate a variable size environment defined by the user.

Angélica González · José Gómez · Miguel Ramón · Luis García
Computers and Automation Department, Universidad de Salamanca, Salamanca, Spain
e-mail: {angelica,marin,luisjaviergs}@usal.es,
miguel.ramon@dgp.mir.es

A. Abraham et al. (Eds.): International Symposium on DCAI, AISC 91, pp. 35–42.
springerlink.com

2 System Description

The system for management and minutiae recognition has been implemented as an application that lets the user interact with digital images of fingerprints of many kinds, including WSQ format designed by the FBI, which is a compressed format specially dedicated to fingerprints;. The application consists of a main program incorporating a number of libraries where all the features of the system are implemented, there is also a communication with a database that stores information of individuals, their fingerprints stored into digital images and the minutiae extracted of the fingerprints.

3 Processing Digital Images of Fingerprints

The application can perform a complete fingerprint image processing; including a large number of filters and operations to carry out treatment and enhancement of the images. It is very important that the quality of the images managed by the system is high, meaning as quality tha the images are not damaged and, therefore, that the ridges and valleys of the fingerprints are well differentiated. Ideally, the dactylogram presents this quality from the source for high rates of success in extracting minutiae; otherwise the filtering functions can overcome some shortcomings of the image source. Among the filtering techniques are include: [1]

- Smooth: This filter repair damaged or low quality fingerprints images, softening gray boundaries of the entire image. The smoothing filter, as others used in the system, can be viewed as a 2D linear filter. The filters in two dimensions, using a convolution matrix whose form is generally 3x3 or 4x4 matrixes where the filter coefficients are stored.
- Medium: A medium-type filter is a nonlinear filter that calculates the median value of replacing each image by the median of the values that surround it in a window. The window size is usually 5x5. This filter has been used in the system to clean the smudges from the edges of the ridges and valleys contained in the image of dactylogram.

4 Extraction of Minutiae in Fingerprints Images

The application enables automated and manual minutiae extraction in a dactylogram. Before the system user can extract the minutiae, the user must manually enter the center of the dactylogram generating a Cartesian axes grid, typically of 0.75 mm, because to 1 mm can be more than 1 minutia point [2,3] this value is configurable by the system user. For automatic extraction some operations are performed in the fingerprint image, which consist of preprocessing, which cleanses the image of its imperfections, getting the last item on the thinning image, at this point the image is prepared to search automatically the characteristics that make it unique from any other.

Fig. 1 Extracting minutiae operations of a dactylogram.

This technique is known as classical extraction [4], it performs a preprocessing of the image before detecting the minutiae. Once done the system look for patterns to identify the fingerprint image, in which the width of the ridges is one pixel. Dactylogram preprocessing operations make the extraction algorithm can work with a wide range of qualities; the price paid is the time to perform this preprocessing. To extract the characteristic points first a sweep of the entire image preprocessed is made to detect those points that are candidate sites by analyzing patterns around them that make up a resizable window, and after each point determines its orientation to identify the subtype.

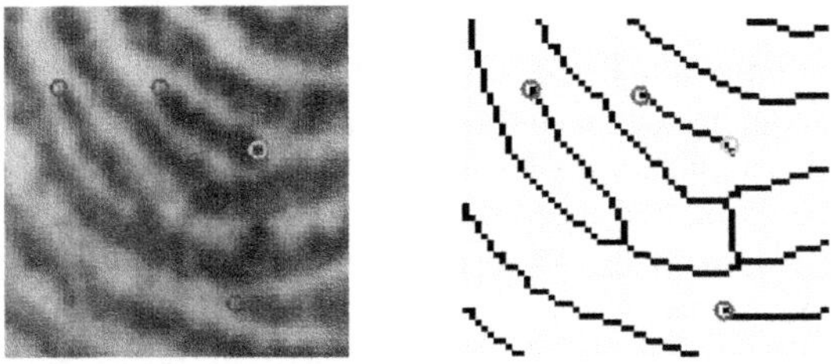

Fig. 2 Minutiae extracted by the system on a section of a dactylogram

The system is able of distinguishing between eighteen different characteristic points, as shown in figure 3.

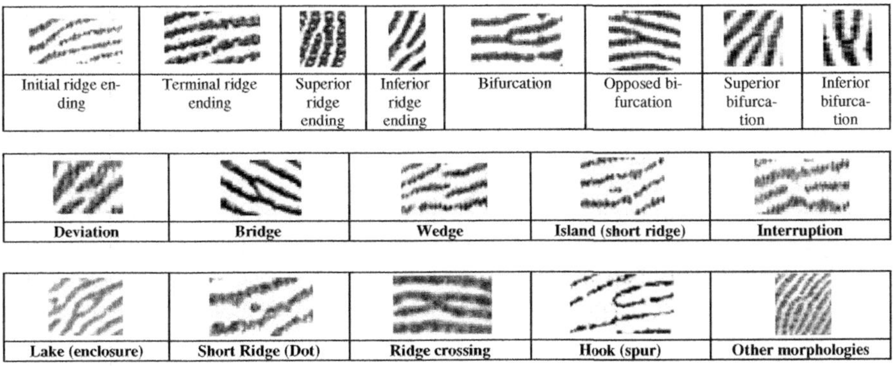

Fig. 3 Morphology of minutiae

Information concerning the minutiae is stored in the system database, the most important aspects that are saved are the type of minutiae, point location within the dactylogram, this location is stored in different scales; in pixels of the image on the grid X and Y axis of the Cartesian grid whose center is identified by the user on the image of dactylogram also calculated and automatically stores information on the number of ridges crossing a imaginary line connecting the minutia with the center of dactylogram.

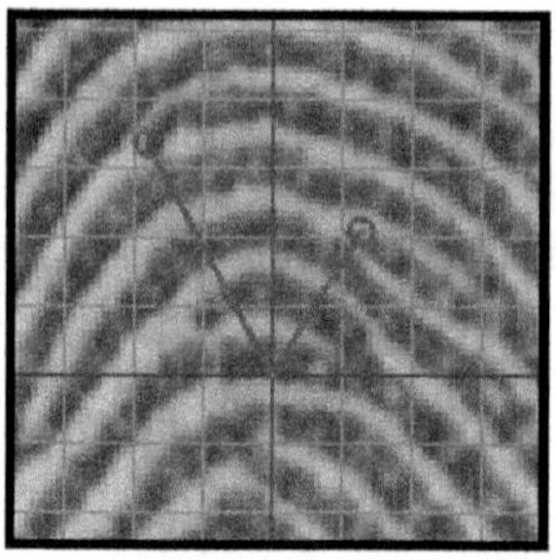

Fig. 4 Minutiae with Cartesian axes and grid lines at the center of a dactylogram

There are not a large number of computer systems that manage and work on alternative methods that are raised in this paper. However, studies [5] that have a similar approach to the application of filters to reduce noise dactylogram image, but about the minutiae extraction using wavelet transformation in analysis of sub windows of the image around the Core Point of the dactylogram. In the system described in this paper we opted for a classical approach of pattern matching, because it was identified the need to work with partial fingerprints where the Core Point could not be present and the range of types of minutiae to be extracted was large.

Comparative minutiae detection algorithms based on the quality of the source image [6] indicate that the ridge valley clarity approach, as used in the system preprocessing of the image fingerprint implemented with local thinned clarity score, have the advantage of that clarity between ridges and the valleys can be calculated by counting the misclassified pixels, while the weakness is in the region of high curvature as in some points of singularity. This extraction technique is among the best results for a wide range of fingerprint image qualities.

5 Generation of Statistics

With the minutiae extraction data stored the system is capable of generating a wide range of statistics that allows further study of the fingerprints processed by the system to obtain relevant information to indicate whether the fingerprints oscillate significantly between men and women, each of the fingers or by classification of the same fingerprint. The computer application is able to elaborate the following statistics:

1. Graphs and tables of the frequencies of different minutiae as the number of ridges that separate them from the central point.

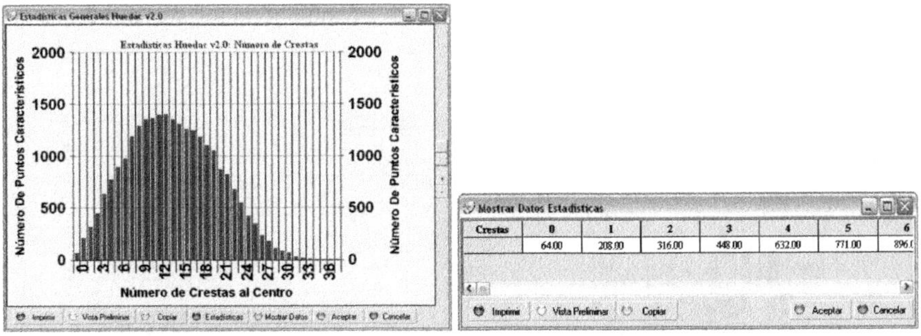

Fig. 5 Graph and frequency data by the number of ridges

2. Graphics and percentage frequency tables of individual minutiae as the number of ridges that separate them from the central point.

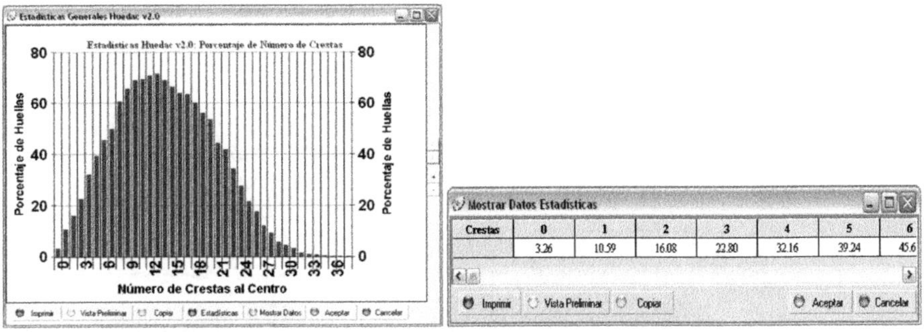

Fig. 6 Graphic and percentage frequency data the number of ridges

3. Graphs and tables of the frequencies of different minutiae by their location in grid.

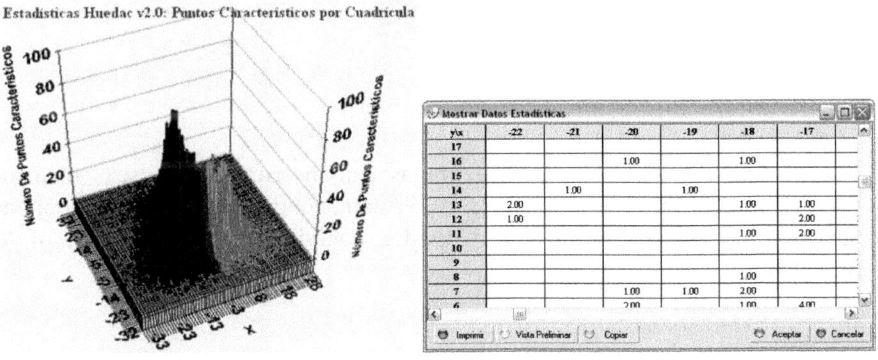

Fig. 7 Chart and data grid frequency

4. Graphs and tables of percentage frequencies of the characteristic points by their location in grid.

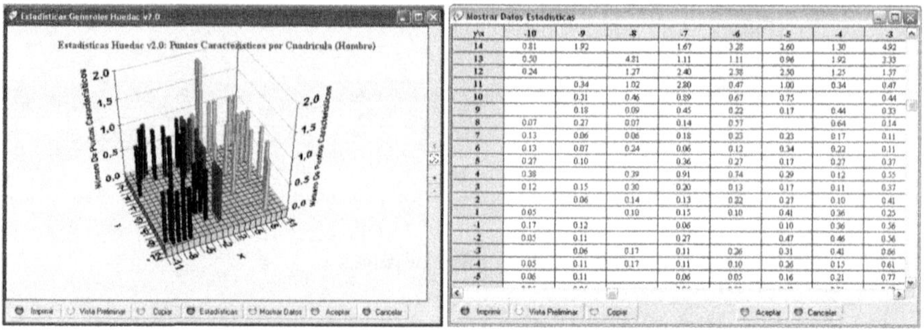

Fig. 8 Graphic and percentage frequency data of the number of ridges

5. Calculation of cumulative frequency of the selected points as they are marking. [7].

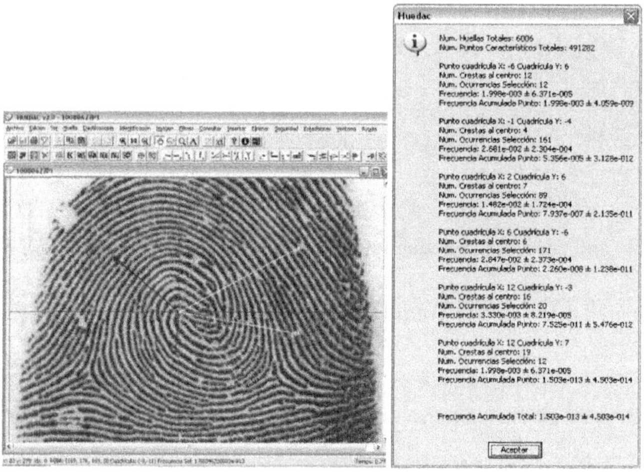

Fig. 9 Cumulative frequencies

Regarding the above tables and graphs can discriminate by sex (man and woman), finger (between 10 fingers), type (plain, central pocket, double loop, accidental) and that sector has been divided dactylogram and for each of the estimated minutiae.

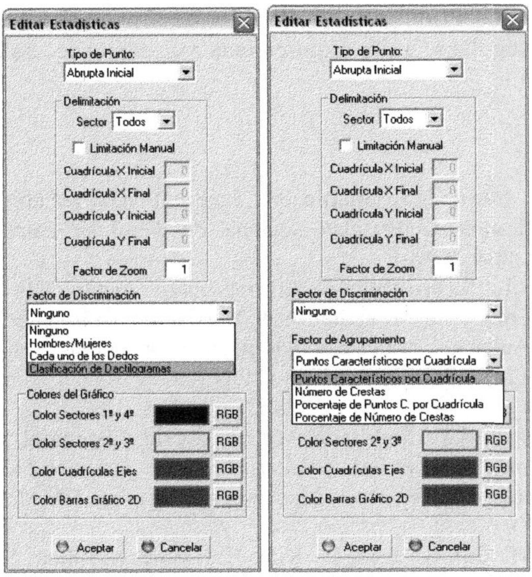

Fig. 10 Dialog for statistical discrimination

6 Conclusions and Lines of Future Work

The most important and complex process of implementation is the extraction of the minutiae of fingerprints; This automatic process is based on a minutiae detection algorithm on the preprocessed image using features of automatic pattern recognition in digital images. As success rates of these automated processes are moderate, especially if dactylogram quality is not high, can be further improved in future work. Besides the characteristic points shown can be manually edited by the user and the centre of dactylogram to make corrections if consider necessary.

The generation of stored data statistics can be considered as the ultimate goal of the system. Thus, the application offer the possibility to make statistical studies o fingerprints stored in the system to obtain important information on whether the characteristic points oscillate significantly between men and women, each of the fingers or according to their classification. You can also calculate the frequency of each of the minutiae and also the cumulative frequencies of a set of selected points, the calculation to ascertain the rates of discrimination in fingerprint identifications. When dactylogram also has a classified, allowing comparison with other fingerprints stored in the database, showing some statistics indicating the degree of similarity between compared fingerprints.

It is recommended further improve of the automatically extraction of minutiae to be more precise, accurate and efficient, here is expected that the use of neural networks is the best option [8] especially in low-quality images, but is also expected that the high success rates cannot be generalized to all fingerprints, since the results are not guaranteed. The evolution of the application also goes for using it in a client-server environment with different users access to a common database, strongly considering the security requirements for remote access [9].

References

[1] Arsuate, G.A., Nasisi, O.H., Martin, M.: Recognition of features in fingerprints for human identification. Universidad Nacional de San Juan. Faculty of Engineering. Automation Institute (1997)

[2] Sclove "The, S.L.: occurrence of fingerprints characteristics as two dimensional processes". Journal of the American Statistical Association 74(367) (1979)

[3] Trauring, M.: Automatic comparasion of finger-ridge patterns. Natura, 938–940 (1963)

[4] The fingerprint identification (Prodac + C Fits), `http://www.ica.es.com/prodac/dossier.htm`

[5] Janjua, F., Javed, M.Y., Sarfraz, N.: ch.19: Hybrid Fingerprint Verification System Based on Fusion of Feature Extraction and Minutiae Detection Strategy. In: 3rd International Conference on Geometric Modeling and Imaging, GMAI 2008, July 9-11, pp. 114–119 (2008)

[6] Jin, C., Kim, H., Cui, X., Park, E., Kim, J., Hwang, J., Elliott, S.: Comparative Assessment of Fingerprint Sample Quality Measures Based on Minutiae-Based Matching Performance. In: Second International Symposium on Electronic Commerce and Security, May 22-24, vol. 1, pp. 309–313 (2009), doi:10.1109/ISECS.2009.59

[7] Study on the frequency of appearance of the minutiae in fingerprints, FRAPUC. Commissioner General of Police Science. General Directorate of Police and G.C. Date of publication (2002)

[8] Bajo, J., De Paz, J.F., Rodríguez, S., González, A.: Hierarchical neural network for clustering and classification. Logic Journal of the IGLP, `http://bisite.usal.es/webisite/?q=en/node/160`, ISSN 1368-9894, ISSN 1367-0751

[9] Pinzón, C.I., De Paz, J.F., Rodríguez, S., Corchado, J.M., Bajo, J.: A hybrid agent-based classification mechanism to detect denial of service attacks. Journal of Physical Agents (JoPHA) 3(3), 11–18 (2009), doi: `http://hdl.handle.net/10045/12528`, `http://bisite.usal.es/webisite/?q=en/node/92`, ISSN: 1888-0258

Associative Watermarking Scheme for Medical Image Authentication

Neveen I. Ghali, Lamiaa M. El Bakrawy, and Aboul Ella Hassanien

Abstract. With the widespread and increasing use of internet and digital forms of image; and the convencience of medical professionals that the future of health care will be shaped by teleradiology and technologies such as telemedicine in general. In addition to the various radiological modalities which produce a variety of digital medical files most often datasets and images. These files should be protected from unwanted modification of their contents, especially as they contain vital medical information. Thus their protection and authentication seems to be of great importance and this need will rise along with the future standardization of exchange of data between hospitals or between patients and doctors. In this paper, an associative watermarking scheme is conducted to perform associative watermarking rules to the images which reducts the amount of embedded data, vector quantization indexing scheme is used to embed watermark for the purpose of image authentication. The vector quantization decoding technique is applied to reconstruct the watermarked image from the watermarked index table. The experimental results show that the proposed scheme is robust. The watermarked images are resistant to severe image processing attcks such as Gaussian noise, brightness, blurring, sharpening, cropping, and JPEG lossy compression.

Keywords: Vector quantization, Association rules, Edge block detection method.

Neveen I. Ghali · Lamiaa M. El Bakrawy
Faculty of Science, Al-Azhar University, Cairo-Egypt
e-mail: {nev_ghali,lamiaabak}@yahoo.com

Aboul Ella Hassanien
Faculty of Computers and Information, Cairo University
e-mail: aboitcairo@gmail.com

A. Abraham et al. (Eds.): International Symposium on DCAI, AISC 91, pp. 43–50.
springerlink.com

1 Introduction

In digital watermarking techniques, some types of watermarks such as logos, labels, trademark, or random sequence, representing the author's ownership, are embedded into the desired digital image. Generally, a registration to the authentication center is necessary, which helps to solve ownership disputes by identifying the owner of the disputed media. If necessary, the embedded watermark in the digital image can be used to verify ownership [11], [4].

In general, any watermarking algorithm consists of three parts as follows [8]: The watermark, which is the information to be embedded into the original image such as text or an image. The embedding algorithm which is the process to include the watermark into the original image. and the extraction algorithm which is the process to recover the watermark information from the watermarked image using the key and with the help of the original image. In this paper the concept of association rules in data mining based on vector quantization (VQ) and Sobel operator are employed to propose a robust watermarking technique for medical images by embedding the fingerprint as a watermark.

The remainder of this paper is organized as follows. Section (II) reviews the related work. Brief introduction of VQ, association rule, and edge block detection are respectively introduced in section (III). Details of the proposed scheme are presented in section (IV). Section (V) shows the experimental results. Finally, conclusions are discussed in Section (VI).

2 Related Work

Wu et al. in [13] presented a novel digital image watermarking scheme based on VQ technique. During the encoding process of the VQ compression technique, the proposed scheme embeds a representative digital watermark in the protected image so that the watermark can be retrieved from the image to effectively prove which party is in legal possession of the copyright in case an ownership dispute arises. In their method, the codewords in the VQ codebook are classified into different groups according to different characteristics and then each binary watermark bit is embedded into the selected VQ encoded block. El-Bakrawy et al., in [4] proposed an associative watermarking scheme which is conducted by the concept of association mining rules and the ideas of VQ and Sobel operator. Unlike traditional watermarking techniques the association rules of the watermark are hidden instead of the watermark itself. The reconstructed images has robustness against aggressive image processing as as cropping, blurring and sharpening. It has effective resistance to noise.

Shih and Wu in [10] presented a technique for embedding the signature image and the fragile watermark into the frequency domain of RONI part of a medical image by using improved genetic algorithms. By compressing the ROI part using lossless compression and the RONI part using lossy compression, they could obtain a higher compression rate. Furthermore, the fragile watermark is embedded to detect any unauthorized modification.

In this paper, the concept of association rules in data mining based on VQ and Sobel operator are employed to propose a robust watermarking technique. It achieves more effective resistance against several images processing such as sharpening, JPEG lossy compression especially in case of adding in Gaussian noise and blurring.

3 An Overview

3.1 Vector Quantization (VQ)

VQ is a simple data compression technique which was first proposed by Gray [14]. In the beginning, an image is segmented into several blocks with the same size, such as 4×4. Each block consists of 16 pixels. These pixels, from left to right and top to bottom, can form a vector $v = \{v_1, v_2, ..., v_k\}$, where k represents number of dimensions. The pixel value of each block is different, so before encoding, representative vectors, called codeword c, should be collected to form a codebook $CB = \{c^i : i = 0, 1, ..., L-1\}$, where L denotes the codebook size and i denotes the index value. The well-known LBG algorithm [5] can be employed to form the codebook. By clustering code words, it finds a representative codeword from each cluster and uses the representative code words to form a codebook. Through Euclidean distance, we can find code words nearest to the input vector and record the index value of each code word. Once the closest codeword for v is found, the index i of the best matching codeword is assigned to the input vector v for the basis of future VQ decoding [14,2].

3.2 Association Rules

Association rule mining, which was first proposed by Agrawal et al. [1], is one of the most important topics in the area of data mining. It has many successful applications, especially in the analysis of consumer market-basket data [9,12].

An association rule is a probabilistic relationship, with the form $X \Rightarrow Y$ between sets of database attributes, where X and Y are sets and termed as itemsets, and $x \bigcap y = \Phi$. It is inferred empirically from examination of records in the database. Such a rule reveals that transactions in the database containing items in X tend to contain items in Y, and the probability, measured as the fraction of the transactions containing X also containing Y, is called the confidence of the rule. The support of the rule is the fraction of the transactions that contain all items in both X and Y [6].

3.3 Edge Block Detection Method

Edge detection is the task of finding the boundaries between the objects that appear in a digital image [7]. Sobel operator is used to detect image edges by calculating partial derivatives in 3×3 neighbourhoods. The reason of using Sobel operator is that it is insensitive to noise and it has relatively small masks than other operator such as Robert operator, two-order Laplacian operator and so on [3].

4 Proposed Scheme

In the proposed method both the original image X with size $A_X \times B_X$ and the watermark W with size $A_W \times B_W$ are divided into non-overlapping blocks with size $k \times k$, and for each block, the codebook C (including Lk^2-dimensional codewords) is utilized to find the closest codeword so as to obtain the index tables of the original image and watermark, X_T and W_T, respectively. The size of X_T and W_T are $(\frac{A_X}{k} \times \frac{B_X}{k})$ and $(\frac{A_W}{k} \times \frac{B_W}{k})$. Subsequently, association rules defined upon 4-itemset are exploited to mine association rules from X_T and W_T, respectively. Then we embed the association rules generated from the watermark into the original image.

4.1 Mining Association Rules of the Original Image and Watermark

For each index in the index tables of original and watermark images the 4-itemset association rules can be illustrated as $(item1, item2, item3) \Rightarrow (item4)$. The first three items are utilized to find the nearest original image rules for the watermark rules, and by changing the fourth item's value of the rule, which is derived from some selected original image blocks, such that the watermark can be embedded. The four items for each block's rule are defined in Algorithm 1.

4.2 Embedding Process

The detailed procedure of hiding association rules W_R in X_R being derived from X_T and W_T is described as follows:

1. The first three items play a role as the matching pattern in X_R and W_R until each rule in W_R has found one matched rule in X_R.
2. By replacing the index of this block with the item4 value of w_r, the purpose for watermarking is successfully achieved.
3. VQ decoding is performed on the watermarked index table, which has been embedded with all the rules in W_R, to reconstruct the watermarked image.

Algorithm 1. The four items for each block's rule

Input Parameters: T, S, Imagearray, ImageCells, Normalizedfactor
Phase-I Embedding :

Calculate the mean of its index and the indices of its neighbouring eight blocks
if the mean value >= T_1 **then**
  Item1 =1
else
  Item1 = 0
end if
Calculate the variance of its index and the indices of its neighbouring eight blocks
if the variance value >= T_2 **then**
  Item2 =1
else
  Item2 = 0
end if
Sobel is applied to do convolution on this block to determine whether its corresponding codeword is an edge block or not
if any value of those two computed values >= T_3 **then**
  this block is an edge block, and Item3 = 1
else
  it is = 0
end if
The item4 value is the corresponding index value indicated in the index table, Where T_1, T_2, T_3 are given threshold.

4.3 Extracting Process

For extracting the watermark from the watermarked image Y. Four keys should be recorded.

1. key1: the set of all selected X_T 's blocks' locations.
2. key2: the set of original indices of these blocks.
3. key3: the MSE values between the codewords of original indices of these blocks, and the codewords of indices of these blocks after embedding.
4. key4: for each element of key2, record all of its corresponding blocks in W_T

Then the detailed process is described as follows:

1. Perform VQ encoding on the test image Y to obtain the index table Y_T.
2. Use key_1 to pick out the watermarked blocks Y_{TW} from Y_T.

3. key_2 and key_3 are used to examine Y as indicated in [9]. If it is treated as a watermarked image goes to 4, Otherwise, Y is not watermarked, and the extraction is terminated.
4. According to key_4, restore each element in key into its corresponding locations on the watermark index table and perform VQ decoding with the above results to reconstruct the extracted watermark

5 Experimental Results

In this paper, PSNR (Peak Signal-to-Noise Ratio) is used to evaluate the difference between the watermarked and the original medical image; we implemented the proposed algorithm over three different kinds of medical images: CT, MRI and interventional image with size 512×512 as given in Figure (1), respectively. NC (Normalized Correlation) is applied to determine the degree of similarity between the original watermark and the extracted watermark, where the watermak image size is 64×64. For convenience, the thresholds T_1 and T_2 are set as the half of the codebook size which is 128. The threshold T_3 applied to determine whether an image block is an "edge block" is set to 70,

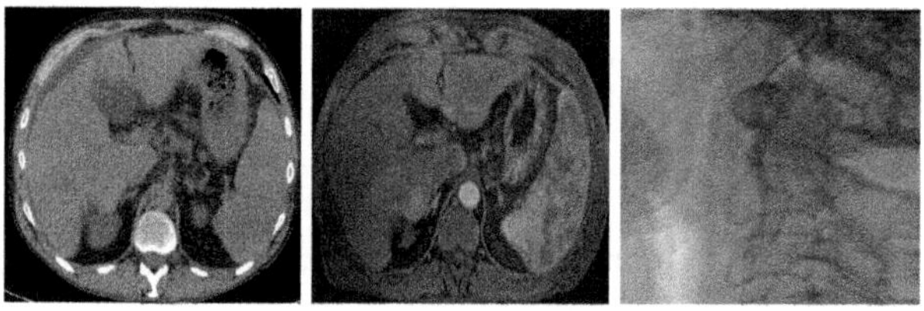

(a) CT Image (b) MRI image (c) Interventional

Fig. 1 Medical Images

Figure (2) shows watermarked images and the corresponding extracted watermarks of the proposed method. Also the well-known image processing software, Photoshop, is applied to perform different image attacks (including JPEG lossy compression (quality level 0) sharpening (3 times), and adding in Gaussian noise (σ= 20), blurring (3 times) on the interventional watermark images. Then NC is used to compare the difference between the original watermark and the extracted watermark. Table (I) shows that no matter which kind of attacks the watermarked images suffer from, extracted watermarks are still similar with the watermark extracted from non-attacked watermarked image.

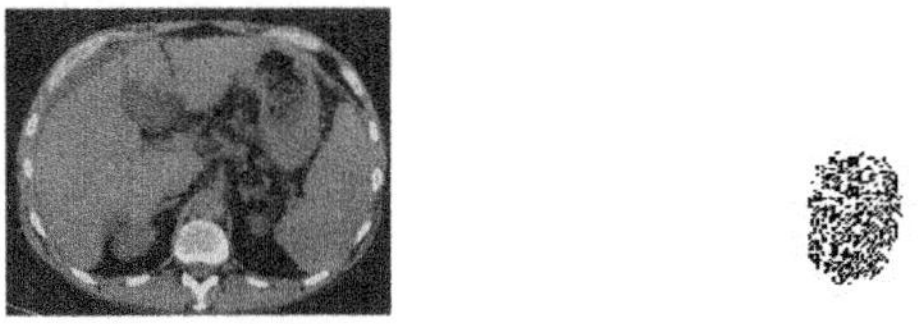

(a) Watermarked CT PSNR = 30.8078(dB) (b) Extracted watermark NC = 0.9962

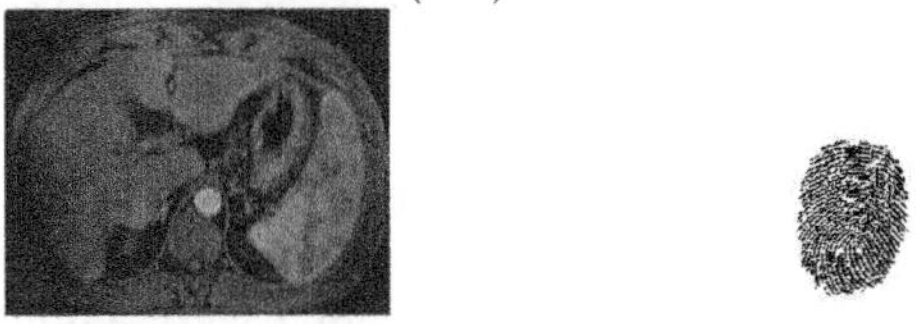

(c) Watermarked MRI PSNR = 30.8620(dB) (d) Extracted watermark NC = 0.9783

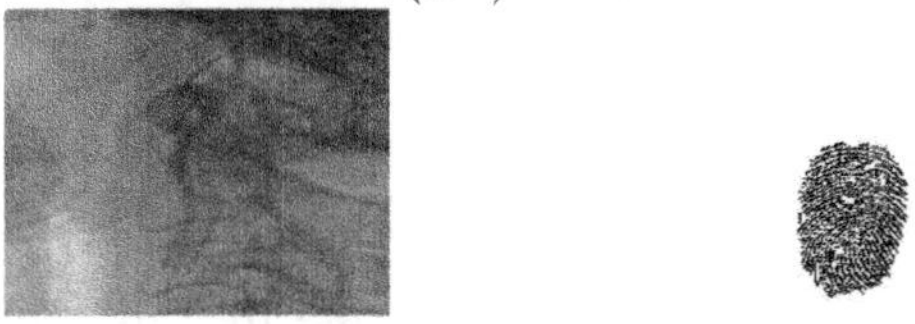

(e) Watermarked Inter. PSNR = 31.8363(dB) (f) Extracted watermark NC = 0.9794

Fig. 2 Watermarked Medical Images and Extracted Watermarks

Table 1 NC values of watermarks extracted from attacked watermarked images of proposed method

Experiments	NC
JPEG lossy compression (quality level = 0)	0.9712
Sharpening three times	0.9708
Adding in Gaussian noise ($\sigma = 20$)	0.9757
Blurring three times	0.9743

6 Conclusions

In this paper, a robust watermarking technique is used depending on vector quantization (VQ) and association rules depending on Sobel operator to detect edge of image as proposed in [4]. This study is implemented to hide biometric data, fingerprint image, over three different types of medical images: CT, MRI and Interventional radiology images. Experimental result shows that the proposed scheme achieves an effective resistance against several images processing such as JPEG lossy compression, sharpening, blurring, and adding in Gaussian noise. Future work in the area should include considering invertible techniques, or ROI techniques if increased robustness is required, and that different watermarks should be applied to different medical image types.

References

1. Agrawal, R., Imielinski, T., Swami, A.: Mining association rules between sets of items in large Databases. In: Proceedings of the ACM SIGMOD Conference on Management of Data, pp. 207–216 (1993)
2. Chen, N., Zhu, J.: Multipurpose audio watermarking algorithm. Journal of Zhejiang University SCIENCE A 4, 517–523 (2008)
3. Dong, Q., Song, C., Ben, C., Quan, L.: A fast subpixel edge detection method using Sobel-Zernike moments operator. Image and Vision Computing 23, 11–17 (2005)
4. El Bakrawy, L., Ghali, N., Hassanein, A., Abraham, A.: An Associative Watermarking based Image Authentication Scheme. Accepted for Publication in ISDA (2010)
5. Linde, Y., Buzo, A., Gray, R.M.: An algorithm for vector quantization design. IEEE Transactions on Communications 28, 84–95 (1980)
6. Ma, W., Wang, K., Liu, Z.: Mining potentially more interesting association rules with fuzzy interest measure. Soft Computing 10, 1–10 (2010)
7. Meinhardt, E., Zacur, E., Frangi, A., Caselles, V.: 3D Edge Detection by Selection of Level Surface Patches. Journal of Mathematical Imaging and Vision 34, 1–16 (2009)
8. Ping, N., Ee, K., Wei, G.: A Study of Digital Watermarking On Medical Image 4(track 14), 2264–2267 (2007)
9. Shen, B., Yao, M., Wu, Z., Gao, Y.: Mining dynamic association rules with comments. Knowl. Inf. Syst. 23, 73–98 (2010)
10. Shih, F., Wu, F.: Robust watermarking and compression for medical images based on genetic algorithms. Information Sciences 175, 200–216 (2005)
11. Tsai, P., Hu, Y., Chang, C.: A color image watermarking scheme based on color quantization. Signal Processing 84, 95–106 (2004)
12. Weng, C., Chen, Y.: Mining fuzzy association rules from uncertain data. Knowledge and Information Systems 9, 24 (2009)
13. Wu, H., Chang, C.: A novel digital image watermarking scheme based on the vector quantization technique. Computers & Security 24, 460–471 (2005)
14. Yang, C., Shen, J.: Recover the tampered image based on VQ indexing. Signal Processing 90, 331–343 (2010)

Static Mutual Approach for Protecting Mobile Agent

Antonio Muñoz, Pablo Anton, and Antonio Maña

Abstract. In terms of the mobile agent paradigm, multi-agent systems represent a promising technology for emerging Ambient Intelligent scenarios in which a huge number of devices interact. Unfortunately, the lack of appropriate security mechanisms, both their enforcement and usability, is hindering the application of this paradigm in real world applications. In this paper, we present a software based solution for the protection of multi-agent systems. Our solution is focused on the cooperative agents model and the core of this concept is the protected computing approach. Finally, one of the most appealing aspects of this approach is based on its user friendly style for agent based system developers who are not security experts.

1 Introduction

In the area of information systems, security is one of the most interesting topics. Recently, with the huge growth in the number of distributed systems, the number of computing attacks has increased and therefore so has the number of protection systems. This paper focuses on mobile agent based systems and the security within them. More specifically, our work deals with static mutual security schemes [8]. First works in software agents arose in the mid 1970s from Carl Hewitt [4]. Hewitt created an agent model (named Actor) which he defined as an autonomous object that interacts and executes concurrently with an internal state and communication capability. Since that initial conception, and due to works developed in Distributed Artificial Intelligence (DAI), a new concept has arisen known as the Multi-Agent System. The main appeal of these systems is that they allow two or more entities to join forces to perform a common task, which is very difficult to complete

Antonio Muñoz · Pablo Anton · Antonio Maña
Computer Science Department, University of Málaga, Spain
e-mail: {amunoz,panton,amg}@lcc.uma.es

A. Abraham et al. (Eds.): International Symposium on DCAI, AISC 91, pp. 51–58.
springerlink.com

individually. Nowadays a huge variation of software agents exists according to their features, abilities or properties.This work addresses the multi-agent systems based on mobile agents, and particularly the security of these systems.

Several protection approaches exist for each of the above points, but this paper will address the related with the protection of agents against the host. We have decided that we do not trust the agencies that host mobile agents when these migrate, producing security risks in the whole multi-agent system. One strategy that helps to solve these security problems, as well as add a higher level of security to the whole system, is based on the protected computing concept [7]. The core of this strategy is based on the idea of dividing the code in two or more mutually dependent parts that will be executed in a trusted processor, while remaining parts can be executed in any other processor, whether trusted or not. In the application of this strategy for the security of multi-agent systems, we have achieved a model in which each agent collaborates with one or more remote agents that are executed in different agencies, trusted or not. Thus a unique successful attack requires the cooperation of every agency in the system, which, in practice, does not make sense. In mutual protection we can differentiate between two schemes. The static mutual protection is the simplest solution fully implemented and described along this paper. And the dynamic mutual protection is an evolved solution more flexible and applicable for any real multi-agent system.

According to protected computing we have developed mutual protection. Furthermore, we have designed and developed tools to carry out this strategy by means of an automatic process. The objective of this tool is that it might be possible to automatically get a secure multi-agent based system on a regular multi-agent system and the mutual protection scheme. This is very useful in testing the efficiency of the system. This paper is organized as follows: in section 2 we review related works and we introduce the MAS (multi-agent system), mobile agents, JADE platform and security schemes. Section 3 presents the main approach of this paper; the automatic generation of a MAS making use of the mutual static strategy. In section 4, we describe the features and architecture of the tools developed, and finally we conclude with some reflections.

2 Related Work

In previous sections we stated that the Protected Computing approach is based on the division of code in two or more parts. Some of these parts will be executed in a trusted processor, but the others will be executed in a regular processor. These divisions are performed in such a way that the execution of the application is not possible without the collaboration of the trusted processor. One of the most important aspects of this technique is the way in which the code is divided. This might be carried out in mutually dependent parts, but it is essential that the public part of code will not be able to be used to get information from the protected one and any communication trace is possible between both parts to get information from the protected part.

A huge number of potential applications exist for this scheme. For instance, we use this idea to protect mobile agents in a MAS where several agents will be sent to different non-trusted agencies to carry out collaborative tasks. It is seriously difficult to warrant the correct execution of the agent and its integrity, chiefly because of the unacknowledged agencies. Thus, the main goal of this scenario is the protection of the agents for potentially malicious agencies. Every agent interacts with one or more remote agents, and these will be executed in different agencies. Then the agents will protect among them and one by one. We previously pointed out that the unique possible attack of this scheme consists on the collaboration of every agency to hack the system, but this case is outside the scope of this paper due to its irrelevance in real applications. Mutual protection strategy presents two different schemes according to the requirements of the system static and mutual strategies.

2.1 Static Mutual Protection

The work presented in this paper is based on this static scheme. Its name is settled by the fact that the protected parts of the agent might be directly included in the agent or in the protector's agents prior to the execution, which main appeal is the increased performance of the system in efficiency terms. However, the system is very restricted and the previous setting of the system is mandatory, thus agents are protected before their execution.

Henceforth we have described the theoretical part of our contribution. However, to achieve a complete solution, a practical basis and tools that implement those concepts are required. We need a platform to develop and execute agents. Due it being the most widely used in the agent development community, we have used JADE[11].

2.2 Dynamic Mutual Protection

The Static Mutual Protection strategy can be successfully applied to many different scenarios. However, there are scenarios where it is not possible to foresee the potential interactions between the agents; where the agents are generated by different parts, or involve very dynamic multi-hop agents. In these cases the Static Mutual Protection strategy will be difficult or impossible to apply. In the Dynamic Protection Strategy each agent is able to execute arbitrary code sections on behalf of other agents in the society. In this strategy each agent includes a public part, an encrypted private part and a specific virtual machine similar to the one described in [7]. This virtual machine allows agents to execute on-the-fly code sections (corresponding to the private parts) received from other agents. The scalability of this scheme is very good since only a few agents (one in most cases) are involved in the protection of any other agent.

2.3 SecureAgent Library

The SecureAgent library consists of a library that implements the static mutual scheme described in previous sections. This library is fully developed in Java and is completely integrated in the JADE platform.

However, the main appeal of this library is the facilities provided in the development of a secure MAS. In fact, when we build a secure MAS system with the properties and functionalities of the static mutual scheme, we only create an object of SecureAgent class. Inheriting agents of this class have behavior descriptions in order to perform the protection tasks. Fortunately it is not necessary to use the library to know the function of each behavior. Another important point relates to the PrivateCode interface, in which only one method is described. In such a way, the part of the private code of every agent is included in the execute() method belonging to the specific class that implements the PrivateCode interface. The code is transparent for the agent, therefore this only executes the code calling to execute() with the appropriate parameters, being unaware of its implementation.

3 Methodology for Agent Protection: Solution

The main goal of this solution consists of facilitating the tasks of protecting agents to the MAS developers, that is, a developer should be able to protect his MAS using the mutual static libraries producing an equivalent version of MAS. However, this task implies recoding some classes and selecting parts of data and instructions to protect. The tasks derived from repeating these steps once or twice are not very complicated, but it is tedious and certainly not efficient. Let us suppose the case that after a couple of hours protecting the MAS we realize that the protected and equivalent version of MAS is very heavy and its efficiency quite low, or even that the security level is not appropriate. We consider the possibility of changing the security setting, studying results and deciding the most appropriate settings for our concrete MAS to be interesting points. For this reason, our tool is focused on speeding up and automating the whole process. Once we have described the starting scenario and the main target we will list the structure related requirements of the system and select the needed tools. An overview of the system scheme is shown in figure 1.

We noticed that the feedback of our application is the unsecured MAS. This system is composed of a set of agents defined by Java classes (. class files). Evidently, the output of this tool is an equivalent MAS in functionality but a secure one, that are a set of java classes. Therefore, the work of this tool is composed of several sequential tasks: read, analyze, modify and create .class files. Thus, we need a tool that allows us to handle these kind of files. We found several tools for this purpose such as BCEL [3], Javassist [1] or ASM [2]. Among these, we choose BCEL due to it being the most popular in the community, more documentation being available, full development in Java and being developed by Apache software foundations, which provides more facilities in the integration.

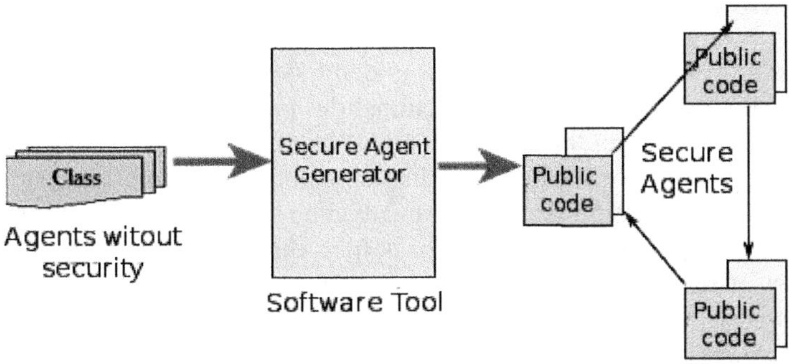

Fig. 1 Secure Agent Generator Tool

3.1 Byte Code Engineering Library (BCEL)

A ".class" file has a quite complex and hard to manage internal structure. A huge number of references and the low level code they have make a it a hard and tedious task to analyze and create. To work with these files we use BCEL library. This consists of an Apache Software Foundation project that provides a wide API, clearly differentiated in three parts:

- A pack containing classes that describe the "static" limitations of class* files. In this case, classes might be used to read and write class files to and from a specified file. It is especially useful to analyze Java classes when the source code is not available for any reason. Finally, the main class of this pack is JavaClass.
- The second pack is to generate or dynamically modify class files. It can be used to add or analyze code from class files, etc.
- The third pack contains a set of code samples and utilities, a viewer for class files and a converter tool from class to HTLM and Jasmin assembler language.

The static component contains several classes to model the internal .class file structure. JavaClass is the main class that is built from a .class file and offers a huge number of functions to browse in different components, fields, methods, local variables, internal classes, and so on. Also, several patterns are provided for the controlling and analyzing of elements, for instance the visitor pattern or the observer pattern. In our case, and especially for the instruction analysis, the visitor pattern has been widely productive. The bytecode instructions hierarchy fits perfectly to the visitor pattern and allows an easy organization of the classes. For instance, a LOAD/STORE instruction is needed to know the field to read or the method to call in an INVOKE instruction. At this point we have described the tool that analyzes the input files, that is the set of unprotected agents and the output is the set of classes that represent secure agents using the SecureAgent library.

4 Architecture

Throughout this section we provide an in-depth description of the most important features and characteristics of the functionalities provided by our tool. It is important to highlight the fact that we are focused on the development of a tool for the automatic generation of secure MAS, implementing the mutual static strategy that as input has a set of agents that compound the non secure MAS.

It is important to note that the feedback files, the set of non secure agents, must fulfil a set of restrictions (preconditions) as described every file must be precompiled and stay in ".class" format; every file must represent a class inherited from jade.core.Agent and internal anonymous classes are not allowed in these files.

Similarly, there are some output conditions to take into consideration that each of these new agents will have a protector agent pre-assigned, which cannot be changed at runtime. Our aim is that the output MAS is equivalent to the input MAS, meaning the behavior of the new MAS with the security incorporated will be exactly the same that the input MAS. The architecture of the system is built using the model view controller pattern. However, the view is a simple graphic user interface to facilitate the use of the tool and we will focus on the data model and in their working phases.

Our tool has three clearly differentiated phases: the loading of original agents, security settings that meet requirements and the final creation of secure agents. Each of these phases was implemented with a set of classes in charge of providing a correct execution. Thus we illustrate the whole process by means of an example. The class contains the code of a non not protected agent, which inherits from Agent class. It is necessary that each agent has associated an agent protector. Following, every phase of the process is described by means of a concrete example.

4.1 Phase I: Load

This is the first phase in the generation of the secure agents process. The goal of this step is to load class files and parse their content. For this purpose, we have selected the set of non secure agents and then we have identified and analyzed all its elements, that is, methods, fields, instructions, internal classes, etc.

In this .class files analysis stage, we have used the static component of the BCEL libraries. Exactly as we previously stated, this part of the API provides the methods to load a .class file and the automatic generation of the structure of a class file [5]. This process is repeated until, for each of the elements from the file class (methods, fields, internal classes, instructions, etc), a modeling object is created and is used to handle it.

Every element generated with BCEL component is a read only one. However, it is important to save information from each of these elements by means of annotations. Thus we created classes that inherit directly from the BCEL classes that include all the useful information for the next phases. In this analysis process we clearly noticed the special case of the instructions, this is a more complex case than the rest of the elements. Inside these classes files there is a section dedicated to class methods. Among other elements inside these methods we found bytecode instructions. These instructions are useless if they are executed separately; they depend on the previous one and the next one.

Once all agents are loaded in the system we progress to the setting phase. This phase is very important because we indicate the degree of security and describe the protection links in it. The information of this phase is very relevant due to it selecting the elements to be protected.

4.2 *Phase II: Setting*

The second phase of this process is the simplest of the three and easiest to implement. This stage controls the specification of security parameters for the creation of the new secure agents. Among the compounding parts of a JADE agent there are two elements to be protected instructions and data (Class fields).

The information needed to determine the security degree is indicated in the percentage of instructions and data to protect. All this information is modeled with a class called SecureAgentConfiguration. Regarding how to indicate the security parameters, we have implemented two separate ways. The first method is needed in order to specify for each of the loaded agents (first phase) the security parameters in the system. This fact implies the selection of every agent and the insertion of the percentage of instructions and selection of fields to protect. This method is a bit more tedious due to the number of agents loaded and the number of testing proofs to perform. For this reason, we have implemented an option that allows us to apply a security template to all agents by means of an XML file. We have developed three basic templates and the possibility to build a customized template. Then it is essential to establish protection links. This task can be automatically performed since this functionality is integrated in the graphic tool. In our case, we only have two agents, then one protects the other and vice versa.

4.3 *Phase III: Creation of Secure Agents*

Finally, we have the secure agent creation phase. Thanks to the previous stages, in this stage we collected all the information needed for this creation. For each of the original agents, at least two new classes may be created, one related to the new secure agent with its public code (data and instructions) and the other with the private code. In addition to these classes, as many internal classes contain the original agent as new more classes will be created. To create these new precompiled classes we make use of the dynamic component of BCEL. This part of the BCEL API will facilitate the creation of the skeleton of class files and, depending on the security parameters set in the previous phase, the original code is inserted in one part or the other.

A new class is created for code protected implementing the PrivateCode interface, but in this new class it is essential to insert protected fields (in our example they do not exist), the “execute()” method that contains the protected code divided by sections. The information to know which section to execute is in the method arguments.

5 Conclusion and Ongoing Work

In this paper we provide a method based on the protected computing approach to protect mobile agents in multi-agent systems. Previous sections show how our contribution provides some graphic tools to easily implement the mutual protection among agents by means of a friendly interface. These graphic tools open up a new field of research that concerns performing a number of analytical and statistical studies about the kind of protection to implement, depending on the system requirements. At this point we have implemented automatic tools for the administration of the static mutual protection scheme. The next logical step is to develop automatic tools for the implementation of the dynamic mutual protection scheme. This proposal is more complicated due to the necessary implementation of a small java virtual machine, but a secure one. We are currently working in this field.

References

1. Chiba, S.: Javassist (Java Programming Assistant). Sun Microsystems, Inc. (2009)
2. OW2 Consortium. ASM
3. Apache Software Foundation. BCEL (Byte Code Engineering Library) (2006)
4. Hewitt, C., Baker, H.: Actors and continuous functionals (1977)
5. Lindholm, T., Yellin, F.: The JavaTM Virtual Machine Specification. Sun Microsystem (1999)
6. Maña, A., Lopez, J., Ortega, J.J., Pimentel, E., Troya, J.M.: A framework for secure execution of software. International Journal of Information Security 3(2), 99–112 (2004)
7. Maña, A., Muñoz, A.: Mutual Protection for Multiagent Systems. In: Third International Workshop on Safety and Security in Multiagent Systems, Hakodate, Japan, p. 37 (2007)
8. Maña, A., Muñoz, A., Serrano, D.: Towards secure agent computing for ubiquitous computing and ambient intelligence. In: Indulska, J., Ma, J., Yang, L.T., Ungerer, T., Cao, J. (eds.) UIC 2007. LNCS, vol. 4611, pp. 1201–1212. Springer, Heidelberg (2007)
9. Nwana, H.S.: Software agents: An overview. The Knowledge Engineering Review 11(3), 205–244 (1996)
10. Pearson, S., Balacheff, B.: Trusted computing platforms: TCPA technology in context. Prentice Hall PTR, Englewood Cliffs (2003)
11. Telecom Italia SpA. JADE (Java Agent Development Framework) (2009)

Dynamic Assignation of Roles and Tasks in Virtual Organizations of Agents

Carolina Zato, Ana de Luis, Juan F. De Paz, and Vivian F. López

Abstract. Nowadays, a common problem that affects the workflow and the results of an entity is the planning and distribution of tasks. Doing this manually implies anticipate workloads and employee characteristics, which is inefficient and almost uncalculated in high dynamic environments. In this paper, a model that generates a planning of tasks, minimizing the resources necessary for its accomplishment and obtains the maximum benefits is presented. Within this proposal, genetic algorithms, queuing theory, and CBR are used in different stages to obtain an efficient distribution. To test the system, the chosen case study that fits the scenario, is the e-Government where an elevated number of tasks must be solved in a precise term using the minimal resources.

Keywords: multiagent systems, virtual organizations, queuing theory, genetic algorithm, scheduling, e-Government.

1 Introduction

Planning and distribution of tasks is a problem that can be encountered in various activities [1] [2]. The goal of these works is optimize the quantity of resources needed, minimize costs and moreover, get the maximum benefits possible. In this aspect, several variables has to be considered such as the high dynamism of the scenario that forces to find a balance between time spent on frequently planning and on implementing plans to adapt the organization. Therefore, it is necessary to create a system that predicts demand for resources so it can perform tasks allocation maximizing benefits and minimizing delays.

Traditionally, the techniques used for planning and resource allocation are performed using exact methods. The exact methods like linear programming [21], nonlinear [13] [15] programming and graph theory, ensure that you get optimal solutions in execution times depending on the number of existing variables. Thus,

Carolina Zato · Ana de Luis · Juan F. De Paz · Vivian F. López
Department of Computer Science and Automation, University of Salamanca
Plaza de la Merced, s/n, 37008, Salamanca, Spain
e-mail: {carol_zato,adeluis,fcofds,vivian}@usal.es

A. Abraham et al. (Eds.): International Symposium on DCAI, AISC 91, pp. 59–68.
springerlink.com

such are not suitable to problems of the NP-Hard type and have difficulties defining constraints and objective functions. It is therefore advisable for such types of problems to use metaheuristics [17] [18] [19] [20] techniques that allow obtaining efficient solutions in reasonable execution times.

This paper proposes a system able to estimate work demands in order to estimate the number of resources needed and according to these claims carry out a work-resources distribution to maximize benefits and minimize delays. The planning model is performed by applying CBR systems [6]. The CBR system integrates into the various stages of reasoning techniques to estimate resources based on queuing theory and planning using genetic algorithms. This planning mechanism is applied to virtual organizations [5] of agents [3] to simulate the behaviour of organizations in planning and allocation of work in the case study selected e-Government, when given the large number of procedures and users of this entity it is essential to have a good system of planning that can do the work successfully and on time, devoting staff just enough to meet the challenges ahead.

This article is divided as follows: section two describes multi-agent systems and planning mechanisms used for assigning dynamic tasks, section three presents the proposed model, sections four and five describe the results obtained and the conclusion

2 Multi-agent Systems

A multi-agent system (MAS: Multi-Agent System) is basically a network of organizations focused on solving problems and working together to find answers to problems that are beyond the individual capabilities or knowledge of each entity [3]. In our case, these entities are CBR-BDI agents, which get its name from the BDI architecture with a CBR reasoning system [6] [7]. A CBR manages cases (past experiences) to solve new problems. The way cases are managed is known as the CBR cycle, and consists of four sequential stages which are recalled every time a problem needs to be solved: retrieve, reuse, revise and retain. Each of the steps of the CBR life cycle requires a model or method in order to perform its mission.

The open MAS [22] should allow the participation of heterogeneous agents, which change over time, with architectures and even with different languages. For this reason, we cannot rely on agents' behaviour, when it is necessary to establish controls on the basis of norms or social rules. For this and because of the characteristics of open environments, new approaches are needed to support the evolution of systems, and to facilitate their growth and run-time updates especially due to the dynamics of open environments. This is one of the reasons that encourage the use of virtual organizations (VO). A VO [5] is an open system designed for grouping, for the collaboration of heterogeneous entities and where there is a separation between form and function that defines their behaviour. The concept of organization is seen as a promising solution to manage the coordination of the agents and control their behaviours and actions.

3 Planning Tasks

Planning problems are usually performed by exact techniques such as linear programming which can be used as in the Simplex algorithm [21], quadratic programming or by using other nonlinear programming techniques such as Lagrange multipliers [13], Kuhn-Tucker [15] or allocation problems in graph theory and application of algorithms such as Ford-Fulkerson [4]. The nature of this type of problems is combinatorial and therefore the time required to find the optimal solution grows exponentially with the number of tasks considered. These problems are included within the area known as combinatorial optimization and in most cases these problems belong to the NP-hard family problems. This situation justifies the application of heuristic algorithms for the search of solution in a reasonable time.

Heuristic algorithms, which are easily generalized to several types of problems, are called metaheuristics. Metaheuristics are smart strategies to design, improve and optimize very general heuristics, giving high performance with respect to traditional heuristics. Among these algorithms are the GRASP algorithm, taboo search, simulated annealing or genetic algorithms. Greedy algorithms (AVM) like GRASP generate solutions iteratively from a randomly generated greedy solution, which is modified to optimize it [17] [16]. Taboo search [18] [19] adjust iteratively a particular solution from the local search of optimal solutions. These algorithms have the problem of local minima and due to this, the simulated annealing algorithms [20] [18] explore alternative solutions to local optimum in a probabilistic way in order to avoid local minima. Genetic algorithms [10] include characteristics of the previous metaheuristics and initially generate multiple random solutions, which are iteratively altered by mutation and crossover operators, and explore solutions locally by mutations to avoid falling into local minima.

In recent years, a number of approaches [12] have been proposed to model and solve several problems of scheduling tasks, with varying degrees of success. Reviewing various comparisons [11] [14] of the efficiency of different metaheuristics methods shows that any single case a method is clearly distinct from another.

4 Proposed Model

The model proposed in this paper focuses on developing a planning mechanism to coordinate the agents included in the VO. The roles of the agents are:

1. **Processor role.** Responsible for carrying out the activities required for each specific task. For this reason, the responsible agent will specialize depending on the type of tasks the system must solve.

2. **Planner role.** Design the overall plan to be implemented by the organization. Sets the number of processor agents and makes the distribution of tasks depending on the role they play. Replan depending on the size of the input queue or inability to accomplish with a plan.
3. **Distributor role.** Distributing tasks according to its completion by the agents and checks that each task is being processed within time limits to serve the plan.
4. **Coordinator.** Performs the general control of the system. Communicate with the platform elements to carry out control actions (connect, disconnect, exceptions, etc.).
5. **Manager role.** This agent manages all the information of the task and communicates to the user.

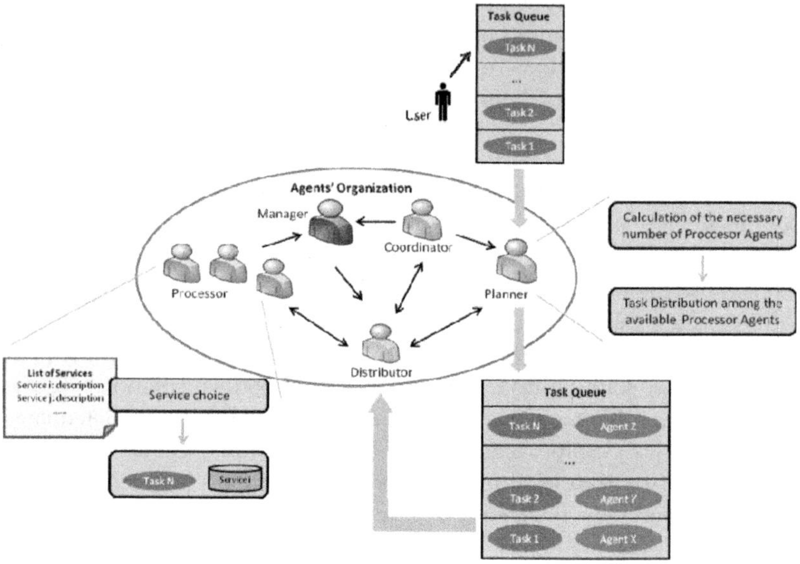

Fig. 1 Outline of proposed model

In the figure 1 the different agents of the system and the interactions among them are represented. In the upper corner of the figure shows the task list that store the activities to carry out in the multi-agent system and in the centre of the image, the agents and the interconnections are showed.

To carry out the planning process, the agent follows the planning model CBR-BDI. The first point in the definition of a CBR-BDI model is the definition of case (1).

$$C = \{t_i / t_i = (ids, idt, b, B, f, p, d, a), i = 1...n\} \quad (1)$$

Where t_i represents the task *i*, *ids* is the identifier of the task type, *idt* the process, *b* benefit, *B* the accrued benefit, *f* date of entry, *p* deadline date, *d* duration, *a* the

ID of the agent that performed the task. The application of the different stages of reasoning is performed as follows:

Recovery. During this stage the most similar cases to the current case cn+1 are retrieved. The most similar cases are those containing more tasks of the same type as the tasks, which are currently queued in the system. The number of cases retrieved is predefined to subsequently implement the adaptation phase. The set of recovered cases are called $C_r \subseteq C$.

Adaptation. During this stage, the recovered information Cr is adapted from cases of memory to generate the entire plan corresponding to the case cn+1. The information recovered is adapted by applying queuing theory and genetic algorithms. On one hand, the recovered information is used to determine the arrival rate and service for each of the tasks and thus using queuing theory we can determine the number of agents needed to run the indicated tasks. Recovered cases are the basis for constructing initial chromosomes in genetic algorithms.

Review. The review stage is performed automatically as the agents are finalizing tasks. The agent updates the duration of the tasks as they are completed and in turn, makes new plans if any notice is received from the processor agents on the inability to complete a plan under time constraints provided.

Learning. The learning phase is limited to store the case when the day is finished. The new case memory C 'is defined as follows: $C' = C \cup c_{n+1}$ To limit the size of cases of memory, some are removed from memory if exceeded a predefined age.

4.1 Dynamic Planning Roles

The number of agents that should be available in the system is estimated dynamically. It is intended that the number of agents suits demand to ensure that the system utilization factor ρ is less than 1. This estimate will be done through the use of queuing theory in a model M/G/s, where the arrival rate follows a Poisson distribution (the most commonly used in similar work [9]), the exponential service and the existence of multiple servers (agents)

The problem of planning multiple tasks can be reduced to the case of planning for a single task of each type. Thus, for each task a planning is performed independently so as to calculate the average waiting time and average queue length independently. The average waiting time and the overall average length is reduced to calculating the average values calculated for each of the tasks. In the case of the M/G/s model where s = 1,2,3, ... is the number of agents and given an arrival rate $\lambda_n = \lambda =$ cte, the service rate when there are n processes is defined by the following equation (2) [8].

$$\mu_n = \begin{cases} n\mu & n = 1, 2, \ldots, s-1 \\ s\mu & n \geq s \end{cases} \tag{2}$$

Where μ represents the average service rate for s available agents. This value depends on both the agents and the machine found.

Assuming that the system is in a stable condition, i.e. it meets the utilization factor $\rho = (\lambda / \mu s) < 1$, the queuing theory [8] allows to calculate the probability that n tasks exists (Pn) in the system, the number of tasks (L) in the system.

To determine the optimal number of agents, an estimation that minimizes the cost function is calculated and it depends on the number of agents used and on the waiting time in the queue. The function is defined in a particular way for each service depending on the actual costs of each agent in the system, though the following benefit function is provided (4).

$$f(L, P_0, \ldots, P_{s-1}, \mu', \overline{p}, \overline{b}) = f_b(L, \mu', \overline{p}, \overline{b}) - k \cdot s \tag{3}$$

$$f_b(L, \mu', \overline{p}, \overline{b}) = \begin{cases} (\overline{p} / u')\overline{b} \cdot s \cdot (1-\rho) & \text{si L} \cdot u' > \overline{p} \cdot s \cdot (1-\rho) \\ L\overline{b} & \text{si L} \cdot u' \leq \overline{p} \cdot s \cdot (1-\rho) \end{cases} \tag{4}$$

Where k is a constant associated with the cost of having an agent working, $\overline{b}$ the average benefit of performing the task, μ' is the average time to complete the task obtained from the service rate, $\overline{p}$ the average time to execute a task. If the conditions of stability are overpass, fb is counted only up to the utilization factor 1. The utilization factor ρ varies according to the new services added to the queue till it reaches the utilization factor of 1.

Following the cost function given in (4), a global cost function is introduced (5) that takes into account the implementation of the various services.

$$f(f_1, \ldots, f_k) = \sum_{j=1}^{k} f_i \tag{5}$$

Where f_i is calculated from equation (4). Because sometimes you might not be given the stability conditions, it is necessary to calculate the terms in order of benefit depending on the type of the task so that when you reach the utilization factor of 100%, the process ends calculating the summation terms. Once the optimization function defined in (5) the maximum value is calculated iteratively starting with number of agents equal to 1, the fixed value is the first local maximum that corresponds to the global maximum.

4.2 Task Assignment

Once the number of starting agents is considered to minimize costs, an allocation of tasks between the available agents has be done. If the system utilization factor does not exceed the value of 1, the distribution of tasks among agents is performed so as to ensure as far as possible that it can perform assigned tasks in case of delays or the time to perform a task increases. It is performed so as to maximize the following function (6):

$$\max \sum_{i=1}^{k} f_i \; where\, fi = \begin{cases} \log(1-|x_i - \overline{x}|) & (x_i - \overline{x}) \geq 0 \\ -\log(1-|x_i - \overline{x}|) & (x_i - \overline{x}) < 0 \end{cases} \tag{6}$$

Where $x_i = t_i - a_{i-1} - c_i$ with t_i the maximum time for completion of the task i, a_{i-1} the cumulative time to perform the tasks i-1 above and finally c_i the time to run the task i, which is customized according to the agent selected and calculated from the average value of previously executed tasks. Minimizing the differences get all the tasks to have a uniform distribution of the remaining time so it gets easier to achieve them.

If the system utilization factor is greater than 1, the aptitude function is redefined to minimize possible losses of the work already done.

$$\min \sum_{i=1}^{k} - B_i \tag{7}$$

As in the previous case to know the value of each Bi it is necessary to establish the order of execution of the procedures. If at the time of completing this task the value of Bi was not taken into account.

The chromosome encoding is performed so that each gene is composed of the elements listed by ti identified in (1). The crossover operator is defined similarly to the multi-junction used in other problems such as TSP. The operator is defined as follows:

- Select a partial route
- Exchange of direct segments of tasks where ID matches
- The exchanges define a series of matches that relates each of the genes of a chromosome with which occupies the same position in the other parent

Mutation operators define various modes that will be executed randomly, and just those mutations that improve the aptitude of chromosomes will be selected. The defined mutation operators are: exchange order of tasks, exchange of assigning contiguous tasks and changing the allocation of a task.

Elitism operator is defined to keep the percentage of efficient solutions in every generation of population and population size as a constant this involves the replacement of parents by the children chromosomes in generations with the exception to remain with elite chromosomes. The roulette selection is the criteria chosen for this.

The initiation of chromosomes is based on the received tasks; each chromosome is initiated with the tasks of the new case. For each task of the chromosome a sequential search is made in the case and assigns the task to an existing agent so that this association is maintained for the remaining tasks.

5 Case Study and Results

The case study presents a society oriented system to process all cases that reach the public administration by electronic means. The implemented system simulates the behaviour of the Planner agent within the VO. In order to assess the mechanism of the proposed planning, taking into account that the final goal is to process all the records within time limits or, if not possible, to maximize the benefit being carried out by different tests. The system was tested with two different records for four simulation Modes: Mode 1 – Without planning, Mode 2 - Calculating the number of agents needed within queuing theory, Mode 3 - Planning (including queuing theory and genetic algorithms), Mode 4 – The whole planning with CBR. In the first case (Test 1) a list of 500 records were entered within a period of 120 minutes and the second case (Test 2) had 1500 records into the system within 210 minutes. Previously it had a memory of 700 cases, based on cases where values were altered randomly following a normal distribution.

Table 1 Results obtained during the simulation

	Number of cases not processed in time		Benefits obtained	
Plan	Test 1	Test 2	Test 3	Test 4
Mode 1	5	25	1100	4100
Mode 2	4	20	1350	4500
Mode 3	2	9	1640	5060
Mode 4	1	5	1720	5290

Table 1 shows that the qualitative leap occurs in both cases by the introduction in the third mode of simulation, with genetic algorithms to carry out the distribution. The introduction of the CBR also produces an improvement in the last mode so as to work with parameters that fit closer to reality due to learning by the system. Test 2 for Mode 4 shows that the number of unprocessed cases on time is higher than Test 1, this is mainly because the recovered arrival rate exceeded the records retrieved from the memory of the CBR.

The benefits obtained in different modes are even more significant. To assign a value to the benefit field for each record a point system is used taking into account various aspects such as the type of record, the economic gain expected, previous work, etc. As expected, the worst benefit is obtained without planning mode. In the second case, the use of queuing theory improves the benefits being able to balance between the number of agents required and ordination benefit and thus assigning priority to improve the figures. In the third mode, as the above chart shows, the introduction of a task distribution based on genetic algorithm is the strong point of planning. Finally, still better benefits are obtained taking into account the adaptation from past cases provided by the CBR.

6 Conclusions and Future Work

Throughout this article a model of VO with a mechanism for planning and distributing tasks in a dynamic environment is presented. A genetic algorithm is introduced as a mechanism for the exploration of the search domain and for optimization has been effective within the proposed problem characterized by its complexity, given the large number of available combinations. Using queuing theory allowed setting the number of agents required for resource optimization. Both methods have been successfully incorporated into the reasoning cycle of CBR, specifically, in the adaptation phase. The results confirm that planning leads to an improvement and by the refinement of the cases included in the reasoning cycle of CBR, it will be possible to achieve a high level of learning and adaptation in a dynamic environment. In the short term, the proposed future work is to explore other heuristics and metaheuristics methods in order to try different formulas for the allocation of tasks, setting an efficiency value to each method and recommending to the user different planning methods associated with its success rate.

Acknowledgments. This work has been supported by the MICINN TIN 2009-13839-C03-03.

References

1. Trippi, R.R., Ash, A.W., Ravenis, J.V.: A mathematical approach to large scale personnel assignment. Computers & Operations Research 1(1), 111–117 (1974)
2. Zülch, G., Rottinger, S., Vollstedt, T.: A simulation approach for planning and reassigning of personnel in manufacturing. International Journal of Production Economics 90, 265–277 (2004)
3. Durfee, E.H., Lesser, V.R., Corkill, D.D.: Trends in Cooperative Distributed Problem Solving. IEEE Transactions on Knowledge and Data Engineering 1(1), 63–83 (1989)
4. Shin, K., Corder, S.: Implementing the ford-fulkerson labeling algorithm with fixed-order scanning. Computers & Operations Research 19(8), 783–787 (1992)
5. Esteva, M., Rodríguez, J., Sierra, C., Garcia, P., Arcos, J.: On the formal specific ations of electronic institutions. In: Dignum, F.P.M., Cortés, U. (eds.) AMEC 2000. LNCS (LNAI), vol. 2003, pp. 126–147. Springer, Heidelberg (2001)
6. Kolodner, J.: Case-Based Reasoning. Morgan Kaufmann, San Francisco (1993)
7. Corchado, E., Assumpcio, M., MBorrajo, M.L.: A maximum likelihood Hebbian learning-based method to an agent-based architecture. Int. Journal of Computer Mathematics. 86(10-11), 1760–1768 (2009)
8. Martín, Q.: Investigación Operativa. Prentice-Hall, Englewood Cliffs (2003)
9. Menasce, D.A.: Trade-offs in designing Web clusters. IEEE Internet Computing 6(5), 76–80 (2002)

10. Yu, J., Buyya, R.: Scheduling scientific workflow applications with deadline and budget constraints using genetic algorithms. Scientific Programming 14(3), 217–230 (2006)
11. Rodríguez, P.: Discusión y Análisis de la metaheurística SN. Depto. Investigación Operativa, InCo, FI, UdelaR. Reporte Tecnico 03-02 (2003)
12. Seda, M.: Mathematical Models of Flow Shop and Job Scheduling Problems. Worl Academy of Science, Engineering and Technology (31), 122–127 (2007)
13. Wan, S., Yang, F., Izquierdo, E.: Lagrange multiplier selection in wavelet-based scalable video coding for quality scalability. Signal Processing: Image Communication 24(9), 730–739 (2009)
14. Pacheco, J.A., Casado, S.: Estudio Comparativo de Diferentes Metaheurísticas para la Resolución del Labor Scheduling Problem. Estudios de Economia Aplicada 21(3), 537–557 (2003)
15. Shi, C., Lu, J., Zhang, G.: An extended Kuhn-Tucker approach for linear bilevel programming. Applied Mathematics and Computation 162(1), 51–63 (2005)
16. Andres, C., Miralles, C., Pastor, R.: Balancing and scheduling tasks in assembly lines with sequence-dependent setup times. European Journal of Operational Research 187(3), 1212–1222 (2008)
17. Kim, K.H., Park, Y.: A crane scheduling method for port container terminals. European Journal of Operational Research 156(3), 752–768 (2004)
18. Porto, S., Kitajima, J.P., Ribeiro, C.: Performance evaluation of a parallel tabu search task scheduling algorithm. Parallel Computing 26(1), 73–90 (1996); Steinhofel, K., Albrecht, A., Wong, C.K.: Two simulated annealing-based heuristics for the job shop scheduling problem. European Journal of Operational Research 118(3), 524–548 (1999)
19. Taillard, E.: Parallel Taboo Search Technique for the Job shop Scheduling Problem. Journal on Computing Science 6, 108–117 (1994)
20. Kolonko, M.: Some new results on simulated annealing applied to the job shop scheduling problem. European Journal of Operational Research 113(1), 123–136 (2009)
21. Kabadi, S., Punnen, A.: A strongly polynomial simplex method for the linear fractional assignment problem. Operations Research Letters 36(4), 402–407 (2008)

Agent Simulation to Develop Interactive and User-Centered Conversational Agents*

David Griol, Javier Carbó, and José M. Molina

Abstract. In this paper, we present a technique for developing user simulators which are able to interact and evaluate conversational agents. Our technique is based on a statistical model that is automatically learned from a dialog corpus. This model is used by the user simulator to provide the following answer taking into account the complete history of the interaction. The main objective of our proposal is not only to evaluate the conversational agent, but also to improve this agent by employing the simulated dialogs to learn a better dialog model. We have applied this technique to design and evaluate a conversational agent which provides academic information in a multi-agent system. The results of the evaluation show that the conversational agent reduces the time needed to fulfill to complete the the dialogs, thereby allowing the conversational agent to tackle new situations and generate new coherent answers for the situations already present in an initial model.

1 Introduction

As we move towards a world where all the information is in the digital domain, it becomes necessary to provide straightforward ways of retrieving it. To achieve this goal it is necessary to provide an effective, easy, save and transparent interaction between the user and the system. Thus, it is important to identify which modality or combination of modalities would be optimal to present the information and interact with the user. To do so, in the last years there has been an increasing interest in simulating human-to-human communication, including the so-called conversational agents in multi-agents system [8]. There is a high variety of applications in

David Griol · Javier Carbó · José M. Molina
Group of Applied Artificial Intelligence (GIAA), Computer Science Department, Carlos III University of Madrid
e-mail: {david.griol, javier.carbo, josemanuel.molina}@uc3m.es

* Funded by projects CICYT TIN2008-06742-C02-02/TSI, CICYT TEC2008-06732-C02-02/TEC, CAM CONTEXTS (S2009/TIC-1485), and DPS2008-07029-C02-02.

A. Abraham et al. (Eds.): International Symposium on DCAI, AISC 91, pp. 69–76.
springerlink.com

which conversational agents can be used, one of the most wide-spread of which is information retrieval. One of the most recent applications of these agents is for the development of e-learning and tutoring systems [7].

Multi-agent systems are designed as a collection of interacting autonomous agents, each having their own capacities and goals that are situated to a common environment. This way, the development of multi-agent systems offers the capability of simulating autonomous agents and the interaction between them. In the literature, there are several corpus-based approaches for developing user simulators, learning optimal dialog strategies, and evaluating conversational agents [10, 9, 6, 3]. The construction of user models based on statistical methods has provided interesting and well-founded results in recent years and is currently a growing research area. A probabilistic model to emulate the user agent can be trained from a corpus of human-computer dialogs to simulate user answers. Therefore, it can be used to learn a dialog strategy by means of its interaction with the conversational agent. A summary of user simulation techniques for reinforcement learning of the dialog strategy can be found in [9].

In this paper, we present a technique to develop a user agent simulator to automatically interact with a conversational agent and generate the dialogs required to learn an enhanced dialog model for a conversational agent. Our user simulation technique is based on a classification process in which a neural network is employed to take into account the previous dialog history to select the next user answer. We have applied this technique to develop a conversational agent which provides academic information in Spanish. The results of the evaluation of the conversational agent show that the conversational agent reduces the time needed to fulfill the different tasks, thereby allowing the conversational agent to tackle new situations and generate better answers for the situations already present in an initial model.

2 Design of an Academic Conversational Agent

The design of our conversational agent is based on the requirements defined for a dialog system developed to provide spoken access to academic information about the Department of Languages and Computer Systems in the University of Granada [1]. To successfully manage the interaction with the users, the conversational agent carry out six main tasks: automatic speech recognition (ASR), natural language understanding (NLU), dialog management (DM), database access and storage (DB), natural language generation (NLG), and text-to-speech synthesis (TTS).

The dialog manager of the the conversational agent has been developed using VoiceXML documents that are dynamically created using PHP. This way, it can adapt the system responses to the context of the conversation and the dialog state, which improves the naturalness of the interaction. For example, the help messages provided by the conversational agent take into account the topic that the user and the agent are addressing at a particular moment. The context is used as well to decide the confirmation strategy to use. In addition, we have implemented a statistical module to automatically select the next system response (i.e, a VoiceXML file by using a model which is learned from a dialog corpus for the task [2].

The information that the conversational agent provides has been classified in four main groups: subjects, professors, doctoral studies and registration. The information that the agent provides for each of these categories is shown in Table 1. As can be observed, the conversational agent must have gathered some data by asking the user about the name of the subjects, the professors, etc. The way in which the user is queried for this information follows in most cases a system-directed initiative.

Table 1 Information provided by the academic conversational agent

Category	**Information provided by the user (names and examples)**		**Information provided by the system**
Subject	*Name*	Compilers	Degree, lecturers, responsible lecturer, semester, credits, web page
	Degree, in which it is taught in case that there are several subjects with the same name	Computer Science	
	Group name and optionally *type*, in case he asks for information about a specific group	A Theory A	Timetable, lecturer
Lecturers	Any combination of *name* and *surnames*	John John Smith Mr. Smith	Office location, contact information (phone, fax, email), groups and subjects, doctoral courses
	Optionally *semester*, in case he asks for the tutoring hours	First semester Second semester	Tutoring hours
Doctoral studies	Name of a doctoral program	Software development	Department, professor
	Name of a course if he asks for information about a specific course	Object-Oriented Programming	Type, credits
Registration	Name of the deadline	Provisional registration confirmation	Initial time, final time, description

As in many other conversational agents, the semantic representation that we have chosen for the task is based on the concept of frame, in which one or more concepts represent the intention of the utterance, and a sequence of attribute-value pairs contains the information about the values given by the user. In the case of user turns, we defined four concepts related to the different queries that the user can perform to the system (*Subject*, *Lecturers*, *Doctoral studies*, *Registration*), three task-independent concepts (*Affirmation*, *Negation*, and *Not-Understood*), and eight attributes (*Subject-Name*, *Degree*, *Group-Name*, *Subject-Type*, *Lecturer-Name*, *Program-Name*, *Semester*, and *Deadline*). The labeling of the system turns is similar to the labeling defined for the user turns. A total of 30 task-dependent concepts was defined:

- Task-independent concepts (*Affirmation*, *Negation*, *Not-Understood*, *New-Query*, *Opening*, and *Closing*).
- Concepts used to inform the user about the result of a specific query (*Subject*, *Lecturers*, *Doctoral-Studies*, and *Registration*).
- Concepts defined to require the user the attributes that are necessary for a specific query (*Subject-Name*, *Degree*, *Group-Name*, *Subject-Type*, *Lecturer-Name*, *Program-Name*, *Semester*, and *Deadline*).
- Concepts used for the confirmation of concepts (*Confirmation-Subject*, *Confirmation-Lecturers*, *Confirmation-DoctoralStudies*, *Confirmation-Registration*) and attributes (*Confirmation-SubjectName*, *Confirmation-Degree*, *Confirmation-GroupName*, *Confirmation-SubjectType*, *Confirmation-LecturerName*, *Confirmation-ProgramName*, *Confirmation-Semester*, and *Confirmation-Deadline*)

A set of 100 dialogs was acquired with the conversational agent by means of its interaction with real users using an initial version of the system [1]. The acquisition process resulted in a spontaneous Spanish speech dialog corpus with 60 different speakers. The total number of user turns was 422. The recorded material has a duration of 150 minutes. Figure 1 shows an example of a dialog extracted from this corpus. Turns with S refer to system turns, and turns with U refer to user turns.

S1: Welcome to the university information system. You can consult information about subjects, lecturers, doctoral studies and registrations.
U1: I want to know information about subjects in Computer Science.
S2: Which subject do you want to consult?
U2: Language Processors I.
S3: Here you can consult the information of the subject. Anything else?
U3: Can you inform me about the schedules of Group D?
S4: I give the information for the Group D of Language Processors I: Monday 9:00-11:00, Wednesday 11:00-13:00. Anything else?
U4: Can you tell me the name of the lecturer?
S5: I give the information for the Group D of Language Processors I. The name of the lecturer is David Griol. Anything else?
U5: No, thank you.
S6: Thank you for using the system.

Fig. 1 Example of a dialog acquired with the interaction of the conversational agent with real users

3 Our Proposal to Develop a User Agent Simulator

Our proposed architecture to provide context-aware services by means of conversational agents is described in [5]. It consists of five different types of agents that cooperate to provide an adapted service. *User agents* are configured into mobile devices or PDAs. *Provider Agents* supply the different services in the system and are bound to *Conversational Agents* that provide the specific services. A *Facilitator Agent* links the different positions to the providers and services defined in the system. A *Positioning Agent* communicates with the ARUBA positioning system to extract and transmit positioning information to other agents in the system. Finally, a *Log Analyzer Agent* generates user profiles that are used by Conversational Agents to adapt their behaviour taking into account the preferences detected in the users' previous dialogs.

The user simulator replaces the user agent in our architecture. This agent simulates the user intention level, that is, the simulator provides concepts and attributes that represent the intention of the user utterance. Therefore, the user simulator carries out the functions of the ASR and NLU modules, i.e., it generates frames in the same format defined for the output of the NLU module.

The methodology that we have developed for user simulation extends our work for developing a statistical methodology for dialog management [4]. The user answers are generated taking into account the information provided by the simulator throughout the history of the dialog, the last system turn, and the objective(s) predefined for the dialog.

In order to control the interaction, our user simulator uses the representation the dialogs as a sequence of pairs (A_i, U_i), where A_i is the output of the dialog system (the system answer) at time i, expressed in terms of dialog acts; and U_i is the semantic representation of the user turn (the result of the understanding process of the user input) at time i, expressed in terms of frames. This way, each dialog is represented by $(A_1,U_1),\cdots,(A_i,U_i),\cdots,(A_n,U_n)$, where A_1 is the greeting turn of the system (the first turn of the dialog), and U_n is the last user turn. We refer to a pair (A_i,U_i) as S_i, the state of the dialog sequence at time i.

In this framework, we consider that, at time i, the objective of the dialog manager is to find an appropriate user answer U_i. This selection is a local process for each time i and takes into account the sequence of dialog states that precede time i, the system answer at time i, and the objective of the dialog $\mathcal{O}$. If the most probable user answer U_i is selected at each time i, the selection is made using the maximization:

$$\hat{U}_i = \underset{U_i \in \mathcal{U}}{\operatorname{argmax}} P(U_i|S_1,\cdots,S_{i-1},A_i,\mathcal{O})$$

where set $\mathcal{U}$ contains all the possible user answers.

As the number of possible sequences of states is very large, we establish a partition in this space (i.e., in the history of the dialog preceding time i).This data structure, that we call *User Register* (UR), contains the information provided by the user throughout the previous history of the dialog. After applying the above considerations and establishing the equivalence relations in the histories of the dialogs, the selection of the best U_i is given by:

$$\hat{U}_i = \underset{U_i \in \mathcal{U}}{\operatorname{argmax}} P(U_i|UR_{i-1},A_i,\mathcal{O})$$

We propose the use of a multilayer perceptron (MLP) to make the assignation of a user turn. The input layer receives the current situation of the dialog, which is represented by the term $(UR_{i-1}, A_i, \mathscr{O})$ in the previous equation. The values of the output layer can be viewed as the a posteriori probability of selecting the different user answers defined for the simulator given the current situation of the dialog. The choice of the most probable user answer of this probability distribution leads to the previous equation. In this case, the user simulator will always generate the same answer for the same situation of the dialog. Since we want to provide the user simulator with a richer variability of behaviors, we base our choice on the probability distribution supplied by the MLP on all the feasible user answers.

A real corpus includes information about the errors that were introduced by the ASR and the NLU modules during the acquisition. This information also includes confidence measures, which are used by the conversational agent to evaluate the reliability of the concepts and attributes generated by the NLU module. This way, an error simulator agent has been designed to perform error generation. This agent modifies the frames generated by the user simulator once the UR is updated. In addition, the error simulator adds a confidence score to each concept and attribute in the frames.

4 Results of the Evaluation

A dialog corpus of 3000 successful dialogs was acquired using the proposed user simulation technique following the same set of scenarios defined for the acquisition with real users. A maximum number of 14 user turns per dialog was defined for the acquisition. A user request for closing the dialog is selected once the system has provided the information defined in the objective(s) of the dialog.

We have considered three dimensions in order to evaluate the initial conversational agent and its evolution once the simulated dialogs are incorporated to learn a new dialog model: high-level features (dialog and turn lengths), dialog style (speech-act frequency and proportion of goal-directed actions), and dialog efficiency (goal completion rates and times). Table 2 shows the comparison of the different high-level measures defined for the evaluation. As it can be seen, after the incorporation of the simulated dialogs there is a reduction in the average number of turns required to fulfill the complete set of objectives defined in the scenarios. This reduction can also be observed in the number of turns of the longest, shortest and most seen dialogs. Figure 2 shows the previously described evolution of the average duration in terms of total dialog turns (*duration*). It also shows the reduction in the number of

Table 2 High-level dialog features defined for the comparison of the conversational agents

	Initial Convers. Agent	**Final Convers. Agent**
Average number of user turns per dialog	4.99	3.75
Number of turns of the most seen dialog	2	2
Number of turns of the shortest dialog	2	2
Number of turns of the longest dialog	14	12

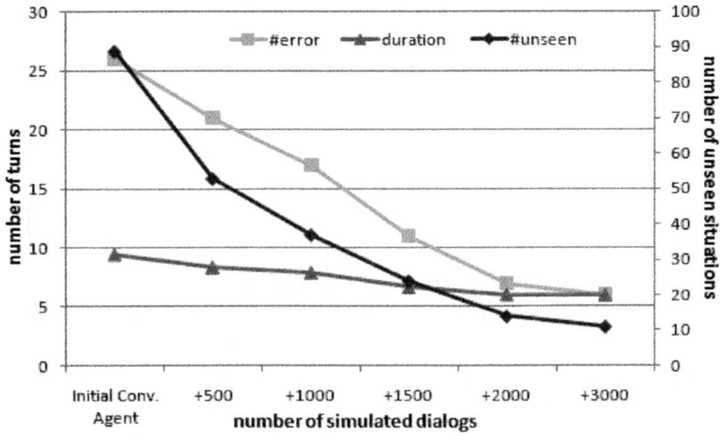

Fig. 2 Evolution of the number of unseen situations, number of errors, and average number of turns

responses provided by the conversational agent which cause a failure of the dialog (*#error*) and the number of unseen situations for which there is not a system response in the dialog model (*#unseen*). As it can be seen from these results, the final conversational agent not only reduces the time required to fulfill each one of the objectives of the dialog, but also it reduces the possibility of selecting an erroneous response.

Finally, Table 3 shows the frequency of the most dominant user and system dialog acts in the initial and final conversational agents. From its comparison, it can be observed that there are significant differences in the dialog acts distribution. With regard to user actions, it can be observed that users need to employ less confirmation turns in the final agent, which explains the higher proportion for the rest of user actions using the final conversational agent. It also explains the lower proportion of yes/no actions in the final agent, which are mainly used to confirm that the system's query has been correctly provided. With regard to the system actions, it can be observed a reduction in the number of system confirmations for data items.

Table 3 Percentages of different types of user [up] and system [down] dialog acts

	Initial Convers. Agent	Final Convers. Agent
Request to the system	31.74%	35.43%
Provide information	20.72%	24.98%
Confirmation	10.81%	7.34%
Yes/No answers	31.47%	28.77%
Other answers	3.26%	3.48%

	Initial Convers. Agent	Final Convers. Agent
Confirmation of concepts and attributes	13.51%	10.23%
Questions to require information	18.44%	19.57%
Answers generated after a database query	68.05%	70.20%

This explains a higher proportion of turns to inform and provide data items for the final agent. Both results show that the final conversational agent carries out a better selection of the system responses.

5 Conclusions

In this paper, we have described a technique for simulating user agents and evaluate conversational agents. Our technique is based on a statistical model which takes the complete history of the interaction into account to decide the next user answer. This decision is modeled by a classification process in which a neural network is used. In addition, the simulated dialogs are used to automatically reinforce the dialog model of the conversational agent. We have described the application of this technique to develop an enhanced academic conversational agent. The results of the evaluation of this agent show that the proposed user simulation methodology can be used not only to evaluate Conversational agents but also to explore new enhanced dialog strategies. As a future work, we are adapting the proposed user simulation technique for its application in more difficult domains in our multi-agent architecture.

References

1. Callejas, Z., López-Cózar, R.: Relations between de-facto criteria in the evaluation of a spoken dialogue system. Speech Communication 50(8-9), 646–665 (2008)
2. Griol, D., Callejas, Z., López-Cózar, R.: Statistical Dialog Management Methodologies for Real Applications. In: Proc. of the 11th Sigdial Meeting, pp. 124–131 (2010)
3. Griol, D., Hurtado, L., Segarra, E., Sanchis, E.: A Statistical Approach to Spoken Dialog Systems Design and Evaluation. Speech Communication 50(8-9), 666–682 (2008)
4. Griol, D., Riccardi, G., Sanchis, E.: Learning the Structure of Human-Computer and Human-Human Dialogs. In: Proc. of the 10th Interspeech Conference, pp. 2775–2778 (2009)
5. Griol, D., Sánchez-Pi, N., Carbó, J., Molina, J.M.: An Architecture to Provide Context-Aware Services by Means of Conversational Agents. In: de Leon F. de Carvalho, A.P., Rodríguez-González, S., De Paz Santana, J.F., Rodríguez, J.M.C. (eds.) Distributed Computing and Artificial Intelligence. AISC, vol. 79, pp. 275–282. Springer, Heidelberg (2010)
6. Lemon, O., Liu, X.: Dialogue Policy Learning for Combinations of Noise and User Simulation: Transfer Results. In: Proc. of the 8th Sigdial Workshop on Discourse and Dialogue, Antwerp, Belgium, pp. 55–58 (2007)
7. Litman, D.J., Silliman, S.: ITSPOKE: An Intelligent Tutoring Spoken Dialogue System. In: Proc. of the 4th HLT/NAACL Conference, pp. 233–236 (2004)
8. McTear, M.F.: Spoken Dialogue Technology: Towards the Conversational User Interface. Springer, Heidelberg (2004)
9. Schatzmann, J., Weilhammer, K., Stuttle, M., Young, S.: A Survey of Statistical User Simulation Techniques for Reinforcement-Learning of Dialogue Management Strategies. Knowledge Engineering Review 21(2), 97–126 (2006)
10. Scheffler, K., Young, S.: Automatic learning of dialogue strategy using dialogue simulation and reinforcement learning. In: Proc. of HLT 2002 Conference, pp. 12–18 (2001)

A Survey on Quality of Service Support on Middleware-Based Distributed Messaging Systems Used in Multi Agent Systems

Jose-Luis Poza-Luján, Juan-Luis Posadas-Yagüe, and José-Enrique Simó-Ten

Abstract. Messaging systems are widely used in distributed systems to hide the details of the communications mechanism to the multi agents systems. However, the Quality of Service is treated in different way depending on the messaging system used. This article presents a review and further analysis of the quality of service treatment in the mainly messaging systems used in distributed multi agent systems. The review covers the issues related to the purpose of the functions provided and the scope of the quality of service offered by every messaging system. We propose ontology for classifying and decide which parameters are relevant to the user. The results of the analysis and the ontology can be used to select the most suitable messaging system to distributed multi agent architecture and to establish the quality of service requirements in a distributed system.

1 Introduction

To make a transparent connection in a distributed system is necessary to hide the details of the communications system to the applications. To make this work, appears the concept of middleware. The scope of middleware is very extensive. Therefore, middleware is commonly defined as an intermediate layer between the application and the communications system that facilitates all aspects related with the connection [1]. The distributed messaging systems (DMS) paradigm has become increasingly popular in recent years to implement the middleware layer. There are a lot of middleware systems based on this paradigm; mainly due to the excellent adaptation of the DMS to provide support to the large number of distributed intelligent multi-agent systems (MAS) architectures.

There are many proposed architectures to cover the middleware in a MAS or in a DMS. From the most used architectures, the proposal of the Foundation for Intelligent Physical Agents (FIPA) [2] is the closest to the MAS, Java Message Service (JMS) [3] and Common Object Request Broker Architecture (CORBA) [4]

Jose-Luis Poza-Luján · Juan-Luis Posadas-Yagüe · José-Enrique Simó-Ten
University Institute of Control Systems and Industrial Computing (ai2),
Universidad Politécnica de Valencia, Camino de vera, s/n. 46022 Valencia (Spain)
e-mail: {jopolu,jposadas,jsimo}@ai2.upv.es

A. Abraham et al. (Eds.): International Symposium on DCAI, AISC 91, pp. 77–84.
springerlink.com

are considered architectures used in MAS but also middleware systems. Finally Data Distribution Service (DDS) [5] is considered mainly a middleware.

Communications in MAS are particularly important; so that, the Quality of Service (QoS) is one aspect that middleware must cover. Because of this, the management of the QoS in a middleware is one of the main objectives in a MAS system.

This article provides that review of systems above cited. The review focuses on the QoS support offered in each system and emphasizes the consequences of using either system, as well as the features needed in future proposals.

The article is organized as follow. Section two describes the main features of the systems considered and discusses the role of each system. Section three contextualizes the systems in the communication process between agents and organizes the QoS features. Section four discusses the results of the previous section. Finally the article reports the conclusions of the study.

2 Messaging Systems Used in Multi Agent Systems

The four systems analyzed (Table 1) are the most supported by standardization organizations and commercial companies. FIPA is a proposal of the organization with the same name. Currently the Institute of Electrical and Electronics Engineers (IEEE) continues the standard. FIPA is used in several agent platforms as JADE [6], JACK [7], ZEUS [8] or Grasshopper [9]. JMS is a part of the Java Platform Enterprise Edition (J2EE) [10] and is used by well known commercial products as ActiveMQ [11] or MQSeries [12]. CORBA is a standard defined by the Object Management Group (OMG) [13] an organization aimed at setting standards for distributed systems and model-based standards. CORBA is widely used for academic and commercial purposes. DDS is a specification of a middleware for distributed systems. The aim of DDS is standardize the programming model for distributed systems. Like CORBA, DDS is an OMG specification but DDS is Real Time oriented and allows the user to specify QoS parameters by means a set of policies. DDS is used in critical systems as aerospace or military products and is supported by robust commercial systems as RTI [14] or OpenSplice [15]

Table 1 Systems compared with the corresponding standardization organization, the main scope or use of the system (agents, middleware or both) and the year of the first version.

System	Standardization organism	Scope	1st version
CORBA	OMG	Middleware → Agents architecture	1991
FIPA	IEEE	Agents architecture	1996
JMS	Sun (Oracle)	Agents architecture → Middleware	2001
DDS	OMG	Middleware	2003

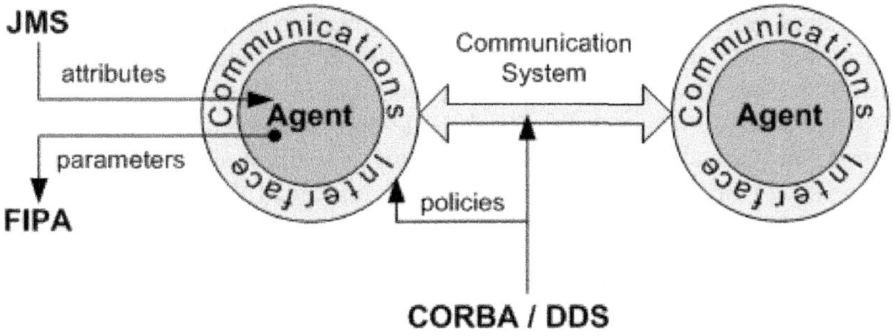

Fig. 1 Location in the MAS communications model of each system analyzed.

CORBA appears in 1991, focused on Object-Oriented Programming (OOP), the programming paradigm used in that time. When the Agent-Oriented Programming model is consolidated appears FIPA. Although is not his main role, CORBA is used as middleware of a great amount of MAS systems, coinciding with the appearance of the agents model. When the concept of middleware is extended, appears JMS. The distributed messaging paradigm is used by JMS and his paradigm is also employed by the DDS model. DDS is the latest model appears and is intentionally a model to be used as middleware.

Figure 1 organizes the scope in the communications process between agents of each system analyzed. JMS is designed to provide to the component the attributes to control the QoS, while FIPA is more focused on the parameters of the agent. CORBA adds the control of the QoS by means the policies, even if the attributes on which it works are more oriented to the communication connection. Finally DCPS is focused on managing the communication of all the distributed system.

3 Quality of Service Supported in Messaging Systems

QoS is a concept that defines a set of parameters to evaluate a service offered. In the field of control architectures, there are many definitions of QoS. From the viewpoint of processing, QoS represents the set of both: quantitative and qualitative characteristics of a distributed system needed to achieve the functionality required by an application [16]. From the communication viewpoint, QoS is defined as all the requirements that a network must meet to message flow transport [17]. In the communication layer, QoS provides temporal parameters like messages delay or easy message flow control like congestion control.

In FIPA, QoS is considered optional, so it is the responsibility of the programmer develops the functions to obtain and manage the QoS parameters. However, [18] proposes 14 relevant QoS parameters. The parameters proposed are similar to the traditionally used in communication systems without taking into account aspects like the message flow or the metadata interchange. Due to FIPA specifies parameters; the model only offers a static idea of the communications.

JMS does not provide QoS parameters. However, different implementations based on the JMS model uses the concept of QoS attributes to manage the communications specified in [19]. In JMS is possible to control 6 attributes, principally specialized in message flow and the temporal characteristics. Nevertheless, JMS don't provide attributes to the handling of the communications faults.

The concept of attributes, offers a dynamic vision of the communications, because some aspects of the message interchange, like the deadline required by the receiver or the deadline that the service can provide, can be configured by the user.

Table 2 QoS areas and parameters included in the systems analyzed.

QoS area	Parameter / Policy	FIPA	JMS	CORBA	DDS
Connection management	Connection delay	Yes			
	Connection errors or liveliness	Yes			Yes
	Connection mode		Yes		
	Connection status	Yes			
	Reconnection			Yes	
Error handling	Error rate / Reliability	Yes			Yes
	Mean up time	Yes			
Message flow	Delivery order			Yes	Yes
	Max hops			Yes	
	Persistence		Yes		Yes
	Priority		Yes	Yes	Yes
	Routing			Yes	
	Synchronization			Yes	
	Topology				Yes
	Transaction type		Yes		
Metadata	Metadata				Yes
	Presentation				Yes
Performance	Bandwidth	Yes			
	Resource limits				Yes
	Throughput	Yes			
Time management	Delay	Yes			
	Delivery mode / Durability		Yes		Yes
	Lifespan				Yes
	Round trip time / Latency	Yes			Yes
	Temporal filter				Yes
	Timeout / Deadline		Yes	Yes	Yes
	Time to live		Yes		
	Timestamp		Yes		

CORBA introduces the concept of QoS policy [20] that works in a communications quality environment. CORBA specifies 13 different policies. The scope of CORBA policies is very wide including policies to manage the message routing. Like CORBA, DDS uses the concept of QoS policy, but DDS defines 22 different QoS policies, covering almost all aspects of the communications. The use of policies by CORBA and DDS offers both visions, static and dynamic, of the communications. A QoS policy can be viewed as a function that returns the state of the communications. Moreover, the QoS policy can be viewed as a function that allows the user to change some aspects of the message interchange.

The middleware isolates the components of the distributed system in time, space and message flow, accordingly, the QoS parameters or policies related with the middleware must contemplate mainly this areas. Table 2 organizes the parameters or policies in the main different QoS areas. Areas are based on the fields covered usually by middleware systems. Detailed definitions of each parameter can be located in the references mentioned above. In a few words, connection management refers to components aspects. Error handling is related with the communications failures. Message flow, performance and time management are classical areas in middleware. Finally, metadata is the additional information that the message syntax can't cover.

4 Analyses and Proposals

As expected, the main areas covered by the QoS parameters are time and message flow management. FIPA focuses on communications aspects as the connection management or the communications performance. JMS is centred on the message flow and time management parameters. CORBA works mainly with message flow, and DDS cover all aspects analyzed.

The parameters most used are priority and deadline. With the priority is possible provide a minimal message flow control in a communications system. In the same way, the deadline provides the minimal time control to messages. These two parameters are the minimum requirements to define the QoS in a communications system. Therefore, in real-time systems is common to find only priority and deadline as QoS parameters.

To increase the QoS support is necessary to provide more QoS parameters, especially to manage the connection among the distributed components. As the number and type of QoS parameters is increased, the complexity of the administration of the system, especially in the middleware, grows. A large number of parameters may be difficult to implement and use the QoS, so the concept of *QoS policy* can help to user with to program the communications with a set o performance constraints.

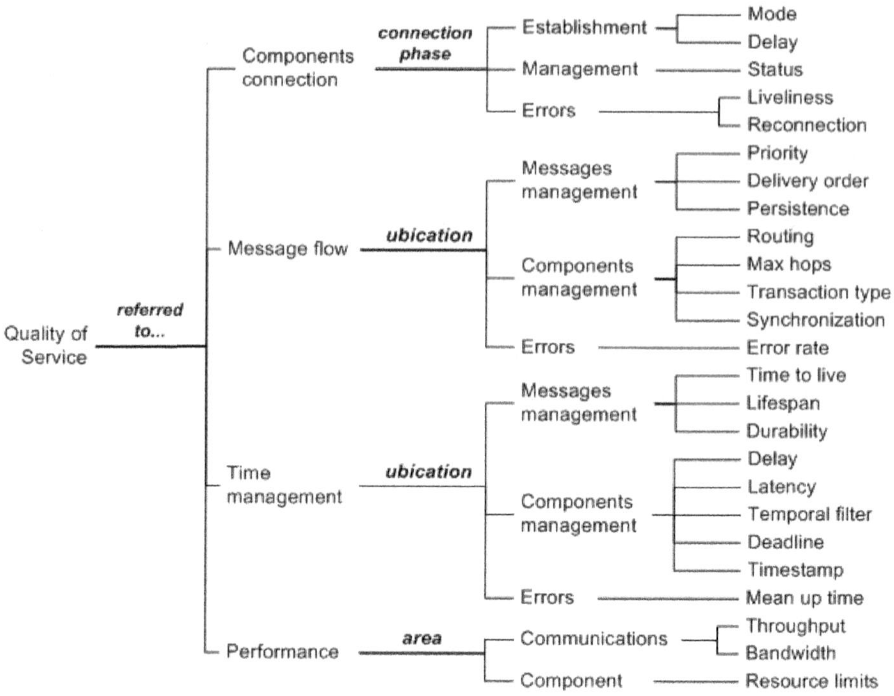

Fig. 2 QoS ontology, where the parameters are organized in function of the system requirements.

It is difficult to determinate the most important or necessary QoS parameters to implement in a distributed MAS; so that, it is necessary organize the parameters. Figure 2 shows an ontology that organizes the parameters in the areas and fields where the QoS parameter is applied. With this ontology, is possible to infer from the type of QoS required which type of QoS parameters are necessary to provide.

The QoS parameters dedicated to manage the components connections, as connection status or liveliness, usually are responsibility of the communications protocol. The performance parameters (throughput, bandwidth or resources management) are highly recommended in all distributed system. Priority, durability, synchronization and deadline, guarantee a minimum QoS in real-time based distributed systems.

The ontology can be organized as a table where the rows describe the main QoS areas (message flow and time management) and columns determine the object to manage (messages or components). Besides, for each QoS area, some parameters can be classified as required while others are only recommender. The table 3 shows the organization described above.

Table 3 QoS parameters recommendations

QoS area	Type of availability	Message management	Component management
Message flow	Required	Priority	Synchronization
	Recommended	Delivery order, Persistence	Transaction type, Routing, Max Hops
Time management	Required	Durability	Deadline
	Recommended	Time to live, Lifespan.	Timestamp, Delay, Latency, Temporal filter

5 Conclusions

On distributed systems, is possible to cover a minimum of the QoS requirements. Depending on the implementation of the system will be more appropriate to use one of the system described in this article. FIPA covers the requirements of the connection on communications systems. In FIPA, other QoS aspects and parameters are the responsibility of the agents. JMS and CORBA provide more QoS parameters than FIPA, JMS is focused on the message flow management, and CORBA offers more parameters to time management, although CORBA diversifies more his parameters.

Finally DDS cover all aspects of the QoS parameters, except the connection management. DDS is the latest standard proposed based on publish-subscribe paradigm, so has taken into account the relevance of the QoS management in the distributed systems. As the complexity of the distributed system grows, the QoS in the middleware is more necessary.

At the time of implementing the QoS in a distributed system, it is necessary to determine the area of communications to manage: the message flow or the time issues. The ontology presented, in combination with the table 3, can help to designed to choose the most appropriate distributed messaging based middleware. To manage a great amount of QoS policies, is recommended use the concept of policies.

From the analysis offered in this article, we can deduce that the QoS support offered by the middleware to the MAS it is increasingly necessary to ensure coherence between the agent communications.

Acknowledgments. The study described in this article is a part of the coordinated project SIDIRELI: Distributed Systems with Limited Resources. Control Kernel and Coordination. Education and Science Department, Spanish Government and European FEDER found. CICYT: MICINN: DPI2008-06737-C02-01/02.

References

1. Gaddah, A., Kunz, T.: A survey of middleware paradigms for mobile computing. Technical Report SCE-03-16. Carleton University Systems and Computing Engineering (2003)
2. Foundation for Intelligent Physical Agents, `http://www.fipa.org/`

3. Java Message Service Specification, http://java.sun.com/products/jms/docs.html
4. Common Object Request Broker Architecture, http://www.corba.org/
5. Data Distribution Service, http://portals.omg.org/dds/
6. Java Agent DEvelopment Framework, http://jade.tilab.com/
7. Agent Oriented Software Pty Ltd., JACK Intelligent Agents: User Guide (1999)
8. Nwana, H., Ndumu, D., Lee, L., Collis, J.: ZEUS: A tool-kit for building distributed multi-agent systems. Applied Artifical Intelligence Journal 13(1), 129–186 (1999)
9. Perdikeas, M.K., Chatzipapadopoulos, F.G., Venieris, I.S., Marino, G.: Mobile Agent Standards and Available Platforms. Computer Networks Journal, Special Issue on 'Mobile Agents in Intelligent Networks and Mobile Communication Systems' 31(10) (1999)
10. Perrone, P.J., Chaganti, K.: J2EE Developer's Handbook. Sam's Publishing, Indianapolis (2003)
11. Apache ActiveMQ, http://activemq.apache.org/
12. IBM WebSphere MQSeries, http://mqseries.net/
13. Object Management Group, http://www.omg.org/
14. RTI Data Distribution Service. RTI corp., http://www.rti.com/
15. OpenSplice DDS. PrismTech Ltd., http://www.prismtech.com
16. Vogel, A., Kerherve, B., von Bochmann, G., Gecsei, J.: Distributed Multimedia and QoS: A Survey. IEEE Multimedia 2(2), 10–19 (1995)
17. Crawley, E., Nair, R., Rajagopalan, B.: RFC 2386: A Framework for QoS-based Routing in the Internet. IETF Internet Draft, 1–37 (1998)
18. Foundation for Intelligent Physical Agents. FIPA Quality of Service Ontology Specification. Doc: SC00094A (2002)
19. Sun Microsystems, Inc. Java(TM) Message Service Specification Final Release 1.1 (2002)
20. Object Management Group (OMG). The Common Object Request Broker Architecture and Specification. CORBA 2.4.2 (2001)

Intelligent Decision Support and Agent-Based Techniques Applied to Wood Manufacturing

Eman Elghoneimy and William A. Gruver

Abstract. A rough mill is a manufacturing facility where loads of lumber of approximate dimensions are cut into components of specific sizes, priorities and qualities. By improving the processes in the rough mill, the cost is reduced and the waste of natural material is decreased. In addition, the rough mill scheduling of components on machines on the chop-line is a challenging problem that cannot be solved by traditional methods because the defects in the wood are not known in advance and the wood sizes are approximate and often inaccurate. In this research, a multi-agent system for decision support and simulation is designed, implemented and evaluated. Scheduling algorithms for the rough mill are implemented using constraint satisfaction and heuristic methods.

Keywords: multi-agent systems, scheduling algorithms, wood manufacturing, decision support, rough mill operations, intelligent distributed systems.

1 Introduction

In a rough mill, bundles of lumber (*jags*) are stored in a warehouse, and then transferred to a ripsaw which cuts the boards into strips. Strips are cut by a chop-saw into components of specific lengths (*cutlist*). Components are then processed to produce manufactured products such as furniture, windows and doors.

Lumber is a natural material which has a variety of defects that are not known in advance. A schedule for cutting the material into defect-free fixed-size components cannot be done in advance since the processing time of each task cannot be calculated in advance. Therefore, scheduling cannot be done using traditional

---

Eman Elghoneimy
Research Associate at Texas A&M University at Qatar. Previously, she was with the School of Engineering Science, Simon Fraser University, Burnaby, BC Canada
e-mail: eman@ieee.org

William A. Gruver
Professor Emeritus in the School of Engineering Science, Simon Fraser University, Burnaby, BC Canada
e-mail: gruver@cs.sfu.ca

A. Abraham et al. (Eds.): International Symposium on DCAI, AISC 91, pp. 85–88.
springerlink.com

methods. Moreover, the schedule has to be dynamically updated; once one task is done, the next is scheduled. In addition to the above challenges, most factories require flexibility for addressing adding or canceling orders, and jammed machines (breakdowns). As factories grow and upgrade its machines and tools, new solutions are required to be flexible, adaptable and expandable.

Current solutions in the rough mill depend on a human operator to create a schedule and dynamically schedule a component on a machine once the previous component is done. Material selection is also based on human operators. Such decisions are rarely optimal, which leads to wasted time and materials.

Agent-based systems [1] are hailed as the next technology for manufacturing systems due to their flexibility, fault-tolerance and expandability. Agents have been used for manufacturing applications ranging from control, scheduling and planning to enterprise operations and supply-chain management [2]. There is no existing agent-based solution that is fit for naturally-defective materials.

2 Agent-Based Decision Support and Simulation System

Fig. 1 shows the architecture of the system. The purpose of this system is to present recommendation for the operators to select material and schedule orders and to view the simulation. An application-specific ontology is developed for defining the concepts and agent actions. JADE framework is used for implementation, as it is compliant with FIPA IEEE standards [3]. There are three categories of agents used in the proposed system: user interface, decision support and simulation agents. This design has a modular architecture where agents represent different physical and logical entities in the rough mill. The system was designed to mimic the functionality of the real-life system. First, a prototype system is verified and validated using JADE sniffer agent, as well as a debug output. Another prototype for two lines of production is implemented using FIPA-PROPOSE negotiation protocol. Finally, the third system RMDSSS (Rough Mill Decision Support and Simulation System) uses the prototype to implement a complete system with real data, simulation and decision support algorithms and two GUIs.

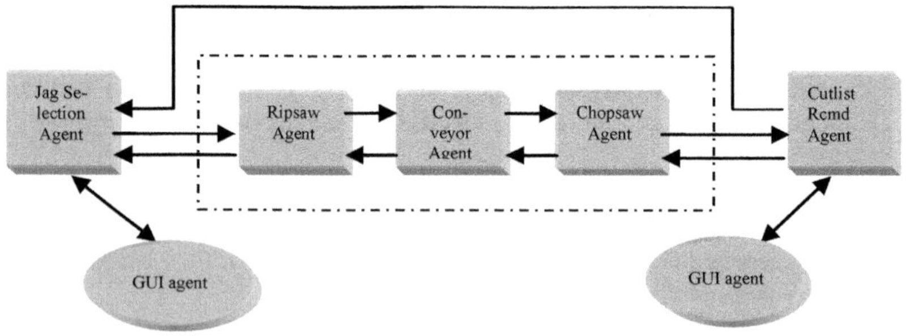

Fig. 1 Rough Mill Decision Support and Simulation System.

The RMDSSS is verified using the JADE GUI (for agent communication) and debug status statements as in prototype evaluation. In addition, the simulation report and the information displayed on the two GUIs and are both used for verification. The simulation is validated against a simulation model, which was used and tested for several years by researchers and rough mill staff.

The advantages of using agent-based design are as follows: (1) modeling rough mill operations, (2) communication using ontology elements, (3) integration of new agents and connecting them to existing agents, (4) expandability by adding more lines or other rough mill operations, and (5) mobility and remote/web monitoring supported by the JADE framework.

The disadvantage of using agent-based systems is the difficulty of the synchronization of a decentralized system. Synchronization of the simulation is done using the conveyor, which mimics the real conveyor behaviour that detects jamming and stops the ripsaw. Simulation time is calculated using bottleneck detection.

3 The Rough Mill Scheduling Problem (RMSP)

The rough mill scheduling problem can be divided into two sub-problems. The first is initial scheduling in which the chop line is empty, so we attempt to schedule as many components as possible on the line at once. The second sub-problem is replacement scheduling where the required quantity of one of the components scheduled on the line is completed, and we need to replace it with another component from the order list.

Natural defects in the wood and inaccurate dimensions result in unknown processing times, therefore traditional scheduling methods cannot be used to solve this problem [4]. The problem can be formulated as a constraint satisfaction and search problem [5].

Backtrack search is implemented and evaluated to discover feasible cutlists. The number of feasible solutions and the runtime grow exponentially when increasing the number of components and machines.

A heuristic method for scheduling is developed. It can be viewed as a best-first search method. Pre-sorted components are assigned to the first valid machine with the smallest range. If the number of components on the list is small, the search is repeated with relaxed parameters. A randomized heuristic method is also implemented that assigns pre-sorted components to the first valid random machine, and n-solutions are presented to the user.

Three replacement scheduling algorithms are also developed to dynamically replace components done on the line with new ones. The first method presents all the feasible replacement components, the second method selects one component, and the third method presents n-components based on heuristics.

The backtrack algorithm is tested on the rough mill data. As the number of components and the number of machines increase, the number of feasible cultists and the runtime increase exponentially. For scheduling of a small test order list on 25 machines, $4.4*10^8$ feasible lists were discovered in 17 hours. This makes this method not useful for large numbers, since it is impossible go through all the nodes of the search tree, as well as the impractical runtime. The results show the complexity of the RMSP, as it can be solved by complete search methods. It is essential to develop other methods that find a reasonable number of cultists in acceptable time. All the heuristic initial and replacement algorithms provide instant results, scheduling 19-20 of the 25 available machines. The results are validated by tracing debug statements, and analyzing the order list and machine information. The randomized heuristic algorithm provides more choices of cultists, some with more components than the heuristic method. Similarly, the heuristic n-replacement provides more choices for replacement components, starting with components of the same length as the component that was done.

4 Conclusion and Future Work

A multi-agent decision support and simulation system is implemented to address the challenges of rough mill operations. A rough mill ontology is designed and used by the agents. The system provides a model, simulation, material selection decision support and scheduling. The problem of scheduling components on the chop-line is addressed. Backtrack, heuristic and randomized heuristic scheduling methods are developed and evaluated using rough mill data. Backtrack provides a large number of feasible lists in a very long run time. The heuristic algorithm is based a best-fit search method provides one feasible cut list, while the randomized heuristic method provides n results, all of which meets the heuristic criteria.

Future research involves extending the agent-based system to add more operations and provide remote or web monitoring. Learning algorithms can be used to improve the simulation agents. Scheduling can gather data from the simulation and improve on existing heuristics. Several local search methods with various objective functions can be implemented and compared using the simulation.

References

1. Wooldridge, M.: Intelligent agents: The key concepts. In: Multi-Agent-Systems and Applications II: 9th ECCAI-ACAI/EASSS, pp. 151–190 (2002)
2. Elghoneimy, E., Gruver, W.A.: Agent-based manufacturing systems: A survey. In: Proc. of the IEEE SMC International Conference on Distributed Human-Machine Systems, pp. 127–132 (2008)
3. Bellifemine, F., Poggi, A., Rimassa, G.: Developing multi-agent systems with a FIPA-compliant agent framework. Software - Practice & Experience 3(3), 103–128 (2001)
4. Pinedo, M.: Scheduling: Theory, Algorithms, and Systems. Prentice-Hall, Englewood Cliffs (2002)
5. Russell, S.J., Norvig, P.: Artificial Intelligence: A Modern Approach. Prentice-Hall, Englewood Cliffs (2010)

Multiple Mobile Agents for Dependable Systems

Ichiro Satoh

Abstract. This paper presents a framework for enabling mobile agents to be organized dynamically and autonomously with two unique compositions and interagent interactions on distributed systems. The first enables an agent to contain other agents inside it and migrate to another agent or computer with its inner agents. It provides a powerful approach to composing and deploying large-scale mobile software. The second enables an agent to define its destination according to another agent's location with several policies to support adaptation on distributed systems. It also introduces several higher-level coordinations between mobile agents, e.g., master-slave and redundancy, which are useful for implementing systems. This paper also describes a prototype implementation of the framework with mobile agent technology and several applications of it.

1 Introduction

Distributed systems are constructed as an organization of software components, which may be running on different computers connected through networks, in coordination patterns, e.g., client-server and peer-to-peer. For example, *MapReduce*, which was introduced by Google [3], is a pattern for computing large data sets on clusters of computers and has been used in cloud computing. Distributed systems need unique coordination patterns, e.g., master-slave and those that are redundant ones. Distributed systems, on the other hand, are dynamic in the sense that computers are dynamically added to or removed from the systems, the network topology is changed, and the requirements of applications running on the systems are changed. Distributed systems need to adapt to changes in them. However, most existing work on distributed systems has assumed the organization of components is statically defined and remains at computers.

Ichiro Satoh
National Institute of Informatics
2-1-2 Hitotsubashi, Chiyoda-ku, Tokyo 101-8430, Japan
e-mail: ichiro@nii.ac.jp

A. Abraham et al. (Eds.): International Symposium on DCAI, AISC 91, pp. 89–98.
springerlink.com

This paper addresses a framework for dynamically deploying and organizing software components in coordination patterns at distributed computers. The framework is unique to other existing work, because it enables software components to define their own policies without any global policies and components are implemented as mobile agents, where mobile agents are autotomous and self-contained programs that can travel from computer to computer under their own control. Mobile agent technology was explored by many researchers fifteen years ago. We believe that the technology can still provide a powerful technique of implementation for distributed systems, particularly adaptive and sustainable ones. However, researchers at the time addressed the mobility of agents and their agents were designed as isolated entities. This is serious because existing mobile agent platforms cannot be used to compose large-scale and dynamic mobile software as a mobile federation between multiple mobile agents.

Our framework offers the notion of self-defined policies to mobile agent technology. This is because the deployment of components is needed to adapt coordinations between components to a dynamic distributed system whose computers may be dynamically added to or removed from the system and networks are (dis)connected. The framework can provide various coordination and deployment patterns among software components, including mobile agents, running on different computers.

2 Basic Approach

The framework presented in this paper introduces two novel compositions of mobile agents, called *containment* and *traction*, and higher-level agent coordination, to support distributed systems. These are introduced as policies between agents.

2.1 Containment Composition of Mobile Agents

The *containment* composition enables each mobile agent to have one or more slots, where each of the slots can contain at most one agent that satisfies their specifications.[1] Like the data types in programming languages, each slot can only hold the agent that has its specified properties inside it. When an agent can have more than one slot, it is called a *container* agent. This composition consists of two policies, called *injection* and *extraction* (Fig. 1). The first means that an agent can enter another agent and the second means that a visiting agent can be left behind by its current agent. Each agent can have actions defined as policies when another agent enters or leaves it. As composition provides a mechanism for dynamically assembling and deploying a group of services at computers, it is useful in supporting the availability and reliability of distributed systems.

[1] Containers cannot become their own descendent agents to avoid to problem of cycle containing.

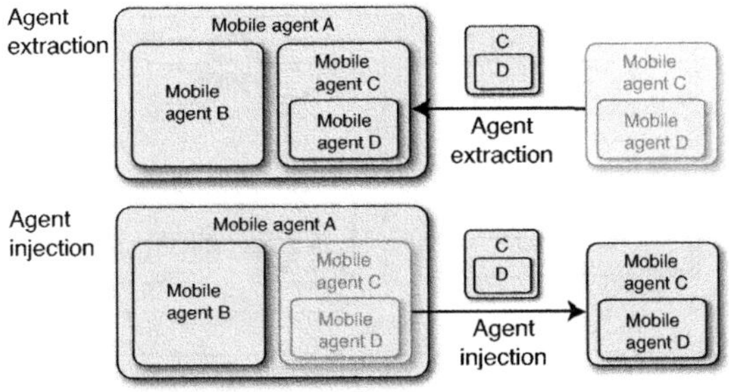

Fig. 1 Containment relationships and migration between agents.

2.2 *Traction Composition*

The framework also enables each agent to explicitly specify a rule, called a *traction*, for agent migration. The current implementation provides two types of tractions, as shown in Fig. 2. The first enables an agent to follow another agent and the second enables an agent to migrate to the source location of another agent. They can put off the deployment of agents for a specified time after other agents have migrated. Therefore, they can be treated as *dynamic friction* between agents and the source locations of the agents. Each traction is declared a relocation between its target agent and another agent. It is defined as a container agent that can have at most one agent inside it. This composition is useful in supporting the confidentiality and integrity of distributed systems.

- If one agent declares a *follow* policy for another, when the latter exists or migrates to a host, the former migrates to the latter's current or destination host.
- If an agent declares a *dispatch* policy for another, when the latter migrates to another host, a copy of the former is created and deployed at the latter's destination host.
- If an agent declares a *shift* policy for another, when the latter migrates to another host, the former migrates to the latter's source host.
- If an agent declares a *fill* policy for another, when the latter migrates to another host, a copy of the former is created and deployed at the latter's source host.

The framework allows each agent to have at most one policy for at most one agent.

2.3 *Higher-Level Coordination of Interagents*

The framework provides two levels of interactions between agents. The first, called *interaction*, offers a meeting place for multiple agents that are contained in the same

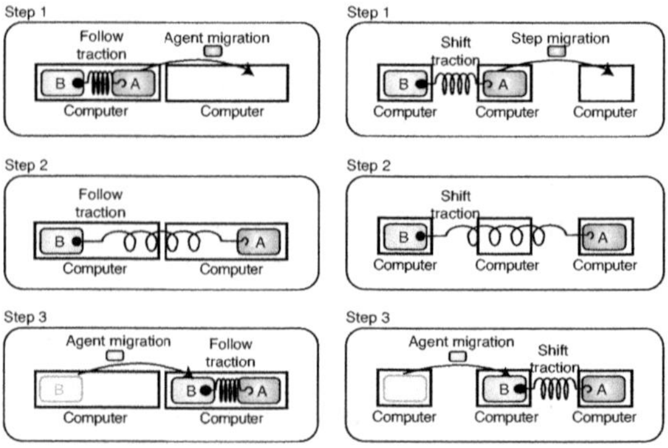

Fig. 2 Agent migration with traction composition.

container agents. Each container agent can define coordination between its inner agents. The current implementation provides two types of built-in coordination:

- **Multicast coordination:** When a container agent receives a message from one of its inner agents or an external agent, it multicasts the message to all its inner agents.
- **Selective coordination:** When a container agent receives a message from one of its inner agents or an external agent, it selects one of its inner agents according to the following three sub-policies and forwards the message to the selected agent.

where the sub-policies are *static assignment* selective coordination forwards messages to a specified inner agent, *random assignment* selects one agent among the inner agents randomly and forwards messages to the selected agent, and *alternate assignment* selects one agent among the inner agents as a master agent in turn and forwards messages to the current master agent.

The second level, called *delegation*, supports two kinds of coordination between agents that are not contained in the same agents. It is defined as a container agent that can have at most one agent inside it and it is useful to organize software components in distributed systems:

- **Master-slave coordination:** When an agent receives a message from an external agent, it forwards the message to its inner agent if the agent is not busy or is present. Otherwise, it forwards the message to a specified external agent.
- **Redundancy coordination:** When an agent receives a message from an external agent, it forwards the message to both its inner agent and a specified external agent.

3 Implementation

This section describes our mobile agent platform for supporting the agent composition and coordination presented in the previous section. It was implemented with

Java language and operated on the Java virtual machine. Each runtime system runs on a computer and is responsible for executing agents at the computer and migrating agents to other computers through networks (Fig. 3). Each system also establishes at most one TCP connection with each of its neighboring systems in a peer-to-peer manner without any centralized management server and exchanges control messages, agents, and inter-component communications with these through the connection. When the life-cycle state of an agent is changed, the runtime system issues certain events to the agent and its descendent agents. We tried to contain the implementation within the framework as much as possible. Each agent in the current implementation of the framework is a collection of Java objects in the standard JAR file format. It has its own name based on the agent hierarchy and a message queue for incoming messages.

Management of agent hierarchy: Unlike other mobile agent platforms, our platform organizes multiple agents hierarchically. Each runtime system manages an agent hierarchy as a tree structure in which each node contains an agent slot and its attributes. Each agent slot contains at most one agent. Each runtime system also corresponds to the container agent that contains all the agents running on it and maintains the root agent between other runtime systems. Therefore, every agent, which is not the root agent, is contained in at most one agent. This framework assumes that each agent is active but subordinate to its container agent. Therefore, each agent has direct control of its descendent agents. That is, an agent can instruct its descendent agents to move to other agents, and serialize and destroy them. No agent has direct control over its ancestral agents.

Management of agent execution: Each agent is provided with its own Java class load, so that its namespace is independent of other agents in each runtime system.[2] Therefore, even when two agents are defined from different classes whose names are the same, the runtime system disallows agents from loading other agents's classes. To prevent agents from accessing the underlying system and other agents, the runtime system can control all agents in its agent hierarchy, under the protection of Java's security manager. Each agent can have one or more activities, which are implemented by using the Java thread library. Furthermore, the runtime system maintains the life-cycle of agents: initialization, execution, suspension, and termination. When the life-cycle state of an agent is changed, the runtime system issues certain events to the agent and its descendent agents. The system can impose specified time constraints on all method invocations between agents to avoid being blocked forever.

Agent migration: When an agent is moved inside a runtime system, the agent and its inner agents can still be running. When an agent is transferred over a network, the runtime system stores the state and the codes of the agent, including the agents contained in it, into a bit-stream formed in Java's JAR file format, which can support

[2] The identifier of each agent is generated from information consisting of its runtime system's host address and port number, so that each agent has a unique identifier in the whole distributed system.

digital signatures for authentication. The current system basically uses the Java object serialization package for marshaling agents. The package does not support the capturing of stack frames of threads. Instead, when an agent is saved or migrated, the runtime system issues events to it and all its descendent agents to invoke their specified methods, which should be executed before they migrate, and it then suspends their active threads and migrates them to the destination.

Traction composition management: The traction composition of agents is managed by each runtime system without any centralized management server. Each runtime system periodically advertises its address to the others through UDP multicasting, and these runtime systems then return their addresses and capabilities to the system through a TCP channel.[3] When an agent migrates to another runtime system, the destination sends a query message to the source of the visiting agent about agents that declare policies to the moving agent.

Higher-level agent coordination: Most interactions between agents in object-oriented systems within a computer can be covered by three primitives: event passing, method invocation, and stream communication. Our framework enables these primitives to be available in partitioned systems on different computers. Achieving syntactic and (partial) semantic transparency for remote interactions requires the use of proxy objects that have the same interfaces as the remote agents. The framework introduces such objects, called *references*, to track possibly moving targets and to interact with the these through the three primitives. Each runtime system offers an RMI mechanism through a TCP connection. It is implemented independently of Java's RMI because Java's RMI lacks any mechanisms for updating references for moving agents. Each runtime system can maintain a database that stores pairs of identifiers of its connected agents and the network addresses of their current runtime systems. It also provides agents with references to others that belong to the same application federation. Each reference enables one agent to interact with another it specifies, even if the agents are on different computers or move to others.

Current Status and Performance: A prototype implementation of this framework was constructed with Sun's Java Developer Kit version 1.5 or later versions. The implementation provided graphical user interfaces to operate the mobile agents. These interfaces allowed us to easily load and migrate mobile agents via full drag-and-drop operations.

Although the current implementation was not constructed for performance, we evaluated that of several basic operations in a distributed system where eight computers (Intel Core Duo 2 1.8 GHz with MacOS X 10.5 and J2SE version 5) were connected through a fast Ethernet. The cost of agent migration in an agent hierarchy was measured to be 3 ms, including the cost of checking whether the visiting agent was permitted to enter the destination agent or not. The cost of agent migration between agents allocated on two computers was measured to be 28 ms. The

[3] We assumed that the agents comprising an application would initially be deployed at runtime systems within a localized space smaller than the domain of a sub-network.

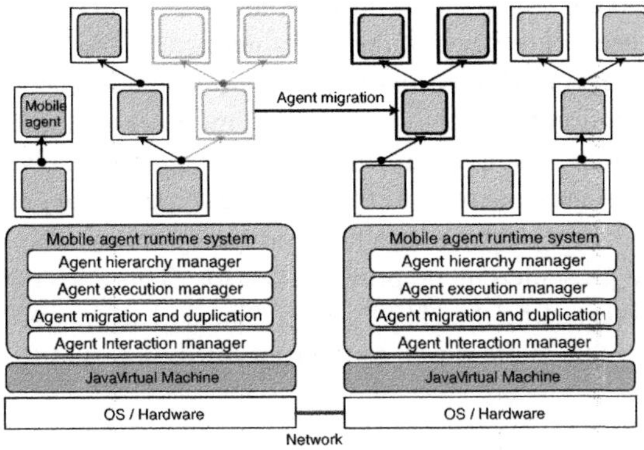

Fig. 3 Basic structure of runtime system.

moving agent was simple and consisted of basic callback methods and contained two inner agents. Its data capacity was about 7 Kbytes (zip-compressed). The cost of agent migration included that of opening TCP-transmission, marshaling the agents, migrating the agents from their source computers to their destination computers, unmarshaling the agents, and verifying security.

4 Experience

This section presents several agent organizations to illustrate how the framework works based on our experience.

4.1 Tracing Mobile Agents

When an agent tries to visit or communicate with another agent, the latter agent may have moved to another computer. Our runtime systems do not offer mechanisms for communicating between agents contained in different container agents, which may be running on a computer. Instead, the framework introduces *forwarder* agents, to support inter-agent communications between agents. Forwarder agents are also used for tracking the current locations of moving agents. Immediately before each agent migrates to its destination, i.e., another agent or computer, it can explicitly leave an agent, called a forwarder agent, corresponding to the master-slave coordination policy. When each forwarder agent receives messages or other agents that want to visit the moving agents, it automatically transfers the message or visiting agents to the destination according to this policy.

4.2 Dynamic Deployment for Duplicated Servers

We constructed a fault-tolerant HTTP-based server to illustrate the utility of the policies presented in this paper. An agent for the HTTP-based server makes a clone of it and creates two container agents corresponding to the redundancy coordination policy. The agent itself enters one of the container agents and the clone agent enters the other container agent. This occurs where the container agents corresponding to redundancy coordination forward their receiving messages to their target agents and one another. Therefore, when one of the container agents receives a message, the target agent and its clones can receive the message. After the agent duplicates itself, the cost of deploying its clone in another runtime system is about 180 ms for the distribution system presented in the previous section. This does not include the cost of terminating and restarting the HTTP server. The cost of forwarding a message is about 22 ms, where this is measured as the round-trip time and the message has no value.

4.3 Mobile Agent-Based MapReduce System

MapReduce is a framework for computing certain kinds of distributable problems using a cluster consisting of a large number of computers [3]. The original MapReduce consists of two elements: a master node and workers and is processed in three phases:

- *Map* phase: The master node divides a problem into smaller sub-problems and then distributes those to worker nodes.
- *Worker* phase: Each worker node processes that smaller problem, and passes the answer back to its master node.
- *Reduce* phase: The master node then takes the answers to all the sub-problems and combines them.

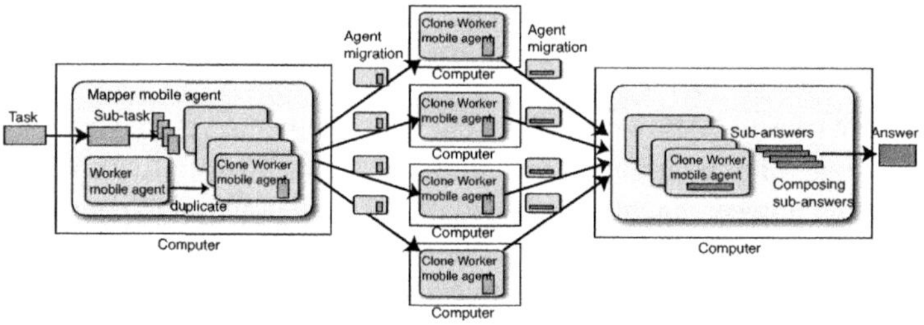

Fig. 4 Mobile agent-based MapReduce system

We tried to implement the MapReduce pattern by using three kinds of mobile agents, *Mapper*, *Worker*, and *Reducer*, with our framework (Figure 4). The *Mapper* agent is a container agent corresponding to the master node in the MapReduce pattern. It supports multicast coordination and contains at least one worker agent inside it. When it receives a task message from the external system, it forwards the message to all its worker agents. If the task can be divided into smaller sub-tasks by using a function specified by users like the original MapReduce, the agent forwards sub-tasks to worker agents. Each *Worker* agent migrates from the mapper agent to its target computer to process its own task by using its internal program after being deployed at the worker node. When it finishes the task, it migrates into the reducer agent. The *Reducer* agent is a container agent that can receive worker agents. To collect the results from the worker agents, it can strand its inner worker agents until it can satisfy specified conditions, e.g., the number of worker agents.

Since the mapper and reducer agents are still mobile, we can change the computers at which the agents are located in addition to worker agents, unlike these in the original MapReduce pattern. We evaluated our mobile agent-based MapReduce system with ten computers (Intel Core Duo 2 1.8 GHz with MacOS X 10.5 and J2SE version 5), where the first was for the mapper agent, the second was for the Reducer agent, and the remainders were eight worker agents. After the mapper agent received the task, the cost of collecting the answers from the eight worker agents was about 780 ms, where the worker agents' programs for processing the sub-tasks were null functions. Since map, reduce, and worker agents were mobile, the system could dynamically distribute them over a distributed system unlike that in the original MapReduce system.

5 Related Work

Numerous mobile agent systems have been released, e.g., Aglets [5], Mole [8], and Telescript [9]. Several mobile agent systems, e.g., Telescript, have introduced the concept of places in addition to mobile agents. Although places are agents that can contain mobile agents and places inside them, they are not mobile. Our mobile agent system, on the other hand, allows one or more mobile agents to be dynamically organized into a single mobile agent, and thus we do not have to distinguish between mobile agents and places. Therefore, a distributed application, particularly a mobile application that is complex and large in scale, can easily be constructed by combining more than one agent. Of these, the FarGo system introduces the notion of a dynamic layout for distributed applications [4] in a decentralized manner. This is similar to our relocation policy in the sense that it allows each agent to have its own policy, but it is aimed at allowing one or more agents to control a single agent, whereas ours aims at allowing one agent to describe its own migration. This is because our framework treats agents as autonomous entities that travel from computer to computer under their own control. This difference is important, because FarGo's policies may conflict if two agents can declare different relocation policies for one single agent. Our framework is free of conflict because each agent can only declare a policy to relocate itself but not that for other agents.

6 Conclusion

This paper described a framework for dynamically organizing multiple mobile agents for computing. It is unique to existing mobile agent systems because it provides three mechanisms for organizing multiple mobile agents. The first enables a mobile agent to contain another mobile agent inside it and migrate to another mobile agent or computer with its descendent agents. It is useful in developing large-scale mobile software from a collection of mobile agents. The second supports higher-level coordination between mobile agents, e.g., master-slave and redundant approaches, which are useful in distributed computing. The third enables mobile agents to be dynamically replaced by other agents without any coordinations between agents. We designed and implemented a prototype system for the framework and demonstrated its effectiveness in several practical applications.

References

1. Babaoglu, O., Meling, H., Montresor, A.: Anthill: A Framework for the Development of Agent-Based Peer-to-Peer Systems. In: Proceeding of 22th IEEE International Conference on Distributed Computing Systems (July 2002)
2. Baumann, J., Radounklis, N.: Agent Groups in Mobile Agent Systems. In: Proceedings of Conference on Distributed Applications and Interoperable Systems (1997)
3. Dean, J., Ghemawat, S.: MapReduce: Simplified Data Processing on Large Clusters. In: Proceedings of OSDI 2004, pp. 137–150. USENIX (2004)
4. Holder, O., Ben-Shaul, I., Gazit, H.: System Support for Dynamic Layout of Distributed Applications. In: Proceedings of International Conference on Distributed Computing Systems (ICDCS 1999), pp. 403–411. IEEE Computer Society, Los Alamitos (1999)
5. Lange, B.D., Oshima, M.: Programming and Deploying Java Mobile Agents with Aglets. Addison-Wesley, Reading (1998)
6. Paun, G.: Computing with Membranes: An Introduction. Springer, Berlin (2002)
7. Satoh, I.: MobileSpaces: A Framework for Building Adaptive Distributed Applications Using a Hierarchical Mobile Agent System. In: Proceedings of International Conference on Distributed Computing Systems (ICDCS 2000), April 2000, pp. 161–168. IEEE Computer Society, Los Alamitos (2000)
8. Strasser, M., Baumann, J., Hole, F.: Mole: A Java Based Mobile Agent System. In: Tschudin, C.F., Ryan, M. (eds.) MOS 1996. LNCS, vol. 1222, Springer, Heidelberg (1997)
9. White, J.E.: Telescript Technology: Mobile Agents, General Magic (1995)

A Multi-agent System for Resource Management in GSM Cellular Networks

Jamal Elhachimi and Zouhair Guennoun

Abstract. This paper presents a new experience in designing and developing a multi-agent system for managing frequency resources in a Regional Access Network (RAN) in GSM system (Global System for Mobile Communications). In our approach, a group of agents are distributed in the network with each regional network overseen by a supervisor agent; i.e. combining a cooperative agent to each cell called the station agent that handles the assignment of a frequency. Within a Radio Area Network – RAN and at each step, an agent is elected by all its neighbors: The election is based on empirical rules for calculating the degree of separation of an agent, the degree of saturation and the improvement claimed by the neighbors for an assignment. The elected agent assigns the smallest frequency in the spectrum that meets all its constraints. In the case of a non permitted assignment, the agent may be served by a neighboring RAN, through a mechanism of cooperation between supervisor agents of both RANs. All RANs are handled in a localized region regardless of the operating band. Our multi-agent system has been implemented in JADE, a well-known multi-agent platform based in JAVA [4]. Experimental evaluations using standard benchmarks of frequency assignment problems show that this approach can find optimal solutions and exact solutions for some instances of these problems and the results obtained are equivalent to those of current methods using simulated annealing, constraint satisfaction/optimization techniques, or neural networks. These results show that our approach is more efficient in terms of flexibility and produces an excellent degree of optimality in terms of flexibility, autonomy and resource requirements.

Keywords: Multi-agent systems, Frequency assignment problem, constraint optimization techniques, JADE.

1 Introduction

Recent demand for mobile telephone service has been growing rapidly. At the same time, the electromagnetic spectrum or frequencies allocated for this purpose

Jamal Elhachimi · Zouhair Guennoun
Laboratory of Electronics and Telecommunications (LEC) - Mohammadia School of Engineers (EMI), Rabat, Morocco
e-mail: jamal_elhachimi@hotmail.com, zouhair@emi.ac.ma

A. Abraham et al. (Eds.): International Symposium on DCAI, AISC 91, pp. 99–106.
springerlink.com

are very limited and getting more and more scarce. This makes solving the frequency assignment problem (FAP) more and more critical. The main goal of this problem is to be able to obtain an efficient use of the scarcely available radio spectrum on a network. The available frequency band is assigned into channels (or frequencies) which have to be allocated to the transceivers installed in each base station of the network. Both these components have an important role in the definition of this problem. This work is focused on the concepts and models used by the current cellular frequency planning, while it considers the following three conditions as the electromagnetic compatibility constraints as in [1] and [2]:

- Co-Channel Constraint: the same frequency cannot be assigned co-channel constraint (CCC) where the same channel cannot be assigned to certain pairs of radio cells at the same time.
- Adjacent Channel Constraint (ACC): similar frequencies cannot be simultaneously assigned to adjacent cells.
- Co-Site Constraint (CSC): any pair of frequencies assigned to the same cell must have a certain gap.

The goal is to find a frequency plan that satisfies the above constraints using a minimum number of frequencies. An optimal solution is sought to facilitate the subsequent addition of new links [3]. This problem has been undertaken in various ways: Some approaches, based on mathematical programming techniques and Heuristic techniques, can be used to solve the problem to optimality but computational time is usually excessive. These approaches have been found less efficient and more costly towards the rapid emergence of the connected devices with wireless links and the current trends of a telecommunication network to use effectively and fully radio resources and multimedia applications.

Connect at best anywhere, anytime and with any network, Customize the more powerful features stimulated by the increasing consumers' demand, Find solutions for the mobile business, and Tend toward several access technologies whose assignment is local and continuously and independently updated, rending impossible any overall control. Hence there are needs to a highly dynamic adaptable distributed system architecture based on a paradigm of distributed agents for modeling and simulating a complex system. This architecture involves different entities interacting with each other and with their neighborhood in order to maintain real-time reactions when sudden changes happen.

This paper is structured as follows. In the next section we provide the state of the art of frequency assignment problem in GSM mobile networks. The mathematical formulation is described in section III. Section IV presents our intelligent resolution approach based on a distributed multi-agent system. The results of the experiments are analyzed in Section V. Finally, conclusions are in the last section.

2 The State of the Art of Frequency Assignment Problem

This problem arises in GSM phone networks where the network is subdivided into cell areas. Each cell is covered by base station transceivers (BTSs) and a number of transmitters (or TRXs) whose locations are known. Therefore, transceivers are the main element to be considered. A BTS can be viewed as a set of TRXs, which are organized in sectors. The TRX is the physical equipment responsible for providing the communication link between the mobile terminal and the network. The frequency assignment problem arises because the number of available frequencies (or channels) to be assigned to each TRX is very scarce. Therefore, the available frequencies need to be reused by many transceivers of the network [1] and [6]. The reuse of these frequencies can compromises the quality of the service (QoS) of the network. Hence, it is extremely important to make an adequate reuse of these frequencies to several TRXs, in such a way the total sum of interferences occurring in the network is minimized. Consequently, it becomes extremely important to quantify the interferences provoked by an assignation of a frequency to a TRX and its influence on the remaining TRXs of the other sectors of the network. To quantify this value an interference matrix is used, called compatibility matrix and denoted C or M [4]. Each element, M(i,j), of M represents the degradation of the network quality if sector i and j operate with the same frequency value. This represents the co-channel interferences. Moreover, adjacent-channel interferences need to be considered. Adjacent-channel interferences occur when two TRXs, in two different sectors, operate on adjacent channels (i.e., when one TRX operates on channel f and the other on channel f+1 or f-1). Therefore, the interference matrix plays an import role in the computation of the cost function and the formulation of the FAP problem, which aims to minimize the sum of interferences occurring in the network (see Eq. 1).

3 Mathematical Formulation

Let $T=\{t_1, t_2,\dots,t_n\}$ be a set of n transceivers (TRXs), and let $F_i=\{f_{i1},\dots, f_{ik}\} \subset N$ be the set of valid frequencies that can be assigned to a transceiver $t_i \in T$, $i=1, \dots, n$ (the cardinality of F_i could be different to each TRX). Furthermore, let $S = \{S_1, S_2, \dots, S_m\}$ be a set of given sectors (or cells) of cardinality m. Each transceiver $t_i \in T$ is installed in exactly one of the m sectors and is denoted as $s(t_i) \in S$. It is also necessary to consider the interference matrix, M, defined as: $M = \{(\mu_{ij}, \sigma_{ij})\}$ of cardinality m*m. The two elements μ_{ij} and σ_{ij} of a matrix entry $M(i,j) = (\mu_{ij}, \sigma_{ij})$ are numerical values greater than or equal to zero and they represent respectively, the mean and standard deviation of a Gaussian probability distribution used to quantify the interferences ratio(C/I) when sector i and j operate on a same frequency. Therefore, the higher the mean value is, the lower interferences are, and thus it will have a superior communication quality. A solution to the problem lays in assigning to all the TRXs (t_i) a valid frequency from its domain (F_i), in order to minimize the following cost function:

$$C(p) = \sum_{t \in T} \sum_{u \in T, u \neq t} C_{sig}(p,t,u) \qquad (1)$$

Where C_{sig} will compute the co-channel interferences (C_{co}) and the adjacent-channel interferences (C_{adj}), for all sectors s_t and s_u, in which the transceivers t and u are installed, that is, s(t) and s(u), respectively. p∈ $F_1 \times F_2 \times ... \times F_n$ denotes a solution (or frequency plan), where p(t_i)∈ F_i is the frequency assigned to the transceiver t_i. Moreover, $\mu_{s_t s_u}$ and $\sigma_{s_t s_u}$ are the interference matrix values at the entry M(s_t,s_u) for the sectors s_t and s_u. In order to obtain the C_{sig} cost from equation 1, the following conditions are considered:

$$\begin{cases} K & if \quad s_t = s_u, |p(t) - p(u)| < 2 \\ C_{co}(\mu_{s_t s_u}, \sigma_{s_t s_u}) & if \quad s_t \neq s_u, \mu_{s_t s_u} > 0, |p(t) - p(u)| = 0 \\ C_{adj}(\mu_{s_t s_u}, \sigma_{s_t s_u}) & if \quad s_t \neq s_u, \mu_{s_t s_u} > 0, |p(t) - p(u)| = 1 \\ 0 \quad otherwise & \end{cases} \tag{2}$$

Where K is a very large value defined in the configuration files of the network. The K value makes it undesirable to allocate the same or adjacent frequencies to TRXs that are installed in the same sector. In our approach, this restriction was incorporated in the creation of the new solution (frequency plan) produced by the algorithm. Therefore, we assure that the solution does not have this severe penalty, which causes the most undesirable interferences as shown in [1].

4 Resolution Approach and Development

The approach comes in the form of a group of distributed agents in the network where each regional network is overseen by a supervisor agent, combining an agent to each cell called a station agent.

4.1 Station Agent

An agent can be defined as a computer system located in an environment and which can act autonomously and flexibly to achieve the objectives for which it was designed. To each station l_i is associated an agent A_i responsible of assigning a value f_i in its domain D_i. Two data are sufficient to characterize the agent in outside its environment:

1-The frequency value of the corresponding station - The agent chooses among the values in the frequency domain corresponding to this station: For each A_i in A, the f_i is in D_i, where D_i is the frequency domain of l_i.

2-The difficulty of an agent defined as a quantitative measure that reflects the current status of this agent. It is the decision criterion used to choose an agent as the elected agent. This measure is expressed in the form of two essential and sufficient entities that are the degree of separation and the degree of saturation. These two entities are intuitively and experimentally determined. For any agent A_i in A, we note D(A_i) the degree of separation of the corresponding link l_i as the sum of the incident constraints values to stations: D(A_i) = { $\sum_{i \neq j} C_{ij}$, C_{ij} in C }

The degree of saturation at step p is determined from the banned intervals for those stations that are not yet assigned. It can be deduced by the number of unsatisfied constraints. For A_i in A, NIS(A_i) is the number of unsatisfied constraints with its value f_i: NIS(A_i) ={ $\sum_{i \neq j} C_{ij}$ for each A_j such as f_j != 0 and C_{ij} != 0}

At step p and for each A_i in A, D_SATp(A_i) = NIS(A_i) ; The agent who has the greatest degree of saturation will be considered as the most on difficulty.
Each agent operates in a physical environment that is its frequency domain. Even though, for several agents, these domains may be identical but not shared. Similarly an agent has an unshared copy of constraints that allows it to be independent of other agents. The social Environment consists of all neighbors of an agent from which it has only a partial view; i.e., it knows about its neighbors only their values and their difficulties, but it has no idea about its neighbors' constraints, views, and domains. The communication is performed by sending messages and a mailbox is associated with each agent that stores the received messages from other agents [7].
The neighborhood of an agent A_i is defined by all agents connected by a constraint to this agent:

For each A_i in A, V(A_i) = {A_j in A / C_{ij} in C , C_{ij} != 0 }

Any change of view leads to an immediate update of the state of constraints. The agent will be in a consistent state at any time.

4.2 Behavior

The behavior of an agent takes place in three phases:

Step 1: Determine V(Ai); V(Ai)={Aj$\in$ A (j$\neq$ i)/ Cij$\neq$0}; Calculate D(Ai);
For all Aj$\in$ V(Ai) (j$\neq$ i) {Ai sends D(Ai) to Aj; Ai Receives D(Aj) ; **End For**
If { $\forall$ Aj$\in$ V(Ai), D(Ai) > D(Aj)} **then** Ai is elected ; Ai: fi $\leftarrow$ f such as
f = min {fi, $\forall$ fi$\in$ Di / $\forall$ Aj$\in$ V(Ai) such as fj $\neq$ 0 and |fi – fj| >Cij (Cij is true)};
Ai sends fi to all Aj$\in$ V(Ai) ; Ai deactivates; goto step 3;
Else {Receives fj}; goto step 2;
Step 2: Calculate D_SATp(Ai); //the degree of saturation on step p
For all Aj $\in$ V(Ai) such as fi=0 (j$\neq$ i): **do** Ai sends D_SATp(Ai) to Aj ;
Ai receives D_SATp(Aj) ; **End For**
If { $\exists$ Aj $\in$ V(Ai) / D_SATp(Aj) > D_SATp(Ai)} goto Step 2;
Else If { $\exists$ Aj$\in$ V(Ai) / D_SATp(Aj) = D_SATp(Ai)}
If {D(Ai) > D(Aj)} then Ai is elected ;
Ai : fi$\leftarrow$f such as f = min {fi, $\forall$ fi$\in$ Di / $\forall$ Aj$\in$ V(Ai)
such as fj $\neq$ 0 and |fi – fj| > Cij (Cij is true)}; Ai sends fi to all Aj$\in$ V(Ai) ;
Ai deactivates; go to step 3; **Else** Goto Step 2;
Else Aj is elected ; Ai : fi $\leftarrow$ f such as f = min{fi, $\forall$ fi$\in$ Di / $\forall$ Aj$\in$ V(Ai)
such as fj $\neq$ 0 and |fi – fj| >Cij (Cij is true)}; Ai sends fi to all Aj$\in$ V(Ai);
Ai deactivates; goto step 3;
Step 3: Exit; // Elimination of the agent

4.3 Supervisor Agent

First, the supervisor agent is in charge of the cooperation between other neighbors RANs supervisor agents. Second, the supervisor agent oversees the management of assignments by:

- Initializing agents associated to stations: called station agents.
- Sending all RAN data associated to station agents: associated Frequency Domain, re-use matrix, stations rentals.
- Holding and collecting responses (until triggering a timeout). In case of a non permitted assignment within its RAN, the agent may resort to another supervisor agent.

The supervisor agent can communicate with other resources outside of its frequency domain through a cooperation procedure similar to all supervisor agents of various RANs. In the case of a blockage, a Taboo search is performed on the overall allocation to achieve an optimal allocation of all stations of the associated RAN. This part will be considered in details in our forthcoming publications.

5 Result

For testing our approach we have used the Philadelphia problems as a benchmark. The Philadelphia problems have been used widely in previous researches including [1]and[5]. These problems are formulated based on an area in Philadelphia, Pennsylvania-The network consists of 21 cells as shown in Figure 1.

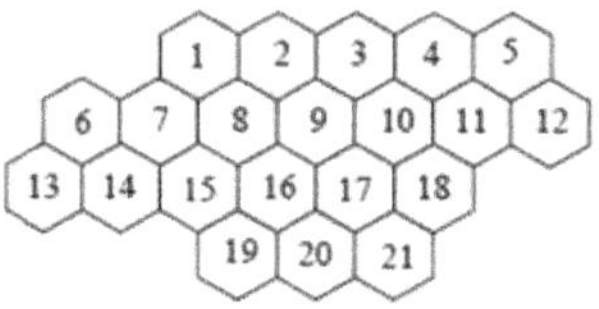

Fig. 1 Cellular Geometry of Philadelphia Problems

Our experiments were conducted on an Intel Pentium M (with 512MB of RAM). There are many variations for setting constraints and demands and several competing teams of researchers have worked on the same instances of problem. We present in Table 1 the parameter setting used both in some approaches and adopted by our evaluations. In this table, "Nc" means the square of required distance for co-channel constraints, assuming that the distance between adjacent cells is 1. For example, if Nc=12, while cell 1 and cell 5 can use the same frequency (the distance is 4), cell 1 and cell 4 cannot (the distance is 3). "acc" represents the separation required for adjacent channel constraints, and "cii" represents co-site constraints. The demand vectors used in the table are as follows:

Case 1: (8 25 8 8 8 15 18 52 77 28 13 15 31 15 36 57 28 8 10 13 8)
Case 2: (5 5 5 8 12 25 30 25 30 40 40 45 20 30 25 15 15 30 20 20 25)

Case 3: (16 50 16 16 16 30 36 104 154 56 26 30 62 30 72 114 56 16 20 26 16)
Case 4: (32 100 32 32 32 60 72 208 308 112 52 60 124 60 144 228 112 32 40 52 32).

Table 2 shows the results obtained with our approach. We consider the theoretical lower-bounds as it is represented in [6]. We use the best solution obtained so far. Ours results are compared with results of the best three methods, from seven reported methods. The three methods are respectively a constraint satisfaction method (CS), a neural network method (NN), and a simulated annealing method (SA). The last row in the table shows our results.

Table 1 Philadelphia Specification

Instance	Nc	acc	cii	Demand vector
P1	12	2	5	Case 1
P2	7	2	5	Case 1
P3	12	2	7	Case 1
P4	7	2	7	Case 1
P5	12	2	5	Case 2
P6	7	2	5	Case 2
P7	12	2	7	Case 2
P8	7	2	7	Case 2
P9	12	2	5	Case 3
P10	12	2	5	Case 4

Table 2 Comparison of solution quality

Instance	Lower bounds	CS	NN	SE	MAS
P1	427	427	427	460	427
P2	427	427	427	447	427
P3	533	533	536	536	533
P4	533	533	533	533	533
P5	258	258	283	283	258
P6	253	253	270	270	253
P7	309	309	310	310	309
P8	309	309	310	310	309
P9	856	856	...	...	856
P10	1714	1714	...	...	1714

As shown in the Table 2, our algorithm obtains optimal solutions for all instances, and obtains exact solutions for other problem instances. Moreover, our method reaches better or equivalent solutions compared with existing methods.

6 Conclusion

In this paper we have been interested in the problem of frequency allocation in the context of a dynamic deployment. After offering an overview and a suitable formulation of the resource allocation problem we have made an intelligent approach based on the AMAS theory (Adaptive Multi-Agent System), which has shown a profit to different RANs, through a general distribution of a unified spectrum between regional networks ensured by a system of distributed agents in different locations of real or virtual network and through a simple communication between these agents. The set of constraints is solved using the criteria of generic problems and decentralized data through a cooperative system of resolution. Experimental evaluations using standard benchmark problems showed that for most of the problem instances, our approach can find better or equivalent solutions compared with existing optimization methods. These results imply that our approach shows a rapid convergence to an optimal solution and presents a good perspective of managing the radio spectrum that can be a relevant strategy and effective management of resources in the future generation networks.

References

[1] da Silva Maximiano, M., Vega-Rodríguez, M.A., G\'{o}mez- Pulido, J.A., Sánchez-Pérez, J.M.: A Hybrid Differential Evolution Algorithm to Solve a Real-World Frequency Assignment Problem. In: Proceedings of the International Multiconference on Computer Science and Information Technology (IMCSIT 2008), Polskie Towarzystwo Informatyczne (Poland), pp. 201–205. IEEE Computer Society, USA (2008)

[2] Gamst, A., Rave, W.: On frequency assignment in mobile automatic telephone systems. In: Proc. GLOBECOM 1982, pp. 309–315 (1982)

[3] Dupant, A.: Etude d'une meta-heuristique hybride pour l'affectation de fréquences dans les réseaux tactiques évolutifs. Université des sciences et Techniques du Languedoc (Octobre 2005)

[4] Bellifemine, F., Poggi, A., Rimassa, G.: Developing multi-agent systems with JADE. In: Castelfranchi, C., Lespérance, Y. (eds.) ATAL 2000. LNCS (LNAI), vol. 1986, pp. 89–103. Springer, Heidelberg (2001)

[5] Luna, F., Blum, C., Alba, E., Nebro, A.J.: ACO vs EAs for Solving a Real-World Frequency Assignment Problem in GSM Networks. In: GECCO 2007, London, UK, pp. 94–101 (2007)

[6] Mishra, A.R.: Radio Network Planning and Opt. In: Fundamentals of Cellular Network Planning and Optimisation: 2G/2.5G/3G...Evolution to 4G, pp. 21–54. Wiley, Chichester (2004)

[7] Sivarajan., K.N., McEliece, R.J., Kethcum, J.W.: Channel Assignment in Cellular Radio. In: Proceedings of 39th IEEE Vehicular Technology Society Conference, pp. 846–885 (1989)

MISIA: Middleware Infrastructure to Simulate Intelligent Agents

Elena García, Sara Rodríguez, Beatriz Martín, Carolina Zato, and Belén Pérez

Abstract. Nowadays there is a clear trend towards using methods and tools that can help to develop multiagent systems (MAS). This study presents a multiagent based middleware for the agents behavior simulation. The main challenge of this work is the design and development of a new infrastructure that can act as a middleware to communicate the current technology in charge of the development of the multiagent system and the technology in charge of the simulation, visualization and analysis of the behavior of the agents. It is a key element when considering that MAS are autonomous, adaptive and complex systems and provides advances abilities for visualization. The proposed middleware infrastructure makes it possible to visualize the emergent agent behaviour and the entity agent. It also allows visualization of the interaction between the agent and the environment.

Keywords: Multiagent systems, Simulation, JADE, Repast.

1 Introduction

Nowadays, simulation is used for several purposes ranging from work flow to system's procedures representation. Simulation can be defined as the representation of the operation or features of one process or system through the use of another. The current research lines that attempt to answer the question of which technique is best for different purposes of simulation are gaining importance. The contribution from agent based computing to the field of computer simulation me-diated by ABS (Agent Based Simulation) is a new paradigm for the simulation of complex systems that require a high level of interaction between the entities of the system. Possible benefits of agent based computing for computer simulation include methods for evaluation of multi agent systems or for training future users of the system [5]. Many new technical systems are distributed systems and involve complex interaction between humans and machines, which notably reduce their usability. The properties of ABS makes it especially suitable for simulating this kind of systems. The idea is to model the behaviour of the human users in terms of software agents.

Elena García · Sara Rodríguez · Beatriz Martín · Carolina Zato · Belén Pérez
Computers and Automation Department, University of Salamanca, Salamanca, Spain
e-mail: {elegar,srg,eureka,carol_zato,lancho}@usal.es

A. Abraham et al. (Eds.): International Symposium on DCAI, AISC 91, pp. 107–116.
springerlink.com

However, it is necessary to define new middleware solutions that allow the connection on ABS a simulation software.

This paper describes the results achieved towards a multiagent-based middleware for the agents' behavior simulation. The middleware, called MISIA (*Middleware Infrastructure to Simulate Intelligent Agents*), allows simulation, visualization and analysis of the agent' behavior. The main contribution of this paper is the design of a new infrastructure that make it possible to provide these capabilities. MISIA makes use of technologies for the development of multiagent systems known and widely used, and combines them so that it is possible to use their capabilities to build highly complex and dynamic systems. On one hand, it is JADE [9],the most widely used platform for based software agents middleware. On the other hand, it is Repast (Recursive Porous Agent Simulation Toolkit) [10], a free and open-source agent-based modeling and simulation toolkit.

Our second contribution is the reformulation of the FIPA protocol used in JADE [9], achieving several advantages: (i) development of a new framework that provides independence between the model and visualization components; (ii) improvement on the visualization component that makes it possible to use the concept of "time", essential for simulation and analysis of the behavior of agents; (iii) and improvements to the user capabilities to which several tools were added, such as message visualization, 2D (and future 3D agents), analysis behavioral, statistics, etc.

Both contributions resulted in the first middleware infrastructure to simulate intelligent agents with visualization, simulation and analysis capabilities.

The article is structured as follows: Section 2 makes a review of agent-modeling toolkits and presents the challenges for simulated multiagent systems. Sections 3 introduces a description of the middleware specifically adapted to the simulation of multiagent systems within dynamic environments (MISIA). Finally, some results conclusions are given in Sections 4 and 5.

2 Background

Agents and multiagent systems are adequate for developing applications in dynamic, flexible environments. Autonomy, learning and reasoning are especially important aspects for an agent. These capabilities can be modelled in different ways and with different tools [15]. Open MAS should allow the participation of heterogeneous agents with different architectures and even different languages [16][4]. The development of open MAS is still a recent field of the multiagent system paradigm and its development will allow applying the agent technology in new and more complex application domains. Thanks to the contribution from agent based computing to the field of computer simulation mediated by ABS is obtained benefits like methods for evaluation and visualization of multi agent systems or for training future users of the system [4]. There are existing work on agent simulation study using commercial-offthe- shelf simulation packages with built-in agent-based modeling and BDI (Belief-Desire-Intention) behaviour architecture [11], modeling detailed complex human behaviours.

Mainly there are two ways for visualizing multiagent systems simulation: the agents interaction protocol and the agent entity. In the former, it is visualized a sequence of messages between agents and the constraints on the content of those messages. On the other hand, the latter method visualizes the entity agent and its iteration with the environment. Most software programs, such as JADE platform [2][9] and Zeus toolkit [3], provide graphical tools that allow the visualization of the messages exchanged between agents.

The toolkits MASON [6], Repast [7][10] and Swarm [12] provide the visualization of the entity agent and its interaction with the environment. The Swarm [12] is a library of object-oriented classes that implements the Swarm conceptual framework for agent-based models and provides many tools for implementing, observing, and conducting experiments on ABS. The MASON [6] is multiagent simulation library core developed in Java. It provides both a model library and an optional suite of visualization tools in 2D and 3D. The Repast [10] is a free and open-source agent-based modelling and simulation toolkit. There are other works like Vizzari et al. [14] where is developed a framework supporting the development of MAS-based simulations based on the Multilayered Multiagent Situated System model provided with a 3D visualization. We did not adopt this framework because would add a non-desired complexity to our system. We chose the Repast toolkit because, when the project started, it was one of the few to offer 3D visualization feature, as well as being simple and having good documentation. Moreover, the Repast system, including the source code and is available directly from the web. Repast seeks to support the development of extremely flexible models of living social agents, but is not limited to modelling living social entities alone. Repast is differentiated from other systems since it has multiple pure implementations in several languages and built-in adaptive features such as genetic algorithms and regression [8].

3 MISIA Middleware

The most well-known agent platforms (like Jade [9]) offer basic functionalities for the agents, such as AMS (Agent Management System) and DF (Directory. Facilitator) services; but designers must implement nearly all organizational features by themselves, like simulation constraints imposed by the MAS topology. In order to model open and adaptive simulated systems, it becomes necessary to have an infrastructure than can use agent technology in the development of simulation environments.

The framework presented in this paper is called MISIA (*Middleware Infrastructure to Simulate Intelligent Agents*). It is a middleware infrastructure that allows to model JADE multiagent systems with the possibility of being represented in Repast. The main concept introduced in this environment is the notion of time in JADE, which means it is possible to render in real time the events into Repast. One of the main differences between JADE and Repast is that in JADE, there not exists the concept of time as such, and the agents interact each other based on changes or events that occur in the execution environment. However, Repast has a time unit : the tick, which is what sets the pace and allows simulations.

On the other hand, agents in the JADE context are implemented based on FIPA standards. This allows to create multiagent systems in open environments, which is not possible within Repast. These differences are what MISIA solved, integrating these two environments and achieving a working environment for creation and simulation of multiagent systems more powerful and versatile.

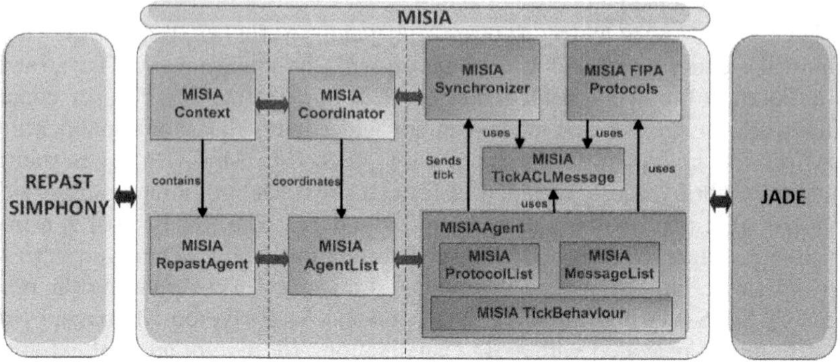

Fig. 1 Functional structure of MISIA

MISIA consists of three principle components or layers: the upper layer is the contact with Repast, the intermediate layer, whose main goal is the interconnection of the two platforms (JADE and Repast), and the bottom layer, which enables JADE supports the notion of time. Next, we will proceed to explain each functional block independently, but to better understand the overall operation, it is necessary to take in mind the following ideas: It is neccesary to syncronize JADE to work simultaneously to Repast. This is achieved by keeping the JADE agents informed about the tick of the simulation they are involved. Moreover, agents are informed when a tick is elapsed. To obtain versatile simulations, it is necessary that all events occurring in JADE are rendered instantly on Repast. The minimum unit of time is the tick, thus, the idea is that every JADE agent can perform functions in a tick (must be simple actions, such as sending a message, receiving or reestablishment of their state) and once finished, they can be updated in Repast. This must occur during the course of all ticks, which are getting updated in real time all events.

The bottom layer of the framework is which connects JADE, and is divided into four functional blocks: (i) **MISIAAgent,** is the extension of JADE agent. Performs the same functions, but adapting them to the presence of ticks. It consists of a number of features to manage the time in JADE. These functions are detailed in the following subsection. (ii) **MISIATickACLMessage.** JADE messages are used for communication between agents. MISIAAgent agents communicate between them with MISIATickACLMessage messages. MISIATickACLMessage is the extension of JADE ACL message that incorporates the concept of time. It includes aspects such as the tick where to send the message, and the delay that the message has when achieves its destination. In JADE, the messages exchanged between agents are sent and arrive instantly, but in real life, that is not the case. It aims to simulate and view the evaluation of the system as time passes, and to achieve this, it is necessary that messages are not instant, but must have a shipping time and a different reception time. (iii) **MISIAFIPAProtocols.** As discussed above, JADE implements FIPA standards, which, among other things, specify multiple communication protocols. These define a series of patterns that respond to different types of communication that two or more agents can perform. The objective is to adapt FIPA protocols defined in JADE with Repast ticks. Features of this module will be detailed in Subsection 3.2. (iv) **MISIASynchronizer** is a JADE agent that acts of notificator. It is responsible for notifying the MISIAAgent when a tick goes by. Is the system clock synchronization. When a tick goes by, MISIASynchronizer is notified in order to notify MISIAAgents. It is made through MISIATickACLMessage messages with a special performative.

The top layer is the contact with Repast. Contains two functional blocks, which are: (i) **MISIARepastAgent**. Each MISIAAgent existing in the system will be represented by a MISIARepastAgent in the context of Repast. This means that for every agent that we want to have on the system actually have to create two: a MISIAAgent agent running on JADE , and its respective MISIARepastAgent released on Repast. It can be seen as follows: a logical agent, and two physical agents. MISIARepastAgents have an important role: they cannot updated their status until their respective MISIAAgents does not end with all the work they need to perform during that tick. This is a very important aspect, since it is the characteristic of the framework as a system in real time. (ii) **MISIAContext** has two important objectives. One is to establish the synchronism in the execution. When a tick goes by, lets know MISIASynchronizer agent that it is necessary to notify MISIAAgent agents that following tick happened. The other goal of this module is to incorporate new agents MISIARepastAgent that entry in the context of the Repast simulation. For each new MISIAAgent that appears in the system, MISIAContext will create their respective MISIARepastAgent and will added it to the simulation environment.

Finally, the intermediate layer is divided into two functional blocks, and its goal is to join adjacent layers. These modules are: (i) **MISIAAgentList**, as its name implies, stores all agents in the system at a given time. It plays an important role because it enables communication between a MISIAAgent and their respective MISIARepastAgent, and vice versa. The diagram shows two-way information flows ranging from MISIARepastAgent to MISIAAgentList and MISIAAgentList to MISIAAgent. These flows are representing that communication, that union between the two physical agents, to confine a logical agent. (ii) **MISIACoordinator** coordinates communication between the two adjacent layers. It is necessary the presence of a coordinator to maintain synchronism between both layers. Thanks to MISIACoordinator, MISIAContext can notify the occurrence of a tick to MISIASynchronizer, and MISIASynchronizer can assure that its purpose is served to MISIAContext, reporting that all MISIAAgent received tick. This kind of communication is necessary to maintain full synchronization between the two platforms. Also shown in the diagram two flows between MISIAContext and MISIACoordinator, between MISIACoordinator and MISIASynchronizer, and represent the flow of ticks and the synchronism.

3.1 JADE Adapted to the Notion of Time

The adaptation of JADE to support the notion of time is the most important and complex feature of our proposal. It was necessary to redefine a series of classes of JADE agents. The new capacity autonomously manages the receipt of ticks and maintains synchronism with Repast, so that it abstracts all these aspects and provides flexibility to the final programmer who uses this framework. Broadly, the sequence of steps that occurs for a tick is: (i) The occurrence of a tick is generated on Repast, as it is the platform that has this ability. MISIAContext is notified of this and, through MISIACoordinator can notify MISIASynchronizer of its tasks. (ii) When MISIASynchronizer receives the notice of MISIAContext, alerts all MISIAAgent about the new tick. As MISIASynchronizer is an agent, it can communicate with MISIAAgent agents through MISIATickACLMessage messages. That's how it notifies them the new tick, sending a message with special semantics to each agent. (iii) Once MISIASynchronizer has sent all the special messages notification of tick, it must wait until all MISIAAgent answer to ensure that all received the notification. Moreover,it sends a ACK message to maintain a strong synchronism between two sides. (iv) When a MISIAAgent receives a tick by MISIASynchronizer carries out various actions. First it sends the messages has to ship in this tick. MISIAAgent agents have a special queue for sending messages. Below is a comparative picture of how is shipping in JADE, and what was the change made in the framework for adaptation over time.

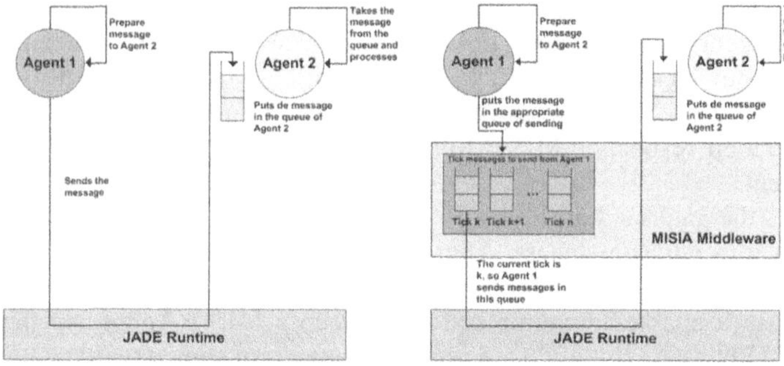

Fig. 2 (a) Sending a message in JADE. (b) Sending a message incorporating MISIA

In JADE, sending a message is instant, the programmer gives the order, and the message is sent ipso facto. MISIAAgent agent does not it in this way. It has a messaging queue MISIATickACLMessage, so that when a programmer gives the order to send a message, the message is automatically inserted in the scheduled queue with the tick which it must be sent. Thus, when a tick goes by, MISIAAgent immediately sends messages that are queued labeled with the current tick. Subsequently, it adds to the agent the protocols needed in this tick. This idea will be further detailed in the following subsection. After performing this task, executes a method designed to make the final programmer overwritten. The aim of this method is to have a function similar to *step ()* Repast method in JADE, which is automatically executed when a tick went by. This function is very convenient to the developers, because, being called on every tick, it is possible separate the actions of the agent in order to carry out them on a specific tick. Finally, after performing all tasks, notifies its agent Repast counterpart, MISIARepastAgent, to make the necessary changes in the simulation environment. Once all MISIARepastAgent agents have been updated in the context of the simulation, this tick went by, and Repast proceed with the change to the next tick.

3.2 *Redefinition of FIPA Protocols*

JADE has a number of implemented FIPA protocols, which help the programmer. With these classes, it abstracts the developer from having to prepare messages to be sent, sending, or to manage the reception of them, among other things. In this framework has been re-implemented FIPA protocols defined in JADE to support the notion of time. These communication protocols JADE defines two roles, which starts the conversation (Initiator role) and which is envolved in the conversation (Responder role). The Initiator agent role will begin by the conversation by sending a message to the recipient. Therefore, it follows the logic developed with the message queue. When a MISIAAgent agent wishes to follow a communication protocol in a given tick, just add the protocol of communication to the agent in the tick established. Therefore, one of the functions of MISIAAgent agent after receiving a tick is to add communication protocols. The rest of communication for sending and receiving messages is re-implementing, recording different behaviors that make the different functions of the protocols. The novelty is that these new behaviors support MISIA modules redefined for JADE , such as support MISIATickACLMessage messages or the ability to respond to a message in a certain tick, without being immediately.

An example reimplemented is the FIPA-Request protocol, which is like follows: the agent with Initiator role sends a request to agent with Responder role. Responder replies, accepting or rejecting the request, and immediately returns to answer the agent with Initiator role informing the result (if the request was made correctly, or there was a problem). With the new definition by MISIA of this protocol, it is possible to send messages during the tick chosen. In this case, MISIA only redefines the role Responder. The Initiator is not necessary because it only sends a message to the beginning.

In the case of the Responder role, must send two messages, as discussed above. So, MISIA provides to programmers two handles, like JADE; one to send the first message, and another to send the second one, abstracting from all the system logic that is to managing ticks.

Below is a fragment of code in Java where it shown how a behavior is reimplemented to manage the arrival of the request by the agent with Initiator role. In this example, *handleMISIARequest* is the procedure that the final developer overwrites to provide the message he want to send in response.

```
registerPrepareResponse(new OneShotBehaviour(){
   public void action() {
         //Get DataStore to obtain the request message
      DataStore ds = getDataStore();
      ACLMessage requestMessage = (ACLMessage) ds.get(REQUEST_KEY);
      TickACLMessage agreeMessage = null;
      try {  agreeMessage = handleMISIARequest(requestMessage);
       } catch (Exception e) {}
   //If the message isn't null, send
   if (agreeMessage != null)     jadeAgent.MISIASend(agreeMessage);    }
```

4 Experimental Results

It has been developed a case study using this middleware to create a multiagent system aimed at facilitating the employment of people with disabilities, so it is possible to simulate the behavior of the agents in the work environment and observe the agents actions graphically in Repast. This is a simple example that defines four jobs, which are occupied by four people with certain disabilities. Every job is composed of a series of tasks. Agents representing the workers have to do them, and according to their capabilities, carry out the assignment with varying degrees of success. Performing various simulations, and seeing the evolution in time, the results can be assessed to determine what would be the most suitable job for each employee. In addition, taking into account the capabilities of Repast, it is possible to make a collection of data generated from each simulation and exported to external applications such as MatLab, for different studies on them. Below is an example of the execution of this case study.There are two ways for visualizing multiagent systems simulation: the agents interaction protocol and the agent entity. MISIA provides the capabilities visualize the sequence of messages between agents and the entity agent and its iteration with the environment. The union of these two platforms involves having a highly efficient environment for the creation of multiagent systems, getting the benefits of JADE to create the systems, as is the use of FIPA standards; and also the visual representation and extraction of simulation data to different applications provided by Repast.

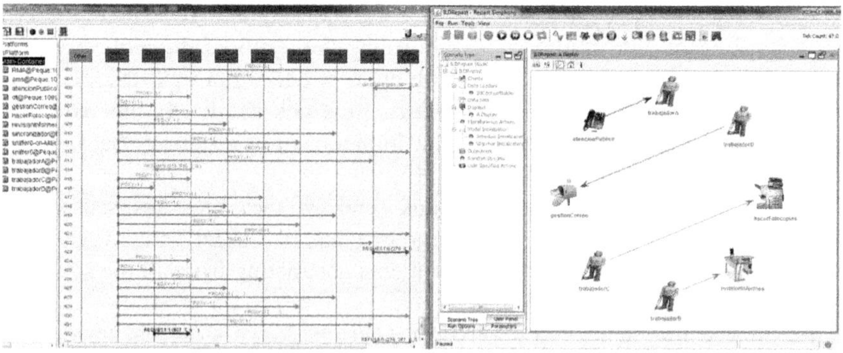

Fig. 3 Case Study MISIA

4 Conclusions and Future Works

Simulation is a helpful tool for understanding complex problems. Therefore, the simulation of multiagent systems in several levels of details and the emergent behavior is fundamental for analyzing the systems processes. In this study, a list of basic concepts and advances is presented for the development of simulated multi-agent systems. We showed in detail the visualization and simulation infrastructure for developing the MAS behavior simulators. MISIA allows simulation, visualization and analysis of the behavior of agents. With the MAS behavior simulator it is possible to visualize the emergent phenomenon that arises from the agents' interactions.

The proposed visualization system also suggests further developments. One of them is make the agent representation more photo-realistic. A 3D agent visualization in more levels of details showing the interaction them would make the system complete and realistic. Another future work is to improve interactivity with the user. We would like to allow the user to visualize the agent state and its simulation individually. We also want to improve the interactivity by means of allowing the interaction of the specialists with the live execution besides the basic functionalities such as play, pause, stop and increase/decrease the speed, by means of putting some substances in the position and observing the emergent behavior. It would allow the self-organization optimization and the proposal of new hypotheses. Even more: generation of reports about the information visualized during the simulation process in several levels of detail, which could increase the comprehension about the process. MISIA is the ideal framework for this purpose.

Acknowledgments. This work has been partially supported by the MICINN project TIN 2009-13839-C03-03.

References

[1] Agent Oriented Software, Ltd., ACK Intelligent Agents-Agent Practicals, 4 (2004)

[2] Bellifemine, F., Caire, G., Poggi, A., Rimassa, G.: Jade a white paper. EXP in Search of Innovation 3(3), 06–19 (2003)

[3] Collis, J.C., Ndumu, D.T., Nwana, H.S., Lee, L.C.: The zeus agent building tool-kit. BT Technol Journal 16(3) (1998)

[4] Corchado, E., Pellicer, M.A., Borrajo, M.L.: A MLHL Based Method to an Agent-Based Architecture. International Journal of Computer Mathematics 86(10, 11), 1760–1768 (2008)

[5] Davidsson, P.: Multi agent based simulation: Beyond social simulation. In: Moss, S., Davidsson, P. (eds.) MABS 2000. LNCS (LNAI), vol. 1979, pp. 97–107. Springer, Heidelberg (2001)

[6] Luke, S., Cioffi-Revilla, C., Panait, L., Mason, S.K.: A new multiagent simulation toolkit. In: Proceedings of the 2004 SwarmFest Workshop (2004)

[7] North, M.J., Howe, T.R., Collier, N.T., Vos, J.R.: The repast symphony runtime system. In: Proceedings of the Agent 2005 Conference on Generative Social Processes, Models, and Mechanisms (2005)

[8] North, M.J., Collier Nicholson, T., Vos Jerry, R.: Experiences Creating Three Implementations of the Repast Agent Modeling Toolkit. ACM Transactions on Modeling and Computer Simulation 16(1), 1–25 (2006)
[9] JADE, Java Agent Development Platform, http://JADE.tilab.com
[10] Repast, http://repast.sourceforge.net/repast_3/index.html
[11] Shendarkar, A., Vasudevan, K., Lee, S., Son, Y.-J.: Crowd Simulation for Emergency Response using BDI Agent based on Virtual Reality. In: Proceedings of the 2006 Winter Simulation Conference, pp. 545–553 (2006)
[12] Swarm, http://www.swarm.org
[13] Foundation for Inteligent Physical Agents. FIPA Agent Management Specification. Disponible en,
http://www.fipa.org/specs/fipa00001/SC00001L.html
[14] Vizzari, G., Pizzi, G., da Silva, F.S.C.: A framework for execution and visualization of situated agents based virtual environments. In: Workshop Dagli Oggetti Agli Agenti, pp. 22–25 (2007)
[15] Wooldridge, M., Jennings, N.R.: Agent Theories, Architectures, and Languages: a Survey. In: Wooldridge, M.J., Jennings, N.R. (eds.) ECAI 1994 and ATAL 1994. LNCS, vol. 890, pp. 1–22. Springer, Heidelberg (1995)
[16] Zambonelli, F., Jennings, N.R., Wooldridge, M.: Developing Multiagent Systems: The Gaia Methodology. ACM Transactions on Software Engineering and Methodology 12, 317–370 (2003)

Secure Communication of Local States in Interpreted Systems

Michael Albert, Andrés Cordón-Franco, Hans van Ditmarsch, David Fernández-Duque, Joost J. Joosten, and Fernando Soler-Toscano

Abstract. Given an interpreted system, we investigate ways for two agents to communicate secrets by public announcements. For card deals, the problem to keep all of your cards a secret (i) can be distinguished from the problem to keep some of your cards a secret (ii). For (i): we characterize a novel class of protocols consisting of two announcements, for the case where two agents both hold n cards and the third agent a single card; the communicating agents announce the sum of their cards modulo $2n+1$. For (ii): we show that the problem to keep at least one of your cards a secret is equivalent to the problem to keep your local state (hand of cards) a secret; we provide a large class of card deals for which exchange of secrets is possible; and we give an example for which there is no protocol of less than three announcements.

1 Introduction

Interpreted systems and security. A well-known abstract architecture for distributed systems such as multi-agent systems is that of an *interpreted system* [4]. An interpreted system is a collection of global states, where a global state is an n-tuple of local states (one for each agent, given n agents). Each processor/agent only knows its local state, and there is public knowledge among all agents of the set of global states constituting the system. This creates a setting wherein we can investigate two of the agents sending messages to each other, with the intention to communicate information that remains a *secret* from other agents. The worst case scenario for security is when all messages are public. In that case, the protocol executed when sending messages can be modelled as successive *model restrictions* of the given interpreted system. We investigate two cases: that of interpreted systems

Michael Albert
University of Otago, New Zealand
e-mail: malbert@cs.otago.ac.nz

Andrés Cordón · Hans van Ditmarsch · David Fernández · Joost Joosten · Fernando Soler
University of Sevilla, Spain
e-mail: {acordon, hvd, dfduque, jjoosten, fsoler}@us.es

A. Abraham et al. (Eds.): International Symposium on DCAI, AISC 91, pp. 117–124.
springerlink.com

where the agents are card players and the secrets concern ownership of cards (an aspect of the agent's local state) (Sections 2 and 3), and that of interpreted systems in general, where the secret is the local state (Section 4).

Example. Alice and Bob, draw a and b cards from a deck of $a+b+c$ cards, and Eve, the eavesdropper, receives the remaining c cards. Alice and Bob wish to communicate their cards to each other by way of public announcements, without informing Eve of any of their cards. The investigation of the generalized problem with card deal size parameters (a,b,c) was inspired by its $(3,3,1)$ instance that was coined in [2] the *Russian Cards Problem*, and that originates with Kirkman [7]. A standard solution for $(3,3,1)$ is as follows. Suppose Alice holds 0, 1, and 2, Bob holds 3, 4, and 5, and Eve holds 6. Alice announces that her hand of cards is one of $012, 034, 056, 135, 146, 236, 245$, i.e., one of the seven hands $\{0,1,2\}$, etc., after which Bob announces that Eve holds 6. Another solution is that Alice announces that she holds one of the five hands $012, 034, 056, 135, 246$, again followed by Bob announcing that Eve holds 6. We can view such solutions as the execution sequences of an underlying protocol. Some general patterns and special cases of card deal sizes (a,b,c) for which two-announcement solutions exist are found in [1], but a complete characterization is not known.

We can relax the constraints for secrecy in the Russian Cards Problem somewhat. Suppose that the eavesdropper may learn single card ownership for Alice and Bob, but just not their entire holding, i.e., the eavesdropper may not learn the card deal. In that case, simpler protocols suffice. In terms of interpreted systems, Alice and Bob attempt to communicate their local state to each other, without Eve learning their local states. Note that, if Eve were to learn the local state of Alice or the local state of Bill, she would learn the entire deal of cards.

A simple way for Alice to communicate her local state to Bob, in the $(3,3,1)$ case, is to announce that she holds one of 012, 034, and 056. In other words, she gives away that she holds card 0, but this does not disclose her whole hand. After that announcement, Bob as before responds that Eve holds 6. We may call Alice's announcement *state safe*, as opposed to *card safe*, above. There is a relation to *bit exchange* problem: is it possible for Alice and Bob to share a secret bit (i.e., the value of a proposition) by public communication. The seminal publication for the latter is [6].

Motivation. We are motivated in our investigations by the ground separating unconditionally secure (also known as information-based) from conditionally secure protocols. The security of the latter are based on computational features: the intractability of some computation, a one-way function between some cryptographic primitives, etc. It is tempting to say that unconditionally secure protocols are abstractions of conditionally secure protocols. But this high and dry ground seems very poor: abstracting away from keys and one-way functions seems to remove the essence from reasoning about protocols, and it is therefore unclear what results for the abstraction have to bear on practical security matters and protocol design. We do not bridge that gap. But anything we do, aims to bridge that gap.

Our slightly less ulterior motive is to design fast unconditionally secure protocols for information exchange in interpreted systems. Within the more specific bounds that we have set, such as card secure protocol or state secure protocols, we aim to find minimum-length protocols, to find the maximum number of bits that can be exchanged, and to analyze multi-agent versions of protocols ('multi-party' in security jargon; with 'multi' as 'more than two') where the intention to securely exchange information about the ignorance and knowledge of other agents (also known as higher-order preconditions in protocol execution) inevitably draws in dynamic epistemic methodology. An additional challenge in that setting is the reconciliation of what may be called more embedded methods with the more abstract logical and combinatorial approaches. Somewhere on the far horizon remains a link with conditional security.

Results. This work contains the following contributions. For the card exchange problem for card deal size $(n,n,1)$ we characterize a novel class of protocols consisting of two announcements. In that case, we treat the set of cards not as a set of (interchangable) labels as in design theory [10], but as set of consecutive numbers $0,1,...,2n$ and employ number theoretical methods and brute force (Haskell). The protocol is simple: both A and B announce the sum of their cards modulo $2n+1$. The method has promising generalizations. Further, we show that state safe is equivalent to bit safe, and provide a large class of card deal sizes (a,b,c) for which bit exchange is possible (this should be seen as a special case of results in [5]). These protocols typically consist of various announcements, without a claim that these are minimal. We also give an example of a bit exchange protocol consisting of three announcements where no solution of two announcements exists.

An extended version, presented at the (informal) ESSLLI 2010 workshop Logics in Security, is available at `http://personal.us.es/hvd/newpubs/fLiS secret1.pdf`

2 Card Deal Terminology and Known Results

The three *agents* Alice, Bob, and Eve are abbreviated as A, B, C. Given a set/*deck D* of $d = a+b+c$ *cards*, their hands of cards A,B,C consist of a,b,c cards. The *card deal* (A,B,C) is the triple of the three hands of cards, and we call this a card deal of *size* (a,b,c). The cards in the deck may be called anything whatsoever, but it is customary to name them $0,1,...,d-1$.

Given that cards are numbers, and that our examples use small numbers, we allow ourselves some abus de langage. Consider size $(3,3,1)$. For hand of cards $\{0,1,2\}$ we write 012 (and the cards in a hand always in this ascending order), and for deal $(\{0,1,2\},\{3,4,5\},\{6\})$ we write 012.345.6.

We can distinguish the information requirement—what A and B are supposed to learn from each other—from the safety requirement—what C is not supposed to learn from the communications taking place between A and B. The information requirement is for A and B to learn all of their cards (and therefore the entire deal of cards). We call an information state satisfying that requirement *state informative*.

The *card safe* requirement is for C to remain ignorant of the ownership of all of A's and B's cards; whereas in *state safe*, the requirement is for C to remain ignorant of the ownership of at least one of those cards (and therefore ignorant about the hand of cards of the other agents, their *local state*).

Protocols to solve these problems consist of a finite number of alternating truthful public announcements by A and B, all of which are informative (trivial announcements are not allowed), and where each announcement consists of a number of alternatives for the hand of cards of the announcing agent. These are not truly restrictive conditions: for a finite number of cards, there are only a finite number of possible card deals, and each informative announcement results in a reduction of these alternatives. In this work we only consider protocols of length two or three.

An information state is represented by a Kripke model for what agents know, 'informative announcement' can be defined as one resulting in a proper model restriction, any complex logical statement that is announced is equivalent to an announcement of a number of alternatives for the actual hand of cards, and all states in (a bisimulation contraction of) that Kripke model are about different card deals [2]. The various safety and information requirements are formulas that can be checked in such a model.

All the following should hold for any deal of cards for which a given sequence of announcements can be made truthfully. An announcement is *card safe / state safe* if it preserves ignorance of C of all cards / some card (the safety requirement). We will normally call them safe, and let the context determine if card safe or state safe is intended. A sequence of announcements is a protocol. A protocol is safe if it consists of safe announcements. A protocol is state informative if A and B know the card deal after termination, i.e., if the information state reached is state informative. A protocol is good if it is safe and state informative.

In [1] some sizes (a,b,c) are listed for which good protocols consisting of two announcements exist, e.g., (a,b,c) such that $a+b+c = p^2+p+1$ for any prime $p \leq a-1$, and $(3,b,1)$ if $b \geq 3$, and $(a,2,1)$ if $a = 0,4 \mod 6$. In a two-announcement protocol the second announcement is always equivalent to B announcing the cards of C. There may be protocols for (a,b,c) but *not* for (b,a,c), e.g., there is a protocol for $(4,2,1)$ but not for $(2,4,1)$.

3 Card Safe Protocols for Size $(n,n,1)$

The five hand solution for the $(3,3,1)$ case is also known under the form of the 'sum modulo number of cards' [8]. For example, when Alice holds 012, she announces that the sum of her cards modulo 7 is 3. There are five hands of cards having that sum: 012, 046, 136, 145, 235. Not all hands in the five hand announcement 012, 034, 056, 135, 246 in the introductory section have the same sum, but subject to the permutation of cards $s(0) = 1, s(1) = 0, s(2) = 2, s(3) = 4, s(4) = 5, s(5) = 6, s(6) = 3$ it can be transformed in the modulo 7 solution. And instead of responding by announcing Eve's card, Bob could equivalently have announced the sum of his cards modulo 7.

In [1] an 18 hand solution for $(4,4,1)$ and a 66 hand solution for $(5,5,1)$ are given, but no general method was known for $(n,n,1)$. In this section we give such a general method for $(n,n,1)$, for $n \geq 3$.

It should be noted that the sum announcement is not always safe. For example, take card deal size $(4,2,1)$. Assume that A holds 0123. It is not (card) safe for A to announce that the sum of her cards is 6. The quadruples summing to 6 are: 0123 0346 0256 1246 1345. If C holds 4, then she learns that A holds 0.

Let $\sum A$ denote the sum of A's cards modulo d, and similarly for other agents. For our purposes we can equate D with the ring $\mathbb{Z}_d$ of d elements, and + to the sum operation defined on that ring. The announcement by an agent of the sum modulo the total number of cards is called the *sum announcement*, and the protocol consisting of A and then B announcing their sum is called the *sum announcement protocol*. First let us note that if $c = 1$, the sum announcement informs the other agent of your cards.

Proposition 1. *If $c = 1$ and A announces the sum of her cards, then B knows A's cards.*

The same argument applies if B announces the sum of his cards, so that:

Corollary 1. *For $(a,b,1)$, the protocol where first A announces the sum of her cards and then B announces the sum of his cards is state informative.*

A direct result from the proof of Proposition 1 is that

Corollary 2. *A good sum announcement protocol for (a,b,c) is also good for (b,a,c).*

As we have seen in Section 2, this is not necessarily the case for other than sum announcement protocols. Now, let us characterize (card) safety. We only summarize the results. Consider the 'pair swap' property:

Pair swap (for A)
For every $x_0 \in \mathbb{Z}_d$ and every deal (A,B,C) such that $x_0 \in A$, there exist $x_1 \in A$ and $y_0, y_1 \in B$ with $x_0 \neq x_1$, $y_0 \neq y_1$, and $x_0 + x_1 = y_0 + y_1$. (1)

Proposition 2. *Suppose that the triple (a,b,c) satisfies pair swap for A. Then, C does not know any of A's cards after $\sum A$ is announced.*

A similar property, (2), must hold for B. An announcement is (card) safe if (1) and (2) hold. To find out when this is the case, we proceed combinatorially, building on results of [3] and [9].

Proposition 3 ([9]). *Let d be a prime. For a set $A \subseteq \mathbb{Z}_d$, denote $S^n(A)$ as the set of all sums $x_1 + \ldots + x_n$ of n distinct elements of A. Then, $|S^n(A)| \geq \min\{d, n|A| - n^2 + 1\}$.*

Proposition 4. *If d is prime and both $2a - 3 + (b-1) \geq d + 1$ and $(a-1) + 2b - 3 \geq d + 1$, then announcing $\sum A$ (or $\sum B$) is card safe.*

Proposition 5. *If $|A| = n \geq 9$ and $A \subseteq \mathbb{Z}_{2n+1}$, then $|S^2(A)| \geq n + 3$.*

This gives us the following

Corollary 3. *For any $n \geq 9$, announcing $\sum A$ is card safe in the $(n,n,1)$ case.*

We also have that

Lemma 1. *For any $3 \leq n \leq 8$, announcing $\sum A$ is card safe in the $(n,n,1)$ case.*

From Corollary 1, Corollary 3 and Lemma 1 we now obtain that

Theorem 1. *For $n \geq 3$, the sum announcement protocol is a good protocol for size $(n,n,1)$.*

Protocols for one announcement. Alice and Bob can announce their sum at the same time, and this is card safe and state informative. So we can shorten the sum announcement protocol into a single announcement protocol. This is an elementary observation, but still remarkable: for the protocols in [1] (and for all other card protocols that of which we know) Bob can only make a specific response *after* Alice's announcement.

Protocols for more than two announcements. For (a,b,c) where $c > 1$, the two announcement protocol of both agents announcing the sum does not work. From A's announcement, B still learns the sum of C's cards, but two cards that are held by A instead of C may also have that sum. It is conceivable that B then makes some other informative response (other than announcing *his* sum of cards!), from which A learns his cards, and may then make yet another announcement informing B of C's cards. In other words, number theory may assist us to find good protocols consisting of more than two announcements. For that, we also need to be more general than just swapping pairs.

From swapping pairs to swapping n-tuples. Interestingly, in the original Russian cards problem for parameters $(3,3,1)$ the swapping pairs argument for showing safety fails. Let us consider the card deal 013.245.6. There is no pair of cards from 013 with the same sum as a pair of cards from 245, for otherwise the remaining cards in each hand would be equal since $0+1+3=2+4+5$ modulo 7. Observe that, however, safety can be easily shown by a swapping triples argument: it suffices to interchange the whole players' hands. Indeed, this is a general fact. Given a card deal of the $(3,3,1)$ case, if the sum of A's cards is different from the sum of B's cards, the swapping pairs argument works. Otherwise, safety can be shown by exchanging the whole hand of both players.

Employing Haskell, we have encountered several other cases where card safety can be shown by a swapping n–tuples argument. We also conjecture a strengthening of Proposition 5. (Details omitted.)

4 Communicating Local States

As said, the models encoding what agents know in a card deal can also be seen as an interpreted system [4], namely where each processor/agent only knows his local state (namely his hand of cards), and where there is public knowledge among all agents of the set of possible global states of the system, where a global state is an n-tuple of local states (given n agents). That a local state consists of several cards is somewhat less relevant from this perspective. The concern of the agents communicating to each other may simply be to keep their local state a secret, but they may not care about each and every of their cards. That is, the protocols should be *state safe*, but not necessarily *card safe*.

In works like [6] the basic building block for secrecy is not a card, or a state, but a *bit*. A bit may be any proposition that the communicating agents wish to share while keeping it a secret from intruders. Given a card deal of size (a,b,c), 'A and B share a secret' means that there is a proposition p such that it is public knowledge (i.e., common knowledge to A, B, and C) that A and B commonly know the value of p but that C remains ignorant of the value of p. A protocol can be called *bit safe* and *bit informative* (or 'a good protocol for bit exchange') if for each initial state of information a sequence of A, B announcements results in an information state with a shared secret. We note that p typically is some factual proposition p (such as 'A holds card 0', 'the deal of cards is 012.345.6', ...), but it can be any proposition, also an epistemic statement; but this is not the situation typically considered in information theory, nor in security protocol analysis. From this perspective, *state informative is bit informative for the proposition describing the deal of cards*; and we note that this is a different proposition in every different state. There are also less obvious correspondences:

Proposition 6. *State safe is bit safe.*

Similarly, one might wonder if bit informative is state informative. As said, state informative is bit informative: the description of a state is a bit. But it is quite possible to share a secret bit without disclosing all your cards. But, if it is possible to share a secret bit, is there then also another protocol to safely disclose all of your cards? We think the answer is yes, but we do not know the answer.

One can show for a large class of (a,b,c) that they are bit safe. Our results can be summarized as follows. Theorem is a straight consequence of Lemma's 2 and 3.

Lemma 2. *If* $a,b > c$, A *and* B *can share a secret after public communication.*

Lemma 3. *If* $a > b = c > 0$ *or* $b > a = c > 0$, A *and* B *can share a secret after public communication.*

Theorem 2. *Let* $a,b > c$, *or* $a > b = c > 0$, *or* $b > a = c > 0$. *Then* A *and* B *can share a secret after public communication.*

Theorem 2 also follows from [5, Theorem 2.1] of which the special case for two agents sharing a secret is that $a+b \geq c+2$. We note that this involves cases where either a or b is smaller than c, unlike our conditions, so their results are more general. However, it is unclear if our (or their) bounds are sharp and if for all other card deal size (a,b,c), no secret can be shared between A and B. There are cases for which no secret can be shared, e.g., $(1,1,1)$.

We have two other interesting results to report. First, if you don't care *which* two agents share the secret, a secret can always be shared (even for $(1,1,1)$!).

Proposition 7. *Given (a,b,c) where there is uncertainty about the card deal, two agents can share a secret.*

Proof. Take any agent i. Let i announce: "I hold exactly one of $\{x,y\}$. The (single!) other agent j for which this also holds now responds: "So do I." Now, i and j share a secret bit. (Namely, the value of the proposition 'i holds card x'.)

Second, bit exchange protocols above may consist of (strictly) more than two announcements. One case is $(2,2,1)$. (Details omitted.)

Proposition 8. *There are (a,b,c) for which good protocols satisfying state safety always require more than two announcements.*

5 Further Research

Of further logical interest is a language of protocols, and a logic to check protocol properties. A promising logic having such features is found in [11].

Acknowledgement. We thank Marco Vervoort for first proving an $n+1$ lower bound version of Proposition 5. We thank the PAAMS anonymous reviewers for their comments. Joost Joosten is currently affiliated to Dept. Lògica, Història i Filosofia de la Ciència, Universitat de Barcelona. Hans van Ditmarsch is also affiliated to the Institute of Mathematical Sciences Chennai (IMSC), India.

References

1. Albert, M., Aldred, R., Atkinson, M., van Ditmarsch, H., Handley, C.: Safe communication for card players by combinatorial designs for two-step protocols. Australasian Journal of Combinatorics 33, 33–46 (2005)
2. van Ditmarsch, H.: The Russian cards problem. Studia Logica 75, 31–62 (2003)
3. Erdös, P., Heilbronn, H.: On the addition of residue classes modulo p. Acta Arithmetica 9, 149–159 (1964)
4. Fagin, R., Halpern, J., Moses, Y., Vardi, M.: Reasoning about Knowledge. MIT Press, Cambridge (1995)
5. Fischer, M., Wright, R.: Multiparty secret key exchange using a random deal of cards. In: Feigenbaum, J. (ed.) CRYPTO 1991. LNCS, vol. 576, pp. 141–155. Springer, Heidelberg (1992)
6. Fischer, M., Wright, R.: Bounds on secret key exchange using a random deal of cards. Journal of Cryptology 9(2), 71–99 (1996)
7. Kirkman, T.: On a problem in combinations. Camb. and Dublin Math. J. 2, 191–204 (1847)
8. Makarychev, K., Makarychev, Y.: The importance of being formal. Mathematical Intelligencer 23(1), 41–42 (2001)
9. da Silva, J.D., Hamidourne, Y.: Cyclic spaces for Grassmann derivatives and additive theory. Bull. London Math. Soc. 26, 140–146 (1994)
10. Stinson, D.: Combinatorial Designs – Constructions and Analysis. Springer, Heidelberg (2004)
11. Wang, Y.: Epistemic modelling and protocol dynamics. Ph.D. thesis, University of Amsterdam, ILLC Dissertation Series DS-2010-06 (2010)

COMAS: A Multi-agent System for Performing Consensus Processes

Iván Palomares, Pedro J. Sánchez, Francisco J. Quesada,
Francisco Mata, and Luis Martínez

Abstract. The need for achieving consensus in group decision making problems is a common and sometimes necessary task in a myriad of social and business environments. Different consensus reaching processes have been proposed in the literature to achieve agreement among a group of experts. Initially, such processes were guided by a human moderator, but afterwards, some proposals to facilitate such a process arose by automating the moderator tasks. However, not many consensus support systems have been developed so far, due to the difficulty to manage intelligent tasks and cope with the negotiation process involved in consensus. This paper aims to present an initial prototype of an automatic consensus support system, developed by using the multi-agent paradigm that provides intelligent tools and capacities to tackle the inherent complexity found in this problem. To do so, we focus on the consensus model considered, the multi-agent architecture designed to develop such a system, and the ontology used for reasoning and communication tasks.

1 Introduction

In group decision making problems (GDM), two or more experts try to reach a common solution about a decision problem. Traditionally, these problems have been solved performing a selection process where the best alternative/s is/are chosen as the solution, without taking into account any previous agreement among experts [9]. This often leads to situations where some experts may consider that their individual opinion has not been taken into account, and therefore they disagree with the achieved solution [19]. To avoid such situations, the idea of carrying out a consensus process prior to the selection process emerged, so that experts express and discuss their preferences to make them closer to each other with the aim of reaching a high level of agreement before making the decision [15]. The

Iván Palomares · Pedro J. Sánchez · Francisco J. Quesada
Francisco Mata · Luis Martínez
Computer Science Department, University of Jaén, Spain
e-mail: {ivanp,pedroj,fqreal,fmata,luis.martinez}@ujaen.es

A. Abraham et al. (Eds.): International Symposium on DCAI, AISC 91, pp. 125–132.
springerlink.com

consensus process has typically been coordinated by a human moderator, who is responsible for process supervision and evaluation of the level of agreement achieved in each consensus round. Regarding the automation of consensus reaching processes (in order to perform them without human supervision), some models have been proposed [11,16], but not so many have been finally developed.

Intelligent agents are software entities capable of carrying out actions in an autonomous way to achieve one or more aims, reasoning about acquired knowledge and exchanging information with other agents and/or the environment. In most contexts and problems, different agents, each one with its specific role and behaviours, must be identified, thus establishing a multi-agent system (MAS). Applications for MAS include areas such as planning [8], industry [1] and, more recently, works related to web services [21].

By considering the aforementioned problem and statements, this contribution aims to present a multi-agent system to support consensus processes (COMAS). This system automates and guides consensus processes by means of a set of different intelligent agents that manage, supervise and control them. This paper is organized as follows: in the next Section, we briefly review consensus GDM problems in general and show the theoretical consensus model used by COMAS in particular. In Section 3, we then describe the MAS architecture of our system. Section 4 briefly describes the ontology employed for agent communication. Section 5 shows an example of the system's performance, and finally, in Section 6 some conclusions and future works are drawn.

2 Consensus Model Description

Group decision-making (GDM) problems may be defined as decision situations where [16]:

1. There is a decision problem to be solved, where a solution must be chosen among a set of alternatives $X = \{x_1, x_2, \ldots, x_n\}$.
2. Two or more experts $E = \{e_1, e_2, \ldots, e_m\}$ participate, having each one his own opinions about the set of alternatives X that the problem considers.
3. Experts try to achieve a common solution.

The process for reaching a solution in GDM problems consists of two steps [18]:

- *Aggregation phase*, that combines the experts' preferences.
- *Exploitation phase*, to obtain an alternative or subset of alternatives as the problem's solution.

The above selection process does not necessarily imply any agreement among the experts, therefore the solution might not be accepted by some of them if they consider that their preferences have not been taken into account in the process [2]. In order to increase the agreement among them, an extra phase is introduced, the *consensus phase*.

A consensus reaching process consists in a discussion process where a group of experts try to achieve an agreement, supported by a human moderator. Many theoretical consensus models have been proposed in the literature [3,4,5,6,12,13]. Figure 1 shows the consensus model implemented by COMAS, based on the main ideas of the models presented in [10,11,16].

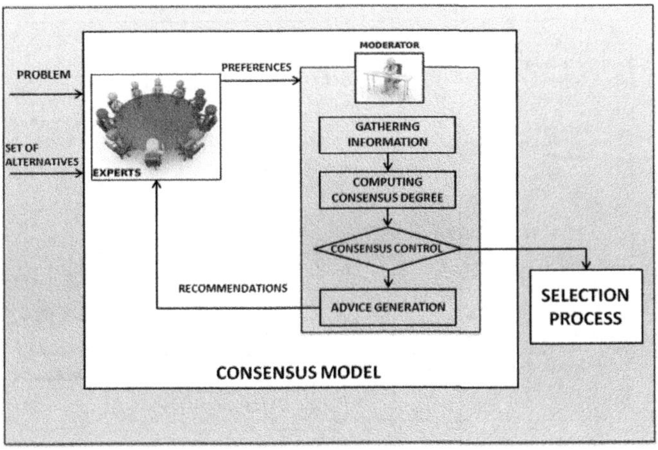

Fig. 1 Consensus reaching model employed in COMAS

Consensus reaching process is seen as an iterative process, where experts provide their preferences about the set of alternatives when each discussion round begins. A brief description of each phase is shown below:

1. **Gathering information.** Experts provide their opinions to the moderator, by means of structures called preference relations, which consist of one matrix by expert P_i, where each element p_i^{lk} represents the degree of preference on the alternative x_l over x_k according to the expert e_i.
2. **Computing consensus degree.** The moderator computes the level of agreement once gathered all the experts' preferences, by means of a similarity measurement. The agreement level is expressed as a value in the interval *[0,1]*, where a value of 1 means total agreement.
3. **Consensus control.** The degree obtained in the previous phase is checked. If it is greater than a consensus threshold given, then the desired agreement has been reached and the consensus process finishes. Otherwise, the process needs more discussion. In addition, a maximum number of rounds is set before beginning the process, so that if it is reached, then the consensus process fails.
4. **Advice generation.** The model suggests how experts should change some of their preferences in order to increase the level of agreement in the following rounds [5,16]. A set of suggestions regarding appropriate changes is generated and delivered to experts.

A further detailed description of the consensus reaching process developed in COMAS can be found in [17].

3 COMAS Multi-agent Architecture

In this section, the multi-agent architecture designed for COMAS is presented. The architecture was developed basing on the FIPA standard[2], and using the JADE agent platform[3]; and is depicted in Figure 2.

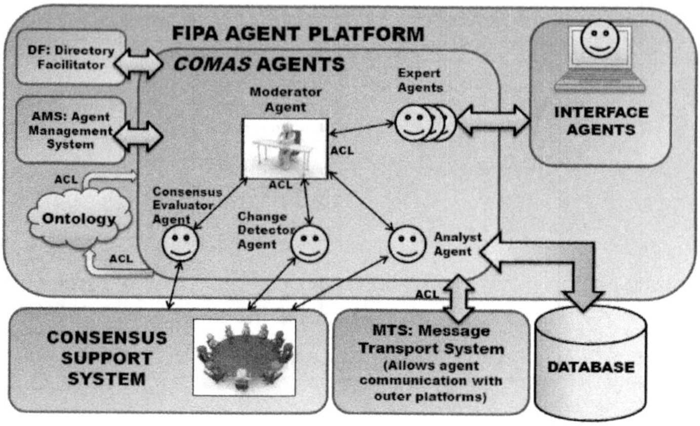

Fig. 2 COMAS architecture

We can consider COMAS a cooperative MAS, where agents work together in order to reach a common objective, i.e., reaching a consensus. Several agent roles are proposed according to the different roles and tasks that can be identified in the previously reviewed consensus model. These roles are described below:

- *Expert Agent:* It represents a human expert, and acts on behalf of him.
- *Moderator agent:* It assumes the human moderator's role, and is responsible for guaranteeing a right development of the overall consensus process.

In addition, because of the complex set of responsibilities initially assumed by the moderator, we have decided to add some specialized agents to support *moderator agent* in various specific tasks in the consensus process:

- *Consensus Evaluator Agent:* It is in charge of computing the consensus degree achieved in each consensus round, and report it to moderator agent.
- *Change Detector Agent:* It is responsible for leading the necessary operations for the *Advice generation* stage in the consensus process.
- *Analyst Agent:* This additional agent assumes the task of storing all the information concerning each consensus process in a database.

Other components provided in our system's architecture are:

- *Interface Agent:* Each expert agent is associated to an interface agent, which allows human experts or users to provide their initial preferences before the

[2] FIPA: http://www.fipa.org
[3] JADE: http://jade.tilab.com

process begins, as well as showing them the results achieved by the consensus process.

- *FIPA-specific components and agents:* FIPA standard (*Foundation for Intelligent and Physical Agents*) provides some utilities for general purpose multi-agent systems, such as the *Directory Facilitator(DF)*, *AMS* (*Agent Management System*), and *MTS (Message Transport System)*.
- *Ontology:* Agents communicate each other by exchanging messages. Therefore, a ontology has been defined for COMAS, so that agents share a common semantics and knowledge about the problem. A detailed description about the system's ontology is given in Section 4.
- *Consensus Support System:* This is the software model containing all necessary Java classes to perform all operations in the consensus process.
- *Database:* It stores and recovers past information about consensus processes.

4 COMAS Ontology

In this section, we show the ontology designed to allow communication among agents under a common language and semantics [7,20].

Since COMAS ontology is based on the idea proposed by Kacprzyk and Zadrozny in [14], its design considers both the necessary concepts for performing consensus processes and those ones used by agents for reasoning about the particular problem's knowledge. Figure 3 shows the components used in our ontology. These components are divided into three categories:

Fig. 3 Ontology components used in COMAS

1. **Concepts:** Symbolic expressions representing objects. They consist of one or more attributes or *slots*. Concepts are not individually used by agents, but they appear as part of predicates or agent actions in ACL messages. Concepts

designed in our system include: *AID, Alternative, Consensus, ExpertAgent, OpAggregation, OpSimilarity, PreferenceRelation* and *Recommendation.*

2. **Predicates:** Expressions about the state of the world, their value would be either true or false. They are commonly used in COMAS as the content of ACL-Inform messages (answer to a request message), and include: *ProvideAssessment, ProvideConsensus* and *ProvideRecommendation.*
3. **Agent Actions:** These expressions are a special case of concepts with an additional semantics, indicating actions that can be performed by agents, and they are usually the content of ACL-Request and ACL-Propose messages in our system. COMAS agent actions include: *JoinConsensus, MakeAssesment, RateConsensus* and *RateRecommendation.*

5 COMAS Performance

In this section, we briefly show some experimental results obtained from a simulation we have carried out on our platform. The simulation consisted in solving a consensus problem with 4 alternatives and 100 experts, by using different consensus measures. The high scalability of COMAS makes possible performing consensus processes automatically with a high number of experts, which hasn't generally been done previously.

Consensus measures are used for measuring the level of agreement among experts, through the use of similarity measures and aggregation operators. We have considered the use of different combinations of these measures for each one of the experiments made. In addition, we consider a consensus threshold of 0.85 and a maximum number of 10 rounds permitted.

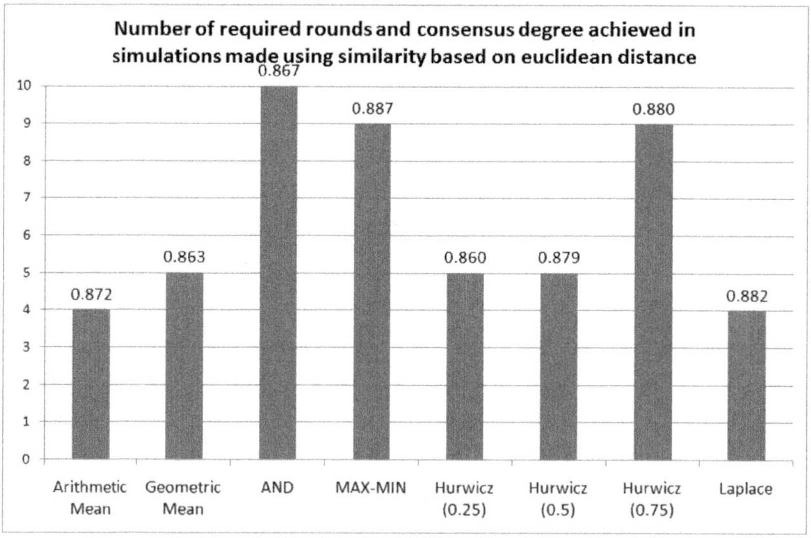

Fig. 4 Results achieved solving a consensus problem through simulation.

Figure 4 shows the convergence towards consensus, by displaying the number of required rounds and the level of agreement achieved, in some of the tests carried out on COMAS by using a *similarity measure based on Euclidean Distance* and different well-known aggregation operators, some of which may be found in [22] for further detail about them.

The overall results obtained from these tests and another ones carried out with more different measures, leaded us to conclude that the use of different consensus measures allows decision makers to use COMAS to solve consensus problems according to their requirements, ranging from situations where a fast agreement is required, to contexts where more discussion among experts is needed.

6 Conclusions and Future Work

Even though consensus processes have been widely studied and many different models were proposed, there are barely real implementations of such a type of system. This contribution has introduced COMAS, a multi-agent architecture to support and automate consensus reaching processes with a high number of experts. We are currently tackling different improvements in COMAS. On the one hand, we are aimed at deploying a user interface based on Web Services, as well as implementing different profiles of 'personality' for experts. On the other hand, we intend to deploy our system in intelligent environments, thus discovering real applications in smart homes.

References

[1] Aldea, A., Bañares-Alcántara, R., Jiménez, L., Moreno, A., Martínez-Miranda, J., Riaño, D.: The scope of application of multi-agent Systems in the process industry: three case studies. Expert Systems with Applications 26(1), 39–47 (2004)

[2] Alonso, S., Herrera-Viedma, E., Chiclana, F., Herrera, F.: Individual and social strategies to deal with ignorance situations in group decision making. International Journal of Information Technology and Decision Making 8(2), 313–333 (2009)

[3] Ben-Arieh, D., Chen, Z.: Linguistic labels aggregation and consensus measure for automatic decision-making using group recommendations. IEEE Transactions on Systems, Man and Cybernetics, Part A: Systems and Humans 36(1), 558–568 (2006)

[4] Ben-Arieh, D., Easton, T., Evans, B.: Minimum cost consensus with quadratic cost functions. IEEE Transactions on Systems, Man and Cybernetics, Part A: Systems and Humans 39(1), 210–217 (2009)

[5] Cabrerizo, F.J., Pérez, I.J., Herrera-Viedma, E.: Managing the consensus in Group decision making in an unbalanced fuzzy linguistic context incomplete information. Knowledge-Based Systems 23, 169–181 (2010)

[6] Fedrizzi, M., Fedrizzi, M., Pereira, R.A.M.: Soft consensus and network dynamics in group decision-making. International Journal of Intelligent Systems 14(1), 63–77 (1999)

[7] García-Sánchez, F., Valencia, R., Martínez, R., Fernández, J.T.: An ontology, intelligent agent-based framework for the provision of semantic web services. Expert Systems with Applications 36(1), 3167–3187 (2009)
[8] Gnansounou, E., Pierre, S., Quintero, A., Dong, J., Lahlou, A.: A multi-agent approach for planing activities in decentralized electricity markets. Knowledge-Based Systems 20(4), 406–418 (2007)
[9] Herrera, F., Herrera-Viedma, E., Verdegay, J.: A sequential selection process in group decision making with linguistic assessments. Information Sciences 85, 223–239 (1995)
[10] Herrera-Viedma, E., Herrera, F., Chiclana, F.: A Consensus Model for Multiperson Decision Making with different Preference Structures. IEEE Transactions on Systems, Man and Cybernetics, Part A: Systems and Humans 32(3), 394–402 (2002)
[11] Herrera-Viedma, E., Martínez, L., Mata, F., Chiclana, F.: A consensus support system model for Group decision-making problems with multigranular linguistic preferente relations. IEEE Transactions on Fuzzy Systems 13(5), 644–658 (2005)
[12] Kacprzyk, J., Nurmi, H., Fedrizzi, M.: Consensus under Fuziness. International Series in Intelligent Technologies, vol. 10. Springer, Heidelberg (1997)
[13] Kacprzyk, J., Zadrozny, S.: On a concept of a consensus reaching process support system based on the use of soft computing and web techniques. Computational Intelligence in Decision and Control 1, 859–864 (2008)
[14] Kacprzyk, J., Zadrozny, S.: Soft computing and web intelligence for supporting consensus reaching. Soft Computing 14(8), 833–846 (2010)
[15] Martínez, L., Montero, J.: Challenges for Improving Consensus Reaching Process in Collective Decisions. New Mathematics and Natural Computation 3(2), 203–217 (2007)
[16] Mata, F., Martínez, L., Herrera-Viedma, E.: An adaptive consensus support system model for Group decision-making problems in a multigranular fuzzy linguistic context. IEEE Transactions on Fuzzy Systems 17(2), 279–290 (2009)
[17] Mata, F., Sánchez, P.J., Palomares, I., Quesada, F.J., Martínez, L.: COMAS: A Consensus Multi-Agent based System. In: Proceedings of the International Conference on Intelligent Systems Design and Applications (ISDA 2010), pp. 457–462 (2010)
[18] Roubens, M.: Fuzzy Sets and decision analysis. Fuzzy sets and Systems 90(2), 199–206 (1997)
[19] Saint, S., Lawson, J.R.: Rules for Reaching Consensus. A modern approach to decision making. Jossey-Bass, San Francisco (1994)
[20] Staab, S., Studer, R.: Handbook on Ontologies. International Handbooks on Information Systems. Springer, Heidelberg (2005)
[21] Wang, S., Shen, W., Hao, Q.: An agent-based web service workflow model for interenterprise collaboration. Expert Systems with Applications 31(4), 787–799 (2006)
[22] Yager, R.R.: Applications and extensions of OWA aggregations. International Journal on Man-Machine Studies 27(1), 103–132 (1992)

Distributed Fuzzy Clustering with Automatic Detection of the Number of Clusters

L. Vendramin, R.J.G.B. Campello, L.F.S. Coletta, and E.R. Hruschka

Abstract. We present a consensus-based algorithm to distributed fuzzy clustering that allows automatic estimation of the number of clusters. Also, a variant of the parallel Fuzzy c-Means algorithm that is capable of estimating the number of clusters is introduced. This variant, named DFCM, is applied for clustering data distributed across different data sites. DFCM makes use of a new, distributed version of the Xie-Beni validity criterion. Illustrative experiments show that for sites having data from different populations the developed consensus-based algorithm can provide better results than DFCM.

Keywords: Clustering Data, Distributed Clustering, Consensus Clustering.

1 Introduction

Data clustering is a fundamental conceptual problem in data mining, in which one aims at determining a finite set of categories to describe a data set according to similarities among its objects [6]. In particular, fuzzy clustering algorithms aim at finding a fuzzy partition of a data set $\mathbf{X} = \{\mathbf{x}_1, \mathbf{x}_2, \ldots, \mathbf{x}_N\}$, with N objects described by n-dimensional feature or attribute vectors $\mathbf{x}_j$, into a certain number k of fuzzy clusters. A fuzzy partition can be represented by means of a $k \times N$ partition matrix $\mathbf{U} = [u_{ij}]_{k \times N}$ where u_{ij} denotes the membership degree of the jth object to the ith cluster. A probabilistic fuzzy partition requires that the membership degrees of each object to all clusters must sum up to one, i.e.:

$$\mathbf{U} = [u_{ij}]_{k \times N};\ u_{ij} \in [0,1];\ \sum_{i=1}^{k} u_{ij} = 1; \forall j \in \{1, \ldots, N\} \tag{1}$$

Lucas Vendramin · Ricardo J.G.B. Campello · Luiz F.S. Coletta · Eduardo R. Hruschka
Departament of Computer Sciences, University of Sao Paulo, Sao Carlos, Brazil
e-mail: `vendra@grad.icmc.usp.br`,
`campello,luizfsc,erh@icmc.usp.br`

A. Abraham et al. (Eds.): International Symposium on DCAI, AISC 91, pp. 133–140.
springerlink.com

The literature on fuzzy clustering is extensive. Several fuzzy clustering algorithms with different characteristics and for different purposes have been proposed and investigated [1, 5]. Some of the most well-known algorithms are the *Fuzzy c-Means* (FCM) [2] and the *Fuzzy c-Medoids* (FCMdd) [7], which is a variant of FCM able to deal with relational data. These algorithms, however, have been originally conceived to process centralized data only. Nevertheless, there are scenarios in which the data are (or can be) distributed among different sites. In these scenarios, the goal of a clustering algorithm consists in finding a structure that describes the distributed data without the need of data and processing centralization. One of these algorithms, named *Distributed Fuzzy c-Means* (DFCM) in this paper, is a parallel version [12] of FCM that assumes that the data sites were generated from the same population and can produce precisely the same FCM results as if the data were centralized. A different approach to distributed fuzzy clustering, which is aimed at describing non-centralized data coming from different populations, have been introduced in [11, 9]. Essentially, the goal of this approach is to reconcile the structure obtained from objects in a local data site with the available FCM-like results from other data sites — by means of a consensus clustering solution.

This paper presents an extension of the consensus-based algorithm to distributed fuzzy clustering described in [11, 9]. A procedure to automatically estimate the number of clusters with the DFCM clustering algorithm, by using a new, distributed version of the Xie-Beni validity criterion, is also provided. Finally, illustrative examples indicate scenarios where the consensus process can be successfully applied.

2 Consensus Clustering

Let $\mathbf{X}[ii] = \{\mathbf{x}_1[ii], \ldots, \mathbf{x}_{N_{ii}}[ii]\}$ be a data set composed of N_{ii} objects, $\mathbf{x}_j[ii]$. Then, consider a scenario in which there is a distributed collection of $p+1$ such (possibly confidential) data sets, each of which is associated with k_{ii} fuzzy clusters. Finally, recall that FCM-like algorithms represent clusters by means of prototypes. The goal of a user in data site $\mathbf{X}[0]$ is to reconcile a fuzzy partition $\mathbf{U}[0] = [u_{ij}[0]]_{k_0 \times N_0}$ induced from data set $\mathbf{X}[0]$ and its prototypes, with the fuzzy partitions $\mathbf{U}[0|ii] = [u_{ij}[0|ii]]_{k_{ii} \times N_0}$ induced from $\mathbf{X}[0]$ and prototypes coming from other data sites $\mathbf{X}[ii]$ $(ii = 1, \ldots, p)$ [11, 9] [1].

Note that all induced partition matrices $\mathbf{U}[0|ii]$ have the same number of columns, but not necessarily the same number of rows. Therefore, a consensus partition $\mathbf{U}_{k \times N_0}$, with k clusters, cannot be obtained directly from the induced partition matrices. The proximity matrix is a viable alternative to deal with this problem. Given a probabilistic fuzzy partition matrix $\mathbf{U}[0|ii]$, i.e., a partition that meet the requirements in (1), we can calculate the proximity between objects i and j ($Prox_{ij}$), in such a way that $Prox_{ij} \in [0, 1]$, as:

[1] Matrices $\mathbf{U}[0|ii]$ are calculated with data from $\mathbf{X}[0]$ and cluster prototypes provided by the iith data site, $\mathbf{X}[ii]$, using the standard FCM procedure to calculate a partition matrix.

$$Prox_{ij}[0|ii] = 1 - \frac{1}{2}\sum_{l=1}^{k_{ii}}(u_{li}[0|ii] - u_{lj}[0|ii])^2 \tag{2}$$

By calculating proximity values for all pairs of objects we get a proximity matrix denoted by $\mathbf{Prox}[0|ii]$. Note that after the transformation from induced partition matrices ($\mathbf{U}[0|ii]$) to induced proximity matrices ($\mathbf{Prox}[0|ii]$), for $ii = 0,\ldots,p$, we obtain matrices with precisely the same dimension, given by $N_0 \times N_0$, which is determined by the number of objects in data site $\mathbf{X}[0]$ (where consensus will take place). The main idea behind the consensus fuzzy clustering approach considered here is to build a partition matrix $\mathbf{U}$ whose proximity matrix, given by $\mathbf{Prox}(\mathbf{U})$, is as close as possible to the induced proximity matrices $\mathbf{Prox}[0|ii](ii = 0,\ldots,p)$. Formally, one aims at minimizing (with respect to $\mathbf{U}$) the following objective function:

$$V = \sum_{ii=0}^{p} \alpha_{ii} \sum_{i=1}^{N_0-1} \sum_{j=i+1}^{N_0} (Prox_{ij}(\mathbf{U}) - Prox_{ij}[0|ii])^2 \tag{3}$$

where $\alpha_{ii} \in [0,1]$ is a weight that can be assigned by the user to reflect the reliability or importance of the iith data site.

Note that the update of the partition matrix elements should be controlled so that the properties described in (1) be satisfied. An approach to satisfy those properties is to use a constrained nonlinear optimization method like the *Interior Point* method [3] — in our case using the partial derivatives of V in Eq. (3) with respect to each element u_{ij}. The formulation of the optimization problem is given by:

$$\begin{aligned}
\min_{\mathbf{U}} \quad & V = \sum_{ii=0}^{p} \alpha_{ii} \sum_{i=1}^{N_0-1} \sum_{j=i+1}^{N_0} (Prox_{ij}(\mathbf{U}) - Prox_{ij}[0|ii])^2 \\
\text{s.t.} \quad & \sum_{l=1}^{k} u_{lj} = 1, \; j = 1,\ldots,N_0 \\
& 0 \le u_{lj} \le 1, \; l = 1,\ldots,k, \; j = 1,\ldots,N_0
\end{aligned}$$

where N_0 is the number of objects in data site $\mathbf{X}[0]$, k is the (user-defined) number of clusters (rows) in the consensus partition $\mathbf{U}$, and $p+1$ is the number of data sites.

This process can also be applied when data sites, called subspaces, are described by the same objects but different attributes. In this case, each subspace finds a partition matrix of the same objects (columns). Since these matrices are built from the same objects, proximity matrices with the same dimensions ($N_0 \times N_0$) can be directly computed from them. By doing so, a consensus process can, analogous to that described above, be applied to find a global structure that represents all subspaces.

3 Estimating the Number of Clusters

The consensus process requires a priori knowledge of the number of clusters. However, the most natural number of clusters in a real data set is usually unknown a priori by the user. A widely known and simple approach to get around this drawback consists of getting a set of data partitions with different numbers of clusters and then select that particular partition that provides the best result according to a quality criterion [6]. The procedure in which the clustering algorithm is run M times for each number of clusters ranging from a given k_{min} to a given k_{max} is referred to here as OMR (*Ordered Multiple Runs*). At the end of this procedure there will exist $M \times (k_{max} - k_{min} + 1)$ partitions that must be evaluated according to the quality criterion adopted (see Section 4). We refer to as OMR-FCM, OMR-DFCM, OMR-FCMdd, and OMR-C when the clustering algorithms used are FCM, DFCM, FCMdd, and consensus (Section 2), respectively.

OMR-FCM and OMR-FCMdd are used to generate partitions of the local data sites to be further combined by means of consensus. OMR-DFCM and OMR-C, by their turn, are each used to generate a partition for the global (distributed) data.

It is important to notice that the partition matrix resulting from OMR-C may contain "virtual" clusters that do not represent any object, i.e., no object in the data set belongs to a higher extent to a given (virtual) cluster than to other clusters. In these cases, virtual clusters can be eliminated by removing its row of the partition matrix $\mathbf{U}$. When eliminating all membership values of a virtual cluster, it is necessary that the constraints imposed by (1) be met. One simple alternative is to normalize each column of $\mathbf{U}$. A more conceptually interesting alternative is to provide the reduced resulting matrix $\mathbf{U}$ back to the optimization process so that it can be readjusted.

4 Distributed Version of Xie-Beni Validity Criterion

In order to evaluate partitions in OMR procedures, we can use as a quality criterion the well-known Xie-Beni (XB) fuzzy clustering validity index, defined as [15]:

$$XB = \frac{\sum_{i=1}^{k}\sum_{j=1}^{N} u_{ij}^{m}||\mathbf{x}_j - \mathbf{v}_i||^2}{N \min_{l,s} ||\mathbf{v}_l - \mathbf{v}_s||^2} \tag{4}$$

where $\mathbf{v}_i$ denotes the ith cluster's prototype, and m is the *fuzzyfication* exponent used by FCM and DFCM, typically $m = 2$. This criterion can be directly used by OMR-FCM and OMR-FCMdd to evaluate local partitions. A distributed version of this criterion, able to evaluate global partitions of distributed data, can be derived from Eq. (4). Let the data be distributed among $p+1$ data sites, each of which with N_{ii} objects, and consider that each data site communicates the XB value respective to a partition of its N_{ii} objects (denoted by $XB[ii]$). Then, it is straightforward to write a distributed version of this criterion as:

$$XB_D = \frac{\sum_{ii=0}^{p}\alpha_{ii}\sum_{i=1}^{k}\sum_{j=1}^{N_{ii}} u_{ij}^{m}||\mathbf{x}_j[ii] - \mathbf{v}_i||^2}{\sum_{ii=0}^{p}\alpha_{ii}N_{ii}\min_{l,s}||\mathbf{v}_l - \mathbf{v}_s||^2} = \frac{\sum_{ii=0}^{p}\alpha_{ii}XB[ii]N_{ii}}{\sum_{ii=0}^{p}\alpha_{ii}N_{ii}} \tag{5}$$

where α_{ii} is the importance of the iith data site (see Eq. (3)). $XB_D = XB$ if $\alpha_{ii} = 1 \forall ii$.

Notice that the use of a validity index such as XB_D to evaluate a global partition of distributed data implicitly presumes that the data sites come from the same population, which is precisely the assumption made by DFCM. So, XB_D will be used by OMR-DFCM. This assumption, however, is not made by the consensus algorithm, which presumes different populations. For this reason, the objective function V of the consensus algorithm itself (Eq.(3)) will be used as a quality criterion by OMR-C.

5 Experiments

We have run the FCM/FCMdd algorithms with the default exponent value $m = 2$. XB and XB_D were used by OMR-FCM/OMR-FCMdd and OMR-DFCM procedures, respectively[2]. Finally, the weights for the data sites in the consensus algorithm were set evenly to one, i.e., $\alpha_{ii} = 1, \forall ii$.

5.1 Illustrative Example I - Same Feature Space

This experiment has two data sites (DS), each of which with $N = 40$ distinct objects described by the same $n = 2$ attributes. We are assuming that the clustering structure obtained locally, at a given data site, can be refined from the clustering structure present in another data site.

The OMR-DFCM procedure was run with $k_{min} = 2, k_{max} = 8$, and $M = 20$. The result of this procedure, a partition with four clusters (circles, squares, diamonds, and triangles), is illustrated in Figs. 1(a) and 1(b). Individual analyses of Figs. 1(a) and 1(b) show that the clusters hardly reflect the local structures. An overall clustering result, obtained by assuming that the data distributed across different data sites are from the same population, can be locally detrimental. Note that objects at the bottom of Fig. 1(a) were divided into two clusters (diamonds and triangles). This is due to the existence of objects in that location in the other data site (Fig. 1(b)). This division can be detrimental to the clustering process in DS1, since there are no objects where the new cluster was identified. In fact, one may prefer a single cluster instead of two clusters in that location.

In these cases, in which the data sites come from different populations, the use of a clustering algorithm that contributes to the improvement of the partitions obtained locally may be desirable. To that end, the OMR-C procedure can be used. We ran the OMR-FCM procedure at each data site individually with the same configurations above for k_{min}, k_{max}, and M. The solutions obtained by OMR-FCM procedure applied individually to each data site are illustrated in Figs. 1(c) and 1(d), and were used for the consensus process. The OMR-C procedure was then applied independently at each data site (considering it as the local data site $\mathbf{X}[0]$) within the same interval for the number of clusters ($k \in [2,8]$). The solution that provided the lowest objective function value, when DS1 and DS2 were considered as the local data sites, is illustrated in Figs. 1(e) and 1(f), respectively.

[2] Similar results were obtained when the Fuzzy Silhouette Criterion [4] was used.

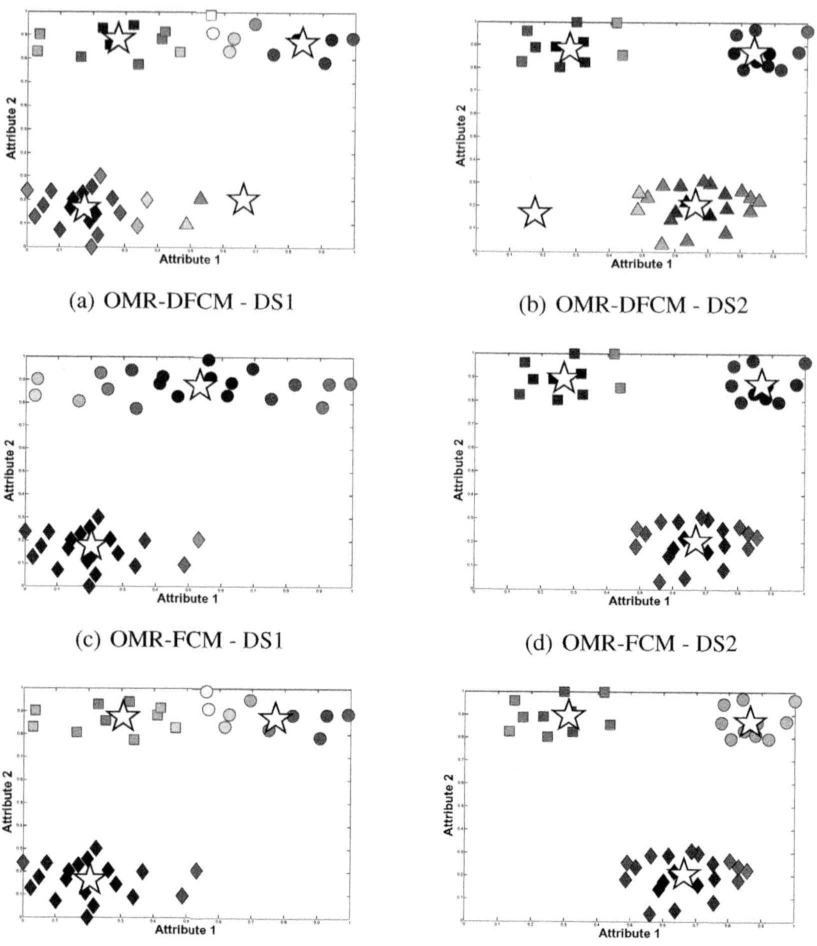

(a) OMR-DFCM - DS1 (b) OMR-DFCM - DS2

(c) OMR-FCM - DS1 (d) OMR-FCM - DS2

(e) OMR-C - DS1 (f) OMR-C - DS2

Fig. 1 Best result obtained by each procedure for both data sites: (a) (b): OMR-DFCM, (c) (d): OMR-FCM locally, and (e) (f): OMR-C – centers are represented as stars.

Note that the consensus process allowed the identification of three clusters at DS1. The objects that are illustrated at the top of Fig. 1(e) were divided into two clusters. This is due to the presence of two clusters in this region identified by DS2 (Fig. 1(d)). In addition, the consensus process has found only one cluster at the bottom of Figs. 1(e) and 1(f), which can be more desirable since there is no evidence of another cluster when evaluating each DS locally. This result illustrates a scenario where consensus can be beneficial to each data site while a distributed clustering algorithm can result in an undesirable structure.

5.2 Illustrative Example II - Different Feature Spaces

This experiment consists in clustering images described by different descriptors. The images were obtained from [14]. Five collections of images were selected, each of which has 111 images of a certain physical object, thus producing a total of 555 images of five different objects[3]. Four descriptors described in [10] were used to build four 555×555 relational matrices (R_1,R_2,R_3, and R_4) with pair distances among all objects (images). Each relational matrix describes the data in a different subspace. The OMR-FCMdd algorithm can be readily applied to relational data if one uses a relational validity criterion, i.e., a criterion that requires a relational data matrix only. A relational version of XB (Section 4) was presented in [13] and will be used here to evaluate the quality of each solution provided by OMR-FCMdd.

Since we have five collections of different images, it is possible to compare the clustering results with the expected solution with five clusters, e.g., using the Jaccard and Corrected Rand external indexes [6]. The higher their values the more similar to the expected partition an obtained partition is. OMR-FCMdd was applied individually to each subspace (i.e., relational matrix) with $k_{min} = 2, k_{max} = 8$, and $M = 50$. In addition, this algorithm was also applied to a single, concatenated matrix $R^+ = R_1 + R_2 + R_3 + R_4$ to produce a single result from the different descriptors of the images (subspaces). Finally, OMR-C was also applied to get a consensus among the results produced by OMR-FCMdd when applied independently to each subspace (R_1,R_2,R_3, and R_4). The results are displayed in Table 1.

Table 1 Number of clusters, Jaccard value, and Corrected Rand value obtained by OMR-FCMdd procedure in each subspace individually, by OMR-FCMdd in the concatenated space R^+, and by consensus (OMR-C).

	$\mathbf{R_1}$	$\mathbf{R_2}$	$\mathbf{R_3}$	$\mathbf{R_4}$	$\mathbf{R^+}$	**OMR-C**
No. of Clusters	7	2	2	6	4	6
Jaccard	0.7788	0.2833	0.2895	0.5182	0.5151	0.7841
Corrected Rand	0.8483	0.1928	0.2046	0.6066	0.5815	0.8487

Although none of the results exhibited the expected number of clusters, 5, the consensus procedure produced a better solution (evaluated by Jaccard and Corrected Rand) compared with OMR-FCMdd when applied either individually to each subspace or using the concatenated relational matrix. The consensus solution is only slightly better than the one obtained from one of the subspaces individually (R_1). However, notice that, in practical applications, there are no expected solutions to be used to compute the Jaccard and Corrected Rand indexes and, so, the user does not known which subspace would provide the best clustering result standalone.

[3] The selected collections are labeled as 422, 656, 792, 915, 959 in [14].

6 Conclusions

In this paper we extended the consensus clustering approach for distributed data described in [9, 11] with (i) automatic detection of the number of clusters and (ii) a new formulation to the optimization problem based on a continuous and differentiable objective function. In addition, a variant of the parallel Fuzzy c-Means algorithm [12], here named DFCM, which is capable of estimating the number of clusters — by using a new, distributed version of the Xie-Beni validity criterion — was also introduced. Our illustrative experiments showed that for sites having data from different populations the developed consensus-based algorithm can provide better results than DFCM. The study of the computational efficiency of the proposed algorithm is an interesting venue for future work.

Acknowledgements. We thank CNPq and FAPESP for their financial support.

References

1. Babuška, R.: Fuzzy Modeling For Control. Kluwer Academic, Dordrecht (1998)
2. Bezdek, J.C.: Pattern Recognition with Fuzzy Objective Function Algorithms Pl.P (1981)
3. Byrd, R.H., Gilbert, J.C., Nocedal, J.: A trust region method based on interior point techniques for nonlinear programming. Math. Progr. 89, 149–185 (2000)
4. Campello, R.J.G.B., Hruschka, E.R.: A fuzzy extension of the silhouette width criterion for cluster analysis. Fuzzy Sets and Systems 157, 2858–2875 (2006)
5. Höppner, F., Klawonn, F., Kruse, R., Runkler, T.: Fuzzy Cluster Analysis: Methods for Classification, Data Analysis and Image Recognition. John Wiley & Sons, Chichester (1999)
6. Jain, A.K., Dubes, R.C.: Algorithms for Clustering Data. Prentice-Hall, Englewood Cliffs (1988)
7. Krishnapuram, R., Joshi, A., Nasraoui, O., Yi, L.: Low-complexity fuzzy relational clustering algorithms for web mining. IEEE Trans. Fuzzy Systems 9, 595–607 (2001)
8. Krishnapuram, R., Keller, J.M.: A possibilistic approach to clustering. IEEE Trans. Fuzzy Systems 1, 98–110 (1993)
9. Loia, V., Pedrycz, W., Senatore, S.: Semantic web content analysis: A study in proximity-based collaborative clustering. IEEE Trans. Fuzzy Systems 15, 1294–1312 (2007)
10. Lux, M., Chatzichristofis, S.A.: Lire: Lucene image retrieval-An extensible java CBIR library. In: Proc. of the 16th ACM International Conference on Multimedia, pp. 1085–1088 (2008)
11. Pedrycz, W., Hirota, K.: A consensus-driven fuzzy clustering. Pattern Recognition Letters 29, 1333–1343 (2008)
12. Rahimi, S., Zargham, M., Thakre, A., Chhillar, D.: A parallel fuzzy c-mean algorithm for image segmentation. In: Fuzzy Information - Processing NAFIPS 2004, pp. 234–237 (2004)
13. Sledge, I.J., Bezdek, J.C., Havens, T.C., Keller, J.M.: Relational generalizations of cluster validity indices. IEEE Trans. Fuzzy Systems 18, 771–786 (2010)
14. Intelligent Sensory Information Systems, Amsterdam library of object images (aloi) (2010), `http://staff.science.uva.nl/~aloi/`
15. Xie, X.L., Beni, G.: A validity measure for fuzzy clustering. IEEE Trans. Pattern Analysis and Machine Intelligence 13, 841–847 (1991)

Resource Sharing in Collaborative Environments: Performance Considerations

Roberto Morales, Norma Candolfi, Jetzabel Serna, David A. Mejía, José M. Villegas, Juan I. Nieto, and Manel Medina

Abstract. Current collaborative environments are now largely populated of a great diversity of heterogeneous mobile devices. In environments such as m-health, the employment of pervasive devices represents an important communication link in gathering information that can be accessed or provided by other shared resources immerse in the environment. Particularly in medical environments, resource sharing middlewares can give rise to an endless number of potential applications to support medical specialists in collaboratively gathering vital health information of a patient. However, the applicability of this technological support which highly depends on the performance provided by the middleware communication channel has not been analyzed. Hence, in this paper, we provide initial results based on a series of experiments aimed at demonstrating the applicability and robustness of resource sharing in hospital work scenarios where, due to its popularity Bluetooth has been adopted by resource sharing middlewares as the main communication channel.

Keywords: Resource sharing, collaborative environments, m-health, performance

1 Introduction

The continuously increasing integration of wireless technology in our daily lives has led to a wide range of collaborative environment applications. Since resource sharing is one of the most common activities in these kind of environments, applications

Roberto Morales · Jetzabel Serna · Manel Medina
Technical University of Catalonia, Computer Architecture Department, Jordi Girona 1-3, 08034 Barcelona, Spain
e-mail: {rmorales,jetzabel,medina}@ac.upc.edu

Norma Candolfi · David A. Mejía · José M. Villegas · Juan I. Nieto
Autonomous University of Baja California, Center of Engineering and Technology, 69042 Valle de las Palmas, México
e-mail: {damejia,ncandolfi,jmanuel,jnieto}@uabc.edu.mx

A. Abraham et al. (Eds.): International Symposium on DCAI, AISC 91, pp. 141–149.
springerlink.com

supporting resource sharing could greatly benefit the individual or cooperative work of users in a collaborative environment, such as m-heatlh. Devices in collaborative environments (e.g. mobile phones, sensors, PDAs) possess each, a great variety of resources. The employment of resource sharing middlewares give devices the opportunity of i) dynamically use resources to improve their capabilities, or ii) dynamically add new resources. For example, devices can aggregate additional resources to improve sound or video quality, by using higher definition speakers, or by placing video with higher resolution in a bigger display, respectively. Devices can augment its capacity with remote storage through the aggregation of additional resources. These capabilities could be exploited, in m-health environments for instance, by using additional external displays to make a collaborative diagnosis of a X-ray image (when physicians occasionally meet in common areas and discuss a particular case). Along with the aforementioned benefits, resource sharing middlewares also convey important challenges, for example the capability of maintaining access to the available resources by the allocation of proper communication channels, which is still a major challenge and an important area of research. Therefore, to achieve successful collaboration, resource sharing middlewares must be able of selecting and managing appropriate communication channels in order to provide a satisfying QoS, and allow different forms of communication exchange, such as device synchronization or information streaming. Furthermore, considering that nowadays pervasive environments are greatly populated with Bluetooth enabled devices, the present study evaluates Bluetooth as the main communication channel for resource sharing in hospital work scenarios, which is particularly suitable due to its unique characteristics, i.e., low cost and low power radio technology. Thus, the main goal of this paper is to perform a series of experiments in order to evaluate the real capacities and behavior of Bluetooth for resource sharing middlewares applied in the collaborative environment of a hospital, where users own conventional mobile devices that often do not support more than one connection, and that, due to the high diversity of included Bluetooth versions, are not able to sustain an adequate performance.

1.1 Outline

This paper is organized according the following structure: Section 2 highlights the basics of resource sharing and introduces the "hospital work" collaborative scenario. The initial experimental evaluation is described in Section 3 and Section 4 presents a discussion of the experimental results. Section 5 overviews the related work and finally the main conclusions and future work directions are drawn in Section 6.

2 Resource Sharing: A Hospital Work Case Study

The great computing diversity in collaborative environments, enable users to share resources and work together to perform a large set of environment-dependent tasks. Current middleware approaches supporting resource sharing and how can hospital work benefit from applying them into the daily activities are explained next.

2.1 Middleware Support for Resource Sharing

Resource sharing allows devices to dynamically interact and share their available resources. Performing these tasks in a transparent manner and advantaging the great communication integration provided by wireless technologies require a general architecture. At present, middlewares presented in [11], [12] or [9] have demonstrated the possibility of extending devices' capabilities by using external resources. For example, using a mobile phone as a mouse to control an external device, employing an internal accelerometer of a device as a joystick, or playing a song using external audio resource. Nevertheless this concept has not been applied yet to m-health.

2.2 Applying Resource Sharing to m-Health Scenarios

Resource sharing can be exploited in a wide set of applications but especially in those involving collaborative work, providing a powerful tool to enhance users' collaborations. The next scenarios present different interactions in hospital work supported by resource sharing, assuming that all collaborating users own a mobile device with resource sharing collaborative applications.

Scenario 1: A physician is in a patient's room when he realizes through the localization application that the medical specialist is walking close to his position and decides to ask him about an X-ray image, and automatically connects his PDA via Bluetooth to the public shared display. Figure 1a represents the physician (master node 01) streaming an image to the shared display (client node). Looking the X-ray results in the public shared display they begin the discussion.
[Physician]: *What do you think about this?*
[Medical Specialist]: *Mmm, I think that he has (disease's name).*
[Physician]: *Are you sure? Because he has (the physician explains the symptoms).*
[Medical Specialist]: *Yes, I do. I had a patient with similar symptoms.*
[Physician]: *Are you sure? I think there could be some differences.*
The medical specialist through his handled (master node 02) accesses remotely to the patient's health record and the image is streamed to the public shared display (client node) which shows the physician an X-ray of that patient.
[Medical Specialist]: *Look, this is the X-ray of the patient I mentioned before.*
[Physician]: *You are right. Which is the treatment for this type of disease?*
The medical specialist accesses his office's PC to retrieve information from a medical guide related to the medication, and transfers it to the physicians' handheld.

Scenario 2: A medical specialist and a couple of physician residents meet in the hallway and discuss about the medical procedures that one of the residents must perform on two patients. Their handheld devices connect via Bluetooth and the Voice Detection Agent [8] starts monitoring the beginning of the interaction for recording it. As depicted in Figure 1b, the device initiating the interaction, i.e., the "client node", automatically selects a "master node" from neighboring devices (the best available audio resource). While the medical specialist starts explaining the procedure of each patient, the master node captures and transmits the audio to the client. In turn, the client node (requestor) processes and sends the information (audio, ids, timestamps, etc.) via WiFi to be stored by the server. The medical specialist completes the explanation, then, the resident goes to the bed ward and reviews the medical record of each patient. At this time, the resident is confused about which procedure must be performed to each patient. Thus, the resident accesses his handheld device to access the recorded information and clarifies his doubts.

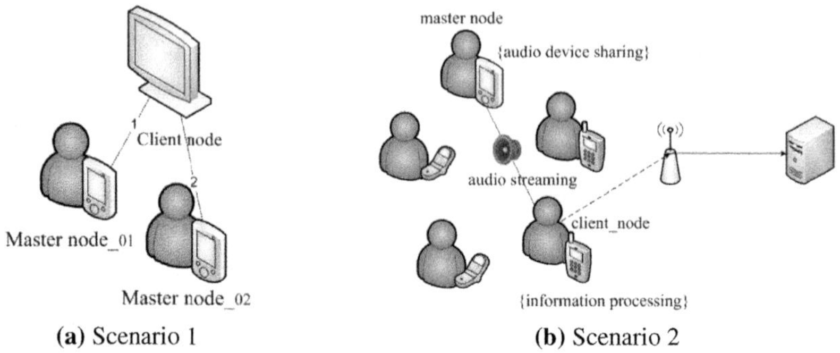

Fig. 1 Resource sharing in m-health scenarios: a) Streaming information to a shared display, b) Gathering audio information from the best available audio resource.

3 Experimental Evaluation

In order to simulate the introduced case study, this section presents a set of initial experiments based on a typical everyday hospital work scenario.

3.1 Test-Bed

In the scenarios 1 and 2 (medical staff visualizing an image in a public shared display and using a shared audio resource) the collaborative interactions are based on a point-to-point connection (a master and a client node at the time). However, to stress the capabilities of Bluetooth and explore other possible scenarios, the experiments included connections up to 5 client nodes, thus providing an insight of maximum number of collaborative users (medical staff) sharing a resource with an acceptable performance degree. In this scenario, nodes were configured with six Anycom USB-200/250 Bluetooth devices (a BT master node and five BT clients). In addition, assuming that, in the scenarios distances between devices are variable, to measure the speed rate among different distances a Motorola A780 smartphone and a Laptop equipped with a 3Com USB Bluetooth dongle were used, and for the test-bed configuration the following parameters were considered:*1)* a transmission packet size of 32k, *2)* L2CAP connection links, and *3)* Anycom proprietary Bluetooth stack and Linux BlueZ. To transmit data a Java client/application was developed and measurements were analyzed with the Colasoft Capsa Bluetooth packet analyzer. Finally, to cope with the requirements of the scenario 2, where the medical staff interactions are recorded (audio streaming via Bluetooth) and transmitted to the server (processed audio transmission via WiFi) and taking into account that both use the same frequency spectrum of ISM band, it is expected that the nodes performance is adversely affected. To confirm this, experiments measurements were performed twice: *1)* with an active WiFi network and, *2)* disabling the WiFi network.

3.2 Experimental Setup

Performance measurements supporting a variable number of nodes. The main objective of this experiment, is to evaluate the communication performance (transmission rate) experienced in a point-to-point collaborative interaction (e.g. a PDA adding a shared resource such as a display), and additionally, the maximum number of collaborative users supported by resource sharing (medical staff sharing a resource to a same master node). Technically the experiment consists of the transmission of packets during a 60s period from a master node to a maximum number of five slave nodes. Packet size and distance between nodes is constant (in a radio of 3mts), and slave (client) nodes are added every 60s, just after each run until the maximum number is reached. During the execution of the experiment packet transmission is monitored by the Colasoft sniffer.

Speed rate measurements considering distance among devices. This experiment aim at measuring performance among different distances to better understand the maximum range to be considered when selecting available shared resources, e.g. the maximum distance allowed to detect a shared resource or to stream an image from a mobile device (PDA) to a public shared display with an acceptable performance degree. The experiment consists on connecting two devices and sending 32k of data at distances of 1, 4, 9 and 15 meters maintaining a minimal link reliability. Devices were configured with a maximum transmission unit (MTU) of 64k, in which devices can transmit enough amount of data without losing performance. Both devices have a class 2 Bluetooth chip, which allows transmissions up to 10 meters with a 721kbit/s transmission speed [3].

4 Experimental Results

This section provides a set of results obtained from the experiments and discusses under what conditions can Bluetooth be applied to the scenarios.

4.1 Measurements

Experiment 1. Figure 2a shows how the master node's performance is affected during packet shipping, the more client nodes added, the less number of shipped packets. This is mainly due to packet retransmission; in other words, while adding

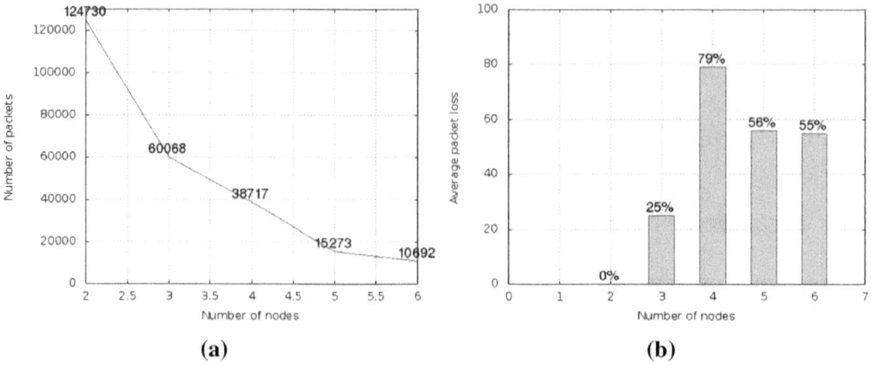

Fig. 2 BT performance: a) number of shipped packets, b) packet loss ratio.

nodes to the piconet, the server receives a retransmission request, and because of the division of the transmission channel, it sends a lower number of packets in the same time window. Figure 2b shows how packets loss increases while nodes are added. Within the experiment, packets were numbered to determine whether all of them arrived in the correct sequence, however, a high percentage of delayed packets occurred. When a node detected a lost packet, it had to wait for its assigned time slot to request for the packet retransmission, but, if the measurement time was reached (60secs), nodes waiting for packet retransmission had to discard the request.

Experiment 2. There were conducted ten communication executions at distances of 1, 4, 9 and 15 meters with direct line-of-sight (i.e. without obstacles). Table 1 shows the results average and its standard deviation. The first three results are very consistent obtaining an average speed of 42Kb/s, while in the fourth case (15m) the distance between devices began to affect the communication speed due to packet loss and retransmission. Figure 3 shows the communication delays at each distance, including a final test executed with a distance of 30m. Note that the distances of 15m and 30m are beyond the Bluetooth chip specifications.

4.2 Results Discussion

Initial results prove that Bluetooth can be particularly suited for scenarios involving point-to-point connections among a master and a slave, e.g. a physician using an external audio or display resource (scenarios 1 and 2 respectively). According to the definition of the second scenario (WiFi coexistence), it has been observed from previous results [10] that, between both measurements (with and without WiFi) there exists no significant difference, determining that BT and WiFi can perfectly coexist, and that, BT demonstrates to be adaptable and efficient enough to the particular needs of the stage. Results also sustain that, Bluetooth only supports a combination of piconets formed of a maximum number of 3 client nodes (slaves), meaning that,

Table 1 Speed rate average.

Distance (meters)	Speed rate average	Standard deviation
1	42.73	1.08
4	42.03	1.37
9	41.11	3.24
15	26.73	11.94

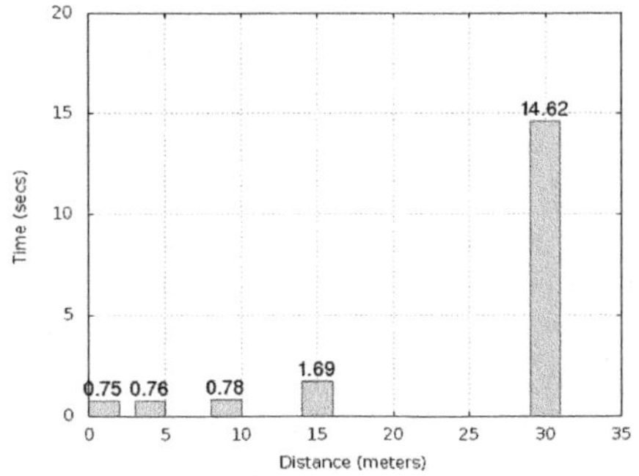

Fig. 3 End-to-end communication delay.

its applicability in resource sharing is reduced to this number of slaves, therefore in collaborative work involving more than 4 medical staff sharing a resource all of them to the same node e.g. a physician transmitting a image to more than 3 residents (shared displays) simultaneously, due to packet loss rate and transmission delays, the performance degree does not meet the QoS requirements. Nevertheless, results obtained in previous research (hospital field research) [8] show that in the 87% of the cases the number of participants in impromptu collaborations was up to 3 (52% of 2 and 35% of 3), and most interactions (79%) were placed by the residents and medical specialists both associated to a same patient involving only one shared resource. Also, analyzing the duration of interactions in a hospital, the study shows an average of 50 interactions per day, where 47% lasted less than 1 min, 44% between 1 and 5 min and 9% more than 5 min, and, the connection time among resource sharing enabled devices was approximately 2644.8ms. Although experiments confirmed the possibility of transmitting data at greater distances (15m and 30m), the response time is slow, what makes it inadequate especially to those resources needing real time information. Note that, results have also corroborated that Bluetooth connections support reliable data transmission rates and response times in a distance range of 10m, enabling a proper selection of available resources on this radio, nonetheless, middlewares must always consider selecting those that are closer, in order to avoid an inadequate power consumption. Hence, we can conclude that even with some limitations, Bluetooth can still be successfully applied in hospital work scenarios.

5 Related Work

Interesting proposals analyzing other Bluetooth features/scenarios that can also be supportive in our research are presented next. Authors of [4] presented an analysis of resource sharing in terms of resource discovery, they evaluate scatternets formed by a large number of devices (between 5-50) via simulation. Their experiments consisted in dynamic updates to provide device's status information (resources' availability). In [1] authors presented a Bluetooth analysis in a highly mobile environment, the experiments consisted in a centralized fixed node broadcasting messages to near devices while moving, they concluded that most of the problems arise because of the time constraints, and the establishment of an stable connection originated by a moving device. Authors of [6] evaluated the Bluetooth's capability to share resources in a collaborative mode, experiments with different nodes densities (2-18) showed that, the best number of nodes needed to successfully share a file with a waiting time of 110secs was six. Other research works focused on the effects caused by the coexistence of different wireless technologies, authors of [5, 14] reported a series of experiments that measured the interference degree between BT and WiFi, their results showed that with BT's 1.2 version the communication was affected by the WiFi's signal. However, this did not happen in our experiments since the version of BT was 2.0+EDR which includes an improvement called Adaptive Frequency Hop (AFH) that avoids collisions in the transmission channel. Nevertheless, authors of [7] remarked that, in a scenario with high mobility this improvement is not enough to avoid interferences. Considering that, our scenarios consisted in interactions based on a minimum time period taken place in common areas, it was possible to avoid this interference, nonetheless this cannot be totally neglected due to the great heterogeneity of collaborating devices (devices with different BT versions). Finally, it is important to remark that because of the highly collaborative nature of the work that takes place in environments such as hospitals [2], this kind of studies are useful for the broad acceptance of technology as an additional working tool. Medical and support staff have the opportunity to benefit from using shared resources available in the environment to provide accurate and timely diagnoses [13].

6 Conclusions and Future Work

This paper presented an analysis of the applicability of resource sharing in hospital work, through the evaluation of Bluetooth as the main communication technology in resource sharing middlewares. Initial results have demonstrated that Bluetooth is able to meet the QoS requirements of this particular environment, where up to 4 Bluetooth devices are able to communicate with an acceptable performance degree, even in scenarios involving distances that are beyond the Bluetooth specification. Hence, Bluetooth can be considered as the main communication channel in hospital work environments, without hindering the possibility of other technologies to coexist, such as WiFi. To stress the capabilities of resource sharing, future work will aim at extending the CARM middleware [9] to enable devices to simultaneously share various resources and addressing additional collaborative scenarios and evaluations (e.g. a mobile device sharing audio and display simultaneously).

References

1. Aiello, M., de Jong, R., de Nes, J.: Bluetooth broadcasting: How far can we go? An experimental study. In: Pervasive Computing (JCPC), December 3-5, pp. 471–476 (2009)
2. Bardram, J.E., Christensen, H.B.: Pervasive Computing Support for Hospitals: An overview of the Activity-Based Computing Project. IEEE Pervasive Computing 6(1) (2007)

3. Bluetooth SIG, http://www.bluetooth.com
4. Brodt, A., Wobser, A., Mitschang, B.: Resource Discovery Protocols for Bluetooth-Based Ad-hoc Smart Spaces: Architectural Considerations and Protocol Evaluation. In: Proceedings of the 2010 Eleventh International Conference on Mobile Data Management (MDM 2010), pp. 145–150. IEEE Computer Society, Washington, DC, USA (2010)
5. Ferro, E., Potorti, F.: Bluetooth and Wi-Fi wireless protocols: a survey and a comparison. IEEE Wireless Communications 12(1), 12–26 (2005)
6. Le Bourdon, X., Couderc, P., Banatre, M.: Spontaneous Hotspots: Sharing Context-Dependant Resources Using Bluetooth. In: Proceedings of the International Conference on Autonomic and Autonomous Systems (ICAS 2006). IEEE Computer Society, USA (2006)
7. Lee, U., Jung, S., Chang, A., Cho, D., Gerla, M.: Bluetooth-based P2P Content Distribution to Mobile Users. IEEE Transaction on Vehicular Technology, 356–367 (2010)
8. Mejía, D.A., Favela, J., Morán, A.L.: Understanding and supporting lightweight communication in hospital work. IEEE Transactions on Information System in Biomedicine (2010)
9. Morales, R., Otero, B., Gil, M.: Mobile Resource Management for a Better User Experience: An Audio Case Study. In: 4th Symposium of Ubiquitous Computing and Ambient Intelligence, UCAmI (2010)
10. Nieto, J., Candolfi, N., et al.: Bluetooth Performance Analysis in Wireless Personal Area Networks. In: Proceedings of the 2009 Electronics, Robotics and Automotive Mechanics Conference (CERMA 2009), pp. 38–43. IEEE Computer Society, Washington, DC, USA (2009)
11. Rellermeyer, J.S., Riva, O., Alonso, G.: AlfredO: An architecture for flexible interaction with electronic devices. In: Issarny, V., Schantz, R. (eds.) Middleware 2008. LNCS, vol. 5346, pp. 22–41. Springer, Heidelberg (2008)
12. Roy, W., Trevor, P., et al.: Dynamic Composable Computing. In: Proceedings of the 9th Workshop on Mobile Computing Systems and Applications, ACM, Napa Valley (2008)
13. Sharmin, M., Ahmed, S., Ahamed, S.I., Haque, M.M., Khan, A.J.: Healthcare aide: towards a virtual assistant for doctors using pervasive middleware. In: Fourth Annual IEEE International Conference on Pervasive Computing and Communications Workshops, PerCom Workshops 2006, March 13-17, p. 6, p. 495 (2006)
14. Shuaib, K., Boulmalf, M., Sallabi, F., Lakas, A.: Performance analysis co-existence of IEEE 802.11g with Bluetooth. In: Second IFIP International Conference on Wireless and Optical Communications Networks, WOCN 2005, March 6-8, pp. 40–44 (2005)

Models for Distributed Computing in Grid Sensor Networks

Buddika Sumanasena and Peter H. Bauer

Abstract. Using local state space models, a method for distributed information processing in rectangular grid sensor networks is presented. Non-linear systems and signal processing algorithms are represented in a local state space model. Then the local state space model is implemented on the sensor network. The local state space models used are generalizations of Fornasini-Marchesini and Givone-Roesser models for linear time-invariant 3-D systems. Realtime implementation issues of the said method are also discussed.

Keywords: Grid sensor networks, Distributed signal processing, Fornasini-Marchesini model, Givone-Roesser model, 3-D systems, nonlinear systems.

1 Introduction

Wireless sensor networks consisting of a large number of sensor nodes, collaborating to accomplish a common objective, have emerged recently as a candidate for a wide variety of applications. Some applications require sensor nodes to be deployed in a regular grid. Application of grid sensor networks for contaminant propagation detection and structural health monitoring is discussed in [15] and [9] respectively. Other application areas that often prefer grid or mesh topology include agriculture and environmental monitoring. Coverage and connectivity of grid sensor networks in the presence of node failure has been studied in [13]. Sensor deployment strategies, robustness against deployment errors, reliability, routing schemes and network capacity limits of grid sensor networks have been studied in [11, 19, 1, 2, 3, 4].

Buddika Sumanasena
Department of Electrical Engineering, University of Notre Dame
e-mail: msumanas@nd.edu

Peter H. Bauer
Department of Electrical Engineering, University of Notre Dame
e-mail: pbauer@nd.edu

A. Abraham et al. (Eds.): International Symposium on DCAI, AISC 91, pp. 151–158.
springerlink.com

Employing distributed schemes for information processing in sensor networks can yield significant benefits in terms of scalability, bandwidth and energy consumption. Furthermore, applications requiring *local actuation in response to a local detection* [8] are best supported by such distributed algorithms, yielding minimum response delays.

Methods for distributed information processing in grid sensor networks are discussed in this paper. For notational convenience it is assumed that the sensor network is two dimensional. However, models discussed can be readily extended to sensor grids of higher or lower dimensionality. It is assumed that the processing algorithm at a given node incorporates measurements of multiple nodes sampled over multiple sampling instances. Therefore the system is 3-dimensional where two dimensions are spacial and the other is temporal.

Methods to implement 3-D linear systems in grid sensor networks in a completely distributed manner are discussed in [5, 16]. Methods discussed in [5, 16] are based on Givone-Roesser(GR) and the Fornasini-Marchesini(FM) local state space models for 3-D systems. Stability of distributed 3-D systems implemented on grid sensor networks using GR and FM models is studied in [14].

In general, unless the processing algorithms at every node are linear, the resulting 3-D system is non-linear. Non linear state space models are proposed for distributed information processing in grid sensor networks in this work. Methods presented in [5, 16] to implement 3-D linear systems in grid sensor networks is extended to incorporate non linear processing algorithms. The method proposed in this work, for implementing 3-D systems in grid sensor networks is really a formulation of a centralized system in a distributed form.

1.1 Contribution

Using the GR and FM models, a method for distributed implementation of linear algorithms in grid sensor networks was presented in [5, 16]. Using non-linear time variant state space models, aforementioned method is extended to non-linear processing algorithms in this work.

1.2 Outline

Non-linear state space models for casual 3-D systems are presented in section II. Application of the state space models for information processing in grid sensor networks is also discussed in section II. Discussion in section II is extended for non-causal systems in section III. Concluding remarks are given in section IV.

2 State Space Models for Non-linear Causal 3-D Systems

Non-linear state space models for causal 3-D systems and their implementation are discussed in this section. Let the sensor measurement and the output, at node (n_1, n_2)

at time t, be denoted by $U(n_1,n_2,t) \in \mathbb{R}^p$ and $Y(n_1,n_2,t) \in \mathbb{R}^q$ respectively. Let the state of the node (n_1,n_2) at time t be $X(n_1,n_2,t) \in \mathbb{R}^n$. Let the sensor network be of size $N_1 \times N_2$.

2.1 The Non-linear GR Model For 3-D Systems

The Givone-Roesser model for 3-D linear systems can be extended to non-linear systems as follows:

$$\begin{bmatrix} x^h(n_1+1,n_2,t) \\ x^v(n_1,n_2+1,t) \\ x^t(n_1,n_2,t+1) \end{bmatrix} = f_{(n_1,n_2,t)}(X(n_1,n_2,t),U(n_1,n_2,t))$$
$$Y(n_1,n_2,t) = g_{(n_1,n_2,t)}(X(n_1,n_2,t),U(n_1,n_2,t)) \qquad (1)$$

Here,$X(n_1,n_2,t) = (x^{h^T}(n_1,n_2,t), x^{v^T}(n_1,n_2,t), x^{t^T}(n_1,n_2,t))^T$. Vectors $x^h \in \mathbb{R}^a$, $x^v \in \mathbb{R}^b$ and $x^t \in \mathbb{R}^c$ are called the horizontal, vertical and temporal state vector components respectively. Functions $f_{(n_1,n_2,t)} : \mathbb{R}^n \times \mathbb{R}^p \to \mathbb{R}^n$ and $g_{(n_1,n_2,t)} : \mathbb{R}^n \times \mathbb{R}^p \to \mathbb{R}^q$ are in general non-linear. For a sensor network of size $N_1 \times N_2$, $n_1 \in [0, N_1-1]$ and $n_2 \in [0, N_2-1]$.

2.2 The Non-linear FM Model For 3-D Systems

The FM model for 3-D linear systems can be extended to non-linear systems as follows:

$$X(n_1,n_2,t) = f_{(n_1,n_2,t)}(X(n_1,n_2,t-1), X(n_1,n_2-1,t), X(n_1-1,n_2,t), U(n_1,n_2,t-1), U(n_1,n_2-1,t),$$
$$U(n_1-1,n_2,t))$$
$$Y(n_1,n_2,t) = g_{(n_1,n_2,t)}(X(n_1,n_2,t),U(n_1,n_2,t)) \qquad (2)$$

Functions $f_{(n_1,n_2,t)} : \mathbb{R}^n \times \mathbb{R}^n \times \mathbb{R}^n \times \mathbb{R}^p \times \mathbb{R}^p \times \mathbb{R}^p \to \mathbb{R}^n$ and $g_{(n_1,n_2,t)} : \mathbb{R}^n \times \mathbb{R}^p \to \mathbb{R}^q$ are in general non-linear. The state vector X of the FM model of a system is not in general the same as the state vector of the GR model of the same system. The state vector of the FM model cannot in general be interpreted as being comprised of horizontal, vertical, and temporal state vector components. The same notation X is used for the state vectors of both the local state space models to be consistent with the notation used in literature.

2.3 Implementation in a Sensor Network

Local state space models (1) and (2) can be implemented with communication only between the adjacent nodes of the grid. This is a key advantage in using them for distributed signal processing in grid sensor networks.

2.3.1 Implementation Using the Non-linear GR Model

The following operations are performed by each node (n_1, n_2) at the time slot t.

- Receive semi-state vectors $x^h(n_1, n_2, t)$ and $x^v(n_1, n_2, t)$ from nodes $(n_1 - 1, n_2)$ and $(n_1, n_2 - 1)$ respectively.
- Use equation (1) to compute $x^h(n_1 + 1, n_2, t)$, $x^v(n_1, n_2 + 1, t)$, $x^t(n_1, n_2, t + 1)$ and the output $Y(n_1, n_2, t)$
- Transmit $x^h(n_1 + 1, n_2, t)$ and $x^v(n_1, n_2 + 1, t)$

Figure 1(a) illustrates the operation of nodes and communication of semi-state vectors between nodes.

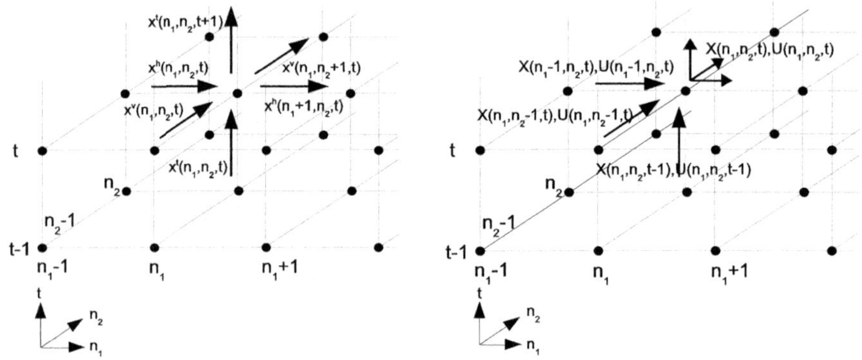

(a) Roesser model based implementation (b) FM model based implementation

Fig. 1 Communication of state vectors between nodes in the network

2.3.2 Implementation Using the Non-linear FM Model

In a sensor network implementation, the following operations are performed by each node (n_1, n_2) at the time slot t.

- Receive state vectors $X(n_1 - 1, n_2, t)$ and $X(n_1, n_2 - 1, t)$ and input vectors $U(n_1 - 1, n_2, t)$ and $U(n_1, n_2 - 1, t)$.
- Use equation (2) to compute $X(n_1, n_2, t)$ and the output $Y(n_1, n_2, t)$.
- Transmit $X(n_1, n_2, t)$ and $U(n_1, n_2, t)$

Figure 1(b) illustrates the operation of nodes and communication of state vectors between nodes.

2.4 *Realization*

The problem of realization is, given a processing algorithm, to select the functions $f_{(n_1,n_2,t)}$ and $g_{(n_1,n_2,t)}$ such that state space models (1) or (2) realize the said algorithm. If the processing algorithm is linear and invariant of time and node the resulting system would be linear and space time invariant. Realization algorithms for the case of linear space time invariant systems have been studied widely in the multidimensional system theory. Algorithms to realize 3-D rational transfer functions in the GR model are given in [10, 6, 12, 21, 17]. Algorithms to realize 2-D and 3-D rational transfer functions in the FM model are given in [7, 22, 20, 18]. However to the best of the authors knowledge realization of non-linear systems in local state space models has not be investigated in the literature even for restricted cases.

2.5 Real-Time Implementation Issues

State vector components $x^h(n_1+1,n_2,t)$ and $x^v(n_1,n_2+1,t)$ of nodes (n_1+1,n_2) and (n_1,n_2+1) are evaluated at node (n_1,n_2) at time slot t. These components are required at time slot t by the nodes (n_1+1,n_2) and (n_1,n_2+1) to perform their computations. In a non-linear FM model based implementation the state vector $X(n_1,n_2,t)$ is evaluated at the node (n_1,n_2) at time slot t. It is required by nodes (n_1+1,n_2) and (n_1,n_2+1) at time slot t to perform their computations. A real time implementation using either model thus requires data transmission with zero time delay which is impossible[1]. There are two options to work around this problem.

One is to allow a time lag between nodes which means the system is not real time. This could be a good solution for small sized sensor networks. The other option is to restrict system models such that zero time delay data transmission is not required in the spatial dimensions to perform computations. This would limit the type of systems that can be implemented.

2.6 Delayed Response Implementation

In the implementation, the problem is that nodes (n_1+1,n_2) and (n_1,n_2+1) do not receive semi-state vectors $x^h(n_1+1,n_2,t)$ and $x^v(n_1,n_2+1,t)$ at time t to perform their computations at time t. In a non-linear FM model based implementation nodes (n_1+1,n_2) and (n_1,n_2+1) do not receive the state vector $X(n_1,n_2,t)$ at time t.A simple solution is to allow those nodes to do the computations they should do at time slot t at time slot $t+1$ instead. This means for each distance unit in either spatial direction there is a time lag of one time slot (one distance unit is equivalent to the distance between two nodes). This time lag could be significant in a moderate sized sensor network.

The time lag can be reduced significantly if the computation front of nodes propagates more than one distance unit in a sampling interval $[t,t+1]$. Let the computation front propagate d distance units along either spatial axis in one time slot. The maximum time delay in computing the output is $\lfloor\frac{N_1+N_2}{d}\rfloor$ for a sensor network of

[1] Note that, performing computations with small time delay at nodes (n_1+1,n_2) and (n_1,n_2+1) does not solve the problem in general since this delay accumulates as computations proceed over the sensor network.

size $N_1 \times N_2$ [16]. If $d > N_1 + N_2$ output at every node at time slot t can be computed within the interval $[t, t+1]$.

2.7 *Real Time Implementation*

2.7.1 Using the Non-linear GR Model

Let state vector components $x^h(n_1+1, n_2, t)$ and $x^v(n_1, n_2+1, t)$ be independent of state vector components $x^h(n_1, n_2, t)$ and $x^v(n_1, n_2, t)$. Then node (n_1, n_2) requires state vector components $x^v(n_1, n_2, t)$ and $x^h(n_1, n_2, t)$ only to compute $x^t(n_1, n_2, t+1)$. Since the state vector component $x^t(n_1, n_2, t+1)$ is used only by the node (n_1, n_2) a node can compute it in the time interval $(t, t+1)$. Thus the node (n_1, n_2) does not require state vector components $x^h(n_1, n_2, t)$ and $x^v(n_1, n_2, t)$ at time t to perform computations. So there is no need for data communication with zero time delay. This restriction to system model makes a real time implementation possible, but limits the impulse responses the system could have.

2.7.2 Using the Non-linear FM Model

Let state vector $X(n_1+i, n_2+j, t)$, where $i > 0$ and $j > 0$ and $i+j = 2$, be independent of $X(n_1, n_2, t)$ and $U(n_1, n_2, t)$. Then nodes (n_1+1, n_2) or (n_1, n_2+1) do not have to wait until they receive $X(n_1, n_2, t)$ and $U(n_1, n_2, t)$ to transmit information required by nodes (n_1+2, n_2), (n_1+1, n_2+1) and (n_1, n_2+2) to perform their computations. Nodes nodes (n_1+1, n_2) and (n_1, n_2+1) can compute their state vectors once they receive $X(n_1, n_2, t)$ and $U(n_1, n_2, t)$ in the time interval $(t, t+1)$.

3 State Space Models for Non-linear Non-causal 3-D Systems

Systems representable by models (1) and (2) are necessarily first octant causal. Spatially non causal systems can be represented as a combination of four systems causal in each of the four quadrants of the spatial plane.

3.1 *The Non-linear GR Model For 3-D Non-causal Systems*

The following state transition model, where $(\alpha, \beta) \in \{-1, 1\} \times \{-1, 1\}$:

$$\begin{bmatrix} x^h_{\alpha\beta}(n_1+\alpha, n_2, t) \\ x^v_{\alpha\beta}(n_1, n_2+\beta, t) \\ x^t_{\alpha\beta}(n_1, n_2, t+1) \end{bmatrix} = f_{\alpha\beta(n_1,n_2,t)}(X_{\alpha\beta}(n_1, n_2, t), U(n_1, n_2, t)) \tag{3}$$

can be used to represent a 3-D system causal in any of the four spatial quadrants by appropriately selecting α and β. A combination of four systems of the form (3), each causal in one of the four quadrants of the spatial plane, can be used to represent a non-causal 3-D systems. The four systems are combined or coupled using the output equation as follows:

$$Y(n_1,n_2,t) = g_{(n_1,n_2,t)}(X_{11}(n_1,n_2,t), X_{1-1}(n_1,n_2,t), X_{-1-1}(n_1,n_2,t), X_{-11}(n_1,n_2,t), U(n_1,n_2,t))$$

3.2 The Non-linear FM Model For 3-D Non-causal Systems

A 3-D system causal in any of the four spatial quadrants can be represented by:

$$X(n_1,n_2,t) = f_{\alpha\beta(n_1,n_2,t)}(X_{\alpha\beta}(n_1,n_2,t-1), X_{\alpha\beta}(n_1,n_2-\beta,t), X_{\alpha\beta}(n_1-\alpha,n_2,t), U(n_1,n_2,t-1)$$
$$, U(n_1,n_2-\beta,t), U(n_1-\alpha,n_2,t)) \quad (4)$$

by appropriately selecting α and β where $(\alpha,\beta) \in \{-1,1\} \times \{-1,1\}$. A combination of four systems of the form (4), each causal in one of the four quadrants of the spatial plane, can be used to represent a non-causal 3-D systems. The four systems are combined or coupled using the output equation as follows:

$$Y(n_1,n_2,t) = g_{(n_1,n_2,t)}(X_{11}(n_1,n_2,t), X_{1-1}(n_1,n_2,t), X_{-1-1}(n_1,n_2,t), X_{-11}(n_1,n_2,t), U(n_1,n_2,t))$$

4 Conclusion

A local state space model based approach for computing in grid sensor networks was presented. It can be used to implement any general signal processing algorithm on a sensor network in a completely distributed manner. Implementation of the model requires only communication between adjacent nodes in the grid sensor network. Therefore it is highly suitable for distributed signal processing in grid sensor networks.

References

[1] AboElFotoh, H., Iyengar, S., Chakrabarty, K.: Computing reliability and message delay for cooperative wireless distributed sensor networks subject to random failures. IEEE Transactions on Reliability 54(1), 145–155 (2005)

[2] AboElFotoh, H.M.F., Elmallah, E.S., Hassanein, H.S.: A flow-based reliability measure for wireless sensor networks. International Journal of Sensor Networks 2(5/6), 311–320 (2007)

[3] Akbar, A., Mansoor, W., Chaudhry, S., Kashif, A., Kim, K.: Node-link-failure resilient routing architecture for sensor grids. In: The 8th International Conference Advanced Communication Technology, Phoenix, USA, pp. 131–135 (2006)

[4] Barrenechea, G., Beferull-Lozano, B., Vetterli, M.: Lattice sensor networks: capacity limits, optimal routing and robustness to failures. In: Proceedings of the 3rd International Symposium on Information Processing in Sensor Networks, Berkeley, USA, pp. 186–195 (2004)

[5] Dewasurendra, D.A., Bauer, P.H.: A novel approach to grid sensor networks. In: 15th IEEE International Conference on Electronics, Circuits and Systems, Malta, pp. 1191–1194 (2008)

[6] Fan, H., Xu, L., Lin, Z.: A constructive procedure for three-dimensional realization. In: Proceedings of the 6th World Congress on Intelligent Control and Automation, Dalian, China, pp. 1893–1896 (2006)

[7] Fornasini, E., Marchesini, G.: Doubly-indexed dynamical systems: State-space models and structural properties. Mathematical Systems Theory 12(1), 59–72 (1978)
[8] Hu, W., Bulusu, N., Jhan, S.: A communication paradigm for hybrid sensor/actuator networks. International Journal of Wireless Information Networks 12(1), 47–59 (2005)
[9] Huang, Y., Loewke, K., Schaaf, K., Nemat-Nasser, S.: Localized SHM with embedded sensor network. In: Proceedings of the 5th International Workshop on Structural Health Monitoring, Stanford, pp. 1554–1561 (2005)
[10] Kanellakis, A.J., Paraskevopoulos, P.N., Theodorou, N.J., Varoufakis, S.J.: On the canonical state-space realization of 3-d discrete systems. IEE Proceedings on Circuits, Devices and Systems 136, 19–31 (1989)
[11] Leoncini, M., Resta, G., Santi, P.: Analysis of a wireless sensor dropping problem in wide-area environmental monitoring. In: Proceedings of the 4th International Symposium on Information Processing in Sensor Networks, Los Angeles, California, pp. 239–245 (2005)
[12] Manikopoulos, C.N., Antoniou, G.E.: State-space realization of three-dimensional systems using the modified cauer form. International Journal of Systems Science 12(21), 2673–2678 (1990)
[13] Shakkottai, S., Srikant, R., Shroff, N.: Unreliable sensor grids: Coverage, connectivity and diameter. In: Proceedings of IEEE INFOCOM, San Francisco, pp. 1073–1083 (2003)
[14] Sumanasena, M.G.B., Bauer, P.H.: Stability of distributed 3-d systems implemented on grid sensor networks. IEEE Transactions on Signal Processing 58(8), 4447–4453 (2010)
[15] Sumanasena, M.G.B., Bauer, P.H.: Distributed m-d filtering for wave front detection in grid sensor networks. In: Proceedings of the 20th IASTED International Conference on Parallel and Distributed Computing and Systems, Orlando, Florida, pp. 423–429 (2008)
[16] Sumanasena, M.G.B., Bauer, P.H.: A Roesser model based multidimensional systems approach for grid sensor networks. In: 43rd Asilomar Conference on Signals Systems and Computers, Pacific Grove (2009)
[17] Sumanasena, M.G.B., Bauer, P.H.: Realization using the FM model for implementations in distributed grid sensor networks. Accepted for 49th IEEE Conference on Decision and Control (2010)
[18] Sumanasena, M.G.B., Bauer, P.H.: Realization using the Roesser model for implementations in distributed grid sensor networks. Accepted for 49th IEEE Conference on Decision and Control (2010)
[19] Xu, K., Takahara, G., Hassanein, H.: On the robustness of grid-based deployment in wireless sensor networks. In: Proceedings of the 2006 International Conference on Wireless Communications and Mobile Computing, Vancouver, Canada, pp. 1183–1188 (2006)
[20] Xu, L., Wu, Q., Lin, Z., Xiao, Y.: A new constructive procedure for 2-d coprime realization in Fornasini-Marchesini model. IEEE Transactions on Circuits and Systems I 54(9), 2061–2069 (2007)
[21] Xu, L., Fan, H., Lin, Z., Bose, N.K.: A direct-construction approach to multidimensional realization and lfr uncertainty modeling. Multidimensional Systems Signal Processing 19(3-4), 323–359 (2008)
[22] Xu, L., Wu, L., Wu, Q., Lin, Z., Xiao, Y.: On realization of 2d discrete systems by Fornasini-Marchecini model. International Journal of Control, Automation, and Systems 3(4), 631–639 (2008)

Virtualizing Grid Computing Infrastructures into the Cloud

Mariano Raboso, Lara del Val, María I. Jiménez, Alberto Izquierdo, Juan J. Villacorta, and José A. de la Varga

Abstract. This paper shows how virtualization techniques can be introduced into the grid computing infrastructure to provide a transparent and homogeneous scientific computing environment. Today's trends in grid computing propose a shared model where different organizations make use of a heterogeneous grid, frequently a cluster of clusters (CoC) of computing and network resources. This paper shows how a grid computing model can be virtualized, obtaining a simple and homogeneous interface that can be offered to the clients. The proposed system is implemented on a system named *virtual grid*. Both cloud computing infrastructure and grid computing technology used, are freely available to all users.

Keywords: grid computing, cloud computing, virtualization, scientific computing environments, message passing interface.

1 Introduction

Increasing demand of computer resources for scientific research has been a strong motivation for the community to develop a wide variety of high performance computing infrastructures (HPC). Huge supercomputer resources are not always available to small research groups, usually limited by restrictive budgets, deploying tasks not suitable for these systems. Therefore, the local resources available must be optimized and shared. Initiatives, such as the European Grid Infrastructure (EGI) [1], aim to develop a sustainable grid infrastructure for all European researchers.

Grid computing systems are powerful solutions for the research community to process computation intensive applications. A grid computing system is a grid of

Mariano Raboso · José A. de la Varga
Facultad de Informática, Universidad Pontificia de Salamanca. Compañía 5, 37002 Salamanca, Spain
e-mail: `mrabosoma@upsa.es`

Lara del Val · María I. Jiménez · Alberto Izquierdo · Juan J. Villacorta
Departamento de Teoría de la Señal y Comunicaciones e Ingeniería Telemática, Universidad de Valladolid, E.T.S.I. Telecomunicación, Paseo Belén 15, 47011 Valladolid, Spain

A. Abraham et al. (Eds.): International Symposium on DCAI, AISC 91, pp. 159–166.
springerlink.com

parallel and distributed computing resources, working together towards a single goal. The grid provides high global computational power if convenient parallelization and concurrency issues are attained.

For building grid computing systems, a great collection of middleware has been provided with projects such as the Open Message Passing Interface (OMPI) [2], which is used worldwide. Affordable hardware infrastructures such as personal computers and network switching resources can be used to develop a high performance computing system capable of performing high intensive calculus operations demanded by the community.

On the other hand, these systems often lack the desirable uniformity of computing resources and network connections: bandwidth and latency [3][4]. Clusters of clusters [5][6] are clear examples of this resultant topology. Resource groups may be geographically distributed on different sites connected among a variety of WAN network technologies such as point-to-point links, MPLS, Frame-Relay or ATM.

In order to efficiently use the global computing infrastructure, many schedulers and algorithms have been proposed [7][8][9][10]. They usually depend on variables such as network distance, latency or bandwidth, to minimize the effect of distance and to integrate non-uniform computing resources.

Recent developments in virtualization have made it possible to virtualize not only computer resources, but also storage and network resources. A new paradigm, "the cloud", has become a cutting-edge technology for data centers and other IT infrastructures. A cloud computing system is an approach to computing, based on on-demand efficient use of aggregate resources, self-managed and consumed as a service.

A grid computing system can take advantage of cloud computing systems. We have developed an infrastructure combining grid and cloud services that provide transparent virtual grids to the clients. Although the grid is shared, the clients will use their own virtual grids with physical computing nodes.

Section 2 and 3 describe the cloud and grid computing developed infrastructures. The new infrastructure, *virtual grid*, is described in section 4. Finally, some discussion is made describing the advantages of the developed system.

2 Cloud Computing Scenario

The cloud computing infrastructure is based on Ubuntu Enterprise Cloud from Canonical [11]. It uses Eucalyptus technology to manage virtual machine images, fully compatible with Amazon's Elastic Compute Cloud (EC2) [12]. Two reasons make this technology a good choice. It is freely available to users, and easy to migrate towards a hybrid or even a public cloud, keeping the same virtual machine images.

For this purpose, a private cloud has been developed using Faculty resources from two universities in Salamanca (UPSA) and Valladolid (UVA). They are connected by RedIris network, an infrastructure that interconnects all the university and research institutes in Spain.

Figure 1 shows the cloud computing system being developed in Salamanca and Valladolid:

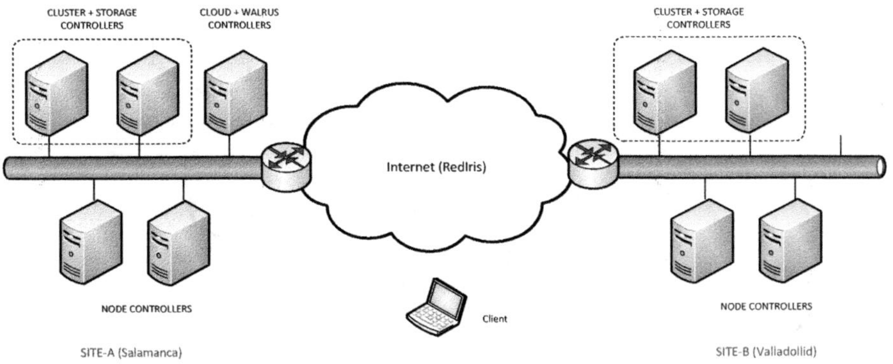

Fig. 1 Cloud computing system architecture.

For network tuning and simulation purposes, two point-to-point WAN interface cards (Cisco Asynchronous/Synchronous WIC) have been installed in the lab routers. It makes possible to experiment with different encapsulation patterns and so with different bandwidth and latency [13][14]. Cloud efficiency can also be measured when virtual machine instances are run in different locations.

A unique master server may implement the controller: storage, walrus, cluster and cloud controllers; two computers run the node controller feature. Remote sites are independent clusters requiring local cluster and storage controllers so virtual machines can be run locally. If needed, storage, cluster, walrus and cloud controllers may be run on the same machine if there is only one cluster.

Node controllers can be connected to a private subnet, but services must be accessible through the public network.

Client access can be made from any where by accessing several interfaces that are offered to run instances (virtual machines). The simplest way is using a plugin for a navigator, but if accurate control is required, instances may be run by the shell command line with the requirement that the machine must be running EC (Eucalyptus) tools.

There are plans to develop a web server integrating all the services that will be required to connect the HPC (grid computing) system.

3 HPC Infrastructure

A grid computing system is developed using the universal and widely used Message Passing Protocol to perform the HPC system [15]. Open MPI middleware [2] and Ubuntu Linux have been selected to enable a grid system that experimentally integrates three computers. One of these computers acts as the master node, being responsible for running jobs and file system sharing.

The rest of the nodes only run jobs assigned by the master. All the nodes are connected to the same network infrastructure, as in the cloud computing system.

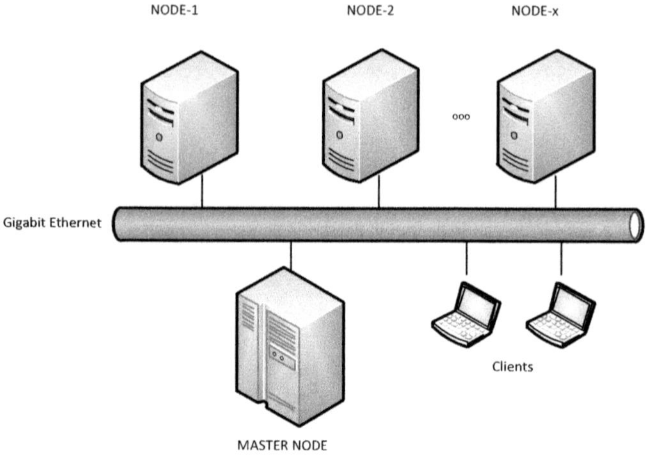

Fig. 2 HPC system developed at the networking lab in Salamanca

The topology is being expanded to integrate the existing computers in other labs. Almost 200 computers can be used as nodes while they are not being used for regular classes and during idle periods of time. Therefore an interesting computing power can be obtained if all machines are efficiently used.

Software running on the grid is developed to simulate digital signal processing applications, specifically acoustic beam forming applications in security, surveillance, biometrics and radar. These applications are computation intensive and usually need a huge quantity of computer resources.

Moreover, these kinds of applications usually take advantage of parallel array processing, so serious effort must be made to (re)write software in the proper way. The HPC system is designed to serve as a stable computing platform for research groups at corresponding universities.

If computing resources (nodes) are located in several sites, then bandwidth and latency communication issues must be considered. To address this problem, several solutions have been adopted that minimize these negative effects. They usually involve schedulers and partitioners [9] that represent the network topology using graphs. As a result, parallel jobs run on the most appropriate machines.

4 Virtualizing the HPC System

In order to batch their jobs, HPC users must connect to the master node. Some effort has been made to offer a standard and uniform interface, so hardware and software component details are transparent for users. Gridsolve and Netsolve platforms [16] provide a middleware that enable users to solve complex scientific problems, while using simple interfaces from their preferred Scientific Computing Environments (SCE) [17]. Interfaces for Matlab, IDL, C and Fortran are included in the latest releases.

Another solution for sharing the grid computing system can be addressed. The main goal is to manage local virtual grids to serve client requests. These requests are made to the cloud computing system via a standard interface. The response consists of creating a new virtual machine using a previously configured master server image. Once the new master is running, the clients will manage their own grid with the computing resources (nodes) assigned.

The infrastructure developed is named *virtual grid*. Figure 3 shows the system and virtual machine creation process:

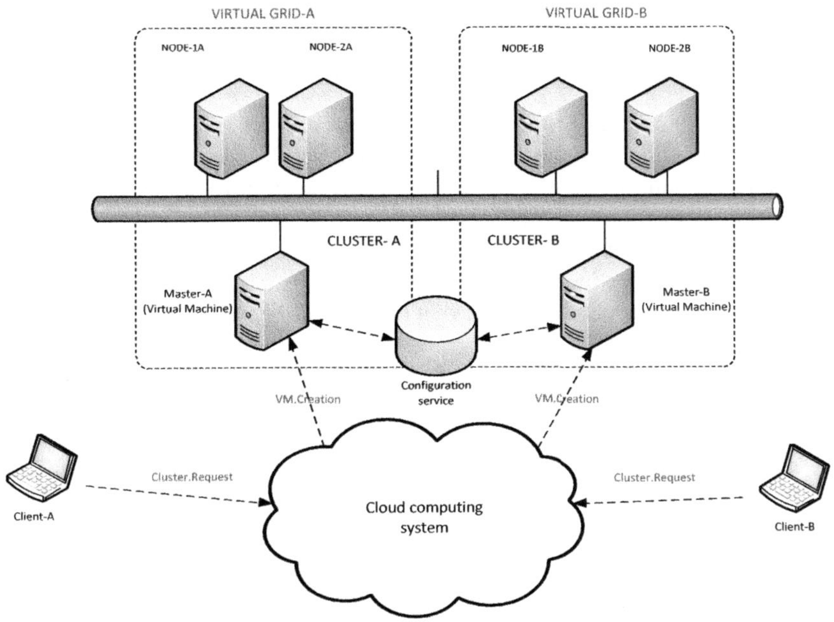

Fig. 3 Virtual grid system architecture and VM creation

When clients want to gain access to the grid, they make a request to the cloud system. The cloud then runs an instance of a master virtual machine that downloads and self-configures the grid with the subnet, nodes and policies assigned. As a result, the clients own their grid with a master node and a set of computing nodes assigned.

Customizing the master virtual machines implies the existence of a configuration server. Once the virtual machine is running, it connects to the server and downloads a specific configuration. The configuration includes:

- Node assignment: number of nodes and subnet number for VLAN isolation.
- Job policy algorithms for the nodes. For example a round robin.
- Time limit or other constraints such as licensing.

5 Virtual Grid Advantages

Virtualizing the grid computing system introduces several advantages that must be considered:

- Uniformity. Every client sees their own grid. Different clients use independent grids, so collateral effects can be minimized or even null.
- Customizable grids. It is possible to offer customized grids by deploying specific images for the virtual master machines. Different configurations may be offered. For example:

- Special hardware configurations: number or computing capacity of nodes.
- Priority jobs, requiring priority access, more resources, ...

- Efficient assignment of local grid infrastructure, users running jobs from different sites, avoiding bottlenecks caused by bandwidth or latency issues.
- Flexible accounting. It is possible to extend accounting plans from the cloud computing system to offer different quality of service. It is possible to pay per use, depending on time or resources used.
- Improved security. Independent grids isolate user jobs running in the grid. Subnet assignment can help achieve a more secure environment.

The key advantage is that every client uses its own grid, so client preferences can be easily customized adding new services to the configuration server. Security is implemented using certificates and managing private and public keys. As a result, only authorized clients may access to the cloud nodes.

6 Conclusions

Traditional high performance computing systems based on grid computing can be improved by using cloud computing infrastructures. They can be built by virtualizing the master node, a key component of the grid.

A new infrastructure for HPC named *virtual grid* has been developed using this paradigm, by virtualizing the master node responsible for running the user jobs. This virtualizing process has numerous advantages such as security, flexibility and efficient use of resources.

Moreover, further techniques for grid computing optimization can be applied to enhance the system.

The system is now under development and is expected to be completed before the end of 2011. Tests made have shown the solution is valid, although some problems have been identified:

- Virtual machines in the cloud run slow, so hardware infrastructure must be improved. This problem affects to the master, so the cloud computing technology must be revised. UEC private cloud technology can be low-level configured in order to obtain better results.

- The number of virtual machines is also reduced, as only two virtual machines can be started per cloud node (one core processor). Every virtual machine can manage one grid system.

Current infrastructure for WAN access is also being testing as they involve using a public infrastructure (RedIris), requiring administrative permissions. Therefore connections are being tested using the WAN interfaces installed on the lab routers. After the necessary tests we will connect the two universities to the cloud.

Future work will involve improving the configuration server to efficiently use the grid computing nodes. Although current services implementation includes basic services for node assignment, we have considered additional services exploiting the cloud public interface available. Other cloud technologies different from UEC must also be tested, in order to obtain better results in terms of computing efficiency and customization.

In order to present a unique interface for accessing to all services, a web platform is also planned. Using this web, clients will be able to start virtual machine instances (new master machines) and perform the needed configuration, avoiding the cloud command shell inconveniences.

Acknowledgements

This project has been partially supported by grant 10MLA-IN-S08EI-1 from the Pontificia University of Salamanca. We also want to acknowledge José María Sánchez-Aguilera's effort, who worked hard to develop the hardware infrastructure.

References

[1] European Grid Infrastructure, `http://www.egi.eu`
[2] Open MPI: Open Source High Performance Computing, `http://www.open-mpi.org`
[3] Das, S.K., Harvey, D.J., Biswas, R.: Parallel processing of adaptive meshes with load balancing. IEEE Transactions on Parallel and Distributed Systems 12(12), 1269–1280 (2001)
[4] Das, S.K., Harvey, D.J., Biswas, R.: Latency hiding in dynamic partitioning and load balancing of grid computing applications. In: Proceedings of First IEEE/ACM International Symposium on Cluster Computing and the Grid 2001, pp. 347–354 (2001)
[5] Pant, A., Jafri, H.: Communicating efficiently on cluster based grids with MPICH-VMI. In: 2004 IEEE International Conference on Cluster Computing, September 20-23, pp. 23–33 (2004)
[6] Lechuga, T.A., de Toledo, M.B.F., Capretz, M.A.M.: An infrastructure for executing WS-BPEL workflows in a Cluster of Clusters. In: 2010 IEEE Symposium on Computers and Communications (ISCC), June 22-25, pp. 681–686 (2010)
[7] Gallet, M., Marchal, L., Vivien, F.: Allocating Series of Workflows on Computing Grids. In: 14th IEEE International Conference on Parallel and Distributed Systems, ICPADS 2008, December 8-10, pp. 48–55 (2008)
[8] Sanguandikul, N., Nupairoj, N.: Implicit information load sharing policy for grid computing environment. In: The 8th International Conference on Advanced Communication Technology, ICACT 2006, February 20-22, vol. 3, p. 5, p. 2054 (2006)

[9] Das, S.K., Harvey, D.J., Biswas, R.: A latency-tolerant partitioner for distributed computing on the information power grid. In: Proceedings 15th International Parallel and Distributed Processing Symposium, April 2001, p. 6 (2001)
[10] de Mello, R.F., Senger, L.J., Yang, L.T.: A routing load balancing policy for grid computing environments. In: 20th International Conference on Advanced Information Networking and Applications, AINA 2006, April 18-20, vol. 1, p. 6 (2006)
[11] Ubuntu enterprise Cloud, http://www.ubuntu.com/cloud
[12] Eucalyptus system: Ubuntu Enterprise Cloud, http://www.eucalyptus.com/products/ubuntu_enterprise_cloud
[13] Matsuda, M., Kudoh, T., Ishikawa, Y.: Evaluation of MPI implementations on grid-connected clusters using an emulated WAN environment. In: Proceedings of 3rd IEEE/ACM International Symposium on Cluster Computing and the Grid, CCGrid 2003, May 12-15, pp. 10–17 (2003)
[14] Frizziero, E., Har'El, Z., Lelli, F., Mandler, B., Maron, G., Molini, P., Pinter, S.S.: Fast Information Transport for an Instrument Enabled Grid. In: IEEE International Conference on e-Science and Grid Computing, December 10-13, pp. 253–260 (2007)
[15] Derbal, Y.: Grid Architecture for High Performance Computing. In: Canadian Conference on Electrical and Computer Engineering, CCECE 2007, April 22-26, pp. 514–517 (2007)
[16] Innovative Computing Laboratory, GridSolve: A system for Grid-enabling general purpose scientific computing environments. Disponible en: http://icl.cs.utk.edu/netsolve/
[17] Dai, Z., Wu, L., Xiao, H., Wu, H., Chi, X.: A Lightweight Grid Middleware Based on OPENSSH - SCE. In: Sixth International Conference on Grid and Cooperative Computing, GCC 2007, August 16-18, pp. 387–394 (2007)

Smart Home Automation Using Controller Area Network

Manuel Ortiz, Manuel Diaz, Francisco Bellido,
Edmundo Saez, and Francisco Quiles

Abstract. In several works, researches have recently proposed the use of a CAN bus (Controller Area Network) as a control network for smart home automation. The use of CAN for the lower network layers has advantages in the field of automation compared with networks based on RS485 due to its multimaster architecture. While compared with other common bus networks such as Ethernet or networks based on token passing, CAN has real time features and ease of implementation and programming of the nodes and therefore, a lower cost. This paper presents the study and evaluation of automation and remote control of an alarm system and HVAC system (Heating, Ventilation, and Air Conditioning) of a home, using CAN as a backbone network. The first version of TUCAN (Tuple Space and CAN) has been also used for the software of the nodes. TUCAN is a data-centric lightweight middleware that provides a CAN bus abstraction, and is based on the concept of Tuple Channel Space. A prototype with four nodes has been used to study the feasibility of the CAN bus as a single network and to evaluate TUCAN middleware.

Keywords: Home automation, home network, distributed control system, Controller Area Network, Tuple Space.

1 Introduction

Today, more and more intelligent buildings are being built for the purpose of increasing the safety and comfort of its occupants. Automation and building management is not only limited to large buildings; more and more homes are being automated by interconnecting the networked electronic equipment. The use of a monitoring network at home allows to monitor and control devices not only within the home, but also remotely using a mobile phone or an internet connection. There

Manuel Ortiz · Francisco Bellido · Edmundo Saez · Francisco Quiles
University of Córdoba, Spain
e-mail: {el1orlom,el1beouf,el1sapee,el1qulaf}@uco.es

Manuel Diaz
University of Málaga, Spain
e-mail: mdr@uma.es

A. Abraham et al. (Eds.): International Symposium on DCAI, AISC 91, pp. 167–174.
springerlink.com

are nowadays many standard home control networks used in home automation as X11, EHS, Batibus, Longworks, Profibus, CAN, etc. [1]. The aim behind this paper is to show that CAN is one of the best options as a single backbone network for a smart home.

The CAN bus is a leader in the automotive industry and machinery, as well as regarding medical equipment. However, in other application areas, especially in the field of automation of small buildings, is not so consolidated, except in the automation of systems such as HVAC (Heating, Ventilation, and Air Conditioning), elevators, etc. [2]. It has been proposed as a backbone network for security systems such as fire detection as in [3] and to direct load control applications in residential areas [4].

The paper is organized as follows. Chapter 2 will show a brief introduction to the CAN bus and discus the advantages over other control networks. In Chapter 3, a description of the TUCAN middleware for application development on the CAN bus is carried out. In Chapter 4, the architecture of the overall automated system will be shown. Chapter 5 will describe the complete application, i.e., the local application on a node using TUCAN middleware. Finally, Chapter 6 will show the conclusions.

2 CAN Overview

The CAN protocol is an internationally standardized protocol suitable for automotive and industrial applications (ISO 11898 for high speed applications [5] and ISO 11519-2 for low speed applications [6]). In addition to the ISO/OSI, the CAN 2.0A and CAN 2.0B specifications are available to CAN controller manufacturers [7]. CAN is a serial bus system which is intended for a loosely coupled system with low information interchange.

A typical CAN network consists of several modules connected to a bus. The information is broadcasted in through the bus. Therefore, every node receives the information and is able to take it. The information interchange is achieved by the use of messages. Messages are named by their identifiers. The identifier designates the information contained in such message, neither the origin nor destination node are included. In addition, the identifier indicates the message priority.

The main features of the CAN bus are:

- Multi-hierarchy, which facilitates the creation of redundant systems.
- High reliability thanks to its sophisticated mechanism of error detection and automatic retransmission of erroneous frames. This ensures data integrity.
- Communication based on the priority of the message, very useful in industrial and critical environments that must meet strict time requirements.
- Broadcast communication. All nodes receive the same information.

The CAN bus is a good alternative as a data link layer over other networks in building management and control. CAN has several advantages over networks based on

RS485, as multimaster architecture and reliability. Regarding the implementation, RS485 networks are only adapted to the master-slave paradigm. CAN allows other programming models, such as producer-consumer or publisher-subscriber.

Likewise, the use of a CAN data link layer offers advantages over other network systems with common-bus architectures that use Ethernet or token passing bus (e.g., ControlNet). These include priority access and a high level of security, which makes the CAN bus communication model predictable and its temporal behavior able to be analyzed using RMA [8]. On the other hand, the implementation of the network interface level sensor or actuator is much easier when using CAN to Ethernet. The CAN network interface can be easily implemented in any low-cost microcontroller.

The disadvantages of CAN compared to these common bus networks are low speed (1 Mbit Max) and the fact that it is not suitable for transmission of large messages. However, in many applications of control, information exchange is small, as in the application shown in this paper. In [9], a detailed discussion of these architectures can be found.

3 TUCAN Middleware Overview

CAN only has defined the physical and link layers. A middleware layer that connects the application with the lower layers is needed to facilitate the development of CAN applications.

TUCAN (TUple space and CAN) is a lightweight data-centric middleware inspired by a Tuple Channel Coordination model space (TCM) [10]. The abstract model of a CAN node based on Tuple Space is shown in Fig. 1. The tasks exchange information by the insertion and extraction of tuples, with a format t=("tag", values), where *tag* is the symbolic name of the tuple and *values* is a list of typed data.

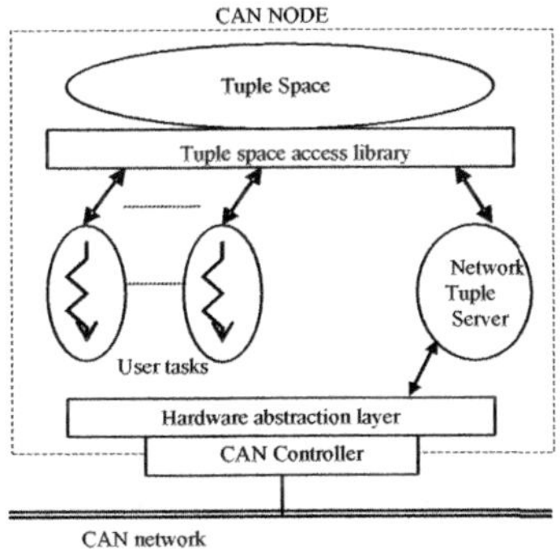

Fig. 1 CAN node Tuple Space.

TUCAN provides an abstraction of a CAN bus to the application layer based on TCM for embedded systems with low processing power and memory, which is a fundamental requirement in consumer electronics. TCM is a model that has adapted to sensor networks with low processing power and memory [11]. TUCAN middleware allows users to create software filters when tuple channels are defined.

A tuple is associated with a CAN message, so that each tuple contains a message and information related to it, so that a priority is assigned to each tuple. Channels are containers of tuples. The channels have attributes that define the channel characteristics (e.g., filter type, timestamp, etc.). Another feature is that the channels are directional. From the point of view of the application, if the application gets information on the channel, then the channel is transmitted and the stored information will be sent by the CAN bus, while if the application extracts information, the channel is received. Messages circulating on the bus are deposited in the reception channels according to the type of filter.

The tasks create the reception channels for the deposit of the messages that are interesting for them. In the case of transmission channels, tuples are deposited and the extracted message is sent to the CAN bus.

We are working on two versions of TUCAN middleware. The simplest one of these is designed to work with an offline identifier assignment and as the channels are created a priori, it is not necessary that the middleware performs the coordination between nodes. This is the version that was used in the application shown in this work and it introduces an overhead similar to the FIFO drivers that are usually used.

4 System Architecture

The objective of the application is the control of a central air conditioning system, an intrusion and fire alarm system of a smart home. In addition to the above discussion of the advantages of using a CAN bus shown in Section 2, in the case of the application at hand, a single CAN network can be used for both systems, minimizing the cost of wiring, due to the fact that bus traffic is small. Messages relating to the alarm system have a higher priority than messages from the HVAC, and this is easily achieved by assigning higher priority identifiers to the alarm system. The nodes related to air conditioning work in coordination and the same happens with alarms, due to the fact that the information is not shared between both systems on our prototype. The modem node is a common element to both systems and performs remote communication with the user. Fig. 2 shows schematically the system architecture.

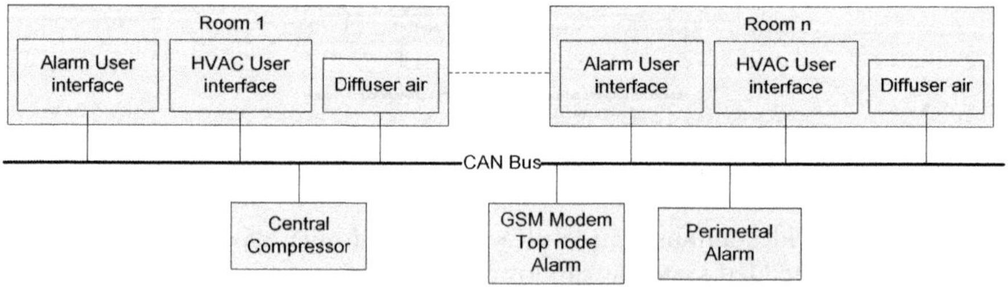

Fig. 2 System Architecture.

In every room, there are three nodes, one for the opening control of the diffuser air, a user node that controls the parameters of the HVAC, and another user node for activation and deactivation of the fire and intruder alarms. One of the nodes performs remote communication with the user. This node is also the top node in the hierarchy of network alarms. The multimaster structure of the CAN bus ensures the scalability of the system regardless of the number of rooms to be checked. New nodes can be connected without modifying the network.

The HVAC system is a fully distributed implementation; hence, the change in the number of rooms does not require a modification in the application of the nodes. In the case of the alarm system scalability, it is more complicated. The alarm system has a distributed and hierarchical structure. The top node needs to know the number of nodes and the type of sensors that is connected to each node.

In this application, the HVAC system is centralized, there is only one compressor control unit. Each room has two physically distributed modules: diffuser air control unit and secondly, user interface and temperature sensor unit.

In this system, the temperature programmed by the user and the room temperature must be known by the diffuser air control unit to regulate the airflow. It must also be known by the compressor control unit in order to power it off, if the programmed temperature has been achieved in every room. The compressor control unit must also know the state of the diffuser air and close it when the compressor machine is in idle state.

5 Overall Application Overview

This section will describe the overall design of the application. The application has been adapted to the "time-driven" programming paradigm which has the advantage of being able to easily design deterministic applications. Each node maintains a regularly updated image of the variables that need to be processed. The network variable interchange follows the producer-consumer paradigm.

Fig. 3 shows the layered structure of the application running on a node. On the node, a cyclic executive is running, as described in [12]. It is a cyclic executive for periodic tasks, which has the restriction that the sum of execution times of all tasks must be less than the duration of a tick. Although it is a strong constraint, in the case of low processing applications, such as those running on these nodes, the tick duration can be easily adjusted.

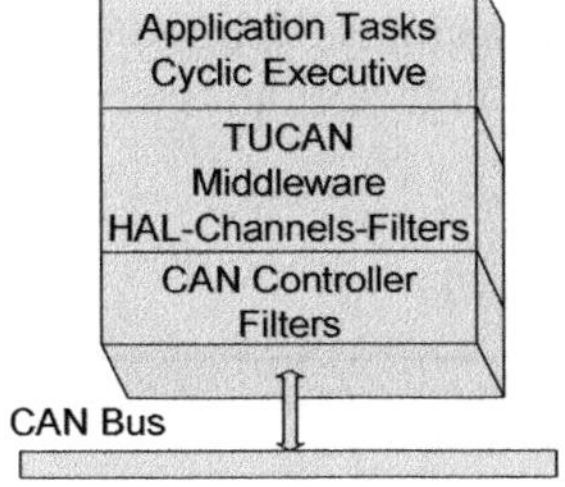

Fig. 3 CAN node layers.

Table 1 shows the CAN 2.0A IDs message and information exchanged on the network. The messages consist of an identifier and a byte of data. The ID refers to the node and to the data byte which contains the information that the node sends to the network. An off-line assignment of identifiers was used, so that the overhead caused by TUCAN is small.

Table 1 Information interchanges in the network.

MsgID (Hex)	**Send**	**Receive** Compre-ssor	Diffuser air	User Air	User Alarm	Modem Air	Modem Alarm
510	Compressor		X	X		X	
52x	Diffuser air	X					
53x	Air-User	X	X				
43x	Alarm-User						X
500	Air-Modem	X					
400	Alarm-Modem				X		

As an example, the following TUCAN primitives used in the compressor node are shown in Fig. 4. The compressor node has three user tasks, each of which uses a channel: *state diffusers*, *air user interfaces*, and *air-modem*. The example shows a part of the state diffusers task. The primitive *createChannel(arg. list)* creates the channel and the primitive *takeTuple(arg. list)* takes tuples deposited by middleware TUCAN according to the filter defined in *createChannel*. *getDatamessage(myTuple)* is a user function.

```
//main task
...
// channel_attibut: FILTER_BASICCAN, OVERWRITING_TUPLE,
//NO_TIME_STAMP, NUMBER_CAN_BUS, etc.
createChannel(CAN20A,channel_attribut);//return channel_ID
...
// end main task
// state diffusers task
...
for i=0 to MAX_ROOM //Diffuser ID=520
        takeTuple(DIFFUSER+i, channel_ID, myTuple);
        if getDataMessage(myTuple) ==OPEN
                then standbyCompressor=FALSE;
end for
...
// end state diffusers task
```

Fig. 4 Compressor Node Application.

To evaluate the use of CAN in the development of smart home applications and the TUCAN middleware, a system with four nodes has been built, as is shown in Fig. 5. A node performs the compressor control, another one is the top node of alarm and modem control, another node that controls the diffuser air, in a single node has been joined to the air conditioning and alarm system user interface. Four prototype boards with an LPC2378 microcontroller running at 12 MHz have been used to implement the nodes. Embedded Development Tools from Keil Company have been used [13].

The TUCAN overhead is small. For example, in compressor node (with three channels) to insert a tuple in the channel from CAN Bus, the worst execution time is 4.6 μs, while the worst execution time in a FIFO driver is 1.3 μs. As regards a message which needs to be filtered and discarded in the node, the worst execution time of TUCAN is 2.2 μs, whereas in a FIFO driver, 1.3 μs are consumed to introduce the message in the buffer and 1.2 μs to remove and discard it. In a FIFO driver, a user task is needed to process the buffer messages.

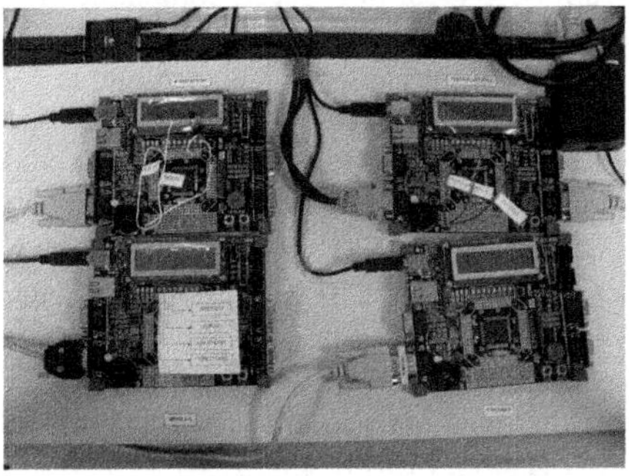

Fig. 5 System prototype.

6 Conclusions

In this paper, we have discussed the advantages of the use of a CAN Network in a smart home, among which, the real-time characteristics of this protocol and its high reliability are included. It has been shown that it is possible to use a single network for different systems that can prioritize the access to the network by simply making an appropriate allocation of the message ID.

We have used both the producer-consumer programming paradigm and "time-driven" paradigm, and this has allowed the development of easily scalable deterministic applications.

The use of TUCAN middleware simplifies the programming of the nodes achieving a real bus abstraction without loss of the features of the CAN bus. TUCAN makes a distribution of messages based on the defined channels and allows the programmer to define filters dynamically without restarting the node. Finally, TUCAN has proven to cause a minimal overhead.

References

1. Kastner, W., Neugschwandtner, G., Soucek, S., Newmann, H.M.: Communication Systems for Building Automation and Control. Proceedings of the IEEE 93(6), 1178–1203 (2005)
2. CAN in Automation (CiA), `http://www.can-cia.org`
3. Lee, K.C., Lee, H.-H.: Network-based fire-detection system via controller area network for smart home automation. IEEE Transactions on Consumer Electronics 50(4), 1093–1100 (2004)
4. Molina-Garcia, A., Torres, R., Munoz, J.L., Encinas, N.: Application of Controller Area Networks to Direct Load Control in Residential Areas. In: 2007 IEEE Lausanne Power Tech., July 1-5, pp. 1970–1974 (2007)
5. International Standard ISO 11519-2, Road vehicles - Low speed serial data communication - Part 2: Low speed controller area network (CAN). ISO Reference number ISO 11519-2 (June 1994)
6. International Standard ISO 11898, Road vehicles- Interchange of digital information - Controller area network (CAN) for high speed communication. ISO Reference number ISO 11898 (November 1993)
7. Robert Bosch GmbH, CAN specification 2.0, Part A and B (September 1991)
8. Davis, R.I., Burns, A., Bril, R.J., Lukkien, J.J.: Controller Area Network (CAN) schedulability analysis: Refuted, revisited and revised. Real-Time Systems 35(3) (April 2007)
9. Lian, F.-L., Moyne, J.R., Tilbury, D.M.: Performance evaluation of control networks: Ethernet, ControlNet and DeviceNet. IEEE Control Systems Magazine 21(1), 66–83 (2001)
10. Diaz, M., Rubio, B., Troya, J.M.: A tuple channel-based coordination model for parallel and distributed programming. Journal of Parallel and Distributed Computing 67(10), 1092–1107 (2007), ISSN 0743-7315
11. Diaz, M., Rubio, B., Troya, J.M.: TCMote: a tuple channel coordination model for wireless sensor networks. In: Proceedings of International Conference on Pervasive Services, ICPS 2005, July 11-14, pp. 437–440 (2005)
12. Dorin, M.: Building "instant-up" real-time operating systems. Embedded Europe (May 2008)
13. Keil Elektronik GmbH, `http://www.keil.com`

Griffon – GPU Programming APIs for Scientific and General Purpose Computing

Pisit Makpaisit and Worawan Marurngsith

Abstract. Applications can accelerates up to hundreds of times faster by offloading some computation from CPU to execute at graphical processing units (GPUs). This technique is so called the general-purpose computation on graphic processing units (GPGPUs). Recent research on accelerating various applications by GPGPUs using a programming model from NVIDIA, called Compute Unified Device Architecture (CUDA), have shown significant improvement on performance results. However, writing an efficient CUDA program requires in-depth understanding of GPU architecture in order to develop a suitable data-parallel strategy, and to express it in a low-level style of code. Thus, CUDA programming is still considered complex and error-prone. This paper proposes a new set of application program interfaces (APIs), called Griffon, and its compiler framework for automatic translation of C programs to CUDA-based programs. The Griffon APIs allow programmers to exploit the performance of multicore machines using OpenMP and offloads computations to GPUs using Griffon directives. The Griffon compiler framework uses a new graph algorithm for efficiently exploiting data locality. Experimental results on a 16-core NVIDIA Geforce 8400M GS using four workloads show that Griffon-based programs can accelerate up to 89 times faster than their sequential implementation.

Keywords: GPU, Accelerating Computing, Automatic translation, CUDA, Parallel Programming.

1 Introduction

Exploiting the powerful and inexpensive graphic processing units (GPUs) for application acceleration has been of interest by researchers. A survey shows that more than 130 literatures from the year 1978 to 2007 present the benefit of the general-purpose computation on graphic processing units (GPGPUs) in broad range of applications[1]. Exploiting the high throughput of GPUs in the computation of artificial neural networks is one of the first early work [1], and is still a

Pisit Makpaisit · Worawan Marurngsith
Department of Computer Science, Faculty of Science and Technology Thammasat University
99 Phaholyothin Road, Pathum Thani, 12121, Thailand
e-mail: haoremixman@gmail.com, wdc@cs.tu.ac.th

A. Abraham et al. (Eds.): International Symposium on DCAI, AISC 91, pp. 175–182.
springerlink.com

challenge to achieve higher level of parallelism in the program [2, 3]. The implementation of GPU-based neural networks has achieved the performance up to 24 times faster than the single-thread CPU implementation [3]. Other applications based on GPGPUs programming have achieved dramatic acceleration for up to 328 times faster than the sequential implementation using a single CPU [4].

The development of programming languages and tools for GPUs plays a key role in the successful of applications accelerating on GPUs. The development can be group in three phases chronologically *i.e.,* using existing 3D-rendering APIs such as OpenGL, using new languages or APIs from non GPU vendors like Brook and Microsoft's Accelerator, and using the extension of languages from GPU venders, *e.g.* NVIDIA's CUDA and AMD's CAL. Among various programming models from GPU venders, NVIDIA's CUDA has been highlighted by its powerful set of library functions to express parallelism on irregular data structures. Che *et al.* showed that GPUs has a potential to support various types of parallel applications with diverse performance characteristics[1][5]. Che *et.al,* have pointed out six performance tuning techniques for CUDA. Most of them related to the requirement for programmers to understand complex concepts of GPGPU programming with CUDA such as the basic properties of the GPUs architecture.

Several languages and tools have been proposed to tackle the complexity and error-prone of implementing CUDA-based GPGPUs applications manually. An example is a new data-parallel programming language for GPU programming called Scout proposed by McCormick *et al.* [6]. In comparison to CUDA, the Scout language simplifies the ways for programmers to express task partitioning and data mapping. However, existing codes have to be re-written to take advantage of the language. A Lee *et al.* proposed the source-to-source compiler framework to transform any C programs augmented by OpenMP shared-memory directives to CUDA codes [4]. Han and Abdelrahman proposed a directive-based language called *hi*CUDA and the compiler framework to transform any *hi*CUDA C-programs to CUDA codes [7]. Both solutions allow programmers to explicitly specify task and data parallelism parts in the C source file and to offload the computation works to GPUs with less effort. However as the OpenMP directives are transformed to CUDA or not supported, programmers cannot combined the benefits of multicore programming using OpenMP with the many core programming using CUDA.

The main objective of this research is to exploit incremental parallelization by using a new directive-based language similar to that of [4, 7], but allows the integration of OpenMP directives with CUDA codes. We propose a new set of APIs for GPGPU programming called *Griffon.* The Griffon APIs comprise a set of compiler directives and a compiler that translates C programs augmented with Griffon directives to the corresponding CUDA programs. Griffon compiler allows programmers to specify parallel regions for CPUs using OpenMP directives and specify the GPUs computation within any OpenMP threads. We measured the performance gain of four Griffon-based programs against their single thread

---

[1] The CUDA applications' speedup ranges from 2.5X to 72X over the single-thread CPU implementations.

implementation for CPU. Two experiments have been performed. The results of the first experiment show that using Griffon we can accelerate three workloads from 22 to 89 times faster than the sequential implementations. Moreover, the results of the second experiment show that using Griffon optimisation on a workload allowed a workload with large shared data to accelerate for up to 3.27 times.

The rest of the paper is organised as follow. The next section introduces the concepts of CUDA programming model. Section 3 presents the overview of Griffon. In Section 4, optimisation techniques used in Griffon are presented. Section 5 shows our experimental results. Section 6 gives the conclusion.

2 The CUDA Programming Model

The Compute Unified Device Architecture (CUDA) is an extension to C programming language for programming on NVIDIA's GPUs [8]. The design of CUDA aims for *scalability*, *i.e.* allowing applications to scale their parallelism up on the increasing number of processor cores. Programmers start with partitioning the problem into several *blocks* or *grid, i.e.* coarse-grain tasks. Each block has *threads,* a finite numbers of independent subtasks. Each grid of threads can be scheduled on any of the available processor cores, in any order. Thus, a multi-threaded CUDA program can be scaling to all CUDA-enabled NVIDIA's GPUs.

```
1:  __global__ void MatAdd(float A[N][N], float B[N][N],
2:                         float C[N][N])
3:  {
4:      int i = threadIdx.x; int j = threadIdx.y;
5:      C[i][j] = A[i][j] + B[i][j];
6:  }
7:
8:  int main()
9:  { ...
10:   dim3 numBlocks = (3, 2);
11:   dim3 threadsPerBlock(1, 8);
12:   MatAdd<<<numBlocks, threadsPerBlock>>>(A, B, C);
13: }
```

Kernel definition (lines 1–7)

Kernel Invocation (line 12)

Fig. 1 An example of assigning blocks to GPUs (its conceptual view shown in Fig.2)

In practice, a workload assigned to all threads is defined as a C function called *kernel* as shown in Fig.1. Every thread executes the kernel once, in a Single Instruction Multiple Data (SIMD) fashion. Programmers have to specify a *grid* of threads to execute the kernel. One grid contains multiple thread blocks organised in two dimensions (*x*, *y*) (see Fig.1 line 10 on how to specify a grid containing six (*3* x *2*) thread blocks). Programmers also have to specify the number of threads in a block that are organised up to three dimensions (*x*, *y*, *z*) (*e.g.* in Fig.1 line 11, each block is specified to have eight threads, (*1* x *8*)). Each thread has a unique *thread ID* that can be accessed via the `threadIdx` variable. The thread ID is normally used for calculating the boundary of data responsible by a thread (as shown in Fig.1 line 4). Once the grid and threads per block have been specified, the kernel function can be invoked by calling the function name and using <<<...>>> brackets to assign the execution space on GPUs (as shown in Fig.1. line 12).

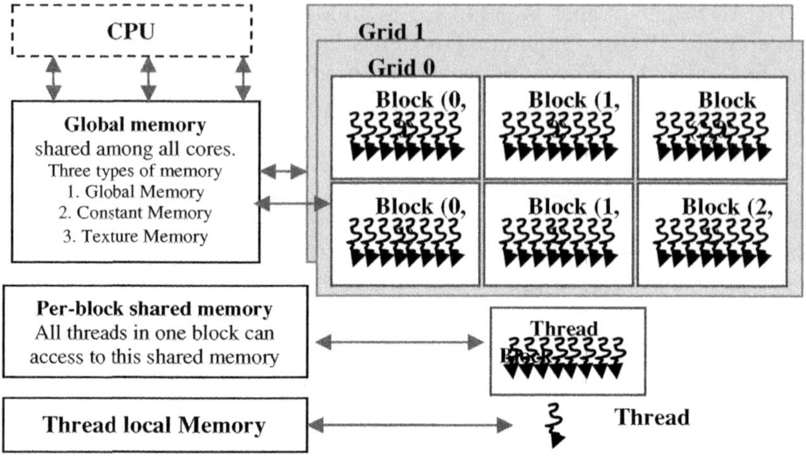

Fig. 2 Memory Hierarchy in CUDA programming model

Fig. 2 depicts the conceptual view and the mapping of memory hierarchy of the CUDA code fragment presented in Fig.1. During execution, CUDA threads may access data from three memory spaces. Each thread keeps its private variables in the *thread local memory*. All threads in the same thread block may communicate by read/write to shared variables stored at the *per-block shared memory*. All grids of threads have access to the *global memory*. When accessing to the shared or global memory, programmers must use suitable synchronisation functions available in CUDA to control multiple accesses to a shared data.

3 Griffon APIs and Compilation Framework

Central to the Griffon is the compilation process that translates Griffon C codes to CUDA-based codes. Figure 3 depicts the overview of Griffon compilation framework and its compilation process. The framework allows programmers to generate a CUDA-based code by specifying several kernels from an existing C program using Griffon directives. Similar to OpenMP, the syntax of Griffon directives comprises construct and clauses (as shown in Fig.4.). There are four groups of directives to facilitate the parallelisation of codes for GPUs. First, the parallel region directive specifies a kernel (as mentioned in Section 2). Programmers use the **`#pragma gfn parallel for`**, immediately following by a 'for' loop which can be parallelised (*i.e.* has no loop carried dependency). Fig. 4 shows an example code fragment of how to parallelise the for loop iterations using the **`parallel for`** directive and its generated code.

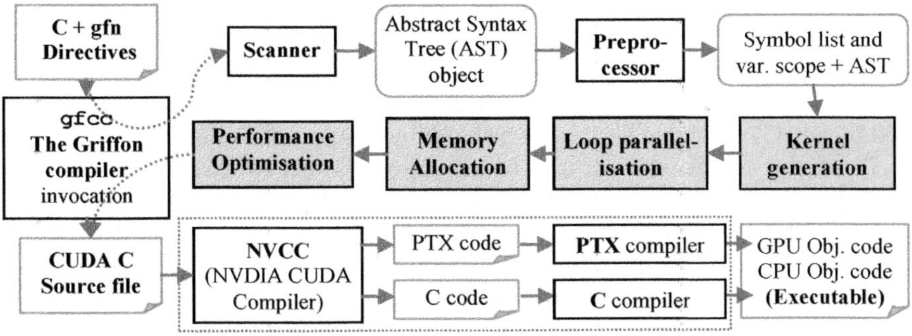

Fig. 3 Overview of the Griffon compiler framework and NVIDIA's compilation process

Second, the GPUs/CPU overlapped computing is defined as a directive. In CUDA, programmers can assign some works to CPU while GPUs are executing their kernels. In Griffon, programmers use the **overlapcompute**()directive to specify the overlapped execution. Third, Griffon also provides a set of synchronization directives and clauses to control the mutual exclusion among threads. These include the **barrier** and **atomic** directives, and the **reduction** clause. Lastly, programmers can control the collaboration of several kernels by using the control flow clause **waitfor**. The clause instructs the compiler to create corresponding dependency graph of tasks of each kernel to identify the group of concurrent tasks.

```
#pragma gfn parallel for
for(i=0;i<N;i+=5)
{
  D[i] = B[i] + A[i]
}
```

```
__global__ void __kernel_0 (int __N)
{
  int __tid = blockIdx.x *
          blockDim.x + threadIdx.x;
  int i = __tid * 5;
  if(__tid < __N) {
     D[i] = B[i] + A[i];
}
```

Fig. 4 Example of Parallel vector addition using Griffon directive and its generated code

4 Optimisation in Griffon

The last step of Griffon compilation process is optimisation. We have observed the performance of the generated codes and noticed four opportunities to maximise their performance. We applied these four techniques in the optimisation step.

First, the code is faster when allocating a kernel to as many threads as possible. This is because CUDA threads executed on GPUs are extremely light weights and cost almost zero overhead. Thus, the Griffon optimiser maximises the number of threads to increase the degree of concurrency.

Second, the number of data transfer is minimised by using data flow analysis. We found that the factor that hurts GPU computing performance most is data transfer operations between main memory on mainboard and global memory of GPUs. The Griffon optimiser alleviates this problem by analysing dependency on shared data. Operations for transferring data from main memory to global memory will be generated by Griffon only when the data will be read by GPUs after any CPUs have modified them. Likewise, data transfer operations from GPU's global memory to main memory will be generated only when the data have been modified by any GPUs and at least one CPU will use them in the future.

Third, the number of data transfer operations issued by a kernel is minimised by finding reusable data in global memory using dependence graph analysis. In this case data that can be reuse, *i.e.* have no read after write (RAW) dependency, do not need to be transferred to main memory. As shown in Fig. 5, the dependence graph shows that variables `A` and `C` in GPU's global memory can be reuse, thus do not need to be transferred. The value of variable `C` written by `K1` will be overwritten by the result from `K2`, thus variable `C` only need to be transferred once at the end of `K2`'s execution.

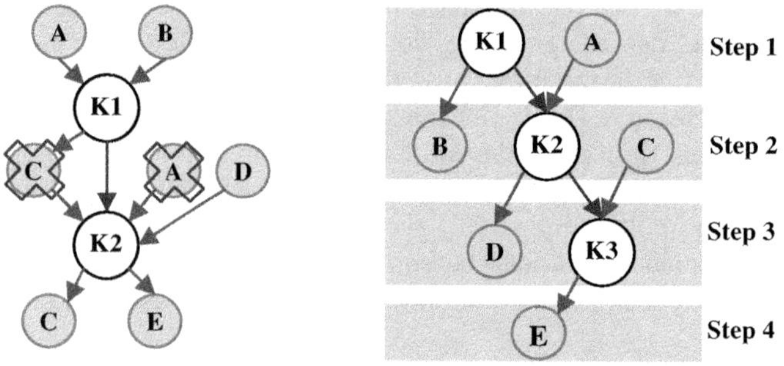

Fig. 5 (left) Dependency graph for data reuse analysis (right) latency hiding technique

Last, the Griffon optimiser hides data access latency by overlapping kernel computation with data transfers by using CUDA's asynchronous data transfer. An example is shown in Fig.5 (right). While the kernel `K1` is executing, variable `A` will be prefetched into GPU's global memory (Step 1). Once `K1` has finished, kernel `K2` will start execution. At the same time variable `C` will be prefetched while variable B is written back to main memory (Step 2), and so on.

5 Experimental Results

Four applications that require high precision of floating point results have been developed in two versions, a sequential version and a Griffon-based version. These are the numerical integration using Trapezoidal Rule (TR), the Sine of vector's elements (SV), the Pi calculation using Monte Carlo method (MC-Pi) and 1D Mean Filter (1D-MF). The sequential and Griffon-generated codes were compiled and executed on a 32-bit Windows Vista machine using Intel Core 2 Duo T9300 2.5 GHz processor, 4GB RAM and a 16-core NVIDIA Geforce 8400M GS 0.8 GHz. The execution results showed that all Griffon-based codes produced the correct results, *i.e.* having the same results as the sequential implementations.

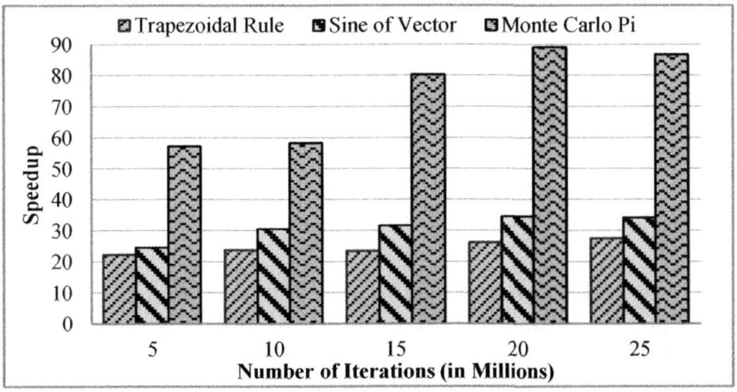

Fig. 6 Speedup of Griffon-based codes in comparison to their sequential implementation

Two experiments have been done. In the first experiment, we aimed to observe how much acceleration the Griffon-generated codes can achieve. The workloads which are computation intensive but contain no large array were selected (TR, SV and MC-Pi descending sort by the number of accesses to shared data). We measured the execution time of these workloads when using 5-25 millions of iterations, and calculated the parallel speedup (see Fig. 6). The results show that all Griffon-generated codes outperform the sequential ones for 22 – 89 times faster. The TR workloads appeared to be suffering from synchronisation to protect shared data. However, the MC-Pi workload gains a surprisingly higher speedup than we expect because of its computation-independent nature which fits well on GPUs.

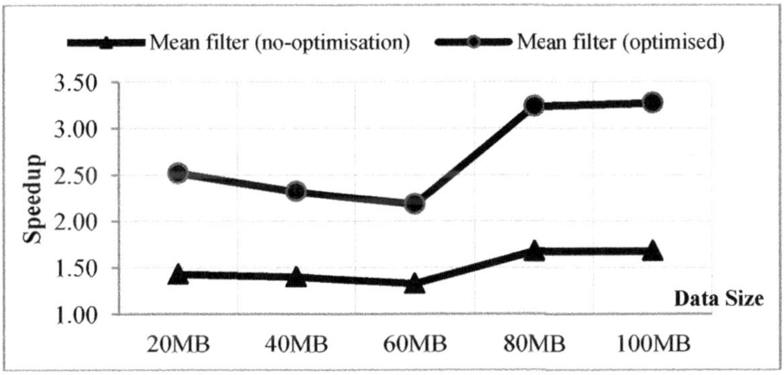

Fig. 7 Speedup of a Griffon-based code using optimisation and no optimisation

In the second experiment, we aimed to test the capability of Griffon optimisation by using 1D-MF, the workload which manipulates a large array. We generated two sets of Griffon-based codes of the workload: using optimisation and non-optimisation (**gfcc -O0**), each set comprises five versions having different working set size (20MB – 100MB). Fig.7 shows the speedup obtained from the workloads. Note that, the original 1D-MF is not easy to parallelise. Thus it requires some structural changes before using Griffon. Although the application did not accelerate aggressively on the GPUs, the results show that using optimisation the application can be accelerate for 2.18 – 3.27 times faster than the sequential ones.

6 Conclusions

We have implemented Griffon, a set of APIs and compilation framework, which allows programmers to develop GPGPU codes for NVIDIA's CUDA environment by using directives. We believe that using Griffon, programmers can rapidly generate GPGPU codes without having to understand CUDA architecture in detail. The results show that the Griffon-based codes could accelerate up to 89X, and using optimisation could gain more speedup from 1-1.5 times. However, current version of Griffon code generation still limited on working with shared data from one scope. Some applications may have to be restructured to fit this constraint. We observed that computation intensive applications with regular access pattern, *e.g.* image recognition, can be good target workloads for improving Griffon. Thus, in future work, we aim to overcome the current limitation and to add powerful features of CUDA to Griffon to achieve aggressive speedup on our target workloads.

References

1. Owens, J.D., et al.: A survey of general-purpose computation on graphics hardware. Computer Graphics Forum 26(1), 80–113 (2007)
2. Scanzio, S., et al.: Parallel implementation of artificial neural network training. In: 2010 IEEE International Conference on Acoustics Speech and Signal Processing, ICASSP (2010)
3. Strigl, D., Kofler, K., Podlipnig, S.: Performance and Scalability of GPU-Based Convolutional Neural Networks. In: 2010 18th Euromicro International Conference on Parallel, Distributed and Network-Based Processing (DPD) (2010)
4. Lee, S., Min, S.-J., Eigenmann, R.: OpenMP to GPGPU: a compiler framework for automatic translation and optimization. SIGPLAN Not. 44(4), 101–110 (2009)
5. Che, S., et al.: A performance study of general-purpose applications on graphics processors using CUDA. Journal of Parallel and Distributed Computing 68(10), 1370–1380 (2008)
6. McCormick, P., et al.: Scout: a data-parallel programming language for graphics processors. Parallel Computing 33(10-11), 648–662 (2007)
7. Han, T.D., Abdelrahman, T.S.: hiCUDA: a high-level directive-based language for GPU programming. In: Proceedings of 2nd Workshop on General Purpose Processing on Graphics Processing Units, pp. 52–61. ACM, Washington, D.C (2009)
8. NVIDIA, NVIDIA CUDA C Programming Guide Version 3.2 (2010)

Efficient Parallel Random Rearrange

David Miraut Andrés and Luis Pastor Pérez

Abstract. Classic shuffling algorithms have linear complexity, but they have the disadvantage of accessing memory with unpredictable patterns, which cause a large numbers of cache misses. In consequence, their execution times are not determined by computation complexity, but by the latency of the memory system. For parallel systems, this penalty gets worse, because of the overheads associated to atomic accesses to data that must be rearranged.

This paper gives an overview of the best known serial and parallel shuffling algorithms, and proposes a new one that minimizes the number of memory accesses and thus, the processors' power consumption. Comparisons among these algorithms and some results are presented for graphic architectures.

Keywords: Parallel Shuffling Algorithms, Graphic Processors, CUDA, GPGPU.

1 Introduction

As technology advances and storage systems grow in capacity, applications that deal with massive volumes of data become more common, even in the field of personal computing, where users expect an interactive response. Since processing power is limited, more and more applications process their data using random sampling to give the user an approximate (and immediate) answer that is gradually refined later on.

Sometimes, this random-sampling based processing involves that some data are accessed and/or processed more than once, without actually improving the final result. An easy way to avoid wasting computing capacity is to plan data access as if the data were disordered, or even reorganize the data according to random patterns to facilitate a better use of the memory hierarchy later on.

David Miraut Andrés · Luis Pastor Pérez
Escuela Técnica Superior de Ingenieria Informática, Rey Juan Carlos University,
e-mail: {david.miraut, luis.pastor}@urjc.es

A. Abraham et al. (Eds.): International Symposium on DCAI, AISC 91, pp. 183–190.
springerlink.com

In this framework, random data permutation has a fundamental role in distributed systems, in areas such as sharing data for load balancing, algorithms where it is interesting to perturb the input so that the worst-case scenarios occur with minuscule probability[14], encryption and security, computer games[2], randomized algorithms[4], etc.

The first publications that explore permutation algorithms date back to the 60s, when Richard Durstenfeld introduced the 235 algorithm designed for computer use. Donald E. Knuth pointed out [10] that a very similar algorithm was proposed many years before by Fisher and Yate[6]. Since then, several variants have been thoroughly studied from the mathematical and implementation points of view, as the *faro shuffle* because of its connection with tricks in card games.

Most of the algorithms proposed so far focus on reducing the computational cost, while ensuring that the $n!$ possible permutations are equally likely to be chosen. Often, researchers do not take into account the cost associated with the process of data movement, which is usually greater than the time invested in the election of the new data positions. As already mentioned, technological changes have brought larger, faster and cheaper storage systems, as well as more powerful processors. Both aspects (memory size and computational power) grow, but not at the same rate. In consequence, the *memory wall* is increasing. To reduce this gap between memory and execution speed and to create the illusion of a fast and large memory space, modern computers implement memory hierarchies, which are based in the principle of locality. Much of the power consumption is due to memory transactions [9], so multilevel memory hierarchies are conceived for reducing it.

Unfortunately, there is little locality in random permutation algorithms (there are not any memory positions more likely to be used at a specific time, and they will not be reused soon). In consequence, the ratio of cache misses is high, and execution times depend on the memory system latency rather than on the methods' computational complexity, even for problem sizes moderately high. Savings in the number of accesses to the data will therefore improve execution times and reduce energy consumption. Memory system designs today tend to mitigate the effects of latency penalty by increasing memory bandwidth (wider buses and cache block sizes), which favours spatial locality. Once again, shuffle algorithms can not take advantage of this feature.

Even though it is always desirable, it is not always necessary for an application to obtain all permutations with equal probability. We distinguish *random rearrange* from shuffle, as the mapping that modifies the order of data in an apparently random way, without necessarily having to be unbiased.

The remainder of this article is organized as follows: Section 2 describes the most classic serial shuffle algorithms. Section 3 shows how they have been adapted to parallel architectures and how they affect memory accesses. Our new and simple parallel *random rearrange* algorithm is described in section 4. Section 5 introduces some numerical results and highlights the performance and scalability of this new algorithm. Last section presents the conclusions that can be reached from this work.

2 Classic Shuffle Algorithms

The simplest or *naïve implementation* of shuffling requires to generate, for each element of the data structure, a pseudo-random real number r between 0 and 1 which is multiplied by the total number of elements N of our data structure. The resulting number j determines the element's position. Collisions often occur, because the new assigned position could be already occupied by a previously shuffled element. These collisions increase the number of memory accesses, and become more common at the end, when there are fewer free places for reallocating elements.

In order to avoid such collisions, another possibility –called *sort-based shuffle*– is to assign each element a random number (within a large range) and then *sort* them according to the assigned numbers. A random permutation is produced by moving each element to its corresponding sorted location. This implementation is easy to program if fast sorting routines are available in the target architecture, and its computational complexity can be as low as $O(n \log n)$. However, this strategy requires a large amount of memory accesses and it can be much more costly than solving the previously mentioned collisions.

The classic Fisher and Yates' algorithm[6] and its variants [5] [13] [7] solve collisions by swapping elements. A random value is assigned to each element within the range of elements that lie ahead. The element indicated by the random number and the current one are exchanged. Therefore, the number of data movements in this algorithm is equal to $2N$, where N is the total number of elements to shuffle.

The next section describes how these algorithms can be extended to be used in parallel architectures.

3 Brief Overview of Parallel Algorithms

Parallel generalizations of the algorithms mentioned above, such as [3] and [1], are common for shared memory multiprocessors. In them, each processor usually prevents other processors from accessing the data elements it is currently processing. Remarkable speed-ups are achieved in comparison to serial algorithms, in spite of the fact that these communication strategies have an additional cost.

Dart throwing algorithms [3] are based on an idea similar to that described in the naïve implementation. In order to reduce collisions, the data structure is iteratively mapped onto a larger space of $d \cdot N$ possible positions, which is compacted later by decreasing d (N is the original problem size). The elements which collide are registered and wait for the next iteration to be mapped onto this space. If, after $c \log\log(n)$ iterations there are elements to be mapped left, the algorithm will restart. In [3] mutex mechanisms are used to prevent collisions, so that if a processor can not access a memory position, it will look for another free slot. Although the process is distributed, the communication cost for the resolution of collisions and synchronization mechanisms is remarkable.

Another family of algorithms, proposed in the context of distributed systems, uses an approach based on distributing the data among p processors and having each one of them accesing only to a portion of the N original elements [8] [12] [3]. They have 3 stages usually: First, each processor randomly chooses a destination processor for each element and sends the element to that processor. After receiving the incoming elements, each processor computes a local random permutation. Finally, the data structure is assembled into a global one. As indicated in [8], the use of resources (memory, bandwidth, computational time and random number generation) is $O(p)$ in each processor, and $O(p^2)$ in total. Both phases of the rearrangement process are susceptible to introduce collisions and extra memory accesses, as in classic parallel shuffle algorithms. In the next section, a new *random rearrange* algorithm that tries to reduce communication overheads and cost is described.

4 Proposed Algorithm

All the algorithms described in the previous sections were designed ensuring that every possible permutation has the same probability of being selected. However, some applications do not require this condition; for them it is enough that the selected permutation is just unpredictable. Practical implementations of the algorithms cited before with standard libraries actually limit the range of possible permutations. A typical example is the implementation of card shuffling with pseudorandom number generators with only 32 bits of internal state, which means that only $2^{32} \approx 4.3 \cdot 10^9$ different sequences of numbers can be produced, a very small fraction of the $52! \approx 8 \cdot 10^{67}$ possible permutations[2].

Indeed, some sets of permutations can be particularly useless. Our proposed algorithm sacrifices some of the possible permutations in order to achieve a greater control of the sequence generation process.

From the point of view of power efficiency, memory accesses are a major cause of energy consumption for both embedded and large scale HPC systems. A lot of optimization efforts have been made, regarding the hardware used, to decrease power consumption. However, energy dissipation is heavily dependent on the software and the algorithms used in computer systems. An additional goal in our method is to reduce power consuption due to processor communication with memory and other processors.

4.1 Linear Congruential Algorithm

Ideally, we would like to have a function that directly maps the original data structure to another, obtained using a random permutation of the original one. Pseudorandom number generators produce deterministic sequences of numbers that appear to be random, but most of the methods give periodic sequences instead. In order to have these sequences seeming random, much of the research effort in the design of these methods has focused on obtaining long repetition periods.

Our strategy is different: we are not interested in long periodic sequences, but in designing a generator of linear congruential generators that provides a straight mapping.

Definition 1. *A linear congruential method produces a sequence* $\{x_{i \geq 0}\}$ *defined recursively by*

$$x_{i+1} \equiv ax_i + c \pmod{m}, \qquad 0 \leq x_{i+1} < m \tag{1}$$

where integers m (the modulus*), a (the* multiplier*), c (the* increment*) and* x_0 *(the* seed*) are chosen such that* $0 \leq a, c, x_0 < m$*. Each different sequence is determined by* (x_0, a, c, m).

We are interested in the tuples (x_0, a, c, m) that:

- Give *purely periodic* sequences, satisfying that $x_{i+d} = x_i \ \forall i$, for a period of d length
- Have a modulus m large enough to cover every element in the data structure as a pseudorandom number in the sequence
- Goes through all the possible values in the range, which is called in the literature *full period length*, so that m can be equal to the number of elements N and coincides with the period d. This feature allow us to save calculations and avoid collisions.

A linear congruential sequence x_i consists of at most m different terms, since each x_i assumes a value in $Z_m = 0, 1, 2, \cdots, m-1$. If a is relatively prime to m (i.e. $a \bmod m = 1$), then given x_{i+1}, congruence (1) can be solved uniquely for x_i since a will have a unique inverse modulo m. Therefore, the sequence x_i is *purely periodic*.

Theorem 1. *To meet the last two desirable conditions given after Definition 1 for full period length sequences, the following conditions must be satisfied[11]:*

- *c is relatively prime to m*
- $a \equiv 1 \pmod{p}$ *for each prime divisor p of m*
- $a \equiv 1 \pmod{4}$ *if m is a multiple of* 4.

In practice, these conditions are necessary but not sufficient; additional calculations must be performed in order to ensure that the sequence covers the entire range of possible elements. Our generator chooses the parameters (x_0, a, c) to get an unsorted list of values that goes over the whole value range between $[0, m)$, where $m = N$, the number of items within the data structure.

4.2 Parallel Scheme

There are many aspects that affect performance for parallel shuffling algorithms. First, in a parallel facility with many heterogeneus processors, each one should receive workload sizes adapted to their computing power. Also, memory collisions, produced while one or several processors are accessing data, should be minimized, because they affect performance and energy consumption. Communication overheads, including those originated by processor setup and synchonization procedures, should be kept as low as possible. And last, the computational burden needed for processing each data element should also be as small as possible.

The next lema allow us to generate any element in the sequence, independently of the previous elements:

Lemma 1. *Let x_i be a linear congruential sequence determined by (x_0, a, c, m). Assuming $a \geq 2$ then*

$$x_{i+k} \equiv a^k x_i + \frac{(a^k - 1)c}{a-1} (mod\, m) \equiv a^{k+i} x_0 + \frac{(a^{k+i} - 1)c}{a-1} (mod\, m) \quad \forall k \geq 0 \qquad (2)$$

In consequence, in a parallel environment, each processor needs to know only the parameters (a, x_0, c) of the generator, the range of elements $[q_{i_1}, q_{i_2}]$ to process, the total number of items m, and pointers to the initial and final data structures (in a shared memory system).

The data dependences implicit to linear congruential sequence generation result in long execution times, even in parallel systems. Nevertheless, the algorithm presented here meets all the requisites mentioned above, thanks to slight variant of Lemma 1. First, each processor p_i can work on a partial subsequence which includes all the input data between positions q_{i_1} and q_{i_2}; iterations can be avoided just by taking its x_0 as q_i position for each element in the sequence (fig. 1). And second, Lemma 1 shows that each processor can compute its assigned subsequence independently of the other ones. This strategy provides therefore a simple, easy to parallelize and collision-free method to perform the *random rearrange* mapping.

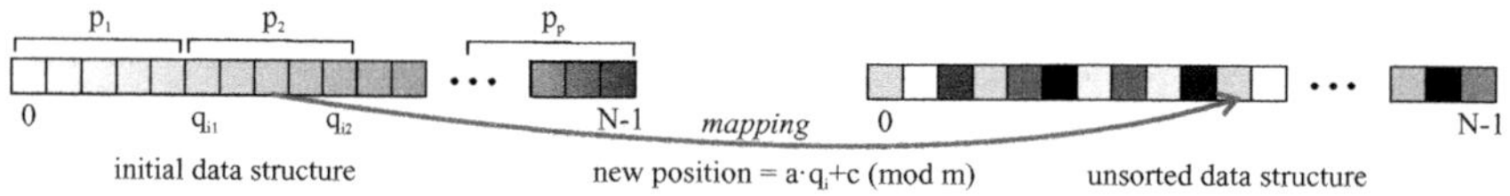

Fig. 1 *Random rearrange* mapping: strategy for dividing data among processors

5 Preliminary Results for Graphics Architectures

Our algorithm and parallel versions of the naïve [3] and Fisher and Yates' shuffle[1] algorithms have been implemented in CUDA to study their behaviour and compare their performance in graphic architectures. Pseudorandom number generation of naïve and shuffle implementations use CURAND library (3.2 RC beta version). The results are presented in figure 2, where average execution times for each of the mentioned algorithms are given for different values of the input data size, N.

We can consider graphics processors as shared memory systems, where each of the millions of threads of their kernels process a small set of elements. We have not implemented *persistent threads*, so communication among processors is limited to atomic operations for accessing parts of the video memory.

The tests presented here were performed on a Xeon server with a Tesla C1060 board, which has 4GB of video memory. In the figure, the time for generating the initial data structure in the CPU was not taken into consideration, although driver

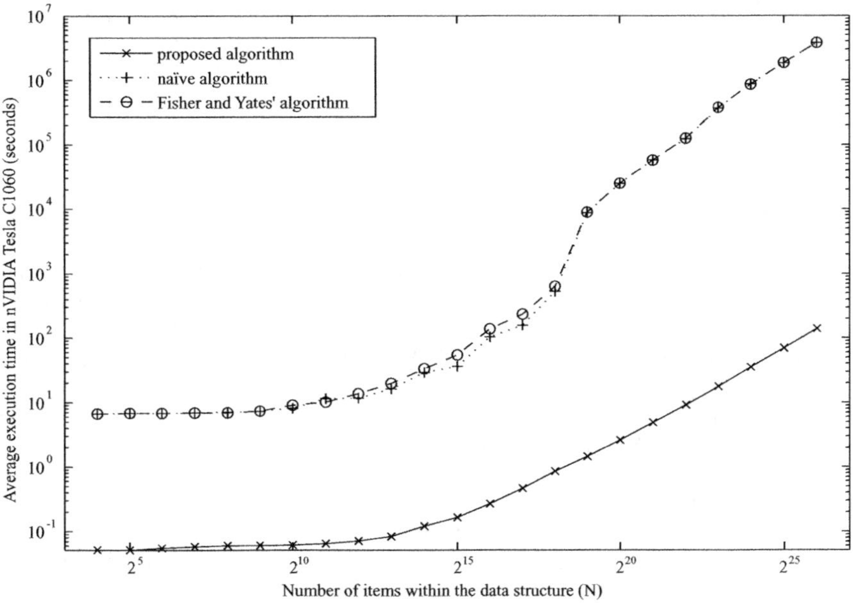

Fig. 2 Average execution time of parallel shuffle algorithms implementations in GPU for different data sizes (both axes are represented in logarithmic scale).

initialization, kernels setup, and the time for data transfers between CPU and GPU were considered.

6 Discussion and Conclussions

The results presented in figure 2 show that even though the three algorithms have linear complexity $O(n)$, inadequate memory access patterns can increase total execution times by several orders of magnitude. In particular, the method proposed here avoids memory collisions as well as the need to exchange elements, not requiring atomic accesses to coordinate the activities of processors on the data set. Also, the proposed method does not need to access memory for pseudorandom number generation. On the other hand, CUDARAND initialization kernels need store a different seed per thread, which takes a significant time in parallel naïve and shuffle algorithms. Globally, this results in a huge difference in execution times and processors' power consumption.

The results achieved so far are encouraging, suggesting that further studies on the relation between permutation space and tuple parameters will result in important improvements on the implementation of randomized algorithms. Finally, many applications areas, such as data mining, will benefit from these improvements.

Acknowledgements. This work was partially supported by the computing facilities of Extremadura Research Centre for Advanced Technologies (CETA-CIEMAT), funded by the European Regional Development Fund (ERDF). The CETA-CIEMAT belongs to the Spanish Ministry of Science and Innovation. This work has been funded by Spanish CICYT, under contracts TIN2007-67188 and TIN2010-21289-C02-01.

References

1. Anderson, R.: Parallel algorithms for generating random permutations on a shared memory machine. In: SPAA 1990: Proceedings of the Second Annual ACM Symposium on Parallel Algorithms and Architectures, pp. 95–102. ACM, New York (1990)
2. Arkin, B., Hill, F., Marks, S., Schmid, M., Walls, T.J.: How we learned to cheat at online poker: A study in software security (September 1999)
3. Cong, G., Bader, D.A.: An empirical analysis of parallel random permutation algorithms on smps. In: Oudshoorn, M.J., Rajasekaran, S. (eds.) ISCA PDCS, pp. 27–34. ISCA (2005)
4. Drineas, P., Drinea, E., Huggins, P.S.: An experimental evaluation of a monte-carlo algorithm for singular value decomposition. In: Manolopoulos, Y., Evripidou, S., Kakas, A.C. (eds.) PCI 2001. LNCS, vol. 2563, pp. 279–296. Springer, Heidelberg (2003)
5. Durstenfeld, R.: Algorithm 235: Random permutation. Commun. ACM 7(7), 420 (1964)
6. Fisher, R.A., Yates, F.: Statistical tables for biological, agricultural and medical research / by Sir R.A.Fisher and F.Yates, 1st edn. Oliver and Boyd, Edinb. (1938)
7. Gries, D., Xue, J.: Generating a random cyclic permutation. BIT Numerical Mathematics 28(3), 509–520 (1988)
8. Gustedt, J.: Efficient sampling of random permutations. Journal of Discrete Algorithms 6(1), 125–139 (2008); Selected papers from AWOCA 2005, Sixteenth Australasian Workshop on Combinatorial Algorithms
9. Hicks, P., Walnock, M., Owens, R.M.: Analysis of power consumption in memory hierarchies. In: Proceedings of the 1997 International Symposium on Low Power Electronics and Design, ISLPED 1997, pp. 239–242. ACM, New York (1997)
10. Knuth, D.E.: The Art of Computer Programming, 3rd edn. Seminumerical Algorithms, vol. 2. Addison-Wesley, Reading (1997)
11. Ripley, B.D.: Stochastic simulation. John Wiley & Sons, Inc., New York (1987)
12. Sanders, P.: Random permutations on distributed, external and hierarchical memory. Information Processing Letters 67(6), 305–309 (1998)
13. Sattolo, S.: An algorithm to generate a random cyclic permutation. Information Processing Letters 22(6), 315–317 (1986)
14. Sintorn, E., Assarsson, U.: Fast parallel gpu-sorting using a hybrid algorithm. J. Parallel Distrib. Comput. 68(10), 1381–1388 (2008)

Cyclic Steady State Refinement

Grzegorz Bocewicz, Robert Wójcik, and Zbigniew A. Banaszak

Abstract. The paper presents a new modeling framework enabling to evaluate the cyclic steady state of a given system of concurrently flowing cyclic processes (SCCP) on the base of the assumed topology of transportation routes, dispatching rules employed, resources and operation times as well as an initial processes allocation. The objective is to provide the rules useful in the course of routing and scheduling executed in SCCP where local cyclic processes interact on the base of a mutual exclusion protocol.

Keywords: cyclic processes, cyclic scheduling, state space, periodicity, dispatching rules.

1 Introduction

Operations in cyclic processes are executed along sequences that repeat an indefinite number of times. In everyday practice they arise in different application domains (such as manufacturing, time-sharing of processors in embedded systems, digital signal processing, and in compilers for scheduling loop operations for parallel or pipelined architectures) as well as service domains (covering such areas as workforce scheduling (e.g., shift scheduling, crew scheduling), timetabling (e.g., train timetabling, aircraft routing and scheduling), and reservations (e.g., reservations with or without slack, assigning classes to rooms) [5], [9], [10], [11], [12]. Such systems belong to a class of so called systems of concurrently flowing

Grzegorz Bocewicz
Koszalin University of Technology, Dept. of Computer Science and Management, Koszalin, Poland
e-mail: bocewicz@ie.tu.koszalin.pl

Robert Wójcik
Wrocław University of Technology, Institute of Computer Engineering, Control and Robotics, Wrocław, Poland
e-mail: robert.wojcik@pwr.wroc.pl

Zbigniew A. Banaszak
Warsaw University of Technology, Faculty of Management, Dept. of Business Informatics, Warsaw, Poland
e-mail: Z.Banaszak@wz.pw.edu.pl

A. Abraham et al. (Eds.): International Symposium on DCAI, AISC 91, pp. 191 – 198.
springerlink.com

cyclic processes (SCCP) [2], [3]. Subway or train traffic can be considered as an example of such kind of systems.

Assumption the subway trains following particular metro can be treated as cyclic processes passing, due to a given timetable, the sequence of stations, allows one to state a question concerning a minimization of the total passenger travel time. So, if passengers travel between two distinguished locations in the transportation network for which no direct connection exists, i.e., transfers become inevitable, the relevant scheduling problem can be stated in the following way. Given a set of metro lines, each one treated as a repeating sequence of stations. Some lines may share the common stations. Given a *headway time* (interval between the trains), i.e., the fixed interval between the trips of a line sometimes called the *period time*. The question considered is: What is a transportation route between two designated terminal stations in the transportation network providing the shortest travel time subject to above mentioned constraints? In other words, a best transportation route of the so called multimodal process, i.e. sharing different lines, is sought.

Many models and methods have been proposed to solve the cyclic scheduling problem [6]. Among them, the mathematical programming approach (usually IP and MIP [12]), max-plus algebra [7], constraint logic programming [1], [2], [3], [4], evolutionary algorithms and Petri nets [8] frameworks belong to the more frequently used. Most of them are oriented at finding of a minimal cycle or maximal throughput while assuming deadlock-free processes flow. The approaches trying to estimate the cycle time from cyclic processes structure and the synchronization mechanism employed (i.e. rendezvous or mutual exclusion instances) are quite unique.

In that context our main contribution is to propose a new modeling framework enabling to evaluate the cyclic steady state of a given SCCP on the base of the assumed processes topology, dispatching rules employed and an initial state. So, the paper's objective is to provide the observations useful in the course of multimodal processes routing and scheduling in systems composed of concurrently flowing cyclic processes interacting between oneself through mutual exclusion protocol.

The rest of the paper is organized as follows: Section 2 introduces to the systems of concurrently flowing cyclic processes. The main definitions clarifying the concept of state space of systems considered is then presented in Section 3. The terms of a cyclic steady state and the corresponding space of cyclic steady states are introduced in Section 4. Conclusions are presented in Section 5.

2 Systems of Concurrent Cyclic Processes

Consider the digraph shown in Fig. 1. The distinguished are three cycles specifying routes of cyclic processes P_1 P_2 and P_3, respectively. Each process route specified by sequence of resources passed on among its execution can interact with other processes through so-called system common resources. So, the process routes are specified as follows:

$$p_1 = (R_6, R_3, R_5), p_2 = (R_2, R_3, R_4), p_3 = (R_1, R_5, R_4), \quad (1)$$

where the resources R_3, R_4, R_5, are shared resources, since each one is used by at least two processes, and the resources R_1, R_2, R_6, are non-shared because each one is exclusively used by only one process. Processes sharing common resources interact each other on the base of mutual exclusion protocol. The possible resources conflicts are resolved with help of assumed priority rules determining the order in which processes make their access to common shared resources (for instance, in case of resource R_4, $\sigma_4 = (P_2, P_3)$ – the priority dispatching rule determines the order in which processes can access to the shared resource R_4, i.e. at first to the process P_2, then to the process P_3, next to P_2 and once again to P_3, and so on).

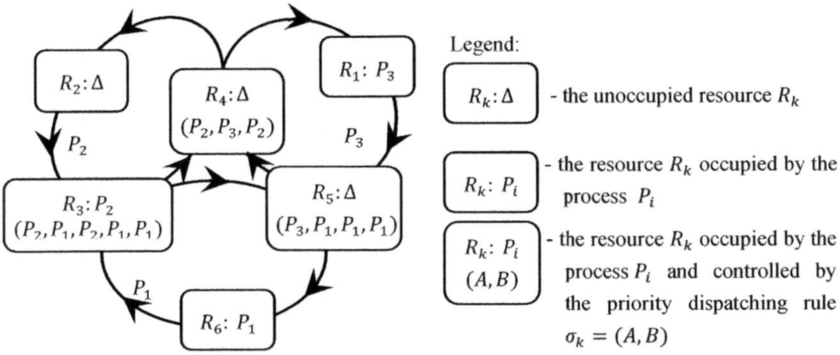

Fig. 1 Process routes structure of SCCP owning three processes

In general case, each process P_i (where $P_i \in P = \{P_1, P_2, \ldots, P_n\}$, and n is a number of processes) executes periodically a sequence of operations using resources defined by the given process route $p_i = (R_{j_1}, R_{j_2}, \ldots, R_{j_{lr(i)}})$, $j_k \in \{1,2, \ldots, m\}$, where $lr(i)$ denotes a length of cyclic process route and m denotes number of resources, and $R_{j_k} \in R$, where $R = \{R_1, R_2, \ldots, R_m\}$.

The time $t_{i,j}$ of operation executed on R_j along P_i, is defined in domain of uniform time units ($\mathbb{N}$ – set of natural numbers, i.e. $t_{i,j} \in \mathbb{N}$). So, the sequence $T_i = (t_{i,j_1}, t_{i,j_2}, \ldots, t_{i,j_{l(i)}})$ describes the operation times required by P_i. To each common shared resource $R_i \in R$ the priority dispatching rule $\sigma_i = (P_{j_1}, P_{j_2}, \ldots, P_{j_{lp(i)}})$, $j_k \in \{1,2, \ldots, n\}$, $P_{j_k} \in P$ is assigned, where $lp(i) > 1$, $lp(i)$ is a number of processes dispatched by σ_i.

In that context a $SCCP$ can be defined as the following quadruple [4]:

$$SC = (\Pi, T, R, \Theta) \quad (2)$$

where: $\Pi = \{p_1, p_2, \ldots, p_n\}$ – the set of local process routes,
$T = \{T_1, T_2, \ldots, T_n\}$ – the set of local process routes operations times,
$R = \{R_1, R_2, \ldots, R_m\}$ – the set of resources,
$\Theta = \{\sigma_1, \sigma_2, \ldots, \sigma_m\}$ – the set of dispatching priority rules.

Let us assume the all operation times equal to a unit operation time (noted as: u.t.), $\forall i \in \{1, \ldots, n\}$, $\forall j \in \{1, \ldots, lr(i)\}$, $(crd_{\,j} T_i = 1\, u.t)$.

The main question concerns of SCCP cyclic steady state behavior and a way this state depends on direction of local process routes as well as on priority rules, and an initial state, i.e. initial process allocation to the system resources. Assuming such a cyclic steady state the next question regarding of travel time along assumed multimodal process route linking distinguished resources plays a pivotal role.

3 State Space

Consider the k-th state S^k (3) composed of: the sequence of processes allocations A^k, the sequence of semaphores (encompassing the rights access to individual resources) Z^k, and the sequence of semaphore indices Q^k:

$$S^k = (A^k, Z^k, Q^k) \tag{3}$$

where $A^k = ({a_1}^k, {a_2}^k, \ldots, {a_m}^k)$ –is the processes allocation in the k-th state (m – is a number of resources occurring in the SCCP). Each element ${a_i}^k$ means the process allotted to the i-th resource R_i in the k-th state: ${a_i}^k \in P \cup \{\Delta\}$, $P = \{P_1, P_2, \ldots, P_n\}$ – the set of processes, ${a_i}^k = P_g$ means, the i-th resource R_i is occupied by the process P_g, and ${a_i}^k = \Delta$ - the i–th resource R_i is unoccupied. In the case considered (see Fig. 1) the processes allocation is specified by the sequence: $A^0 = (P_3, \Delta, P_2, \Delta, \Delta, P_1)$.

$Z^k = ({z_1}^k, {z_2}^k, \ldots, {z_m}^k)$ – is the sequence of semaphores corresponding to the k-th state, where ${z_i}^k \in P$ means the name of the process (specified in the i-th dispatching rule σ_i, allocated to the i-th resource) allowed to occupy the i-th resource R_i. For instance ${z_i}^k = P_g$ means that at a moment the process P_g is allowed to occupy the i-th resource R_i. For the SCCP from Fig. 1 the sequence of semaphores has the following form: $Z^0 = (P_3, P_2, P_2, P_2, P_3, P_1)$.

$Q^k = ({q_1}^k, {q_2}^k, \ldots, {q_m}^k)$ – is- the sequence of semaphore indices corresponding to the k-th state, where ${q_i}^k$ means the position of the semaphore ${z_i}^k$ in the priority dispatching rule σ_i: ${z_i}^k = crd_{({q_i}^k)}\sigma_i$, ${q_i}^k \in \mathbb{N}$ ($crd_i D = d_i$, for $D = (d_1, d_2, \ldots, d_i, \ldots, d_w)$). For instance ${q_2}^k = 2$ means the 2nd position occupied by P_1 in the priority dispatching rule σ_2, where $P_1 = {z_2}^k$. For the SCCP from Fig. 1 the sequence of semaphores Z has the following form: $Q^0 = (1,1,1,1,1,1)$.

The state S^k is feasible only if for any of its co-ordinate $a_i{}^k$ the following conditions hold:

$$\forall_{i\in\{1,2,\dots,n\}}\exists!_{j\in\{1,2,\dots,m\}}\left(P_i = crd_j A^k\right), \tag{4}$$

$$\forall_{i\in\{1,2,\dots,m\}}(crd_i\ A^k \in P\cup\{\Delta\}). \tag{5}$$

The set of all feasible states is called **a state space** $\mathbb{S}$, i.e., $S^k \in \mathbb{S}$.
Consider two feasible states S^k and S^l:

$$S^k = ((a_1{}^k, a_2{}^k, \dots, a_m{}^k), (z_1{}^k, z_2{}^k, \dots, z_m{}^k), (q_1{}^k, q_2{}^k, \dots, q_m{}^k)), \tag{6}$$

$$S^l = ((a_1{}^l, a_2{}^l, \dots, a_m{}^l), (z_1{}^l, z_2{}^l, \dots, z_m{}^l), (q_1{}^l, q_2{}^l, \dots, q_m{}^l)). \tag{7}$$

The state **S^l is directly reachable from the state S^k** if the following conditions hold:

$$\forall_{i\in\{1,2,\dots,m\}}\forall_{j\in\{1,2,\dots,n\}}\left[(a_i{}^k = \Delta) \wedge (a_{\beta_i(P_j)}{}^k = z_i{}^k) \Rightarrow (a_i{}^l = z_i{}^k)\right], \tag{8}$$

$$\forall_{i\in\{1,2,\dots,m\}}\forall_{j\in\{1,2,\dots,n\}}\left[(a_i{}^k = \Delta) \wedge (a_{\beta_i(P_j)}{}^k \neq z_i{}^k) \Rightarrow \left(a_i{}^l \neq P_j\right)\right], \tag{9}$$

$$\forall_{i\in\{1,2,\dots,m\}}[(a_i{}^k = \Delta) \Rightarrow [(z_i{}^l = z_i{}^k) \wedge (q_i{}^l = q_i{}^k)]], \tag{10}$$

$$\forall_{i\in\{1,2,\dots,m\}}[(a_i{}^k \neq \Delta) \wedge (a_i{}^l \neq \Delta) \Rightarrow [(z_i{}^l = z_i{}^k) \wedge (a_i{}^l = a_i{}^k) \wedge (q_i{}^l = q_i{}^k)]], \tag{11}$$

$$\forall_{i\in\{1,2,\dots,m\}}\left[(a_i{}^k \neq \Delta) \wedge (a_i{}^l = \Delta) \Rightarrow [(z_i{}^l = crd_{(q_i{}^l)}\sigma_i) \wedge (q_i{}^l = \gamma_i(q_i{}^k))]\right], \tag{12}$$

$$\forall_{i\in\{1,2,\dots,m\}}\left[(a_i{}^k \neq \Delta) \wedge (z_{\alpha_i(a_i{}^k)}{}^k = a_i{}^k) \Rightarrow (a_{\alpha_i(a_i{}^k)}{}^l = a_i{}^k) \wedge (a_i{}^l = \Delta)\right], \tag{13}$$

$$\forall_{i\in\{1,2,\dots,m\}}\left[(a_i{}^k \neq \Delta) \wedge (z_{\alpha_i(a_i{}^k)}{}^k \neq a_i{}^k) \Rightarrow [(a_i{}^l = a_i{}^k) \wedge (q_i{}^l = q_i{}^k)]\right], \tag{14}$$

where: m – a number of resources, n – a number of processes, $\beta_i(P_j)$ – the index of resource directly proceeding the resource R_i, in the j-th process route p_j, $\beta_i(P_j) \in \{1,2,\dots,m\}$, $\alpha_i(P_j)$ –the index of resource directly succeeding the resource R_i, in the j-th process route p_j, $\alpha_i(P_j) \in \{1,2,\dots,m\}$, $\gamma_i(q_i{}^k)$ – the function defined by (15):

$$\gamma_i(a) = \begin{cases} a+1 & for\ a < lp(i) \\ 1 & for\ a = lp(i) \end{cases} \tag{15}$$

where: $lp(i)$ - the number of processes dispatched by the rule σ_i.

In the case considered (see. Fig. 1) for the state $S^0 = (A^0, Z^0, Q^0)$ such that $S^0 = ((P_3, \Delta, P_2, \Delta, \Delta, P_1), (P_3, P_2, P_2, P_2, P_3, P_1), (1,1,1,1,1,1))$ the only directly reachable state $S^1 = ((\Delta, \Delta, \Delta, P_2, P_3, P_1), (P_3, P_2, P_1, P_2, P_3, P_1), (1,1,2,1,1,1))$ there exists. The graphical illustration of the cyclic steady state space $\boldsymbol{Sc}$ generated by S^0 is shown in Fig. 2.

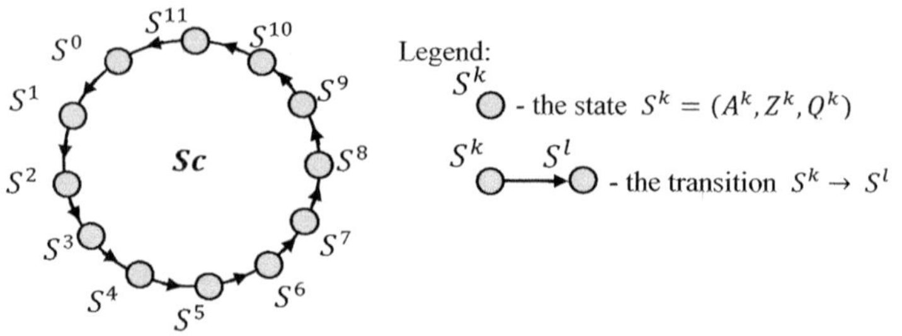

Fig. 2 Illustration of the cyclic steady state space ***Sc*** generated by S^0 in SCCP from Fig. 1.

Note that the transition $S^k \to S^l$ is a state transition function $\delta : \mathbb{S} \to \mathbb{S}$ (16) following conditions (8)÷(14).

$$S^l = \delta(S^k) \tag{16}$$

4 Cyclic Steady State Space

Consider the SCCP and its state space $\mathbb{S}$ (the set of all feasible states defined by (3)). The set $Sc = \{S^a, S^b, S^c, \dots, S^d\}$, $Sc \subset \mathbb{S}$ is called **a cyclic steady state** generated by an initial state $S^a \in \mathbb{S}$ if the following condition holds:

$$S^a \to S^b \to S^c \to \dots \to S^d \to S^a \tag{17}$$

where: $S^a \to S^b$ – the transition defined by (8)÷(14).

In other words a cyclic steady state consists of such a set of states in which starting from any distinguished state it is possible to reach the rest of states and finally reach this distinguished state again. Each cyclic steady state is determined by so called period of cyclic steady state Tc.

A cyclic steady state period *Tc* is defined in the following way: $Tc = \|Sc\|$. Of course, for any $S^k \in Sc$ the following property holds $S^k \xrightarrow{Tc-1} S^k$.

Therefore, searching for a cyclic steady state Sc in a given SCCP can be seen as a reachability problem where for an assumed initial state S^0 the state S^k, such that following transitions $S^0 \xrightarrow{i} S^k \xrightarrow{Tc-1} S^k$ holds, is sought.

Note that cyclic steady state behavior of the SCCP (SSB_{SCCP} for short) follows from assumption that the quadruple (2) has been extended by an initial state $S^0 \in Sc$ and the state transition function δ. So, in general case one may consider the cyclic steady state space defined as sixtuple (18), where SS consists of states belonging to cyclic steady states ($Sc \in SS$).

$$SSB_{\text{SCCP}} = (\Pi, T, R, \Theta, SS, \delta) \tag{18}$$

The graphical illustration of the cyclic steady state space is shown in Fig.3. The multimodal processes can be seen as trajectories passing through different cyclic steady states, e.g. distinguished by the bold line.

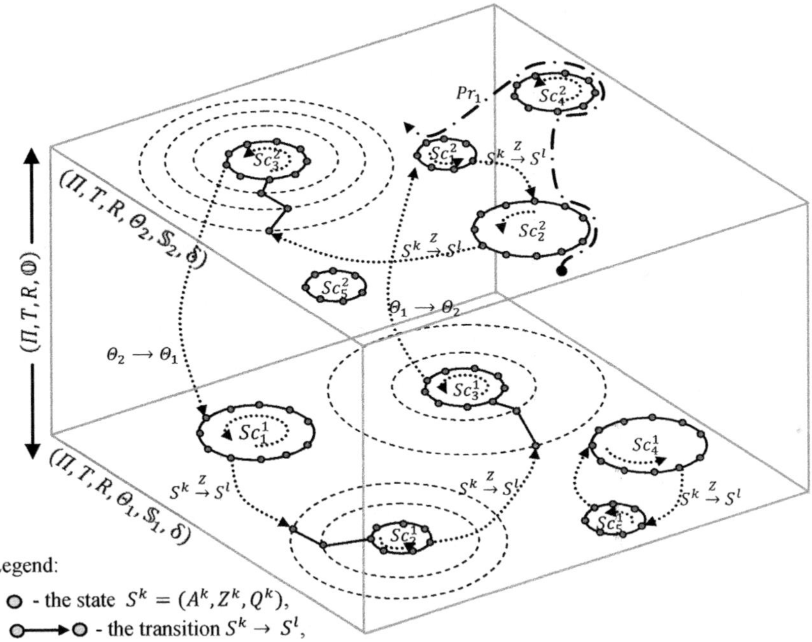

Legend:

- the state $S^k = (A^k, Z^k, Q^k)$,
- the transition $S^k \rightarrow S^l$,
- the j-th cyclic steady state for the i-th set of dispatching priority rules Θ_i,

$S^k \xrightarrow{Z} S^l$ - the transition $S^k \rightarrow S^l$, linking distinguished states S^k, S^l for the given set of dispatching priority rules Θ_i by semaphores Z modification,

$\Theta_i \rightarrow \Theta_j$ - the transition, linking distinguished states S^k, S^l obtained for the sets of dispatching priority rules Θ_i , Θ_j, respectively,

Pr_1 - the multimodal process Pr_1.

Fig. 3 Graphical illustration of cyclic steady states space

5 Concluding Remarks

The approach proposed is based on the system of concurrently flowing cyclic processes concept assuming its cyclic steady state behavior guaranteed by a given set of dispatching rules and assumed set of initial processes allocations. In the general case that means there exists a set of possible cyclic steady states encompassing potential cyclic behaviors of the SCCP at hand. Each cyclic steady state characterized by its cycle time specifies the local processes repeatability, i.e. their periodicity.

In that context, so called multimodal processes that can be seen as processes composed of local cyclic processes lead to two fundamental questions: Does there exist a control procedure (i.e. a set of dispatching rules and an initial state) enabling to guarantee an assumed steady cyclic state (e.g. following requirements caused by multimodal processes at hand) subject to SCCP's structure constraints?

Does there exist the SCCP's structure such that an assumed steady cyclic state (e.g. following requirements caused by multimodal processes at hand) can be achieved? Response to these questions determines our further works.

References

[1] Bocewicz, G., Wójcik, R., Banaszak, Z.: On Undecidability of Cyclic Scheduling Problems. In: Karagiannis, D., Jin, Z. (eds.) KSEM 2009. LNCS (LNAI), vol. 5914, pp. 310–321. Springer, Heidelberg (2009)

[2] Bocewicz, G., Wójcik, R., Banaszak, Z.: AGVs distributed control subject to imprecise operation times. In: Nguyen, N.T., Jo, G.-S., Howlett, R.J., Jain, L.C. (eds.) KES-AMSTA 2008. LNCS (LNAI), vol. 4953, pp. 421–430. Springer, Heidelberg (2008)

[3] Bocewicz, G., Banaszak, Z., Wójcik, R.: Design of admissible schedules for AGV systems with constraints: a logic-algebraic approach. In: Nguyen, N.T., Grzech, A., Howlett, R.J., Jain, L.C. (eds.) KES-AMSTA 2007. LNCS (LNAI), vol. 4496, pp. 578–587. Springer, Heidelberg (2007)

[4] Bocewicz, G., Banaszak, Z.: Cyclic processes scheduling. In: Bzdyra, K., Mleczko, J. (eds.) Applied Computer Science: Production engineering IT-driven concepts, vol. 6(2), pp. 41–70 (2010)

[5] Liebchen, C., Möhring, R.H.: A case study in periodic timetabling. Electronic Notes in Theoretical Computer Science 66(6), 21–34 (2002)

[6] Levner, E., Kats, V., Alcaide, D., Pablo, L., Cheng, T.C.E.: Complexity of cyclic scheduling problems: A state-of-the-art survey. Computers & Industrial Engineering 59(2), 352–361 (2010)

[7] Polak, M., Majdzik, P., Banaszak, Z., Wójcik, R.: The performance evaluation tool for automated prototyping of concurrent cyclic processes. Fundamenta Informatice. Fundamenta Informatice 60(1-4), 269–289 (2004)

[8] Song, J.-S., Lee, T.-E.: Petri net modeling and scheduling for cyclic job shops with blocking. Computers & Industrial Engineering 34(2), 281–295 (1998)

[9] Heo, S.-K., Lee, K.-H., Lee, H.-K., Lee, I.-B., Park, J.H.: A New Algorithm for Cyclic Scheduling and Design of Multipurpose Batch Plants. Ind. Eng. Chem. Res. 42(4), 836–846 (2003)

[10] Trouillet, B., Korbaa, O., Gentina, J.-C.k.: Formal Approach for FMS Cyclic Scheduling. IEEE SMC Transactions, Part C 37(1), 126–137 (2007)

[11] Wang, B., Yang, H., Zhang, Z.-H.: Research on the train operation plan of the Beijing-Tianjin inter-city railway based on periodic train diagrams. Tiedao Xuebao/Journal of the China Railway Society 29(2), 8–13 (2007)

[12] Von Kampmeyer, T.: Cyclic scheduling problems, Ph.D. Dissertation, Fachbereich Mathematik/ Informatik, Universität Osnabrück (2006)

Ant Colony to Fast Search of Paths in Huge Networks

Jessica Rivero, Dolores Cuadra, F. Javier Calle, and Pedro Isasi

Abstract. Nowadays, Ant Colony Algorithms (ACO's) are applied to different domains of application such as automated software testing, computer games or searching of paths. Most of works that use graphs for representing information and ACO algorithm for searching a path are related to small (thousand of nodes and links) and static networks (both position and number of nodes are always fixed). In this paper a new ACO algorithm is presented. The proposal shows a solution to search paths for huge and dynamic networks in a suitable time.

1 Introduction

A huge number of systems that could be found nowadays use graphs as a tool to represent the information with which they work. In some cases, this is a direct transformation, but in others it is required to do an abstraction process to give a meaning both nodes and links which make up the graph. Automatic solution of games are examples of the last case (chess, or draughts, in which each node is an scenario and links which join them are the moves to go from one scenario to another).

In most cases where this kind of representation is done, the main goal is to make an easier handling of information and to perform necessary operations on it to satisfy the requests of users in an effective and efficient way. These requests are habitually solved applying searching algorithms of shortest paths, or paths into a certain quality range.

As it is shown in the next section, there are lot of works trying to find an algorithm capable of solving the problem of searching paths inside graphs. The main problem of these works is that they search paths over small dynamic networks or over huge static networks, but not in huge dynamic graphs which are very important because they could represent some widely used domains as social networks or semantic web. The objective of this paper is to present an efficient and effective algorithm which works with this kind of graphs, giving a path which satisfies a quality range in a limited time, making possible to use it in applications like

Jessica Rivero · Dolores Cuadra · F. Javier Calle · Pedro Isasi
Computer Sciences Department
Carlos III University of Madrid
28911 Madrid, Spain
e-mail: {jrivero,dcuadra,fcalle}@inf.uc3m.es, isasi@ia.uc3m.es

A. Abraham et al. (Eds.): International Symposium on DCAI, AISC 91, pp. 199–208.
springerlink.com

searching persons in social networks, or discovering the procedure which must be followed to solve problems in computer games. This algorithm is presented as an extension of ACO classic algorithms.

To do this, the rest of the paper is organized in the following way: In section 2 current works in the field are going to be presented. After this, the problem is going to be formalised. Once this will be done, a new algorithm is going to be proposed and explained to solve previous problems. In the next section, some experiments over huge network will be shown the good way in which the algorithm works. Finally, some conclusions and future works can be found.

2 Previous Works

Currently the graphs obtained to formalize concepts (nodes) and relations between them (links) are reaching dimensions of hundreds of thousands or millions of nodes and links for various reasons. This is a problem in the cases of searching of paths inside networks, because this search must be solved in a short time (limited time) since in most cases there is an end user waiting for the solution.

Faced with this challenge various studies can be found as the one presented in [4]. This work stores all information of a road network in a file system, and on it a series of routing queries are realised. To do this, authors propose a pre-processing of the information and the use of a modified Dijkstra algorithm, obtaining optimal quality solutions and a short response time. Another work which also stores the information of the network in secondary memory is the one presented in [13] or [14]. In this case, the information is stored in a data base. To obtain answers in short time, they create groups of nodes according to distances and organize them as tree structures.

Other works in the field with a short answer time are [3, 7] (which work with networks of millions of nodes making hierarchies to organize them) or [6] (which does a fragmentation of the network in subnetworks and pre-processes links which are taken into account in each one). The problem is that they store the network in main memory, and by this reason they need optimization processes to store all the network in the minimum space, losing part of the benefits of working with all the information.

Nonetheless, in spite of these works have a short response time and that work with huge networks (none of them use networks with a number of nodes below hundreds of thousands) they expend a lot of time in the pre-processing of the information, making impossible to use them in dynamic networks.

Related to this problem, some of these authors have changed their algorithms to take into account the dynamism in networks [10, 15], but they only consider variations in the cost of links and nodes, but not structural changes. Given these changes, that in cases such as ad-hoc networks or small-world networks [2, 11] can be every few seconds or minutes, it is not advisable to have pre-processing of hours or minutes, because once a path was found it would not be valid because new changes appear during the pre-processing or search.

For this reason, adaptative algorithms are necessary. And because of its own characteristics, ACO [9] could be an option. A proof of this is their use in ad-hoc networks [8] or in vehicle routing over dynamic roads [5].

Nonetheless, ACO has been used in networks of hundreds or thousands of nodes, but never in bigger networks, that is the trend as it was mentioned before.

This problem has already been tackled in [1] which creates an extension of ACO (ACOhg) to solve the problem of search of paths in huge dynamic networks. However, it stores the information in main memory, and in some moments of the search it has to delete some information of the searching process to free space.

In this article another version of ACO is presented to solve such problems (an initial version can be found in [12]). In addition, a secondary storage is going to be used to avoid limitations of main memory such as lost of information during the search as the case mentioned above.

3 Problem Formalization

Once some relevant works in the area have been explained, in this point is going to be presented a formalization of the scenario over which the problem of searching paths wants to be solved.

Let $G(t)=\{N,L\}$ be a connected network at the instant t where:

- $N = \{(x,y) \in \Re \times \Re\}$ is the set of nodes in G.
- $L = \{l_{ij} = (n_i,n_j) \in NxN,\ n_i \neq n_j\}$ is the set of links in G having that $l_{ij} = l_{ji}$.

The weight of each $l_{ij} \in L$ could be defined by the following function:

$$W: L \rightarrow \Re^+ \text{, where } w_{ij} = W(l_{ij})$$

The size of G is equal to the number of elements in N.

G is a dynamic network, such that given two different time instants t_i and t_j where $\Delta t = t_j - t_i \geq 0$, and the state of G in each one ($G(t_i)$ and $G(t_j)$), it is going to happen that $G(t_i) \neq G(t_j)$. This variation with the time could be due to:

- Changes in the weights of links in L: $w_{ij}(t_i) \neq w_{ij}(t_j)$.
- Structural changes: elements in N and L appear and disappear.

The rate of dynamism in G is called K, and it is the percentage of changes in each unit of time (between two consecutive states of G) Δt. Depending of the value of K, the dynamism of G changes, such that if $K>0\%$, G is dynamic.

The services over G are queries to find the path P between whatever two nodes in a time shorter than one fixed by the petitioner ($t_{threshold}$), such that: $P: NxN \rightarrow L' \subseteq L, t_{answer}$, with $t_{answer} \leq t_{threshold}$. Where t_{answer} is the time used to calculate the path.

Depending on the application that wants to be considered to apply the algorithm, the values of the parameters mentioned above are different. Specifically, the values considered in this work make not possible to apply the solutions discussed in the point two of this paper. These values model scenarios with huge dynamic networks and with limited answer time because the end user needs to have an answer as fast as possible. Examples of scenarios with those characteristics are social networks, or on line games. Both cases are online applications, and due to this reason the information in the network is huge, dynamic (all users could change the network with their interaction with it), and it has to be accessible by all the users. These characteristics have the following connotations:

- The value of K is high and Δt is small.
- The size of the network is millions of nodes and has to be visible by all the users.

4 Proposal

To realise the fast search of paths over the scenarios mentioned before, the proposed algorithm joins two different phases:

- A kind of pre-processing with the capacity to be able to adapt to changes in G, but with enough information to help to the second phase. This pre-processing has a difference with respect to approaches in the previous section: it does not change the structure of the graph, it only adds information to help in the searching phase.
- Algorithm ACO with some modifications to give paths faster and adaptable to changes during the search.

4.1 Pre-processing

Let S be a subset of N ($S \subseteq N$) where nodes satisfy a property X (objectives in paths, preferences of the petitioner, nodes which are the center of a cluster, etc), whatever $s_i \in S$ could be interested to have it any path.

Due to this reason a pre-processing around each s_i is going to be realised to reach easier these nodes improving the time to search paths. New definitions appear when this is done:

- Any s_i is going to have associated a value called *Smell* to find it in the network.
- *Initial Smell (m)*: Amount of smell in s_i. It is the greatest smell in the network.
- The smell is going to be spread from nodes s_i creating *areas of smell* (c_k) around each one, such that n areas are going to appear in the network: $C = \{c_1, c_2, \ldots c_n\}$, where $c_k \subseteq N$ and $(c_k \cap c_l) \subseteq \{\emptyset, N\}$.

  Initially, all nodes inside c_i have a *smell* between m and u ($u < smell(n_i) < m$, *with* $n_i \in c_i$), where u is the *Smell Threshold* of the area and it is the smallest value of the *smell* inside the area in the initial time.

- The form in which the *smell* is going to be spread is decreasing its amount as the nodes are further from s_i. Each step *smell* decreases ($k \cdot w_{ij}$), where $k \in \Re(0...,1)$. This parameter is called *Percentage of Deposited Smell.*

With respect to the size of c_k (for $k=\{1,2,...n\}$):

- t=0 (before any request of path): Each c_k has the same size defined by u, k and m. It must be small for two reasons:

  1. To avoid that many changes in G affect the areas being necessary to create them again.
  2. Because X could change and it is necessary that all *smells* can be deleted and created quickly allowing the network to be self-adaptive.

  The form in which this area (c_k) is created is decreasing the smell using the cost of links and the factor k as was explained before. The form of the area is going to be undefined, and its size is going to be fixed by u, because any node inside it must have a smell bigger to that value.

- t>0: In this moment could happen two different events with a direct influence in the size of c_k:

  1. c_k has been used to create the path p_{ij}. In this case, *smell* is going to be assigned to nodes of p_{ij} starting from the nodes with *smell* and decreasing it taking into account the equation (*previousSmell*-$k \cdot w_{ij}$) until the final nodes were reached. With this expansion of *smell*, the size of the area increases.
  2. A change in G happens and it affects to some node n_i with *smell* equal to m' of type c_k. In this case:

     - If the change is out of the initial area of c_k, the nodes which depend of n_i and with *smell* $m'' < m'$ are going to be totally evaporated.
     - If the change is inside the initial area of c_k, all nodes with that kind of *smell* are going to be deleted and the area is going to be created again.

An important thing is that the pre-process does not change the structure of the network over which the query of path is resolved. By this reason, when a restarting of any c_k has to be realized, ants do not have to wait to continue searching. Areas of smell are only extra information to helps ants to search the end node.

4.2 ACO

ACO is going to be adapted to the problem, so the main features are the followings:

- Links are initialized to a fix amount of pheromone.
- Half of ants go from the start node to the end node, and the others go in opposite direction while any of them have smell. If the start/end node has smell, all ants start in the same node (the one without smell).
- A tabu list is used to avoid visiting a node two times during the search.

- When it completes a path, it lays a substance called trail on each l_{km} visited inside p_{ij}. If $\tau_{km}(t)$ is the trail in the link l_{km} at the moment t, then: $\tau_{km}(t+1)=(1-\rho)\cdot \tau_{km}(t)+1/L_{total}$
  Where L_{total} is the length of the path and $\rho\in \Re(0...,1)$ is the dissipation factor.
- It chooses the node n_j to go to with a probability that is a function of the amount of trail present on the connecting edge. The equation of this probability is the following:

$$p(n_i, n_j) = \begin{cases} \dfrac{\tau_{ij}(t)}{\sum_{k\in reachable} \tau_{ik}(t)} & \text{if } j\in \text{reachable from } n_i \\ 0 & \text{if } j\notin \text{reachable from } n_i \end{cases}$$

  Where reachable nodes in tabu list are not included in the formula.
- When all ants finish, all discovered paths to go from n_i to n_j using at least one *smell area* are selected, and the ones with lowest cost are used to expand *smell*.

During the search, each ant acts in the following way when a path is required:

1. Initialisation: A fixed value of pheromone is deposited in all the links of the graph, the start nodes of each ant are established and their tabu tables are emptied. To finish, the value of the parameter that store the length of the found path is set to infinite.
2. Choose the node n_k to move to, with probability $p(n_m,n_k)$, where n_m is the actual node, and insert n_m in tabu list.
3. Once the ant is in n_k, follows Algorithm 1. In it, the method *completePath(tabu)* tries to create a path from the node with smell found and the centre node of its smell area.

```
if (nk=end_node) then
    length=0;
    for    (h=1...(sizeTabu-1))
do
     aux=W(tabu(h),tabu(h+1));
        length=length+aux;
    end;
    if (length<Ltotal) then
        Ltotal=length;
    end if;
    updateTrail(tabu);
    emptyTabu();
    setStartNodeToAnt();
elseif (nk∈ ci) then
    storePartialPath(tabu);
    path = completePath(tabu);
    if (∃path) then
        length=pathLength(path)
    ;
        if (length<Ltotal) then
            Ltotal=length;
        end if;
        updateTrail(tabu);
    end if;
end if;
```

Algorithm 1. Process for each step.

At each step, the execution time is controlled, and if it is bigger than the maximum time, the ant stops. If this condition is not true, then the ant goes again to step two.

5 Case of Study

To show how this algorithm works, a case of study is going to be done. It is divided in two parts due to the two components of the algorithm: Pre-processing and ACO.

These experiments are the first published, and they are centred in showing how the new algorithm works when it is compared with another algorithm and in fixing the values of parameters with influence in it. To do this faster, the experiments are executed over static scenarios. In future papers it is going to be incorporated the dynamic factor using the values obtained in this phase.

The variables considered in the study are the quality of the solution, and the answer time (the time to give the first solution). With respect to the algorithm used to compare the proposal is Dijkstra. This choice was done by different reasons:

- It is the most classical algorithm used in search of paths and the one implemented by all DBMS (Data Base Management System).
- It gives a solution without use any kind of heuristic.

With respect to the environment, it is going to be used a network of 200000 nodes and 600000 links, and the distribution of nodes and the connection between them is randomly generated. This size was selected to guarantee a huge network and the distribution to not benefit to any algorithm.

Paths requested by the users go from one node (randomly selected) to another node which satisfies the property of being the centre of an area with smell.

The number of queries is 1000 (a significant number due to the size of the network) and start nodes are always different. Queries are done one after another.

Once the scenario is explained, the parameters used in the proposal are the ones shown in Table 1.

Table 1 Parameters of the proposed algorithm

Parameter	Value
$t_{threshold}$	800 sec
#ants	500
	0.6
m	1000000
k	100%

The parameter $t_{threshold}$ is fixed to, more or less, the time required by Dijkstra to get a solution.

With respect to the value of u, this parameter it is going to be studied in this paper, and for this reason, three values are going to be compared among them (values are in Table 2).

Table 2 Time to do smell areas

%Nodes with Smell	Time (sec)
1% ($u = 999700$)	11,5
17% ($u = 998600$)	178,3
30% ($u = 998000$)	326,6

The first thing that needs the algorithm proposed in the paper is the pre-processing. This is done in the initialisation time and when changes in the area happen. The time required to do a new area or to repeat that process when changes happen could be seen in Table 2.

The time to create areas of smell increases when the percentage of nodes with smell increases. Nonetheless, in spite of the increase of time, in no case that value is bigger or equal than the one required in others works (which need hours). This is a good result, because the algorithm is able to recover of a change without effect in answer time.

After the pre-processing, ants start to look for the requested paths. In Fig. 1 some results are shown. In it, x-axis represents the number of the executed service, and y-axis represents the time expended by each service to give the first answer to the query (in logarithmic scale), and the quality of the obtained path.

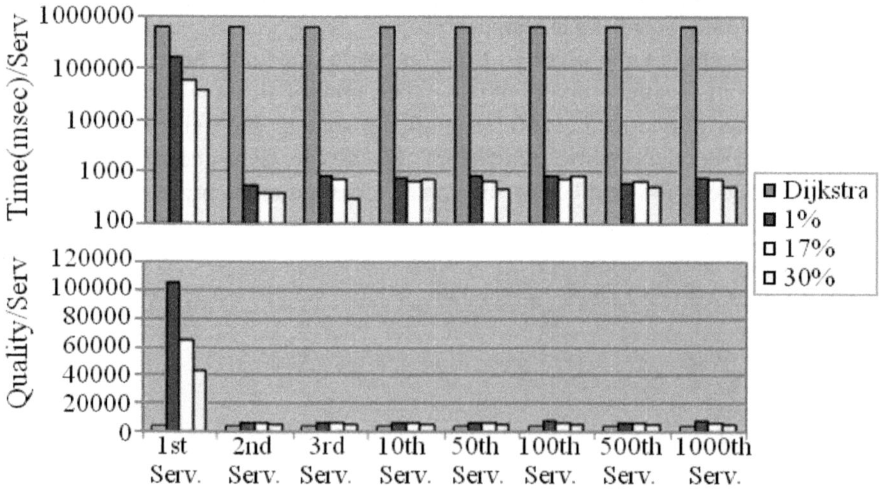

Fig. 1 Comparison between Dijkstra and new ACO.

As could be seen, the first service requires more time, because the first node with smell is far away, but after the new dispersion of smell, the answer time decreases considerably, and in the second service it is reduced to hundreds of milliseconds. In the case of Dijkstra, the required time to give the first answer is always the same, and it is always bigger than the one required by the new algorithm independently of u.

With respect to the quality, it is possible to identify that the quality of paths with the new algorithm is not the optimal, but that is near when more queries are executed. Also, it has to be considered, that results collect the first path obtained by ants. If ants run during more time, the quality of results will improve.

Another thing that can be appreciated in Fig. 1 is an improvement of quality and time as the percentage of nodes with smell rises. However, it is not comparable with the time that is necessary to create the area with smell each time that changes happen in it, and it is better to work with small areas.

6 Conclusions and Future Works

In this paper, a new algorithm has been proposed based on ACO. It uses a pre-processing which helps ants to find paths between nodes faster than other algorithms. This pre-processing is only used to help ants and does not have a negative effect when it has to be repeated.

Experimental results have shown that it is better to have small areas of smell, because they are more adaptable to changes, and because the relation quality-time is more or less the same in all the amounts of smell. This is good, because it permits to have more areas of small size distributed by the entire network instead of only one of big size.

Future works are centred in two ways:

- Improve the algorithm to obtain better solutions.
- Decide the moment in which is better to stop the algorithm to obtain paths with quality closes to the optimal one.

References

1. Alba, E., Chicano, F.: ACOhg: Dealing with Huge Graph. In: Genetic and Evolutionary Computation Conference, London, UK, July 2007, pp. 10–17 (2007)
2. Albert, R., Barabási, A.-L.: Statistical mechanics of complex networks. Reviews of Modern Physics 74, 47 (2002)
3. Bast, H., Funke, S., Matijevic, D., Sanders, P., Schultes, D.: In Transit to Constant Shortest-Path Queries in Road Networks. In: Workshop on Algorithm Engineering and Experiments (ALENEX 2007) (January 2007)
4. Chan, E.P.F., Lim, H.: Optimization and evaluation of shortest path queries. VLDB J. 16(3), 343–369 (2007)
5. De Oliveira, S.M. A study of Pheromone Modification Strategies for using ACO on the Dynamic Vehicle Routing Problem. Doctoral Symposium on Engineering Stochastic Local Search Algorithms (SLS-DS), Brussels, September 3-4 (2009)

6. Delling, D., Holzer, M., Müller, K., Schulz, F., Wagner, D.: High-Performance Multi-Level Routing. In: The Shortest Path Problem: Ninth DIMACS Implementation Challenge. DIMACS Book, vol. 74, pp. 73–92. AMS, Providence (2009)
7. Delling, D., Sanders, P., Schultes, D., Wagner, D.: Highway Hierarchies Star. In: 9th DIMACS Challenge on Shortest Paths (November 2006)
8. Di Caro, G., Ducatelle, F., Gambardella, L.M.: AntHocNet: An Adaptive Nature-Inspired Algorithm for Routing in Mobile Ad Hoc Networks. European Transactions on Telecommunications (ETT), Special Issue on Self Organization in Mobile Networking 16(5), 443–455 (2005)
9. Dorigo, M., Stützle, T.: Ant Colony Optimization. The MIT Press, Cambridge (2004)
10. Nannicini, G., Baptiste, P., Barbier, G., Krob, D., Liberti, L.: Fast paths in large-scale dynamic road networks. Computational Optimization and Applications 45(1), 143–158 (2010)
11. Newman, M.E.J.: The Structure and Function of Complex Networks. SIAM Review 45(2), 167–256 (2003), ISSN 0036-1445
12. Rivero, J.: Fast Search of Paths through Huge Networks. In: Doctoral Symposium on Engineering Stochastic Local Search Algorithms (SLS-DS), Brussels, pp. 3–4 (2009)
13. Sankaranarayanan, J., Samet, H.: Distance oracles for spatial networks. In: Proceedings of the 25th IEEE International Conference on Data Engineering, Shanghai (2009)
14. Sankaranarayanan, J., Samet, H., Alborzi, H.: Path oracles for spatial networks. In: Proceedings of the 35th International Conference on Very Large Data Bases (VLDB), Lyon, France (September 2009)
15. Schultes, D., Sanders, P.: Dynamic Highway-Node Routing. In: Demetrescu, C. (ed.) WEA 2007. LNCS, vol. 4525, pp. 66–79. Springer, Heidelberg (2007)

An Estimator Update Scheme for Large Teams of Learning Automata

Manuel P. Cuéllar, María Ros, Miguel Delgado, and Amparo Vila

Abstract. Learning Automata are stochastic decision-making machines that have been widely used in classification, control, and network routing, between others. Despite their versatility, one of the main drawbacks of these models is the low convergence rate of the learning rules used for the training. Estimator algorithms such as *Pursuit* schemes help to overcome this limitation, although they require a high computer memory cost for their operation. This fact becomes a serious inconvenient when a large set of learning automata collaborate in a team to solve a concrete task, since the memory requirements of these algorithms increases exponentially. In these cases, Pursuit algorithms are ineffective due to memory overflow.

In this work, we address this problem and we propose an estimator algorithm that can be used to train large teams of Learning Automata. The approach uses a similar strategy to Tabu Search algorithms to manage long and short term memory, in order to reduce the memory requirements. The method is applied in classic permutation problems as a test-bed.

1 Introduction

Learning Automata (LA) [7] are adaptive stochastic decision-making machines. A learning automaton is defined by the tuple $<\alpha, Q, R, T>$, where α is the set of decisions or actions available to the automaton, Q is its internal state, R is a set of reinforcement values that are input to the automaton, and T is a learning rule. In a classic scheme, the automaton is connected in a feedback loop with an unknown random environment. At each time instant k, it selects an action from the

Manuel P. Cuéllar · María Ros · Miguel Delgado · Amparo Vila
University of Granada, Department of Computer Science and Artificial Intelligence
E.T.S.I. Informática y de Telecomunicación, C/. Pdta. Daniel Saucedo Aranda s.n.
e-mail: {manupc,marosiz,vila}@decsai.ugr.es,
mdelgado@ugr.es

A. Abraham et al. (Eds.): International Symposium on DCAI, AISC 91, pp. 209–216.
springerlink.com

action-set $a(k) \in \alpha$, and applies $a(k)$ over the environment. Then the environment returns a reward or penalty reinforcement value $\beta(k) \in R$ to the automaton that depends on the suitability $D(a(k))$ of the action selected, and the internal state is updated with the learning rule as $Q(k+1) = T(Q(k), a(k), \beta(k))$. There is a wide variety of LA models depending on the action-set, the environment, and the reinforcement value designs [13, 17]. The most known LA type is the *Finite Action-set Learning Automaton* (FALA) [13], where the action-set $\alpha = \{\alpha_1, \alpha_2, ..., \alpha_r\}$ is discrete and finite. LA systems also depends on the type of the reinforcement values and the environment, and we suggest to read the references [7, 17] for a wider explanation.

The objective of the LA training is to maximize the expectation of receiving a reward from the environment with respect to the automaton action selection [7, 14]. Thus, a single automaton is capable to learn a single optimal action. It is required to use a set of LA in order to solve more complex problems that contain several parameters to be optimized. In this way, the literature offers some interesting approaches to combine a set of LA to achieve the desirable behaviour [2, 16]. The most known is the organization of LA into a team to collaborate and to build the solution [14, 16]. Here, each automaton is autonomous and it selects an action independently of the remaining LA in the team. The solution to the problem is then built from the set of actions selected by the set of automata. Another interesting proposals are to generate hierarchies of LA [3, 2], distributed networks of LA [4], and parallel modules of LA [15]. The organization of the set of LA highly depends on the features of the problem to be addressed, although some approaches have shown a good performance to speed up the learning convergence [3, 2, 15]. The combination of LA has been also studied from the multi-agent systems point of view, due to the capabilities of these models to be implemented as agents [6, 12, 8].

On the other hand, the classic update rules for the internal state of LA are based in reinforcement learning [7, 13]. These techniques have the limitation of low convergence rate to the optimal behaviour of the automata [10, 1]. The estimator algorithms such as *Pursuit* schemes [10, 9, 1] overcome this limitation, although they require a high memory cost. In the case of teams of LA, there are also estimator algorithms to train the team, but they become ineffective when both the number of LA and automata actions increase due to memory overflow. In these situations, some meta-heuristics such as simulated annealing have been proposed to train the team [14]. Other authors have developed specific training methods to solve problems such as the graph colouring problem [18] or stochastic spanning trees [19], between others. More complex algorithms have shown a good performance in tutorial-like systems [11], by mean of simulating a classroom where the automata assume the roles of students and teachers.

The work described in this article focuses on this way. We notice that the power of the estimator algorithms is the capability to maintain in memory the past experiences of the team in order to estimate the reward of each automata action for the future iterations. Thus, a training algorithm for a team of LA should include features of long and short term memory to achieve the optimal action learning, but also considering the memory limitation requirements. In our approach, we include these features in the proposed algorithm inspired by a classic Tabu Search procedure [5]. We believe that permutation problems are a suitable test-bed for our approach, so that we demonstrate the feasibility of our proposal in the Quadratic Assignment Problem. The results obtained are compared with classic linear reinforcement schemes.

The article is structured as follows: Section 2 explains how to use a team of LA to solve permutation problems. Next, Section 3 describes the algorithm proposed to train a large team of LA, and the experiments are shown in Section 4. Finally, Secion 5 concludes.

2 Learning Automata for Permutation Problem Solving

The permutation problems are often formulated as the minimization of a function that depends on a sequence of objects, $x^* = min_x f(x) = min_x f([x_1, x_2, \dots, x_n])$. These objects are usually represented as integer numbers, and the problem is to find the permutation of x that provides the optimal value of $f(x)$. We have selected this type of problems to test our approach since the number of objects in the permutation is usually high and we can build large teams of LA with a high number of actions to solve them.

In this work, a team of FALA is used to address permutation problems. Each automaton is matched with an object, and its action-set contains the available positions in which the object may be located in the sequence. Figure 1 describes the structure of the team with the environment. The automata are sorted randomly in a sequence at each iteration k and then each automaton selects an action from its available action-set, i.e. the positions in the permutation that have been not selected by any other automata at the same iteration. After that, the vector x is built by the *Fuser* module and it is sent to the environment. The reinforcement value follows a *P-model* [7], and the environment returns the reward $\beta(k) = 1$ if $f(x(k)) \leq f(x(k-1))$, or the penalty $\beta(k) = 0$ otherwise. The reinforcement value is then sent to all the automata in the team to update their state and a new iteration starts.

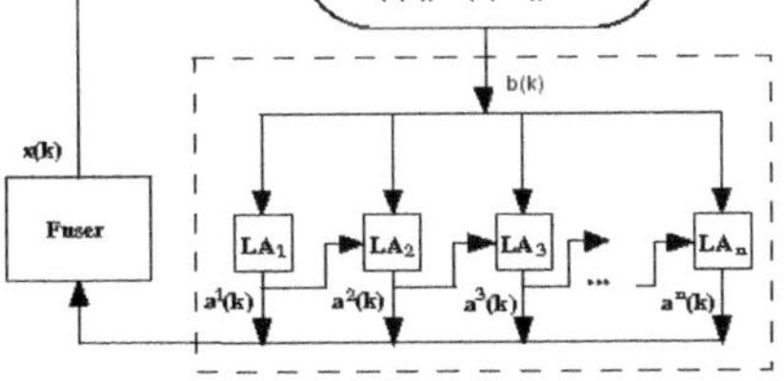

Fig. 1 FALA team to solve permutation problems. At each iteration, the automata are randomly sorted and they select an action in sequence. The actions of all the automata are sent to the Fuser that builds the solution x(k). Then, the environment evaluates x(k) and returns a reinforcement value to the team.

3 Long and Short Term Memory to Train a Large Team of Learning Automata in Permutation Problems

The classic implementation of the internal state of a learning automaton is a distribution probability over the automaton action-set, $Q(k) = P(k) = [p_1(k), p_2(k), ..., p_r(k)]^t$, where $p_i(k)$ is the probability of selection of the action α_i at the k-th iteration [7, 13]. On the other hand, the estimator algorithms define the internal state as $Q(k) =< P(k), D(k) >$, where $D(k) = [d_1(k), d_2(k), ..., d_r(k)]^t$ is the estimation of each automaton action to obtain a reward from the environment [16, 10]. In the case of a team of n LA with equal number of actions, $P(k)$ is a matrix of size $n \times r$ that represents the probability of selection of the automata's actions, and $D(k)$ is a hyper-matrix of size $|\alpha|^n$ [14] to represent the estimation to receive a reward for the team.

The proposed algorithm uses long and short term memory to estimate the probability of obtaining a reward from the environment. Its main advantage is that the memory needs can be adapted to the problem requirements. Our approach is inspired in features of the Tabu Search [5]: The most recent actions selected by the team are saved into the short term memory to avoid unsuitable actions being selected in the next iterations. On the other hand, the actions selected at each iteration and their reinforcement are saved into a long term memory to record the history of the learning and to guide the long term search. If a predefined number of iterations R has been reached with the team providing the same solution, then the algorithm has converged and the probability distributions of the automata are reinitialized considering the long term memory. The main operation of the algorithm is as follows:

1. **Initialization.** Let n be the number of automata in the team. To ease the algorithm description, we assume that all the automata have the same number of actions r. The probability $p^i_j(0) = 1/r$ is initialized for all the automata $1 \leq i \leq n$ and actions $1 \leq j \leq r$, i.e. all the team members are the pure chance automaton. The short term memory of each automaton is initialized to the empty set $S^i_{TM} = \emptyset$, and the values of the long term memory $L_{TM}(0) =< W(0), Z(0) >$ are also initialized to zero. The long term memory is composed of two $n \times r$ matrices W and Z, where $z^i_j(0)$ is the number of times that the i-th automaton in the team selected its j-th action, and $w^i_j(0)$ is the number of times that the environment returned a reward. The value $w^i_j(0)/z^i_j(0)$ provides the probability of obtaining a reward for the i-th automaton if the j-th action is selected. After that, initialize the *best action record* $B = \{b_1, b_2, .., b_n\}$ to the empty set. The set B is the best set of actions returned by the team along all the past learning iterations. Finally, update the current iteration $k = 1$, and the number of iterations in which the team returns the same solution $t = 0$.
2. **Action selection.** The automata in the team, in random sequence, select an action from their action-set. The action selected by the i-th automaton is carried out considering its action probability distribution, i.e. $[p^i_1(k), p^i_2(k), ..., p^i_r(k)]$, and the following criteria:

    a. The automata are allowed to select only the actions that do not remain in their own short term memory S^i_{TM}.

b. An automaton cannot select an action previously chosen by another automaton at the same iteration.
c. The best action $b_i \in B$ can always be selected by the $i-th$ automaton if it was not selected by another automaton at the same iteration, even if $b_i \in S^i_{TM}$.
d. If it is not possible to fulfil all the previous criteria, the best action available in the short term memory is selected.

3. **Short term memory update.** The set of selected actions, $[a^1(k), a^2(k), ..., a^n(k)]$, are added to their respective automaton short term memory S^i_{TM}. If S^i_{TM} is full, then the oldest action is removed before the inclusion of $a^i(k)$.
4. **Evaluation.** The environment evaluates the set of actions returned by the team, $x = [a^1(k), a^2(k), ..., a^n(k)]$, and it provides a reward/penalty reinforcement value $\beta(k) \in \{0,1\}$. If the evaluation $f(x(k))$ is the best one found over all the iterations, then the algorithm updates the *Best action record* as $b_1 = a^1(k), b_2 = a^2(k), ..., b_n = a^n(k)$. Also, if $x(k) \neq x(k-1)$ then t is initialized to $t = 0$. Otherwise, $t = t+1$.
5. **Automata state and long term memory update.** The algorithm updates $z^i_{a_i(k)}(k+1) = z^i_{a_i(k)}(k) + 1$, and $w^i_{a_i(k)}(k+1) = w^i_{a_i(k)}(k) + \beta(k)$. Then update $p^i_j(k+1)$ as in equation 1.

$$p^i_j(k+1) = \begin{cases} max\{p^i_j(k) - \lambda, 0\}; \forall j \neq b_i \\ 1 - \sum^r_{j \neq m} p^i_j(k+1); if j = b_i \end{cases} \tag{1}$$

6. **Reinitialization.** If $t = R$, then it is assumed that a local optimum has been found. The probability distributions of the automata are reinitialized to $p^i_j(k+1) = w^i_j(k+1)/z^i_j(k+1)$. The long term memory is updated to $z^i_j(k+1) = z^i_j(k) - min_j\{z^i_j(k)\}, w^i_j(k+1) = w^i_j(k) - min_j\{w^i_j(k)\}$, and the short term memory of the automata are initialized to $S^i_{TM} = \emptyset$.
7. **Next iteration.** If a stopping criterion is satisfied, then the algorithm stops. Otherwise it is updated $k = k+1$ and returns to step 2.

4 Experiments

In this section, we test our approach over instances of the *Quadratic Assignment Problem* obtained from the *QAPlib* (see *http://www.seas.upenn.edu/qaplib/*), which is a well known minimization permutation problem. The selected problem instances are *scr12* and *esc32a*. The first one is classified as a small problem whose optimal solution provides a fitness of 31410, while the second one is a medium-size instance whose optimal solution has not been found yet, although it is estimated an optimal fitness of 90. On the other hand, the best solution reported for *esc32a* at the *QAPlib* has a fitness of 130. The performance of the approach is compared with the update rules *Continuous Pursuit Reward-Penalty* (*CP_{RP}*) [10], *Discrete Pursuit Reward-Inaction* (*DP_{RI}*) [10] and the *Discrete Pursuit Generalized Algorithm* (*DPGA*) [1]. They have been implemented with a frequency vector to model the expectation of reward for each automaton, as it is done in the classic training where the automata are not in a team, in order to avoid the inclusion of the hypermatrix. All the algorithms were executed under the same conditions: The stopping criterion is to reach 25000 solutions evaluated for *scr12*, and 200000 solutions evaluated for *esc32a*. The algorithms were executed 30 times so that we can make statistical analyses over the results. Table 1 describes the learning rates used for each method and problem, where *LSTMP* is the abbreviation for *Long and Short Term Memory Pursuit Algorithm* and it refers to the proposal in this work. The value λ is the learning rate of each method. These parameters were obtained after a trial-and-error procedure and testing different combinations of values for each item and algorithm.

Table 1 Parameters used for the algorithms

Algorithm	scr12	esc32a
CP_{RP}	$\lambda = 0.0001$	$\lambda = 0.0001$
DP_{RP}	$\lambda = 1/1200$	$\lambda = 1/6000$
$DPGA$	$\lambda = 1/1200$	$\lambda = 1/6000$
$LSTMP$	$\lambda = 0.01, \lvert S^i_{TM} \rvert = 6, R = 3$	$\lambda = 0.01, \lvert S^i_{TM} \rvert = 16, R = 5$

Table 2 shows the average, best and worst fitness obtained by each method in the problems solved. It is verified that the classic algorithms become trapped into local optima solutions, while the $LSTMP$ approach is able to improve the performance of the team. In the case of *scr12*, the algorithm $LSTMP$ has been able to obtain the optimal solution to the problem, while a near-optimal solution has been found for *esc32a*. On average, $LSTMP$ is also able to provide better solutions than the remaining update rules, and also the worst solution found by our proposal is better than using the classic techniques.

On the other hand, the speed of convergence is also improved with $LSTMP$. Figure 2 plots the average evolution of the fitness of the algorithms studied for *scr12* (left) and *esc32a* (right). The short-term memory favours the exploration of the solution space since the recent solutions visited are not considered for selection in the

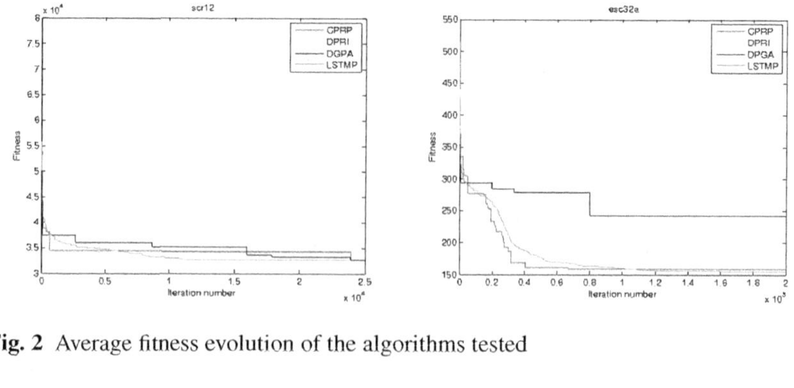

Fig. 2 Average fitness evolution of the algorithms tested

Table 2 Results obtained

Algorithm	Average (scr12)	Best (scr12)	Worst (scr12)	Average (esc32a)	Best (esc32a)	Worst (esc32a)
CP_{RP}	34072	32758	34788	174	158	192
DP_{RI}	34210	32958	35266	179.2	168	206
$DPGA$	33695	32696	34780	267.8	242	278
$LSTMP$	32641	31410	33892	154	144	164

next future iterations. This helps to avoid local optima and to explore other areas of the solution space. On the other hand, the long-term memory is used for the exploitation of promising areas of the solution space when the algorithm converges to a local optima. The combination of these strategies provides a suitable balance between diversity and convergence regarding the solutions visited. A Mruskal-Wallis test was applied to check if there are significant differences between the performance of the algorithms studied, with a 95% of confidence level. The results of the test provided a probability value of 0.4958 for the problem *scr12* and 0.0004 for *esc32a*. These results suggest that there are no significant differences between our approach and the classic methods when the number of LA in the team is small. However, as the size of the team increases, the classic update rules become ineffective and our proposal provides better performance.

5 Conclusions and Future Work

In this work, we have proposed an estimator algorithm for large teams of learning automata. Our goal is to overcome the limitations in computer memory cost required by the classic training algorithms for teams of LA, to achieve a suitable convergence to optimal solutions. We have studied the effects of the inclusion of long and short term memory features within the algorithm. The results obtained suggest that our approach is not only suitable to solve the drawbacks addressed, but also it is capable to overcome local optima solutions during the training. The short term memory is useful for a better solution space exploration, while the use of long term memory allows to exploit promising neighbourhoods. In future works we will explore additional strategies widely used in metaheuristics to achieve a suitable balance between diversity and convergence, such as the hybridation of local search operators with the long and short-term memory features of our approach.

Acknowledgements. This work has been supported by the project TIN2009-14538-C02-01, I+D+I national program, Government of Spain.

References

1. Agache, M., Oommen, B.J.: Generalized pursuit learning schemes: new families of continuous and discretized learning automata. IEEE Transactions on Systems, Man, and Cybernetics, Part B 32(6), 738–749 (2002)

2. Baba, N., Mogami, Y.: A new learning algorithm for the hierarchical structure learning automata operating in the nonstationary s-model random environment. IEEE Transactions on Systems, Man, and Cybernetics, Part B: Cybernetics 32(6), 750–758 (2002)
3. Baba, N., Mogami, Y.: A relative reward-strength algorithm for the hierarchical structure learning automata operating in the general nonstationary multiteacher environment. IEEE Transactions on Systems, Man, and Cybernetics, Part B: Cybernetics 36(4), 781–794 (2006)
4. Beigy, H., Meybodi, M.R.: Utilizing distributed learning automata to solve stochastic shortest path problems. International Journal of Uncertainty, Fuzziness and Knowledge-Based Systems 14(5), 591–615 (2006)
5. Glover, F., Laguna, M.: Tabu search, pp. 70–150 (1993)
6. Lukac, K., Lukac, Z., Tkalic, M.: Behaviour of f learning automata as multicriteria routing agents in connection oriented networks. In: The 12th IEEE International Conference on Fuzzy Systems, FUZZ 2003, vol. 1, pp. 296–301 (2003)
7. Narendra, K.S., Thathachar, M.A.L.: Learning automata: an introduction. Prentice-Hall, Inc., Upper Saddle River (1989)
8. Nowé, A., Verbeeck, K., Peeters, M.: Learning automata as a basis for multi agent reinforcement learning. In: Tuyls, K., 't Hoen, P.J., Verbeeck, K., Sen, S. (eds.) LAMAS 2005. LNCS (LNAI), vol. 3898, pp. 71–85. Springer, Heidelberg (2006)
9. Oommen, B., Agache, M.: A comparison of continuous and discretized pursuit learning schemes. In: Proceedings of 1999 IEEE International Conference on Systems, Man, and Cybernetics, SMC 1999, vol. 4, pp. 1061–1067 (1999)
10. Oommen, B., Lanctot, J.: Discretized pursuit learning automata. IEEE Transactions on Systems, Man and Cybernetics 20(4), 931–938 (1990)
11. Oommen, B.J., Hashem, M.K.: Modeling a student-classroom interaction in a tutorial-like system using learning automata. Trans. Sys. Man Cyber. Part B 40(1), 29–42 (2010), `http://dx.doi.org/10.1109/TSMCB.2009.2032414`
12. Peeters, M., Verbeeck, K., Nowé, A.: Solving multi-stage games with hierarchical learning automata that bootstrap. In: Tuyls, K., Nowe, A., Guessoum, Z., Kudenko, D. (eds.) ALAMAS 2005, ALAMAS 2006, and ALAMAS 2007. LNCS (LNAI), vol. 4865, pp. 169–187. Springer, Heidelberg (2008)
13. Poznyak, A.S., Najim, K.: Learning Automata and Stochastic Optimization. Springer-Verlag New York, Inc., Secaucus (1997)
14. Sastry, P., Thathachar, M.: Learning automata algorithms for pattern classification. Sadhana 24(4-5), 261–292 (1999)
15. Thathachar, M.A.L., Arvind, M.T.: Parallel algorithms for modules of learning automata. IEEE Transactions on Systems, Man, and Cybernetics, Part B 28(1), 24–33 (1998)
16. Thathachar, M.A.L., Ramakrishnan, K.R.: A cooperative game of a pair of learning automata. Automatica 20(6), 797–801 (1984)
17. Thathachar, M.A.L., Sastry, P.S.: Varieties of learning automata: an overview. IEEE Transactions on Systems, Man, and Cybernetics, Part B 32(6), 711–722 (2002)
18. Torkestani, J., Meybodi, M.: Graph coloring problem based on learning automata. In: International Conference on Information Management and Engineering, ICIME 2009, pp. 718–722 (2009)
19. Torkestani, J., Meybodi, M.: Solving the minimum spanning tree problem in stochastic graphs using learning automata. In: International Conference on Information Management and Engineering, ICIME 2009, pp. 643–647 (2009)

Parameter Analysis of a Genetic Algorithm to Design Linear Array Geometries

Lara del Val, María I. Jiménez, Mariano Raboso, Alberto Izquierdo, Juan J. Villacorta, Alonso Alonso, and Albano Carrera

Abstract. This article summarizes several analyses on employing an iterative learning method of the Computational Artificial Intelligence field, a Genetic Algorithm, focused on designing linear arrays. The objective of these analyses is the effectiveness improvement of these evolutive algorithms in this particular problem. The influence of giving certain values to each of the specific parameters of a Genetic Algorithm is characterized. Obtaining the optimal final solution depends on these parameter values. Thanks to this analysis, the Genetic Algorithm is optimized and also the best linear array geometry, based on certain established quality criteria, is found.

1 Introduction

Antennas are the part of the telecommunication systems designed to transmit and receive electromagnetic waves. The correct transmission and reception of a certain desired signal depends mainly on them. There are different types of antennas, but employing a set of antennas, called arrays, is the most appropriate choice, referred to directivity improvement and to better interference rejection. Arrays have the advantage that the beam can be steered electronically.

The array transmitted or received signal can be controlled by varying the power supply amplitude and/or phase of its antennas [1], or by varying its geometry [2].

This article faces the problem of positioning the sensors on the array in such a way that the array performance is the best as possible. So, this article shows an optimization problem, because it is focused on finding an optimum geometry [3]. There are several ways to solve this kind of problems: minimum search methods,

Lara del Val · María I. Jiménez · Alberto Izquierdo · Juan J. Villacorta · Alonso Alonso
Albano Carrera
Departamento de Teoría de la Señal y Comunicaciones e Ingeniería Telemática,
Universidad de Valladolid, E.T.S.I. Telecomunicación,
Paseo Belén 15, 47011 Valladolid, Spain
e-mail: lara.val@tel.uva.es

Mariano Raboso
E.U. Informática, Universidad Pontificia de Salamanca, 37002 Salamanca, Spain

A. Abraham et al. (Eds.): International Symposium on DCAI, AISC 91, pp. 217–224.
springerlink.com

random search methods, etc. In this case an iterative learning method, a Genetic Algorithm (GA) is employed, particularly a continuous GA [3].

GA are employed because they offer several advantages, but their performance and final result are conditioned to the parameters and methods that characterized them. This article analyzes the fitness of employing a continuous GA to solve the problem of designing the geometry of a linear array according to the values of the GA parameters.

2 Array Design

This study shows the search of the array sensor spacing that optimizes the spatial signal filtering [4]. This is achieved by focusing, as much as possible, the main beam of the array beampattern to the desired direction, and by placing nulls on the reception directions of the interference signals.

Main beam width is involved in directivity improvement, arrays are more directive as main beam width becomes narrower. Sidelobe level, that is, the level of the other lobes of the beampattern relative to the main beam level, has an influence on interference cancellation. It is important to be centred on those sidelobes that are adjacent to the main beam, because they can provide detrimental to a good filtering of the desired signal.

This analysis considers that array performance is better as main beam is narrower and as sidelobe level is higher- This performance is also better if the array beampattern has not grating lobes [5]. This analysis has several objectives, so this is a multi-objective optimization problem. The first objective is reached increasing the length of the array, and the second one, reducing this length. So, a compromise solution must be taken.

These objectives are modelled on a fitness function. Array performance is characterized by the quality values that are obtained through this fitness function. This quality value is relating to the main beam width and the sidelobes of a reference array: a uniform linear array (ULA) with 0.5λ sensor spacing. This reference ULA avoids grating lobes appearance and a main beam too wide [6].

Equation (1) assesses the width of the main beam, focused to the direction of interest, and compares it with the reference main beam (BW_0), that is, the main beam width of the reference array:

$$\hat{BW}_{dBi} = 10 \cdot \log\left(\frac{BW_o}{BW}\right) \tag{1}$$

i subscript indicates each pointing direction. M=20 pointing directions, uniformly distributed between -90º and 90º, have been considered.

Regarding sidelobes, those adjacent to the main lobe and grating lobes are the most important ones. So, these sidelobes are quantified, $F(\theta^{max,sl}{}_a)$, and they are weighted according to their level and position, $W(\theta^{max,sl}{}_a)$, showed in Figure 1:

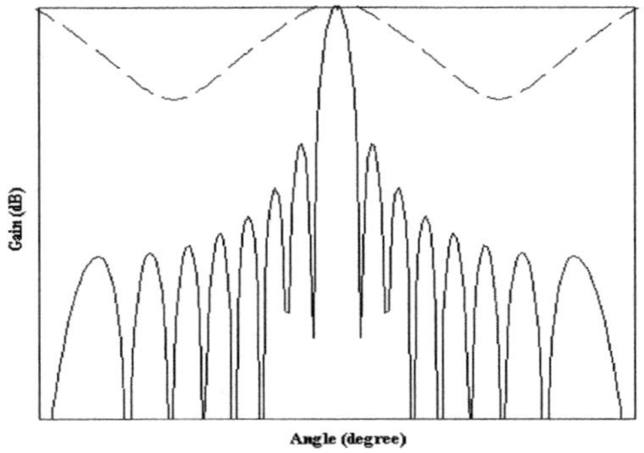

Fig. 1 Beampattern (solid line) and $W(\theta^{max,sl}_a)$ weighting function (dashed line).

$$\Delta_{sl} = \sum_{a=1}^{M-1} F\left(\theta^{\max sl}_{a}\right) \cdot W\left(\theta^{\max sl}_{a}\right) \tag{2}$$

Also this weighted sidelobe value is relative to the reference array:

$$\hat{\Delta}_{sli} = 10 \cdot \log\left(\frac{\Delta_{slo}}{\Delta_{sl}}\right) \tag{3}$$

Both quality criteria, (1) and (3), are included and weighted in a function:

$$c_i = B\hat{W}_i + 13 \cdot \hat{\Delta}_{sli} \tag{4}$$

Finally, array quality is obtained as:

$$c = \frac{1}{M} \sum_{i=1}^{M} c_i \tag{5}$$

So, the array with the highest quality will be the one with the best performance, on average, for all pointing directions.

3 Genetic Algorithm for Linear Array Design

Specific parameters of this kind of algorithms are: population size, selection rate (fraction of individuals that survives from one generation to another) and mutation rate (proportion of mutated genes in a generation). The speed and quality of the obtained solution depend on the values of these parameters.

In this optimization problem, the genes of the GA chromosomes represent the positions of 16 sensors on a linear array ($x_1, x_2, x_3 \ldots x_{16}$). The values of these positions are subjected to several restrictions in sensor spacing: the minimum spacing between sensors is 0.3λ, to be able to put one besides the other, and the maximum spacing is 0.7λ, to restrict the number of grating lobes.

Sensor spacing is randomly and sequentially generated to make up the initial population. Element distribution begins with the first element placed on the origin, and the other elements are placed one by one according to the position of the previous sensor, x_{i-1}, and the randomly calculated distance d: $x_i=x_{i-1}+d$.

Roulette Wheel-Weighting method has been used. In this selection method, each chromosome is weighted according to its quality. So, it is very probable that best chromosomes of the population are part of a new couple of chromosomes. Particularly, *Rank-Weighting* weight assignment technique has been used. This technique establishes weights according to chromosome position on the maintained population, which must be sorted in order of quality [3][7].

On the other hand, the mating method which has been used in this analysis is based on creating one or several descendant's genes from the following expressions: $p_{new1,\alpha}=p_{m\alpha}-\beta[p_{m\alpha}-p_{d\alpha}]$ and $p_{new2,\alpha}=p_{m\alpha}+\beta[p_{m\alpha}-p_{d\alpha}]$, where β is a random number between 0 and 1, and α denotes the position of the gene on the chromosome. After that, a crossover point between genes is established, and from it the other genes of the parents chromosomes are crossed [3][7]. So, the following descendants are created: $descendant_1=[p_{m1},p_{m2},..., p_{new1},..., p_{dn}, ..., p_{dN}]$ and $descendant_2=[p_{m1},p_{m2},..., p_{new1},..., p_{dn}, ..., p_{dN}]$.

Those algorithm executions carried out in the analysis must be similar, only the value of the analysed GA parameter can vary. So, all analyses start from the same parameter values, showed in Table 1. These values are defined taking previous GA studies [7][8] as a starting point.

Table 1 Algorithm initial parameters

Beamforming Parameters	**GA Parameters**	**Value**
Number of iterations	Number of generations	1000
Number of sensors	Number of genes	16
Number of arrays per iteration	Population size	8
Rate of selected arrays on each iteration	Selection rate	0.5
Rate of random changes on sensor positions	Mutation rate	0.03

4 Results of the GA Optimization Analysis

This section shows the results of the analyses on employing GA to design a linear array. Establishing the value of each GA parameter to obtain the best final solution is the objective of these analyses.

4.1 Population Size

Population size represents the number of chromosomes or arrays in each generation. The following sizes have been analyzed: 4, 8, 32 and 48. GA evolution to the optimum solution is showed in Figure 2.

It can be observed that as population size increases, convergence to a better solution is faster. This is due to higher chromosome diversity among the population, which involves higher probability of having chromosomes with higher quality in a particular generation. So, the optimum solution can be reached faster.

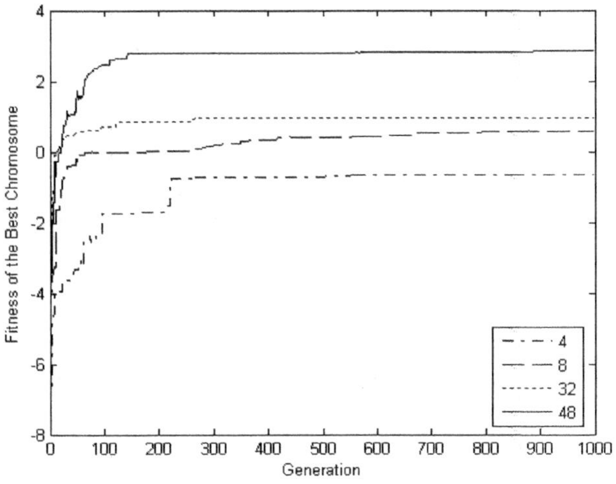

Fig. 2 GA convergence vs. population size

Average fitness evolution for each chromosome generation is shown in Figure 3. Evolution fluctuations are lower with a higher population size, because the importance of one chromosome in the whole population is lower.

The problem of increasing population size is that GA execution time also increases proportionally. Time increase is not a limitation when searching the best linear array design, because this design will remain unaltered once it is found.

4.2 Selection Rate

Selection rate represents the ratio of chromosomes that remain between generations. The following rates have been analyzed: 0.25, 0.5 and 0.75. Obtained results are shown in Figure 4.

With a high rate (0.75), 6 of the 8 chromosomes of each generation remain for the next fluctuations on the convergence function are observed. This is because maintaining more chromosomes, fewer new descendants must be generated, and fewer new combinations must be tested. So, if these new chromosomes quality is not better than the previous one, mutation phase can get quality even worse.

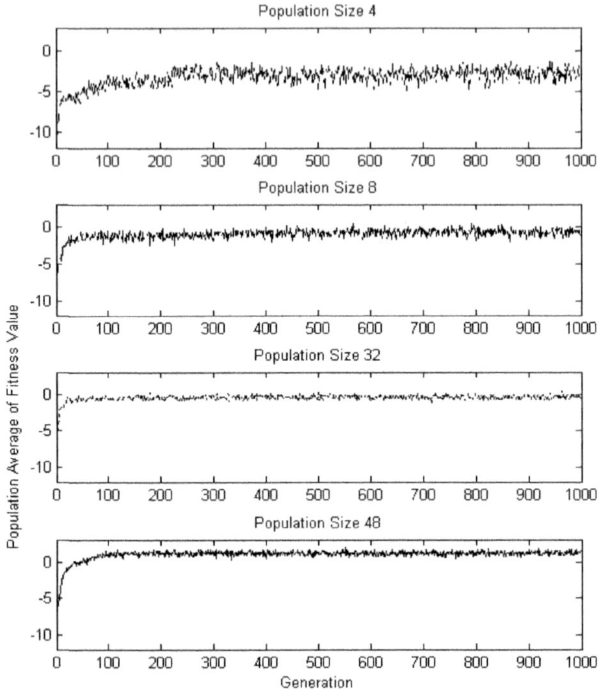

Fig. 3 Average fitness evolution vs. population size

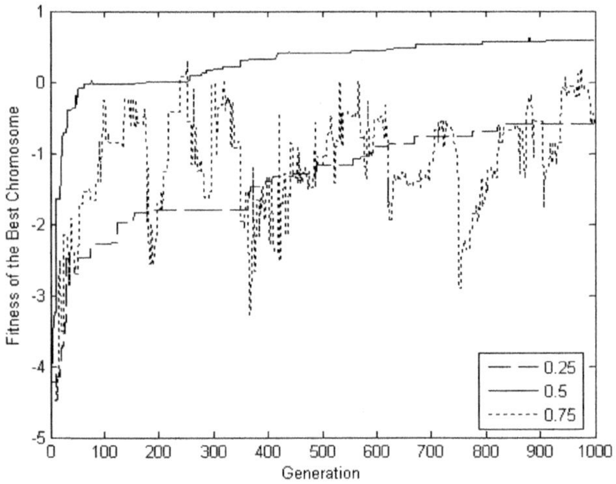

Fig. 4 GA convergence vs. selection rate

Convergence is slower with a low selection rate because creating new chromosomes with a reduced number of them makes the search of the solution slower. Generation worsening between iterations is less probable because new chromosome combinations with higher quality are more probable to be found from more new descendants created from fewer parents.

4.3 Mutation Rate

Mutation rate represents the ratio of sensor positions (genes) of the population that are mutated. The following rates have been analyzed: 0.01, 0.03 and 0.05.

Figure 5 shows that a high rate (0.05) produces big fluctuations on the convergence function. This is because more mutations involve a higher chance of improving and also worsening quality between generations.

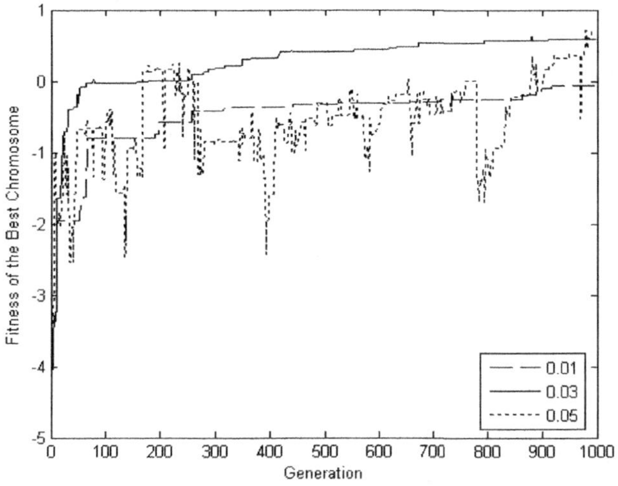

Fig. 5 GA convergence vs. mutation rate

Fluctuations on the convergence function are not caused with a low rate (0.01). In this case, chromosome diversity between generations decreases. So, very few mutations involve a lower chance of worsening all the chromosomes with equal or higher quality than those of the previous generation.

The problem of employing a low rate is that decreasing diversity between generations makes the convergence slower.

After this analysis looking for the suitable values for these three GA parameters, the following values have been chosen: a high population size (48 chromosomes), a middle selection rate (0.5) and a high mutation rate (0.05). With these values the best evolution of a GA that searches the best design of a linear array is obtained.

5 Conclusions

It has been checked that GA final solution depends to a large extent on the values of the GA parameters, not only on the parameters of the problem. That is the reason why whenever a GA is applied to an optimization problem, a previous optimization study must be done to obtain the best solution.

The analysis results show that suitable values for the parameters of a GA employed to find the optimal design of a linear array are: a high population size, a middle selection rat and a high mutation rate.

Future work can be done, studying the interrelation between the three GA parameters that have been analyzed in this paper. The influence of the value of one parameter upon the other parameters can be analyzed. Also the influence of employing different selection and mating methods can be analyzed.

This methodology is also applicable to other array geometries (2D and 3D), and also to other optimization algorithms that are based on natural processes.

Acknowledgments

We gratefully acknowledge Beatriz Lázaro Pastor for the work in Genetic Algorithms that she carried out with our research group.

References

[1] Yan, K.K., Lu, Y.: Sidelobe Reduction in Array-Pattern Synthesis Using Genetic Algorithm. IEEE Transactions on Antennas and Propagation 45(7), 1117–1122 (1997)

[2] O'Neill, D.J.: Element Placement in Thinned Arrays Using Genetic Algo-rithms, Oceans Engineering for Today's Technology, Tomorrow's Perser-vation. Naval Undersea Warfare Center Newport 2, II.301–II.306 (1994)

[3] Haupt, R.L., Haupt, S.E.: Practical Genetic Algorithms, 2nd edn. A Wiley-Interscience Publication, New Jersey (2004)

[4] Van Veen, B.D., Buckley, K.M.: Beamforming: a versatile approach to spatial filtering. IEEE ASSP Magazine, 4–24 (March 1988)

[5] Van Trees, H.L.: Optimum Array Processing. Detection, Estimation and Modulation Theory, Part IV. A Wiley-Interscience Publication, Hoboken (2002)

[6] Kingsley, S.: Understandign Radar Systems. Mc Graw-Hill, UK (1992)

[7] Johnson, J.M., Ramat-Samii, Y.: Genetic Algorithms in Engineering Electromagnetics. IEEE Antennas and Propagation Magazine 39(4), 7–21 (1997)

[8] Michielssen, E., Ranjithan, S., Mittra, R.: Optimal multilayer filter design using real coded genetic algorithms. IEE Proceedings J. 139(6), 413–420 (1992)

Rebeca Through the Looking Glass: A 3D Adventure to Learn to Program

David Miraut Andrés, Ángela Mendoza Mendoza,
Susana Mata Fernández, and Luis Pastor Pérez

Abstract. The popularity of Computer Science and Telecommunication Engineering as intended majors among incoming students has been declining during the last years. Several research groups are joining efforts and creating tools to reverse this dangerous trend for industry in Western countries. "Rebeca through the looking glass" is an educational serious game based on *Alice* -an initiative from Carnegie Mellon University- that tries to approach and to facilitate the teaching of object-oriented programming to young students. Its multilingual interface is an intuitive and visual 3D environment, designed to motivate their curiosity and encourage them to study ICT (Information and Communication Technologies) careers, without having to cope with frustrating syntax errors or enigmatic core dumps. This paper describes the keys of this new development, and our first experiences in the deployment of *Rebeca* in Educational Centers.

Keywords: E-learning, Educational Serious Game, Object-Based Programming.

1 Introduction

20 years ago, students taking their first steps in programming were filled with excitement when writing a simple "Hello World" program. Nowadays, our young people are immersed in technology; not only computers, but any type of programmable electronic devices: phones, music players, game consoles, etc. However, in this age of information -in which we have easy access to technology and the skills of design, modeling and programming are essential to the practice of many professions- young people, paradoxically, are becoming more reluctant to explore the possibilities of the machines and to discover how they are made and how to fully exploit their power.

Escuela Técnica Superior de Ingeniería Informática, Rey Juan Carlos University
e-mail: `{david.miraut,angela.mendoza}@urjc.es`
`{susana.mata,luis.pastor}@urjc.es`

A. Abraham et al. (Eds.): International Symposium on DCAI, AISC 91, pp. 225–232.
springerlink.com

Enrolment in Computer Science and Telecommunication Engineering majors has been steadily declining during the last years in Spain (and many other countries[2]), in spite of the fact that ICT is a promising sector, full of work opportunities.

In some cases, this behavior can be due to a misperception of the creative possibilities of ICT jobs. For other students, that might happen because of the reputation of ICT-related degrees of being very demanding. Both of these ideas are reinforced by the steep slope of the learning curve during the first year of College: most students experience difficulties with the development of abstract thinking, and they get discouraged when they don't see immediately the applicability of the knowledge they learn in object oriented programming courses. The Bologna process brings opportunities that must encourage us to look for new teaching strategies which motivate our students to learn on their own from their very first stages at University.

To overcome some of these problems in programming courses, we have developed a new tool, *Rebeca*, based on *Alice*, an open source software developed by Carnegie Mellon University. *Alice* [4] is an innovative tool designed to teach introductory programming concepts to undergraduates with no previous 3D graphics or programming experience. Both *Alice* and *Rebeca* communicate effectively with young people in the visual language they are familiar with, using a 3D graphics programming environment with a drag and drop interface, where they can make 3D animations and interactive video games while learning to program easily. This is very attractive to students, because the interface is designed to avoid having to face frustrating details related to the syntax of object-oriented programs in their first steps of the learning process. Each object in the virtual world is an object whose behavior can be programmed in a language similar to Java. *Alice* has been very successful in the English-speaking countries, but it has hardly been used abroad, because of its lack of support to other languages besides English. The language barrier is the main stumbling block to introduce *Alice* to young audiences who do not master the English language. Therefore, we focused our efforts on solving this problem and on extending the original software in order to localize it and to make it easily adaptable to any other language. These modifications have required a radical change in the source code and in the programming language itself. This paper describes the keys of the new development, how it can be adapted to any language or region, and our experience in the deployment of *Rebeca* in Educational Centers. Teaching guides, including a collection of problems and solutions, have been written in Spanish with Creative Common licenses which may also be translated to other languages to exploit the potential of this tool, and to show aspects of our profession to young people in an attractive way.

2 Decline of ICT Studies Enrolment in Western Countries

National Institutes of Statistics, like the U.S. Bureau of Labor Statistics [3], are clearly optimistic in relation to ICT business opportunities. However, this increasing demand for ICT engineers is not tuned to the present trend of career choices among young people. This is a real concern for Academia and Industry in Western countries. Many assumptions and conjectures have been made about the reasons that led to the decline in CS enrollment in the early 1990s and again since 2000. Most of the published studies [9] [8] [11] agree that one of the main reasons is the lack of accurate information about ICT professional opportunities among high school students when they are making choices about future careers and appropriate Colleges. This situation becomes aggravated because there are not enough certified high school teachers in CS in Western countries, and High School curricula changes do not encourage an emphasis on Math and Sciences. Creativity and innovation are present in every aspect of ICT professionals' work when searching for new solutions. But young people, in general, do not perceive any excitement about ICT, because computer-related subjects in High Schools tend to cover just basic knowledge of office suites [1]. In order to bring high school students closer to the professional reality, our universities have developed innovative initiatives.

3 New Tools in a New Era

Nowadays, high school students are in close contact with ICT. During their education, they even develop some ICT skills, usually through an informal training in these areas. But -paradoxically- they are more reluctant than ever to take an education in ICT studies and build a career in this field. As mentioned before, our environment and the stimuli we receive from it have clearly changed. Therefore, if we want to raise the student's curiosity and to train good and motivated professionals, it is necessary to change the approach to reach them. We can take a more attractive path, without losing academic rigor, by using a visual language closer to their way of life and to how they create things.

3.1 Rebeca and Alice

"Rebeca through the looking glass" is an educational innovation project based on *Alice*. *Alice* and *Rebeca* allow to easily generate 3D animations to tell stories, and to create interactive games and videos that can be later on shared on the Internet, through a friendly programming environment.

Alice was initially developed as part of a Virtual Reality research project by the Carnegie Mellon Graphics group, led at that time by Randy Pausch. This project has been evolving over 15 years [12], and it has been supported by numerous Departments, teachers, students and companies. *Alice* has been an overwhelming success in the English-speaking community, where this software is used in classes related to Technology by more than a thousand secondary schools [6] [7] [5]). However, their acceptance in other countries has been rather modest. The reason for this difference is that *Alice* is a software created by and for English speakers. It was not designed to be translated into other languages. Additionally, its structure reflects its 15 years of continuous diverse contributions, becoming a project of more than 170,000 lines of code, largely unstructured in 1900 Java, Python and Haskell files. Neither standard internationalization (i18n) Java process, nor traditional debugging techniques can be used in Alice. For this reason and despite it is one of the most requested features, its creators have dismissed the possibility of supporting other languages. This is an additional problem for our students, because they not only face the difficulty of learning to program, but they must also deal with a foreign language. Often, the effort to overcome the language barrier is higher than the one needed to assimilate programming concepts, as our preliminary studies have shown (section 5). "Rebeca through the looking glass" was born as a challenge in response to this need, with the aim that all young people can benefit from the tool that has risen so many vocations overseas.

3.2 *Object-Oriented Programming Learning with Rebeca and Alice*

Learning to program is a hard task for the majority of students, and its complexity is seen as one of the main factors that discourages students in the first year [13].

Rebeca and *Alice* provide mechanisms to overcome typical students' difficulties, such us rigid syntax, unfamiliar structure and the amount of time spent to produce a simple output. The drag-and-drop integrated development environment (IDE) of both *Rebeca* and *Alice* eliminates syntax problems that bedevil first-year undergraduate students. As the student drags-and-drops graphical elements, the source code is constructed and displayed. The student is never permitted to freely edit the code as is the case with most programming development environments. This again shields the student from making syntax errors, but allows them to become familiar with programming language constructs.

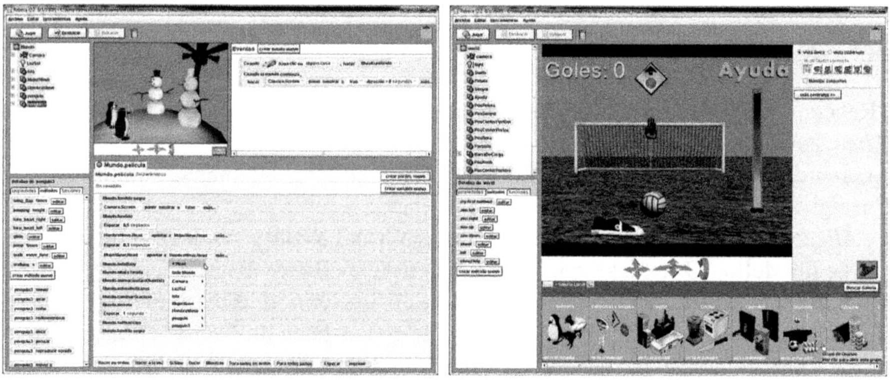

Fig. 1 Rebeca 3D programming enviroment and object gallery.

With *Rebeca* and *Alice*, students can check the status of the program in a much more intuitive way than with a debugger. For example, it is easier to see that an object moves forward rather than backward, than it is to check if a variable has been decremented instead of incremented. Thus, logic errors lead to funny situations that are easily detected. Also, the three-dimensional interface both *Rebeca* and *Alice* present, and the possibilities they offer for creating interactive stories in a collaborative way stimulate the imagination of boys and girls and keep them engaged. Storytelling is a useful metaphor that can help structuring the process of creating programs in a similar way to the animation production process. The design is divided in different stages, in which students sequentially develop storyboards, scripts, dialogs, shots and scenes... to turn the story into pseudocode. At the end of this process, students detail their ideas in methods and functions through the interface. Advanced concepts such as concurrency and parallel algorithms can be explored and studied in a simple way through logical structures specially designed for this purpose, such as *Do together* or *For all together*. Even simple agents can be programmed as characters that interact in the story or with the player. *Rebeca* can be used in High Schools and first year computer programming courses at College. A hybrid approach where *Rebeca* is used at the beginning of a course and Java is teached during the remainder of the semester may be more effective in reaching educational objectives while helping still to motivate and retain students' interest.

4 Rebeca Development

Rebeca has been the result of the joint work of several Graduation Projects offered by the Modeling and Virtual Reality Group at the Rey Juan Carlos University, among which Sergio Ruiz and Irene Montano contributions have been specially relevant. This makes *Rebeca*'s development very special, since it is a piece of software made by and for students with great love and care in details.

Internationalization is the process of designing software so that it can be translated into different languages and regions without the need for further changes to the code. *Alice* has been internationalized and localized as a whole (not just the interface or the tutorials), to allow porting it easily to different regions and languages. Thus, we had to rewrite thoroughly the original software: more than 1800 files have been modified. As a consequence the localization to any language is now immediate. It only requires the edition of a new set of XML resource files that were added to the original project. These text files have pairs key-value, where original Alice expressions are the key and the places for the values are filled in with the translated sentences. In our case the location has been made to Spanish, but thanks to the prior internationalization process, this software can be ported to any other language.

Since many European languages use characters that do not exist in English, it was necessary to modify the set of applications and how they communicate to use an international character set. Again, this feature has required a laborious task of reverse engineering. Carnegie Mellon distributes *Alice* code under a free open license, but it is not documented and current developers do not give support to make modifications to *Alice* version 2 in official forums.

The development of *Rebeca* has allowed us to fix bugs and add new features related to stencil tutorial support, so that the learning experience can be more effective. A teaching guide has also been written in Spanish [10]; this book can be used as an introductory manual for school teachers who want to try *Rebeca* in their courses.

5 First Teaching Experiences and Discussion

Several pilot courses, workshops and competitions of different duration have been carried out in a number of events held in Madrid (Spain) during the last months, such as Juvenalia, fesTICval, Science Week, and different seminars held at the Rey Juan Carlos University. The experience has been very positive in all aspects, particularly thanks to the excellent response of High School students, who stayed in front of their computers even during the breaks. Students did more optional exercises and attend more classes than other groups using traditional tools. Many of them were amazed to discover that programming is a creative discipline, and a 85% of the students considered a good idea to teach *Rebeca* and object-oriented programming in their schools.

During a 3-day ICT Summer Camp for High School students of different years, an experiment with *Alice* and *Rebeca* tools was performed. Half of the students (group A) worked first with *Alice* and then with *Rebeca*, and the other half (group B) worked in reverse order. Both groups had to complete a set of guided tutorials and then create their own animation or small interactive game in pairs that would demonstrate what they had learned. After the experience the students were invited to complete an opinion poll with questions following psychometric Likert scale.

Group A students carried out the tutorials with *Alice* at the beginning, so they had to make an extra effort to understand the vocabulary and follow the tutorials, nearly 45% complete all the tutorials in English; many of them were last year students so their language dependence is not noticeable, but they stated some frustration because of English vocabulary difficulty. All other students left *Alice* before finishing the first stage and they chose to do the tutorials with *Rebeca*. 85% of the participants in group A developed their graphic animation with *Rebeca*, because they considered this tool to be easier and more fun to program.

Group B students were asked to accomplish the first tutorials with Rebeca and all the others with *Alice*. At the beginning, there was a strong motivation to understand and perform the tutorials, all of them agree *Rebeca* allows to learn programming concepts faster. In the second part of the Summer Camp, this group was less interested with *Alice*, even though they knew the structure of the interface of both tools. Only this group had students of first courses; they had serious difficulties to perform tutorials with *Alice* because they could not understand English instructions and programming concepts in a foreign language.

Figure 2 represents some of the variables measured in the survey. We can observe that students in the first years of both groups have a strong dependence on the language. Spanish students decide their professional vocation and what kind of career they want to study in near future during these first courses in High School. Therefore, it is especially important to show them the professional reality of ICT

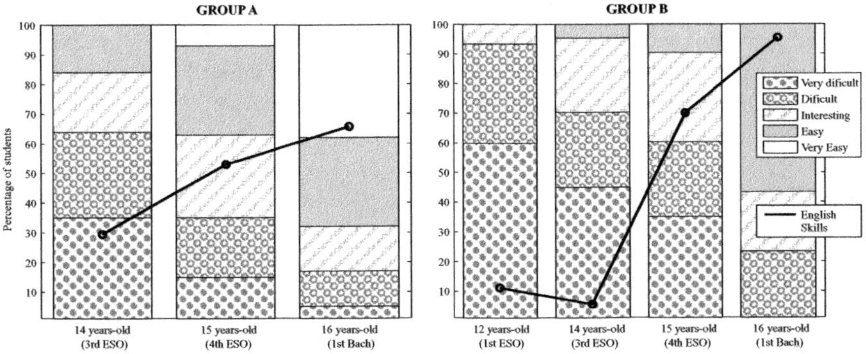

Fig. 2 Students' opinion about *Alice* difficulty in comparison with *Rebeca* (bars) and dependency with Spanish language (bold line).

engineering in a way that the language barrier does not hide the beauty of Computer Science.

The enthusiasm shown by High School students encouraged us to keep improving this tool. In a near future, we will conduct a study to measure the effectiveness of this tool as an aid in first year programming courses between different groups of students, in order to plan future steps in their development.

6 Conclusions and Future Work

The result of this project has been a piece of software completely localized to our language, ready to be ported to any other language and ready for use in teaching. Not only the interface, but also the programming language has been translated. We have designed a programming language similar to Java but with the terminology in Spanish. We have tried to make it as accessible as possible to young audiences, so they can really focus on learning the methodology of object-oriented programming. "Rebeca through the looking glass" and all teaching materials are completely free, and can be downloaded from the official site of the application: `www.gmrv.es/rebeca-es` As soon as possible, the website will also include: a centralized repository where young people can share animations and games made with *Rebeca*, a forum with a *think tank* who will help solve any questions that may arise in the use of the tool and its use in the courses, a blog where we will collect suggestions from teachers and students, and several extensions and plugins to upload animation videos to the most popular social networks.

Acknowledgements

Rebeca through the looking glass is dedicated to all teachers and students of the Rey Juan Carlos and Carnegie Mellon Universities that made it possible. And especially, to Randy Pausch, so that his legacy can reach all young people worldwide.

Rebeca through the looking glass project is funded by VI Investigator Grant Call to Educational Innovation and Improvement of Teaching 2010/11 from Rey Juan Carlos University; including support from Escuela Técnica Superior de Ingeniería Informática in dissemination activities and organisation of *Rebeca*'s events for young people.

References

1. Ali, A., Shubra, C.: Efforts to reverse the trend of enrollment decline in computer science programs. The Journal of Issues in Informing Science and Information Technology 7, 209–225 (2010)
2. Becerra-Fernandez, I., Elam, J., Clemmons, S.: Reversing the landslide in computer-related degree programs. Commun. ACM 53, 127–133 (2010)
3. Career Guide to Industries Bureau of Labor Statistics (2010-11) U.S. Department of Labor. Computer systems design and related services (2010), retrieved from the Bureau of Labor Statistics web site on November 20 (2010)
4. Dann, W.P., Cooper, S., Pausch, R.: Learning to Program with Alice, 1st edn. Prentice-Hall, Inc., Upper Saddle River (2006)
5. Kelleher, C.: Barriers to programming engagement. Advances in Gender and Education 1, 5–10 (2009)
6. Kelleher, C., Pausch, R.: Lowering the barriers to programming: A taxonomy of programming environments and languages for novice programmers. ACM Comput. Surv. 37, 83–137 (2005)
7. Kelleher, C., Pausch, R.: Using storytelling to motivate programming. Commun. ACM 50, 58–64 (2007)
8. Lenox, T., Woratschek, C.R., Davis, G.A.: Exploring declining cs/is/it enrollments. Information Systems Education Journal 6 (2008)
9. Lomerson, W.L., Pollacia, L.: Declining cis enrollment: An examination of pre-college factors. Information Systems Education Journal 4 (2006)
10. Montano, I., Ruiz, S.: Guía didáctica de Rebeca: Aprende a programar con gráficos 3D interactivos, 1st edn., Lulu, Madrid, Spain (2010)
11. O'Lander, R.: Factors effecting high school student's choice of computer science as a major. In: Proceedings of the Symposium on Computers and The Quality of Life, CQL 1996, pp. 25–31. ACM, New York (1996)
12. Pausch, R., Burnette, T., Capeheart, A.C., Conway, M., Cosgrove, D., DeLine, R., Durbin, J., Gossweiler, R., Koga, S., White, J.: Crisis and opportunity in computer science. IEEE Computer Graphics and Applications 15, 8–11 (1995)
13. Sloan, R.H., Troy, P.: Cs 0.5: a better approach to introductory computer science for majors. SIGCSE Bull. 40, 271–275 (2008)

A New Evolutionary Hybrid Algorithm to Solve Demand Responsive Transportation Problems

Roberto Carballedo, Eneko Osaba, Pablo Fernández, and Asier Perallos

Abstract. This paper shows the work done in the definition of a new hybrid algorithm that is based on two evolutionary techniques: simulated annealing and genetic algorithms. The new algorithm has been used to solve the problem of finding the optimal route for a bus in a rural area where people are geographically dispersed. The result of the work done is an algorithm that (in a reasonable time) is able to obtain good solutions regardless of the number of stops along a route.

Keywords: meta-heuristics, simulated annealing, genetic algorithm, demand responsive transport.

1 Introduction

Nowadays public transport systems have some drawbacks to meet the demand for all passengers. The most obvious one is the limitation of the resources. Although each transport system has its own characteristics, there are some limitations shared by all of them: the capacity of the vehicles, the frequency and schedules of the services, and the geographical area of coverage. As a result of this arises the concept of transportation on demand. This concept aims to adapt transportation systems to passenger demand in an efficient manner. Many of the techniques used to solve these problems do not yield an exact solution. This is because the types of problems to be solved are classified as NP-hard [1]. For this reason, heuristics techniques are used for obtaining good approximations.

This paper is divided in 6 sections. Section 2 presents the main types of well known problems related to the work done. Section 3 illustrates the most commonly used strategies in the field of route optimization. Section 4 presents the approach followed to define our hybrid algorithm. Section 5 presents the results of the tests done to validate our algorithm and finally conclusions and future work is detailed.

Roberto Carballedo · Eneko Osaba · Pablo Fernández · Asier Perallos
Deusto Institute of Technology - DeustoTech, University of Deusto,
Avda. Universidades 24, 48007 - Bilbao, Spain
e-mail: {e.osaba,roberto.carballedo,perallos}@deusto.es,
pablo.fernandez@deusto.es

A. Abraham et al. (Eds.): International Symposium on DCAI, AISC 91, pp. 233–240.
springerlink.com

2 Transportation on Demand

Transportation-On-Demand (TOD) [2] is concerned with the transportation of passengers or goods between specific origins and destinations at the request of users. Most TOD problems are characterized by the presence of three often conflicting objectives: maximizing the number of requests served, minimizing operating costs and minimizing user inconvenience. As is common in many combinatorial optimization problems, these objectives are conflicting and it is needed to sort them by importance.

2.1 Well Know Transportation Problems

Most of the problems arisen in transportation on demand topic have similar characteristics, which means that they can be framed as instances of other generic and well know problems. In this section, we present the most common traditional problems in the field of transportation on demand.

Traveling Salesman Problem (TSP) [4]: The Travelling Salesman Problem (TSP) is an NP-hard problem in combinatorial optimization studied in operations research and theoretical computer science. Given a list of cities and their pair-wise distances, the task is to find a shortest possible tour that visits each city exactly once. This type of problem is used as a benchmark for many optimization algorithms.

Vehicle Routing Problem (VRP) [5]: The vehicle routing problem (VRP) is a generalization of the TSP. The aim of the problem is to service a number of customers with a fleet of vehicles. Often the context of this type of problem is related to deliver goods located at a central depot to customers which have placed orders for such goods. Implicit is the goal of minimizing the cost of distributing the goods. Many variants of the VRP are described in the literature [6]. These problems include the addition of variables and constraints. One of the most popular variants includes time windows for deliveries. These time windows represent the time within which the deliveries (or visits) must be made. [7]

Demand Responsive Transport (DRT): Demand Responsive Transport or Demand-Responsive Transit (DRT) or Demand Responsive Service is an advanced, user-oriented form of public transport. It is characterized by flexible routing and scheduling of small/medium vehicles operating in shared-ride mode between pick-up and drop-off locations according to passengers needs. DRT systems provide a public transport service in rural areas or areas of low passenger demand, where a regular bus service may not be economically viable. DRT systems are characterized by the flexibility of the planning of vehicle routes. These routes may vary according to the passenger' needs in real time. This is the type of problem that we used to benchmark the algorithm proposed in this paper.

3 Artificial Intelligence Techniques and Algorithms

In the literature we can find many attempts to find an exact solution to the problems explained in the previous section. For most routing problems is not possible to find the optimal solution, for that reason, there have been a number of strategies to find an acceptable solution, taking care of the basic criteria of the computational complexity: time needed to obtain the solution and consumption of computational resources. This section details the most commonly used techniques for solving the problems explained in the previous section.

3.1 Local Search Algorithms

Most of solution methods begin the resolution process by generating an initial solution that does not have to be correct. From this, and iteratively, these algorithms "search" for a better/good solution. These techniques use an objective function that measures the quality of the solutions obtained during the search process. In this scope we can find the local search techniques:

Simulated Annealing [8]: This is one of the most popular local search techniques. It is based on the physical principle of cooling metal. Using that analogy, it generates an initial solution and the process proceeds by selecting new solutions randomly. The new solutions are not always better than the initial solution, but as time passes and the temperature decreases (the metal becomes stronger), each new solution must be better than previous solutions.

Tabu search [9]: This technique is similar to Simulated Annealing, but with a different approach when selecting the successive solutions. In this case, several memory spaces are used, in which solutions found and discarded during the search process are stored.

Ant Colony [10]: This algorithm simulates a colony of artificial ants working in groups and communicating through artificial pheromones trails. Each artificial ant builds a solution to the problem and the path to reach that solution. When all ants have completed a trip to a solution and all trips are reviewed, the traces are stored. The process of paths constructing is repeated until almost all ants follow the same trip in each cycle.

3.2 Evolutionary Algorithms [11]

These methods include algorithms inspired by the laws of natural selection and the evolution of the animal species. In most cases, an initial population of solutions is defined. This initial population consists of a number of individuals (solutions of the problem.) Then, with the combination and evolution of these individuals, the algorithm tries to get a better solution. The most popular technique in this field is genetic algorithms, which are inspired by the biological evolution of species.

3.3 Hybrid Local Search

This is one of the most used strategies. This approach attempts to solve the problems faced by traditional strategies. To this end, several strategies are combined (usually 2) in a single process. This allows grouping the advantages of each strategy and solving its individual problems. As explained below, this is the approach we used for the design of our algorithm.

4 Proposed Algorithms

Having defined the main types of problems related to route optimization, and techniques used for resolution, we will specify the algorithm that we designed, and the problem we used to validate it. The algorithm we have designed allows us to model and solve any combinatorial optimization problem. Nevertheless, we have defined an instance of a DRT problem, to illustrate the operation and performance of the algorithm.

4.1 Description of Our DRT Problem

To verify our new algorithm, we have defined an instance of a DRT problem. Our problem refers to a bus on demand system. The passengers make requests for travel from one stop to another. There are 15 stops. Five of the stops are mandatory and the rest are optional. The position of all stops is fixed and known, but the passage of buses by an optional stop depends on the passenger demand. The bus will pass the optional stops, if passenger demand exceeds a certain threshold. If the bus does not go through any optional stop, the route between the mandatory stops is always the same. If the bus has to pass more than an optional stop, the route between two mandatory stops should be calculated dynamically to minimize the distance traveled by the bus. The optimization problem we have to solve is based on the calculation of the optimal path between two mandatory stops, through a series of optional stops.

To solve the problem, we designed a hybrid algorithm that combines simulated annealing methods and genetic algorithms. Then we explain the details of each technique separately.

4.2 Simulated Annealing

As explained above, this is a meta-heuristic algorithm based on the physical principle of metals cooling. The most important characteristic of this algorithm are:

Concept of state: A state of a problem, define a specific situation of the problem. This situation is defined by the fundamental elements that make up the problem. In our problem, a state is defined by the order in which the bus travels through the optional stops between two mandatory stops. Then the state of our problem is a path between several stations.

Evaluation function: This function measures the quality of a state. This quality is usually associated with a numerical value that allows us to compare states and determine which is better. In our problem, the evaluation function is the sum of the distances between the stations that make up a state. The evaluation function is the criterion for determining that a solution is better than another.

Successor function: The objective of this function is to obtain a new state based on the current state and the temperature. For this, it takes a random exchange in the order of the stations of the current state, changing the path also. The successor function is designed to create a new state from another. In our problem, the successor function performs a random change of the position of two stops. With this change, a new state is created. This new state represents a new route and it has a new value of evaluation function, usually different from the previous state's value.

As explained previously, the process of simulated annealing algorithm is based on the generation of successor states iteratively. In each iteration, if the value of the evaluation function of the new state is better than the current state value, the successor state becomes the new current state. Otherwise, the successor state will be the new current state with a certain probability that decreases as temperature decreases. Therefore, the temperature is used to select the successor states that do not have a better evaluation function as the new current state.

The temperature function is a mathematical function, which is updated each iteration, and allows controlling the selection of "bad" successor states. In the first iteration, the value of the temperature function will be high and the probability of choosing "bad" successor states will be great, but as the temperature value decreases, the probability of choosing "bad" states, will also decrease.

4.3 Genetic Algorithm

Genetic algorithms are based on the principles of natural selection of species. For this reason, these algorithms work with concepts of chromosomes, genes, genetic combination and mutation.

One of the most important tasks when working with genetic algorithms is the definition of the concept of state. The states of a genetic algorithm (also known as chromosome) are composed by genes. Each gene is a property or a characteristic of the problem. In our problem, a gene represents a stop, and a chromosome is defined by a sequence of stops.

The operation of a genetic algorithm is based on the evolution of an initial population of chromosomes through a series of iterations. The chromosomes evolve through the crossover and the mutation of genes. The basic operation of a genetic algorithm can be defined as follows:

1. Creation of the initial population. In our problem, we create a series of random routes, which represent the initial population.
2. Evaluation of each of the individuals (chromosomes) using an evaluation function. In our problem, this evaluation function is based on the sum of the distance between the stops that define a chromosome.
3. Start an iterative process until it reaches the threshold of generations

4. Selection of the best chromosomes to be parents. The selection process was carried out based on a fitness function.
   4.1. Generation of new chromosomes from the cross between parental chromosomes. The creation of new chromosomes is done using a crossover function.
   4.2. Once new chromosomes are generated, a process of mutation of some genes of the new chromosomes is performed.
   4.3. Selection of chromosomes that form part of the new population. After performing the process of crossover and mutation, the resulting chromosomes are evaluated by fitness function, and the best ones are selected to be part of the new population.
5. Once the process of generating new populations, the solution is the best chromosome of the current population.

Fitness function: This is the function used to measure the quality of the chromosomes. The quality depends on the order of genes, since the value is the sum of the distances between the genes (stops) of a chromosome.

Crossover function: This is the function used to perform the reproduction process. Usually each crossover generates two children. Each child is formed from fragments of each of their parents. In our problem this process is complex, since stations cannot be repeated. This is the simplest reproduction process but there are other ways to make the crossover process.

Selection criteria: The selection criterion is used twice in the process of the algorithm: the selection of the parents of the new population and the selection of the best individuals after a full iteration. There are multiple criteria, from which selected all individuals, even those who selected only the best individuals (according to fitness function). In our algorithm Stochastic Remainder Criteria was used. This selection criterion selects all individuals whose probability of selection is above the average probability of selection of the entire population (according to the value of the fitness function). If this criterion is not reached the target number of individuals to choose, other individuals were selected randomly.

5 Test and Solution Proposed

As indicated above, for the design of our hybrid algorithm, separate versions of simulated annealing and a genetic algorithm have been implemented. In addition, we have implemented a "brute force" algorithm, to find out the optimal solution for small instances of the problem (with few intermediate stops).

With these 3 algorithms, there have been a series of tests to measure the performance of each algorithm and the ability of each one to solve the problem. As a result of these tests, we have obtained several conclusions:

1. The “brute force” algorithm is optimal because it always finds the best solution. Even so, it has the disadvantage that the execution time is unacceptable when the number of stations increases to more than 9 (for a large number of

stations cannot even get a solution). This algorithm cannot be used in a real scenario.
2. The simulated annealing algorithm only finds the optimal solution when the first and last station does not vary during the resolution process. Running time is always the same regardless of the number of stops.
3. In the case of genetic algorithm, the execution time is constant if the number of generations is also constant. An advantage of this algorithm is that the probability of finding a good solution is independent of the number of stops.

After preliminary analysis of algorithms separately, we came to the conclusion that the results of runtime and solution quality were not good. For this reason we decided to combine the two heuristics.

5.1 Our Hybrid Algorithm

Our hybrid algorithm came up with the aim to combine the advantages of genetic algorithms and simulated annealing:

- Rapid and constant execution time (simulated annealing).
- Probability of finding a good solution for the problem instances with many stops (genetic algorithm).

The solution would avoid the main drawback of the two algorithms:

- The solution should be optimal or very close to it.

With all these goals, it thought about making the hybrid. By nature of the two algorithms, it is appropriate to insert the execution of simulated annealing algorithm in the execution of genetic algorithm. That is because the first algorithm is focused on only one solution and the second works simultaneously with different solutions.

Having decided the model of integration, there were two options to do the integration:

- Integrate the simulated annealing in the process of creating the initial population.
- Integrate the simulated annealing in the process of reproduction, right after generating the new population.

After several tests, we concluded that the most effective solution was to apply the simulated annealing algorithm just after the reproduction process. Below is a table showing the results of the tests. The table shows the number of stops, the number of generations used in the genetic algorithm, runtime, and the percentage of times the algorithm finds the optimal solution.

Comparing the proposed alternative with each of the separate algorithms, we can ensure that the execution time is right, regardless of the number of stops. Moreover, in situations where the optimal solution is not found, the average deviation for the optimal solution does not exceed 3% of the value of the optimal solution.

Table 1 Results of the tests.

N. of stations	N. of generations	T. of execution	% of optimal solution
9	5	3 seconds	80%
9	10	5 seconds	100%
10	5	3 seconds	80%
10	10	5 seconds	100%
11	5	3 seconds	80%
11	10	5 seconds	100%

6 Conclusions and Further Work

The work presented is the result of a research project funded by the Basque government. The aim of the project was the optimization of on-demand bus transport systems. Our algorithm is integrated into a Web application that allows passengers to make requests via a mobile device. With these requests, using the algorithm described, we construct the bus route dynamically. In addition, if a request will not be met, the system notifies the passenger the nearest station in which he can take the bus.

During the implementation of the algorithm different software design patterns have been used. This has allowed the generation of a library for modeling and solving problems of route optimization, which may be used in future developments.

At present, we are working on the design of a methodology that facilitates the modeling of route optimization problems to take into account constraints associated with vehicles (capacity and cost of travel) and passenger preferences (time restrictions).

References

1. Garey, M.R., Johnson, D.S.: Computers and Intractability; a Guide to the Theory of Np-Completeness. W. H. Freeman & Co., New York (1990)
2. Jorgensen, R.M., Larsen, J., Bergvinsdottir, K.B.: Solving the Dial-a-Ride problem using genetic algorithms (2004)
3. Applegate, D.L., Bixby, R.M., Chvátal, V., Cook, W.J.: The Traveling Salesman Problem (2006), ISBN 0691129932
4. Dantzig, G.B., Ramser, J.H.: The Truck Dispatching Problem. Management Science 6(1), 80–91 (1959)
5. Pisinger, D., Ropke, S.: A general heuristic for vehicle routing problems (2005)
6. Repoussis, P.P., Tarantilis, C.D., Ioannou, G.: Arc-guided evolutionary algorithm for the vehicle routing problem with time windows (2009)
7. Rutenbar, R.A.: Simulated Annealing algorithms: an overview (2002)
8. Gendreau, M., Hertz, A., Laporte, G.: A tabu search heuristic for the vehicle routing problem (1994)
9. Dorigo, M., Gambardella, L.M.: Ant Colony System: A Cooperative Learning Approach to the Traveling Salesman Problem (1997)
10. Repoussis, P.P., Tarantilis, C.D., Ioannou, G.: An Evolutionary Algorithm for the Open Vehicle Routing Problem with Time Windows (2009)
11. Zhang, L., Yao, M., Zheng, N.: Optimization and improvement of Genetic Algorithms solving Traveling Salesman Problem (2009)

Complications Detection in Treatment for Bacterial Endocarditis

Leticia Curiel, Bruno Baruque, Carlos Dueñas,
Emilio Corchado, and Cristina Pérez

Abstract. This study proposes the use of decision trees to detect possible complications in a critical disease called endocarditis. The endocarditis illness could produce heart failure, stroke, kidney failure, emboli, immunological disorders and death. The aim is to obtained a tree decision classifier based on the symptoms (attributes) of patients (the data instances) observed by doctors to predict the possible complications that can occur when a patient is in treatment of bacterial endocarditis and thus, help doctors to make an early diagnose so that they can treat more effectively the infection and aid to a patient faster recovery. The results obtained using a real data set, show that with the information extracted form each case in an early stage of the development of the patient a quite accurate idea of the complications that can arise can be extracted.

1 Introduction

Machine Learning [1, 2] is a field related to tasks as recognition, diagnosis, planning, robot control, prediction, etc. These concepts involve techniques, such as algorithms for dimensionality reduction as PCA [3], artificial neural networks [4], genetic algorithms [5, 6], fuzzy systems [7] and swarm intelligence [8], which

Leticia Curiel · Bruno Baruque
Department of Civil Engineering, University of Burgos, Burgos, Spain
e-mail: lcuriel@ubu.es

Carlos Dueñas
Complejo Hospitalario Asistencial de Burgos, Servicio de Medicina Interna, Burgos, Spain
e-mail: cjdg@hgy.es

Emilio Corchado
Departamento de Informática y Automática, Universidad de Salamanca, Salamanca, Spain
e-mail: escorchado@usal.es

Cristina Pérez
Complejo Hospitalario Asistencial de Burgos, Servicio de Medicina Interna, Burgos, Spain

A. Abraham et al. (Eds.): International Symposium on DCAI, AISC 91, pp. 241–248.
springerlink.com

investigate complex problems to solve real problems in fields as medicine [9], ecology [10], engineering [11], industrial process [12] and so on.

Endocarditis is a term used to describe a serious infection of the endocardium that can cause severe damage to the inner lining of the heart, to any of the four valves of the heart and to other structures such us the interventricular septum, the chordae tendineae, the mural endocardium, or even on intracardiac devices. The infection can occur in any age and either sex. Usually, the illness is caused by a growth of bacteria on one of the heart valves, leading to an infected mass called "vegetation". It could be classified in:

- Bacterial endocarditis: this is produced when bacteria enter the bloodstream.
- Fungal endocarditis: occurs in people with low resistance to infection, such as those who are taking medications that suppress the immune system.
- Noninfective endocarditis: is a heart inflammation caused by advanced step of cancer or by disorders of the immune system.

According to the American Heart Association (AHA), the infection may be contracted during brief periods of introduction of bacteria in the bloodstream, such as after dental procedures, tonsillectomy or adenoidectomy, examination of the respiratory passageways with an instrument known as a rigid bronchoscope, certain types of surgery on the respiratory passageways, the gastrointestinal tract, or the urinary tract and gallbladder or prostate surgery.

The endocarditis can be diagnosed by many procedures [13, 14] such as transthoracic echocardiography, by transesophageal echocardiography, by Duke Criteria, by autopsy, etc.

Once the illness has been diagnosed a rapid initiation of an adequate therapeutic regimen is important to prevent the patients from severe complications such as heart failure, stroke, kidney failure, septic embolism and various immunological phenomena, variety of systemic signs and symptoms through several mechanisms, including infertility or death.

The main treatment [13, 14] of the infection is through aggressive antibiotics, usually intravenously, which attack the microorganisms. The problem is that the diagnosis of what kind of bacteria originated the infection is based on positive blood culture results with identical microorganisms, which is not an immediate process. So, usually, doctors in many cases have to begin the treatment before knowing the specific bacteria the patient is infected with. Also, antibiotic treatment is sometimes not enough because the valve has been severely damaged and a surgical replacement of the valve is required.

For all these reasons the correct treatment of the patient in the earliest stage as possible is considered as an interesting objective. To help to achieve it, this research proposes the use decision tree [15, 16] techniques to recognize possible complications once the patient is in treatment, helping to identify in advance possible solutions.

The remaining of this paper is organised as follows. Section 2 introduces the decision tree learning techniques used to construct the different classifiers presented. Section 3 describes the dataset used for this analysis; Section 4 shows the experiments and results obtained. Finally, in Section 5, the conclusions are set out and comments are made on future lines of work.

2 Tree Learning Algorithms

Machine Learning [1, 2] deals with algorithms that can construct models to estimate or predict the class to which new cases belong to. One manner to do it is through decision trees [15]. A tree is a leaf node labelled with a class linked to two or more nodes, where each branching node represents a choice between different alternatives. So, to classify instances, an attribute-vector must be presented to the tree and evaluate each of its composing attributes in the corresponding node. To complete the classification process, some tests into the attributes obtained reaching one or other leaf, are carried out.

The inputs of a decision tree consist on a collection of training cases with an expected dependence between variables. Each of the training cases is included into a single class into which the problem to solve is divided. The goal of the decision trees is to learn from these training cases to be able to classify futures instances.

In the following subsections three commonly used systems for induction of decision trees for classification are described: CHAID, ID3 and C4.5.

2.1 Chi-squared Automatic Interaction Detection

Chi-squared Automatic Interaction Detection (CHAID) [17] is a decision tree method useful in exploratory analysis that relates a potentially large number of categorical predictor variables to a single categorical nominal dependent variable.

The method was proposed as a modification of the Automatic Interaction Detector method (AID) [18] for categorized dependent and independent variables.

The algorithm incorporated a sequential merge and split procedure based on a chi-square test statistic and proceeds in steps as follows:

- Cross tabulate the m categories of the predictor with the k categories of the dependent variable.
- Then, find the pair of categories of the predictor which account for the least significant difference on a chi-square test and merge these two categories.
- Repeat the merging process until the chi-square test is significant according to a proposed value.
- Pick the predictor variable whose chi-square is largest and split the sample into $m \leq l$ subsets, where l is the number of categories resulting from the merging process on that predictor.
- Finally, continue splitting, until no "significant" chi-squares result.

2.2 The Iterative Dichotomiser 3

The Iterative Dichotomiser 3 (ID3) [16, 19] is a mathematical algorithm used to generate decision trees. The resulting tree is used to classify new samples. This algorithm consists of constructing a tree from a random subset of the training set. The process must be repeated with the incorrect classifications values while the tree does not classify correctly the remaining cases of the training set.

To achieve this, the algorithm extracts the attribute that best separates the given cases into targeted classes. The algorithm uses the statistical property called "information gain" to choose which attribute is the best at separating training examples. This gain of set S on attribute A is defined as follows:

$$G(S,A) = E(S) - \sum_{v=1}^{t} \frac{|S_v|}{|S|} E(S_v) \tag{1}$$

Where $\sum$ is each value v of all possible values of attribute A; S_v represents a subset of S which attribute A has value v; $|S_v|$ and $|S|$ are the number of elements in S_v and in S, respectively; and $E(S)$ is the information entropy of the subset S expressed by:

$$E(S) = -\sum p(I) \log_2 p(I) \tag{2}$$

Where $p(I)$ is the collection of S belonging to class I.

2.3 The C4.5 Algorithm

C4.5 [20] is an algorithm used to create decision trees and is considering as the successor of the ID3 [16] algorithm developed by Ross Quinlan too. This algorithm works as the same way as its predecessor, ID3, using the information gain (Eq. (1)) to choose the test A that maximizes $G(S,A)$ (Eq. (1)). The problem of using this approach is that it can favour data sets with numerous outcomes. To avoid this, it includes a measure called the "gain ratio" (Eq. (3)) by also taking into account the potential information from the partition itself:

$$P(S,A) = -\sum_{v=1}^{t} \frac{|S_v|}{|S|} \log\left(\frac{|S_v|}{|S|}\right) \tag{3}$$

Finally, the algorithm chooses the test A that maximizes the gain ratio, expressed by:

$$H(S,A) = \frac{G(S,A)}{P(S,A)} \tag{4}$$

2.4 Data Description

The data set has been collected by the Complejo Hospitalario Asistencial Universitario Burgos (Spain) and contains 50 different cases. Those cases contain medical data extracted from the evolution of 50 different patients that were admitted into the hospital and diagnosed with endocarditis.

The following input variables have been considered for the study:

- Patient's age: contains cases ranging from 15 to 89 years old.
- Patient's sex: Male or female.

- Previous valve: Indicates whether the heart valve is native, prosthetic or is a pacemaker.
- Valve type: Indicates the type of infected heart valve: It is discriminated between native valve, prosthetic valve, pacemaker or prosthetic valve with pacemaker.
- Clinical Time: Indicates the time lapse that passed from first symptoms to endocarditis diagnosis (in days).
- Organism: bacteria that causes the infection. Contains more than 10 different types and its variants; such us enterococcus faecalis, enterococcus faecium, Haemophilus parainfluenzae, staphylococcus Lugdunens, staphylococcus parasanguis,...

The output to be predicted is the complications that may occur during treatment. The following complications have been considered:

- Heart failure.
- Cardiogenic shock: worse than heart failure.
- Septic emboli.
- Uncomplicated.

2.5 *Experiments and Results*

The purpose of this multidisciplinary study is the prediction of possible complications once the patient is in treatment of endocarditis.
The dataset considered has 50 different cases: 38 of those cases have been used to train the decision tree and the remaining 12 samples are used to test the model.

In order to get the most adequate classifier to this case, different decision tree learning algorithms have been applied and their results have been compared. This comparison is shown in Table 1.

Table 1 Decision trees results (CHAID, ID3 and C4.5).

		CHAID	ID3	C4.5
Class recall	Uncomplicated	100.00%	100.00%	**100.00%**
	Cardiogenic shock	0.00%	100.00%	**100.00%**
	Septic emboli	0.00%	0.00%	**100.00%**
	Heart Failure	0.00%	50.00%	**50.00%**
Class prediction	Uncomplicated	66.67%	80.00%	**88.89%**
	Cardiogenic shock	0.00%	100.00%	**100.00%**
	Septic emboli	0.00%	0.00%	**100.00%**
	Heart Failure	0.00%	100.00%	**100.00%**
Parameters		minimal size for split 4, minimal leaf size 3, minimal gain 0.1, maximal depth 10 and confidence 0.2	minimal size for split 4, minimal leaf size 2, minimal gain 0.1	minimal size for split 4, minimal leaf size 2, minimal gain 0.1, maximal depth 20, confidence 0.25
Accuracy		66.67%	83.33%	**91.97%**

As shown in Table 1, the best results are obtained with the C4.5 model. C4.5 model is able to predict future cases in a figure close to 92% when the other models that achieve values close to 84% and to 67%. Figure 1 shows the final tree decision model.

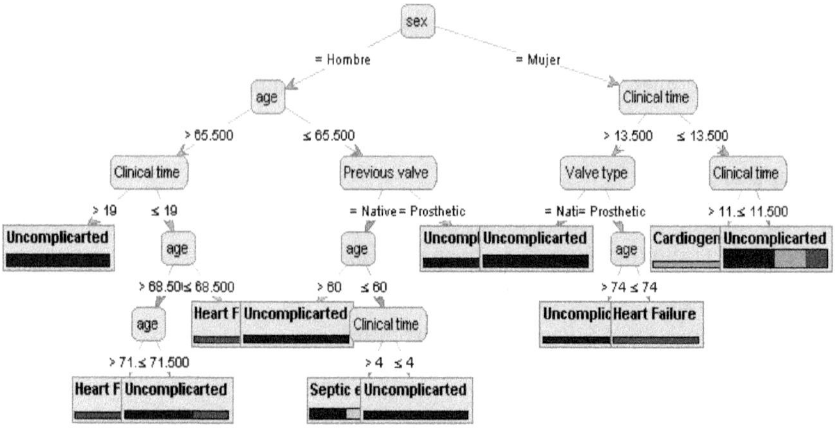

Fig. 1 The C4.5 model

The final model (Fig. 1) shows the structure of the decision tree. It can be noticed that the organism input variable does not appear in the model because it does not affect the classification in a substantial way. As it has been previously mentioned, one of the problems related to the endocarditis treatment is the ignorance of the sorts of bacteria causing the infection, so the identification of a model that is able to classify without this variable is very advantageous.

3 Conclusions and Future Research

The present study describes an ongoing multidisciplinary research in which an application of classical models by means of decision tree algorithms to a medical diagnosis problem has been presented. We have identified the complications with a reasonable degree of accuracy using a relatively quite small amount of samples and attributes. In this application field this means small amount of patients and a low number of medical tests and analyses; which seems as an advantageous feature, being this kind of real data so costly to acquire.

Future work will be focused on the collection and storage of more specific attributes for each patient. Results seem to point to the fact that with more detailed data the medical condition of each patient and enough amount of different patients better results could be obtained. These results may include better prediction of complications based on detailed data obtained from simple tests performed as close to the admission time of the patient as possible.

Another research line may be the use of the information and experience gathered in these experiments for the development of a Case Base Reasoning system

[21] to solve tasks related to the ones presented above. These would be able to handle the incorporation of new information with the treatment and monitoring of the evolution of more patients. They also seem to be more intuitive for medical professionals, which are not used to deal with complex statistical models.

Acknowledgments. We would like to extend our thanks to Complejo Hospitalario Asistencial Universitario de Burgos (SACYL). This research has been partially supported through projects TIN2010-21272-C02-01 from the Spanish Ministry of Science and Innovation and Grupo Antolin Ingenieria, S.A., within the framework of project MAGNO2008 - 1028.- CENIT.

References

[1] Abraham, A., Corchado, E., Corchado, J.M.: Hybrid learning machines. Neurocomputing 72(13-15), 2729–2730 (2009)

[2] Mitchell, T.M.: The Discipline of Machine Learning. Technical Report CMU-ML-06-108, School of Computer Science, Carnegie Mellon University (2006)

[3] Esbensen, K.H., Geladi, P.: Principal Component Analysis: Concept, Geometrical Interpretation, Mathematical Background, Algorithms, History, Practice. In: Brown, S.D., Tauler, R., Walczak, B. (eds.) Comprehensive Chemometrics, pp. 211–226. Elsevier, Oxford (2009)

[4] Herrero, A., Corchado, E., Sáiz, L., Abraham, A.: DIPKIP: A connectionist knowledge management system to identify knowledge deficits in practical cases. Computational Intelligence 26(1), 26–56 (2010)

[5] Lorena, A.C., Ponce, A.C.: Evolutionary design of code-matrices for multiclass problems. In: Soft Computing for Knowledge Discovery and Data Mining, pp. 153–184. Springer, Heidelberg (2008)

[6] Naldi, M.C., Ponce, A.C., Gabrielli, R.J., Hruschka, E.R.: Genetic clustering for data mining, vol. 2, pp. 113–132. Springer, Heidelberg (2008)

[7] Berlanga, F.J., Rivera, A.J., Jesus, M.J., Herrera, F.: GP-COACH: Genetic Programming-based learning of Compact and Accurate fuzzy rule-based classification systems for High-dimensional problems. Information Science 180(8), 1183–1200 (2010)

[8] Das, S., Abraham, A., Konar, A.: Automatic kernel clustering with a Multi-Elitist Particle Swarm Optimization Algorithm. Pattern Recognition Letters 29(5), 688–699 (2008)

[9] Lee, M.Y., Yang, C.S.: Entropy-based feature extraction and decision tree induction for breast cancer diagnosis with standardized thermograph images. Computers Methods and Programs in Biomedicine 100(3), 269–282 (2010)

[10] Baruque, B., Corchado, E., Mata, A., Corchado, J.M.: A forecasting solution to the oil spill problem based on a hybrid intelligent system. Information Sciences 180(10), 2029–2043 (2010); Special Issue on Intelligent Distributed Information Systems

[11] Sedano, J., Curiel, L., Corchado, E., de la Cal, E., Villar, J.R.: A Soft Computing Based Method for Detecting Lifetime Building Thermal Insulation Failures. Integrated Computer-Aided Engineering 17(2), 103–115 (2010)

[12] Sedano, J., Corchado, E., Curiel, L., Villar, J.R., Bravo, P.M.: The Application of a two-step AI Model to an Automated Pneumatic Drilling Process. International Journal of Computer Mathematics 86(10-11), 1769–1777 (2009)

[13] Plicht, B., Erbel, R.: Diagnosis and treatment of infective endocarditis. Current ESC guidelines. HERZ 35(8), 542–548 (2010)
[14] Plicht, B., Janosi, R.A., Buck, T., Erbel, R.: Infective endocarditis as cardiovascular emergency. HERZ 51(8), 987–994 (2010)
[15] Quinlan, J.R.: Learning decision tree classifiers. ACM Computing Surveys (CSUR) 28(1), 71–72 (1996)
[16] Quinlan, J.R.: Induction of Decision Trees. Machine Learning 1(1), 81–106 (1986)
[17] Kass, G.V.: An Exploratory Technique for Investigating Large Quantities of Categorical Data. Applied Statistics 29(2), 119–127 (1980)
[18] Morgan, J.N., Sonquist, J.A.: Problems in the Analysis of Survey Data and a Proposal. Journal of the American Statistical Association 58(3), 415–434 (2010)
[19] Colin, A.: Building Decision Trees with the ID3 Algorithm. Dr. Dobbs Journal (1996)
[20] Quinlan, J.R.: C4.5: Programs for Machine Learning. Machine Learning 16(3), 235–240 (1993)
[21] Aamodt, A., Plaza, E.: Case-Based Reasoning: Foundational Issues, Methodological Variations, and System Approaches. Artificial Intelligence Communications-AICom 7(1), 39–59 (1994)

Pattern Driven Task Model Refinement

Michael Zaki, Maik Wurdel, and Peter Forbrig

Abstract. Task models have been used as a tool to elicit requirements in the early development stages. Moreover, they have recently proved to be a suitable starting point for modeling of interactive processes. During the different development stages several corresponding task models are built. Although every model is just a refined version from the previous one, this refinement process is not a trivial operation as a lot of rules and restrictions have to be respected in order to successfully infer the suitable task model concerning the current level of abstraction. In this paper we aim to assist the developer by giving him the opportunity to move with a given model from one abstraction level to another one in an easier and more seamless way. Thus, we present an approach consisting of some guideline patterns which help the developer to transform a given task model between the different development stages in a more performing and less error-prone manner.

1 Introduction and Background Information

Task models have been used in numerous applications in order to provide realistic information about the user tasks and their relations. Based upon such information the system under construction can be built around the user task world which increases the level of usability, learnability and appropriateness of the system. These advantages are of great interest in ubiquitous computing environments [1], as the main goal behind these environments is to make the information available anywhere, anytime and to give the user a positive impression about the environment.

Knowledge about tasks is especially expedient for interaction development as this is the touching point of user and system. If the UI corresponds with the work processes of the user the system is more suitable. In Fig.1 an example of a task model is shown. Please note that the notation used for our task models is the CTT notation [2]. In this figure the task model of the role presenter is depicted. This model describes the tasks to be executed by a presenter in a conference session. One can notice the temporal operators existing between the tasks. These operators express the precise temporal order in which the actor has to perform the tasks in order to successfully play his role. For example to be able to start a presentation,

Michael Zaki · Maik Wurdel · Peter Forbrig
University of Rostock, Department of Computer Science, Rostock, Germany
e-mail: {michael.zaky,maik.wurdel,peter.forbrig}@uni-rostock.de

A. Abraham et al. (Eds.): International Symposium on DCAI, AISC 91, pp. 249–256.
springerlink.com

the presenter should first introduce himself and that is why the 'enable' temporal operator is chosen to express the relationship between these two tasks. Another example is the 'disable' temporal operator existing between the iterative task 'Next slide' and the task 'Ends presentation' and which means that once the task 'Ends presentation' is executed, then the task 'Next slide' becomes disabled and thus further execution of this task is forbidden. A typical task model contains several task types as we can also notice in the figure. For example, the task 'Next slide' is an abstract task, 'Show next slide' is an application task while the other tasks are user tasks.

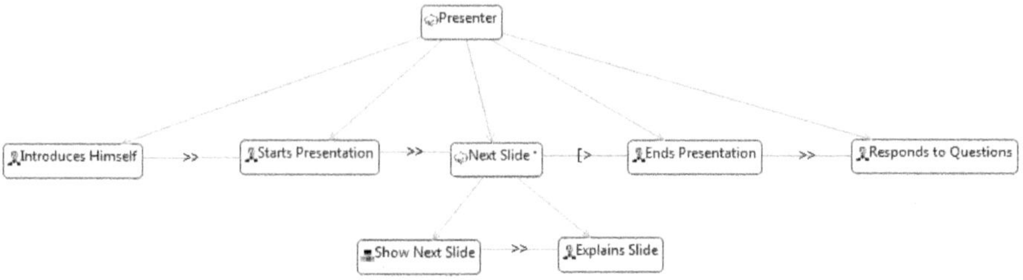

Fig. 1 Task model of the role presenter

During the development life cycle different stages of task modeling exist. First, the user tasks are analyzed and captured via a task model, the so called analysis task model. Sight visits, interviews and questionnaires are typical gathering devices for the analysis task model. The explicit development of a model reflecting the current process helps the developer to keep in mind what the user is actually doing in his day-to-day work and how it is been done. Next, the requirements model is created taking into account the analysis model, the functional requirements and the non functional requirements of the system under construction. As use cases are the standard means for specifying functional requirements consistency checks can also be performed according to the approach of Sinning [3]. Having created such a requirements task model is not sufficient when it comes to lower level design decisions. Therefore, another type of model is proposed which captures such information: the design task model. Each model is usually not created in one step but gradually refined and perfected.

While transferring the task model from the analysis to the requirements level or from the requirements to the design level, structural and behavioral refinement processes are actually taking place. This fact as well as the usage of the so called meta-operators as a tool in order to support behavioral refinement operations are discussed in [4] and [5]. Briefly, structural refinement for a given task can be achieved by adding new subtasks to this task. However, type consistency between the parent task and the new added child tasks must be taken into account while this kind of refinement process is performed. Behavioral refinement consists of assigning meta-operators to tasks in order to decide which tasks may be canceled or deleted in next refinement levels and which tasks must be retained. The structural refinement is not the focus of this paper, as we aim here to stress on the usage of meta-operators in the behavioral refinement process. Having a model in a given level of abstraction according to the actual development stage, the main role of meta-operators is to express which tasks in this model must remain in following refining task models and which tasks may be omitted. Briefly, there are three possible restriction levels that can be assigned to a given task. These levels are expressed using the following three meta-operators:

a) **Shallow Binding operator($\odot$)**: Expresses a mandatory task which has to remain in all following refining models, meanwhile its type can be changed and its subtasks can be removed or changed.
b) **Deep Binding operator ($\otimes$):** Expresses a mandatory task which has to remain with all its subtasks in all following refining models. Furthermore, the tasks types have to remain the same.
c) **Exempted Binding operator ($\ominus$):** Expresses a newly introduced task which did not exist in the first model, but which has to remain with all its subtasks in all following refining models.

The gradual refinement of task models is a challenging task as different information are captured in each type of model and additional information are added gradually. Therefore, in this paper an approach is presented which provides some guideline patterns trying to cover all meta-operators assignments and usage problems. Despite the fact that patterns were first discovered in urban architecture by Christopher Alexander in 1977 [6], their influence has spread and reached the software engineering as well as the HCI area. For example, Tidwell defines a collection of interaction design patterns where the solution is expressed in terms of suggested perceivable interaction behavior [7]. The patterns presented in this paper are useful for domains like ubiquitous computing where task models are highly recommended as a basis and a starting point for any application's development process. While constructing our patterns, we investigated some problems which in order to be solvable, require some few modifications to be realized in the meta-operators rules and constraints. In the next section, these changes are explained in further details. Then in section 3, our task model refining patterns are presented. After that in order to make our ideas more concrete, we discuss an example which illustrates the two major paper contributions, by highlighting the advantages gained by the meta-operators rules modification and also the benefits behind the usage of our patterns within the task model refinement process. Finally, we conclude and we give an overview of related future research.

2 Meta-operators Overriding

A common case that can be encountered by a developer in the modeling process is the existence of a task having a big number of subtasks. We can have the case that just one or few tasks out of the subtasks set have different binding requirements than the others. An inappropriate solution would be to assign a binding operator to every single subtask, however it is also impossible to just assign one binding operator to the parent task, as we have few subtasks for which a different operator is needed. In order to solve such cases, our suggestion is to perform some changes concerning the meta-operators constraints by introducing the meta-operators overriding notion. The idea is simply that for a subtask having its own meta-operator, the binding constraints represented by the meta-operator bound to its parent task are not applicable anymore. In other words, the parent task's assigned operator does not influence this child task because it has been overridden by this child task's own meta-operator. Consequently, a solution for the described case would be to assign one operator for the parent task and different operators to these few subtasks. As a result, fewer operators are assigned for the whole task tree and meanwhile for each task the right binding constraints are expressed which guarantee a successful future refinement process. Within the example presented in Section 4, we present one of the cases where the meta-operators overriding notion's utility is illustrated.

3 Task Model Refining Patterns

As already discussed, the meta-operators help the user in the transformation of task models between the different development stages. However, the assignment of the right operator to every task is not a trivial process as a lot of factors have to be taken into account in order to identify the suitable restriction level to be assigned to the task and thus the corresponding meta-operator. As examples of such factors we can mention the current development phase, the goal behind the task performance and the expected changes occurring to this task in the next refining models. Additionally, the choice of these operators is not the only challenging part during the refining process, as some other problems can encounter the developer while transforming his model from a given level of abstraction to another one. The idea presented here is to use patterns in order to overcome the complexity of the refining operation and to successfully assist the developer in this process. The motivation behind these patterns is to provide solutions for the most common operator assignment problems and perplexing transformation cases. For the sake of brevity, just some of these patterns are presented in further details in the following.

a) Analysis model meta-operators assignment pattern

Name: Analysis model meta-operators assignment
Context: The developer is constructing his analysis model, and so he has to assign meta-operators for some tasks in order to define rules for the further refinement process.
Problem: How to choose the right meta-operator for each task in the analysis level?
Solution: **a)** If for a given task it is sure that no system intervention is required in the requirements level and meanwhile this task including its child tasks are mandatory for the accomplishment of the whole process, bind the task with the deep binding operator and so this binding ensures the consistency of this task in all the following refinement levels. **b)** If a given task has to exist in the following refinement levels and for which an assistance offered by the system under construction is planned, bind it with the shallow binding operator. Consequently, it is guaranteed that the task itself will remain in further refining models but its type may be changed and its subtasks may be removed or modified. **c)** The exempted binding operator does not have to be assigned to any task in the analysis level. **d)** The tasks for which you do not have special constraints, and which can be removed in subsequent refining models, never use the meta-operators.

b) Insertion of new persistent task pattern

Name: New persistent task insertion
Context: The developer is constructing his requirements or design model, and he would like to introduce a new task which did not exist in previous models.
Problem: How to insert a new persistent task in the model?
Solution: Insert the task and bind it with the exempted operator to express that this task was not part of the initial requirements but it has to exist in further refining models.

c) Minority changeable tasks pattern

Name: Minority changeable tasks
Context: The developer is assigning meta-operator to a task which contains a lot of subtasks. This task is of type abstract. The developer wants that just one or few of the subtasks can be changed or omitted while the others have to be persistent.
Problem: How can the developer make most of the subtasks persistent while some can be cancelled? It is always a bad solution to assign a lot of meta-operators.
Solution: Bind the parent task with a deep binding and bind the few subtasks you may cancel or change with the shallow binding.

d) Minority persistent tasks pattern

Name: Minority persistent tasks
Context: The developer is assigning meta-operator to a task which contains a lot of subtasks. The developer wants to prevent just one or few of the subtasks to be omitted or changed in the next refining models, while the other subtasks can be canceled or changed.
Problem: How can the developer make most of the subtasks changeable while some have to remain persistent? It is always a bad solution to assign a lot of meta-operators.
Solution: Bind the parent task by the shallow binding, and bind the few tasks you want to keep unchangeable with the deep binding operator.

4 Patterns Application Example

Let us assume we aim to construct the task model for ATM machine usage in the analysis level. We find that using this machine implies that the user has first of all to identify himself, chooses the language he would like to communicate with and then he can have access to the several services provided. Thus, he is able to withdraw money, display any information concerning his account or transfer money to another account holder. Once we are settled on the tasks needed to present the ATM machine's functionality, we can begin constructing the task model in the analysis level, and now the question is which meta-operator has to be assigned to which task? The role of task model refining patterns is to help the developer finding the answer for this question.

As we are building our analysis model, and in order to successfully assign the meta-operators to the existing tasks, we can have assistance by the analysis model meta-operators assignment pattern. So we have to iterate over the tasks and to apply the solution offered by the pattern. In Fig.2 the corresponding task model for the ATM machine in the analysis level is depicted.

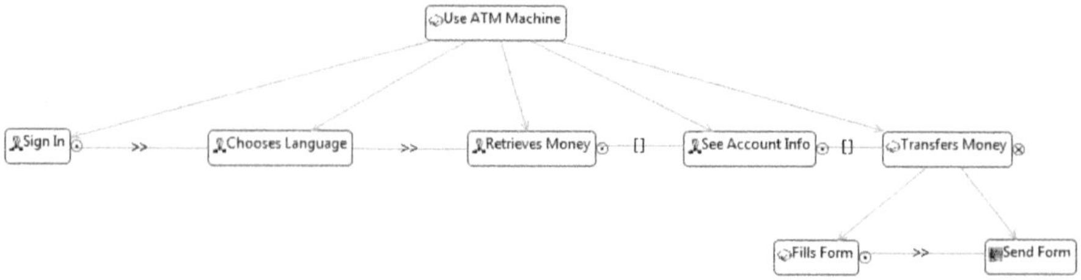

Fig. 2 ATM machine in the analysis level

The tasks "Sign In", "Retrieves Money" and "See Account Info" are mandatory and so they have to remain in following refinement models. However, further changes due to system intervention in requirements level are expected. According to the mentioned patterns, these tasks have to be bound with the shallow binding operator. Also, the task "Chooses Language" is not mandatory for the ATM functionality, and then it may or may not persist in the following refinement models. Thus, following the same pattern no meta-operator should be assigned to this task as we do not have any real constraints concerning its existence. The task "Transfers Money" is an interesting case, in which on the one hand we want to ensure that the task exists with its two subtasks in all following refinement levels and on the other hand we expect to perform some changes to the task "Fills Form" and thus we want to have the opportunity to extend this task in next refinement levels. So "Fills Form" is a child task of "Transfers Money", however they have contradictory binding requirements. Here one can see one of the advantages behind the idea of meta-operators overriding (mentioned in Section 2) because the solution is to bind the parent task "Transfers Money" with the deep binding operator and the child task "Fills Form" with the shallow binding operator. This solution was not realizable or possible using the former deep binding operator semantics and rules, as operators overriding was not allowed.

Having constructed our task model in the analysis level, now we want to transform this model to a valid refined one in the requirements level. Consequently, structural and behavioral refinements are taking place in this process as we can notice in Fig.3. Let us focus on the behavioral refinement and thus the assignment of the meta-operators to the existing tasks in our requirements model. The tasks "Sign In", "Retrieves Money" and "See Account Info" and their subtasks cannot be omitted in further refinement models, as they are mandatory for the functionality of the system under construction. Additionally, no further refinement regarding these tasks due to design level decisions is expected and thus these tasks have to be bound with the deep binding operator. Additionally by having a closer look on the task "Fill Form", we can notice that this task was extended by four different subtasks. On the one hand, for the subtasks "Enter Receiver's Name", "Enter Receiver's Account Number and "Enter Bank Name" no changes are expected in further refining models in requirements or design stages. On the other hand for the subtask "Enter Amount" further changes may occur. For example, this task may be extended in order to let the user be able to choose the currency for the transfer process, and also the system may check first whether the user has a sufficient amount in his account for the transfer operation to be successfully accomplished. Thus, a simple solution would be to assign a shallow binding operator to the subtask "Enter amount" and three deep binding operators for the other subtasks correspondingly. However, this solution is not optimized as an excess usage of meta-operators is not advisable. A better solution is presented by the "Minority persistent tasks" pattern. We have to assign a deep binding operator for the parent task "Fill Form" and a shallow binding operator for the subtask "Enter Amount" and consequently the same binding requirements can be expressed in a better manner using only two operators.

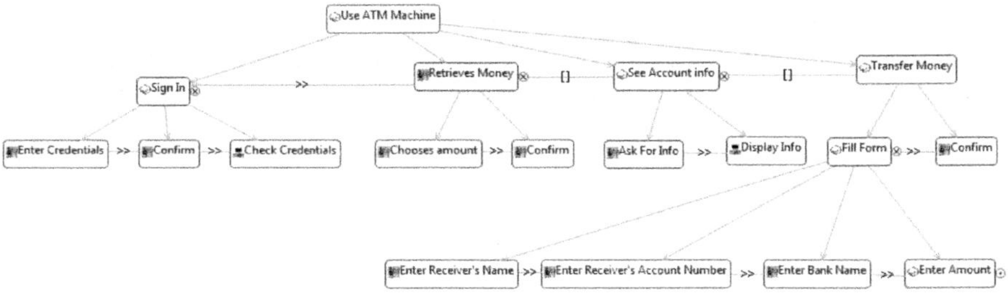

Fig. 3 ATM machine in the requirements level

5 Conclusion

In this paper we discussed briefly the role of meta-operators in the task model refinement operation after motivating the role of task models in nowadays applications and especially for ubiquitous computing environments. Moreover, we gave an overview of the different development stages and the role of each corresponding task model. Then we discussed some few modifications that we applied on the meta-operators semantics by arguing that these changes can provide solutions for contexts and cases which were irresolvable using the former semantics. After that we introduced the usage of patterns into task model refinement process by providing patterns which are able to help the user assigning the meta-operators to the model and also to assist the user to solve the most common encountered problems during the refinement process. Furthermore in order to make our ideas clear and to highlight the benefits behind our patterns, we presented an example of a task model for an ATM machine and we illustrated the way the developer can choose the suitable meta-operators for this model in the analysis and requirements level using some of our patterns. Additionally, within the same example we made clear how the meta-operators overriding notion is useful to express some situations. In the future we plan to provide pattern libraries embracing some concrete solutions for the transformation of the most recurrent situations from the analysis to the requirements level and from the requirements to the design level. Using these libraries the developer will be able to transfer a given model from one abstraction level to another one by just loading the suitable patterns out of these libraries.

References

1. Weiser, M., Gold, R., Brown, J.S.: The origins of ubiquitous computing research at PARC in the late 1980s. IBM Systems Journal (1999)
2. Paterno, F., Meniconi, C., Meniconi, S.: ConcurTaskTrees: A diagrammatic Notation for Specifying Task Models (1997)
3. Sinning, D., Chalin, P., Khendek, F.: Consistency between Task Models and Use Cases. In: DSV-IS 2007 (2007)
4. Wurdel, M., Sinnig., D., Forbrig, P.: Task Model Refinement with Meta Operators. In: DVS-IS (2008)
5. Wurdel, M., Sinnig, D., Forbrig, P.: Task-based Development Methodology for Collaborative Environments. In: TAMODIA/HCSE (2008)
6. Alexander, C., Ishikawa, S., Silverstien, M.: A Pattern Language. In: Towns, Buildings, Construction, Oxford University Press, Oxford (1977)
7. Tidwell, J.: Interaction Design Patterns (1998)

Feature Reduction of Local Binary Patterns Applied to Face Recognition

Juan Carlos García and Francisco A. Pujol

Abstract. In recent years, Local Binary Patterns have proved to be a powerful local descriptor for microstructures of images, having been introduced in many facial recognition systems and intelligent environments. In this work, we present the implementation of a face recognition method based on the use of Local Binary Patterns. We used data mining tools to get a smaller feature vector and thus improve the computational cost of the system. The implementation was tested with the Color FERET database, obtaining a recognition rate of 94% and reducing 75% the original feature vector dimension.

1 Introduction

A facial recognition (FR) system consists of recognizing a biometric sample taken as a face picture. General interest in FR systems has grown considerably over the last decade [1]; among others, in recent years one of the most successful algorithms uses the so-called *Local Binary Patterns* (LBP).

The basic LBP operator was introduced by Ojala et al. [2]. It labels the pixels and threshold each neighbourhood of 3×3 pixels with the central pixel value; thus, the gray value of each pixel in the neighbourhood, g_p, is compared to the gray value of the central pixel, g_c; the LBP label for the central pixel of each region is obtained as:

$$LBP_P(x,y) = \sum_{p=0}^{P-1} s(g_p - g_c)2^p \quad \text{, where } s(x) = \begin{cases} 1 & x \geq 0 \\ 0 & x < 0 \end{cases} \tag{1}$$

This operator was extended to different neighbourhoods [2]. To do this, a circular neighbourhood around the central pixel is used, where the position (x,y) of g_p is

Juan Carlos García · Francisco A. Pujol
Dept. Tecnología Informática y Computación,
Universidad de Alicante, P.O. Box 99, E-03080, Alicante (Spain)
e-mail: juancar.garciavillena@gmail.com, fpujol@dtic.ua.es

A. Abraham et al. (Eds.): International Symposium on DCAI, AISC 91, pp. 257–260.
springerlink.com

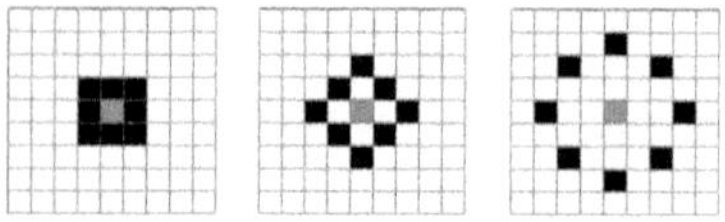

Fig. 1 Three circular neighbourhoods (P,R): (8,1), (8,2), (8,3).

determined by $(-R\sin(2\pi p/R), R\cos(2\pi p/R))$, being R the ratio between pixel p and the central pixel. An example is shown in Fig. 1.

In [3], Ahonen proposed a face recognition system based on a LBP face representation. An input image is divided into R small non-overlapping regions of the same size. Histograms of LBP codes H^r, with $r = \{1, 2, \ldots, R\}$, are calculated over each region and concatenated into a single histogram representing the face image. For classification, χ^2 dissimilarity measure is used.

Although the original algorithm had a high recognition rate (around 94%), this method presents two main problems: it needs a large feature vector and the assignment of the weights to each face region was not much precise. As a consequence, the aim of this work is to try to overcome these problems.

The rest of the paper is divided as follows: Section 2 shows our proposal for improving the method and, next, Section 3 presents the results of the implemented experiments. Finally, some concluding remarks are shown in Section 4.

2 Feature Reduction for LBP

In this work, we have used the Color FERET database [4, 5]. We used the sets *fa* and *fb*, corresponding to 843 individuals with frontal images. We first detected faces and normalized images to a size of 130×150 pixels.

Next, face features must be extracted. We propose to split face images applying three different masks, dividing faces into 7×7, 15×14 and 21×21 regions. The first mask was used in Ahonen's original work [3], and it generates a vector of 2301 features per image, since each image is divided into 39 regions, containing 59 LBPs (58 uniform and 1 non-uniform). A LBP is uniform if the bit transitions contained in its binary value are not greater than 2. These patterns are used to find the pixels that belong to flat areas, contours, corners, etc.

To reduce feature vectors, we have used the data mining tools WEKA[1] and RapidMiner[2]. Using a training set of 200 images, we noticed that the first 14 patterns are the most significant patterns for the recognition tasks. Therefore, if we also add a pattern to label all the non-uniform patterns, a total amount of 15 LBPs will be obtained. Thus, when using Ahonen's mask, our proposal will result in the feature vectors to have 585 elements, reducing by 75% the feature vectors dimensions.

[1] `http://www.cs.waikato.ac.nz/~ml/publications.html`
[2] `http://rapid-i.com/content/view/181/190/`

Then, as it seems clear that each face region contributes in a different way for the recognition process, we shall use the weighted χ^2 measure for obtaining similarities among faces, defined as:

$$\chi_w^2(S,M) = \sum_{r,i} w^r \frac{(S^r(i) - M^r(i))^2}{S^r(i) + M^r(i)}, \quad (2)$$

where w^r is the weight for region r, and S and M correspond to the sample and the model histograms. Fig. 2 shows the weights assigned for the 7×7-region mask.

Fig. 2 Assigned weights: black (w_1^r=0), dark gray (w_2^r=1), gray (w_3^r=3) and white (w_4^r=4).

Finally, the k-nearest neighbours algorithm (k-NN) is used to classify each image.

3 Results

Our first test (see Table 1) has been to obtain the recognition rate for the three proposed masks. The test was performed with $P = 8$ and there was no assignment of weights to the face regions. That is, each region contributes in the same way to χ_w^2.

Table 1 Results for $P = 8$, with and without using specific weights.

Radius (R)	1	2	4	6	8	10	12	14	16
Mask 1 (7×7 regions)	80%	84%	**90%**	89%	87%	85%	84%	81%	81%
Mask 2 (15×14 regions)	84%	86%	**90%**	89%	87%	85%	84%	80%	81%
Mask 3 (21×21 regions)	86%	87%	**90%**	89%	86%	84%	82%	81%	80%
Weighted Mask 1	83%	87%	**94%**	93%	92%	90%	89%	86%	85%

The results are very similar for the first 3 masks; thus, the highest recognition rate is 90% for $R = 4$. It must be also noticed that the recognition rate does not increase much as the dimension of the feature vector increases; i.e., using more regions to obtain the histograms does not implicate a higher recognition rate.

In addition, the last results in Table 1 –where the computed weights are considered– have been performed only for Mask 1. As shown, the assignment of different weights to each face region improves significantly the recognition rate, achieving a 94% of correct results for $R = 4$. There is also an improvement for the rest of radius.

Using the data mining tools, we have reduced significantly the dimension of the feature vector. Table 2 shows a comparison between the dimensions of the feature vectors for both the original method and our proposal, considering the weighted versions of both algorithms.

Table 2 A comparison between methods.

Method	No. of LBPs	No. of regions	Feature vector dimension	Recogn. rate
Original method	59	39	2301	94-97%
Mask 1	15	39	585	94%

As shown, in spite of reducing the feature vector, which is obtained from the LBP histograms, our system has achieved high recognition rates. That is, we obtained a vector of 585 features, a reduction of 75% compared to the originally required features (2301). However, the recognition rate only decreased by 3% (94%), at most. This is a significant decrease of computational cost and will allow our proposal to be implemented in real-time applications, where the computation time is a key factor.

4 Conclusions and Future Works

In this work, we have presented a method to reduce the dimension of the LBP histograms for face recognition, using data mining tools. As shown, we have reduced significantly the computational complexity of the method, since the dimension of the histograms have been reduced by 75%, getting a recognition rate of 94%.

Future works aim to design new methods to obtain higher recognition rates with small feature vectors. We shall also design different face masks to get better results.

References

1. Li, S., Jain, A.: Handbook of Face Recognition. Springer, New York (2005)
2. Ojala, T., Pietikäinen, M., Mäenpää, T.: Multiresolution gray-scale and rotation invariant texture classification with local binary patterns. IEEE Trans. Pattern Analysis and Machine Intelligence 24, 971–987 (2002)
3. Ahonen, T., Hadid, A., Pietikäinen, M.: Face recognition with local binary patterns. In: Pajdla, T., Matas, J(G.) (eds.) ECCV 2004. LNCS, vol. 3021, pp. 469–481. Springer, Heidelberg (2004)
4. Phillips, P.J., Wechsler, H., Huang, J., Rauss, P.: The FERET database and evaluation procedure for face recognition algorithms. Image and Vision Computing 16, 295–306 (1998)
5. Phillips, P.J., Moon, H., Rizvi, S.A., Rauss, P.J.: The FERET Evaluation Methodology for Face Recognition Algorithms. IEEE Trans. Pattern Analysis and Machine Intelligence 22, 1090–1104 (2000)

Anti-Icing Decision Support System Based on a Multi-agent System and Data-Mining

David Martínez Casas, José Ángel Taboada González,
Juan Enrique Arias Rodríguez, and Sebastián Villaroya Fernández

Abstract. This paper presents a system to improve decision support to prevent the ice formation or snow accumulation on the road surface (anti-icing). The main part of the system has the task of predicting the weather parameters [1] involved in the problem. To do that, the system uses data-mining techniques for the prediction process and uses a multi-agent system that is responsible for controlling the flow of information throughout the system and decides the most appropriate actions in each case to address the problem.

Introduction

In recent years, there has been a growing interest among road authorities in getting predictions for ice formation on roads, since proper decisions about road salting require accurate predictions of ice formation some time ahead; valuable warnings of road ice could be issued to the public as well.

In order to survey the road conditions, networks of road stations measuring meteorological parameters in the road and possibly additional information about surface have been established in many places.

The information provided by the measuring systems mentioned is very useful to use in a system that helps us to take the decision to spray anti-icing agent to keep the road clear of ice or snow, minimizing the amount of salt (NaCl) required improving economic efficiency and preserving the environment.

The present paper describes a system based on the application of various data-mining [2] [3] techniques to predict relevant meteorological parameters and classify the future state of the road. We use the paradigm of intelligent multi-agent systems [4] [5] for road ice forecast in short time periods and heavily localized areas where the intelligent agents implement different data-mining algorithms. The model is briefly described in section 2. Experiments and the associated results are described in section 3. Final remarks and conclusions are presented in section 4.

David Martínez Casas · José Ángel Taboada González · Juan Enrique Arias Rodríguez
Sebastián Villaroya Fernández
Laboratorio de Sistemas Departemento de electrónica y computación
Universidad de Santiago de Compostela

A. Abraham et al. (Eds.): International Symposium on DCAI, AISC 91, pp. 261–264.
springerlink.com

Description of the System

The developed system makes use of input data obtained from the meteorological sensors and stored in a database. This data is fed into the system to determine the actions that must be carried out on the control of the spray system flux as well as information on the data and decisions to highway control centers.

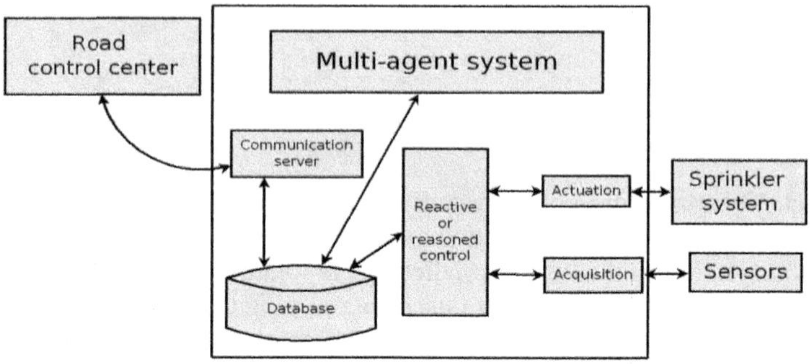

Fig. 1 Global architecture of the system.

We build a system based on multi-agent paradigm and data-mining. Some agents implement data-mining algorithms to determine the actions that are necessary to keep the road surface "clean" of ice or snow. Other agents take charge in the inter-agents communications [6] and decisions, and the data access. To do this, use the data provided by the measurement stations located near the area of interest and stored in a database. The overall system that integrates the multi-agent system is shown in fig. 1. The main element of the global system is a multi-agent system (MAS). The MAS has the following sub-systems.

1. Validation sub-system. It is responsible for validating the input data to ensure a maximum quality of the same. A set of agents, one for each weather parameter input, is responsible for applying heuristic rules to determine if the input data is valid or not. This information is used by other sub-systems.
2. Prediction sub-system. It is responsible to make a numerical prediction of some meteorological parameters of interest for decision making such as air temperature, relative humidity or temperature of road surface. The algorithms used to generate predictive models are such as linear regression, regression trees and neural networks that are trained using historical data stored in the database. A manager prediction agent is responsible for retrieving the answers provided by the prediction agents and choose the best one based on the value of the square root error on the results.
3. Classification sub-system. It is responsible for classifying the future state of the road surface to determine whether or not a preventive action may be taken. The sub-system consists of a set of agents that implement different data-mining algorithms based on any one of the following paradigms: decision trees, decision rules, neural networks and instance-based systems. A manager classification

agent is responsible for retrieving the answers provided by the classification agents and determine the final answer to make the decision to take action or not over the salt spread system. The best final answer can be obtained based on various strategies which the most simple is majority odd vote over the answer (Ice / No_Ice).
4. Actuation sub-system. The actuation system makes use of the historical data, numerical predictions and road state classification to determine the amount of salt needed to maintain the road clean before a bad situation occurs (anti-icing). Also, if an ice / pavement bond is formed the actuation system calculates the amount of salt needed to unbound (de-icing).

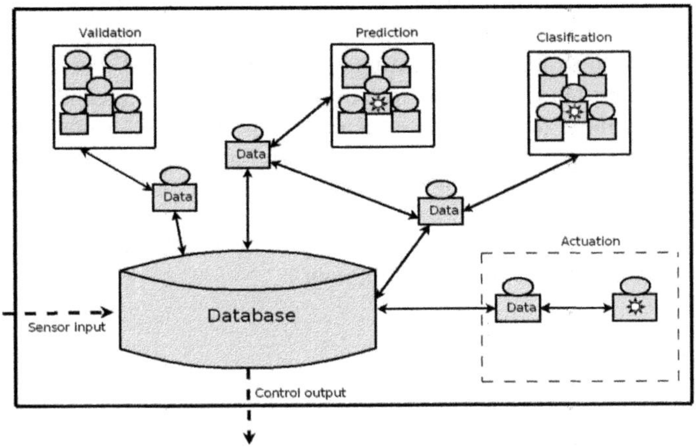

Fig. 2 MAS structure.

The operation of the system is described as follows: "Data" agents periodically check if there are new data in the database. In the case of new data, it enables the validation system to check if the data is valid. If the data is valid, prediction and classification systems come into action, run their processes and store the results in the database. The actuation system queries the database and determines the actions to be taken depending on the situation described by the historical data and the result of the prediction and classification process.

Experiments and Results

The system is fully developed but we have no data to validate de "actuation" system. To test the other systems (validation, prediction and classification) use data from RWIS (Road Weather Information System) of EEUU interstate highway for the period between 14/12/2005 and 19/01/2006 (winter) with a period of ten minutes between two consecutive measures.

The next table shows the results obtained to 30 minutes forecast (time necessary to operate the sprinkler system) for the prediction and the classification. To the state of the road classification process consider two possible outputs (Ice, No Ice) indicating whether or not the use of the sprinkler system is necessary.

	MAE	RSE		Ice	No Ice
Relative humidity (%)	0.78	1.50			
Air temperature (ºC)	0.40	0.63		96.2%	94.8 %
Surface temperature (ºC)	0.37	0.53			

30 min. forecast results.

Conclusions

The results show that the system has enough prediction effectiveness in a large number of cases in which we can act as anti-icing (a strategy to prevent the formation of ice/pavement bond) instead of de-icing (a strategy to allow the ice/pavement bond to form and treating it until the bond is broken). This saves a significant amount of salt and therefore improves the economic and environmental benefits. Furthermore the multi-agent paradigm provides robustness and scalability to the system so as to facilitate easy incorporation of new data mining algorithms adding new intelligent agents to the system.

Acknowledgments. This works has been support by the "Xunta de Galicia" project 07TIC011E in collaboration with "Level Telecom".

References

[1] Casas, D.M., González, J.Á.T., Rodríguez, J.E.A., Pet, J.V.: Using Data-Mining for Short-Term Rainfall Forecasting. In: Omatu, S., Rocha, M.P., Bravo, J., Fernández, F., Corchado, E., Bustillo, A., Corchado, J.M. (eds.) IWANN 2009. LNCS, vol. 5518, pp. 487–490. Springer, Heidelberg (2009)
[2] Orallo, J.H.: Introducción a la minería de datos. Prentice Hall, Madrid (2007)
[3] Araujo, B.S.: Aprendizaje automatico: conceptos básicos y avanzados: aspectos practicos utilizando el software Weka. Prentice Hall, Madrid (2006)
[4] Wooldridge, M.: An Introduction to Multiagent Systems. John Wiley & Sons, Chichester (2002), ISBN 0 47149691X
[5] Botía, J.A., Gómez-Skarmeta, A.F., Valdés, M., Padilla, A.: METALA: A Metalearning Architecture. In: Fuzzy Days 2001, pp. 688–698 (2001)
[6] FIPA (Foundation for Intelligent Physical Agents), `http://www.fipa.org`

Analysis of XML Native Databases for E-Health Applications

Isabel de la Torre Díez, Francisco Javier Díaz-Pernas,
Míriam Antón-Rodríguez, Mario Martínez-Zarzuela,
David González-Ortega, and José Fernando Díez-Higuera

Abstract. XML databases allow to process and organize information in an optimal and simplified way for the final user. In this paper, XML databases advantages and disadvantages for e-health are analyzed. Included results have been obtained through theoretical and practical research, using similar environments as those that can be found in real situations, practical examples, and with the opinion of clinical staff. In the last section, a dissertation on the potential of XML databases for future applications is presented.

Keywords: Databases, standardization, metadata, organization, XML.

1 Introduction

One of the oldest fields in which telecommunications have a greater potential is telemedicine. The possibility of improving the diagnostics capabilities and treat diseases from a remote point has attracted the attention of many people. Now that the technology has been matured enough to offer broadband and mobile technologies, telemedicine is able to provide services such as telediagnosis, remote care and medical consulting or remote surgery among others.

Even with the recent advances of today's telemedical systems, some difficulties for their complete implementation frequently arise: the need for device interoperability and the correct management and processing of the generated data is a critical task and systems have to be designed so that it is possible to update the systems to incorporate the newest technology. For this reason, it is required that, when implementing a new system, scalability data format interchange is guaranteed.

For information storage and processing, databases are required. These databases allow performing four different tasks: create, read, update and delete (CRUD) information [1]. However, different databases perform these tasks in a

Isabel de la Torre Díez · Francisco Javier Díaz-Pernas · Míriam Antón-Rodríguez
Mario Martínez-Zarzuela · David González-Ortega · José Fernando Díez-Higuera
Departamento de Teoría de la Señal, Comunicaciones e Ingeniería Telemática. Universidad de Valladolid, Paseo de Belén, 15. Campus Miguel Delibes. 47011 Valladolid, Spain
e-mail: `{isator,pacper,mirant,marmar,davgonz,josdie}@tel.uva.es`

A. Abraham et al. (Eds.): International Symposium on DCAI, AISC 91, pp. 265–272.
springerlink.com

different way. User's frontend applications will provide different features and behavior depending on the database chosen to store the data.

For many years, the most extended and accepted databases were relational databases, such as MySQL or Microsoft Access. However, as Elliotte Rusty Harold said "If your only tool is a hammer, all your problems will look like nails….." [2]. Certainly, storing and processing data were only possible in the existing platforms, base on relational assignements.

At this point in 1998, was when the XML format appeared [3], introducing a real and effective alternative to the previous databases architecture, which were very limited due to the restrictions in the kind of input and output data types, and in the speed at which large amounts of data could be manipulated.

Since that moment, the continuous evolution of the technology behind XML has driven a great advance towards the interoperability using different types of information, and XML has evolved to a format that is capable to contain the rest of the formats and standardize information streams between old devices using non-uniform models [4], solving the problems of relational databases and providing a more flexible organization.

Along this document, the data storage model in native XML databases is explored, as well as the implications of these databases in telemedicine, where efficient data storage and processing is really important.

Aforementioned CRUD tasks are common in every database application, thus are so in XML native databases. Internal storage of the information, searches, deletions, XML conversions…etc, are processes that can be done in different ways. For this reason, different applications are able to provide a greater interoperability with other formats, while others are oriented to adapt the information to XML before it is stored [5].

In this paper, the capability of XML databases is analyzed. Although the tests performed are limited to specific scenarios, certain similarities and differences against traditional storage models, where fields of data are disposed along tables in rows and columns, will be discussed.

Finally, a comparison among both types of databases for providing telemedicine services is given, and a dissertation on the future direction of the technology related to database management is provided.

2 Methodology

With the aim to prove the evolution suffered in database systems, we decided to check out the behavior and response capability of two different platforms: a traditional relational database, such as Microsoft Access and a XML platform as those proposed in [2]. In this last case, due to the economical restrictions of the research project, the tests were performed using Open Source environments, which provide the same benefits as their privative counterparts, but free of charge. For this purpose, based on their popularity, documentation and web support eXist, Xindice, dbXML, or the Berkeley dbXML databases can be employed.

On the assumption that in the study of relational and XML databases, it will be find that the latter are more efficient, we decided to use the worst XML database in terms of functionality and performance [5]. This way, a Worst Case Scenario for comparing both databases was set. The chosen platform was the eXist database.

For the relational database, we opted to use Microsoft Access, provided in the suite Microsoft Office. This platform was chosen. As the main objective of this study is measuring the performance of the databases in telemedicine, tests were made using non-real patients data, but simulating the necessities of medical clinics.

Tests were performed on a commodity computer equipped with a double core CPU running at a 3.8GHz clock frequency and containing 2 GB of RAM and a hard disk of 250 GB. The operating system was Windows XP Service Pack 3. As previously stated in [5], it is useful using a standard computer like this for the tests, with the aim of reducing the impact of factors which could alter the results obtained in the tests and easing the management and control of the platforms to be used.

After Microsoft Office suite installation, the Microsoft Access database application is available. Microsoft Access tools were used to import and adapt a telemedicine system. Main issues during this process included:

- The inclusion of certain tables that, as is explained in the results section, is necessary for storing information that it is exclusive in the sanitary environment.
- These tables have to be linked to the table Ficha.
- It is necessary to include some data in every table so that searches are possible.
- Finally, some forms, searches, macros and modules were created in Visual Basic so that the input, search and output tests could be performed. Results are explained in the next section.

In order to perform databases comparison, the open source, java-based eXist platform was used. After installation, database configuration and interaction can be done through a web-based application or through a stand-alone Java application.

In order to perform the comparison, it is necessary a XML database managed by eXist. This database has been developed in an easy way, including an enough amount of metadata so that the system can be tested. Following the indications of [6] and [7], these archives were employed:

- A DTD archive to delimit the structure of information blocks in the database. As it is stated in [6], this is not necessary for the correct behavior of the system. However, this archive is included so that XML functioning is better understood.
- A main XML archive to store the information with the structure of labels and tags previously determined in the DTD archive.
- A CSS modeling archive. Although this is again optional, it is possible to use a CSS style sheet or even Javascript or a XLST style to model and present the data contained in the XML document [7].
- The tests included data searches and data inputs.

3 Results

3.1 *Relational Database*

3.1.1 Data Input

In a relational database such as Microsoft Access, data input and output can be done in different ways:

- Filling every field in the tables using the Microsoft Access program. This is a laborious task and, frequently, redundant and not efficient.
- Using a data manager programmed for the database, which can be included in the same program, or
- Using the option provided by the operating system to indicate the data source through Open Database Connectivity (ODBC) and then use the database from a remote application programmed in Java, for example.

For clarity, we decided to use the second option, so programming a Form like the one showed in Fig. 1 was needed. When the complexity of the searches increases it might be necessary to use Visual Basic.

For data input efficiency, this database presents two inconveniences:

Firstly, it is difficult to implement field relationships in an efficient way, so that they can be updated and extra work is not needed from the end user for data input.

Secondly, different data have to be included in different tables, so that the database grows with the variety of information that has to be classified (this happens in the example with tables Cardiology and Radiology, which are different from the table Ficha). In the next section we will see that this induces serious problems to find data and classify the information coherently.

3.1.2 Data Search

In the previous section, Forms were used for data input. For data search, queries are needed. In Microsoft Access queries can be done as showed in Fig. 2.

Once these queries have been developed, it is easy to identify the inefficiency of the relational model, in which it is necessary to obtain data from different tables to obtain specific information about different areas in telemedicine.

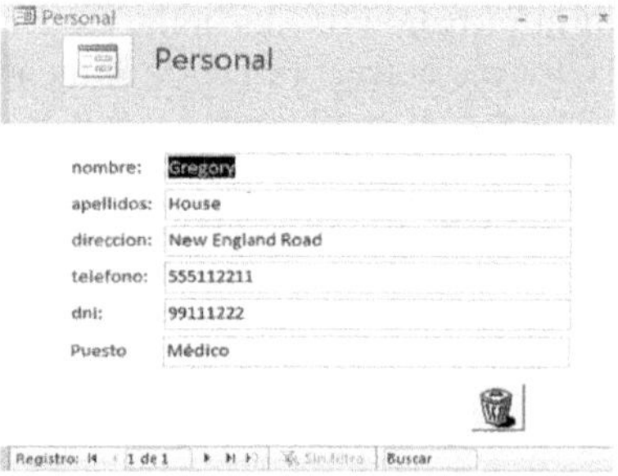

Fig. 1 Example of form to introduce new staff.

Fig. 2 Searching patients.

3.2 *Native XML Database*

3.2.1 Data Input

One of the main advantages of using a XML native database is data format interoperability and the interface for data input is determinant to achieve this objective. However, in the eXist platform we could not find any Web or Java interface which could be used to easily ask for the data that have to be introduced in the database in XML format.

For this reason, data input is tedious in the first moment, because the data conversion gateway is not present and then it is necessary to blindly trust in labels and introduced data integrity and concordance. This cannot be guaranteed by the program.

However, this gateway for data transformation and introduction could be developed using a Web form, some JavaScript controls and the adjustment of the XML labels previously defined. This group of elements is known as xForms [8], and easily resolves the introduction and transformation of data to the XML format. For the final user, data introduction is presented as a regular standard form.

Data input interfaces are critical for the whole system. As it has been explained, data introduction in XML involves a previous conversion of the information using external agents. However, this weakness has an easy solution, previously indicated here and in [5]. This way, data conversion is transparent for the user, making input process very easy for the final user.

With respect to the operation of the generated input interface, it is worth to mention that the data introduced natively using the XML format was correctly processed by the eXist platform. Also, the tests involving the development of simple forms in the format XForm were satisfactory, allowing the introduction of new data in an easy way.

3.2.2 Data Search

The language supported for XML native database searching is mostly xQuery [9], which is in turn based in xPath [10]. In the case we are dealing with, eXist has two interfaces for data searching: using Web and Java.

Nevertheless, and as it happened in the case of relational databases, we could not find a Web (or Java) interface inside eXist that could facilitate data search inside the files. A Web interface is missed (necessary for the medical staff not familiar with the XML technology). This interface would be easily implemented and it is indispensable before these databases can be implemented in the correspondent hospitals.

Different tests were performed, based on the search examples given in [6]. Every test provided a satisfactory result, as it is showed in Fig. 3(a). Also, once these searches were performed, the results obtained in XML were easily reproducible in any program. In this report, the result of these searches has been presented in a web navigator, so that the different elements found are easily identified, as it is shown in Fig. 3(b).

We found especially tedious to learn the xQuery language to perform the searches. Is not difficult to modify the existing searching examples in the included SandBox in eXist, although a more intuitive interface is desirable for the final user to facilitate performing queries in real environments.

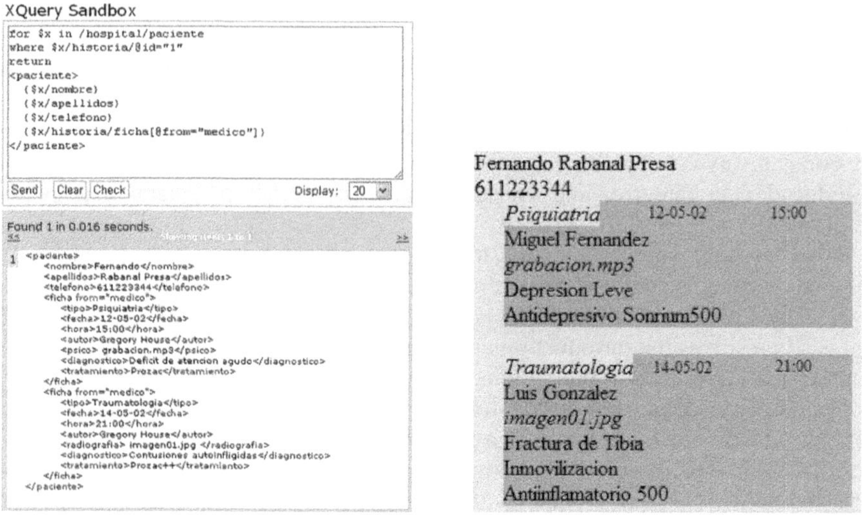

(a) (b)

Fig. 3 Searching information and patient medical records with xQuery and XML (a), and in Web format (b).

4 Discussion

In this section, we discuss the features that differentiate the relational databases from the native XML databases. This way, we will introduce some well-known concepts like information searching capability, ease of use/ usability, and some other less known concepts such as the internal codification of the information or data conversion, so that they can be included in the database.

First of all, it is worth to mention that using XML data storage, the information is stored in a much regular way, as it was contrasted during the tests. The size in bytes to store the information is increased due to this regularization and, therefore, more hardware storing resources are needed and efficient algorithms for data compression are needed. However, relational databases do much more efficient storage of data, although they need some data redundancy (what increases the size of data in memory), so that it is possible to access some information from other information (tables and databases relationships). However, the multidimensionality acquired recently by these systems, together with the large amount that can be stored, makes that this redundancy is at least as large in size as in the case of the XML native databases, although in this case the increment in storage requirements does not result in an increment in the regularization of the information.

On the other hand, data storage can be done in different ways: through logic tables, text format or any other predetermined format. While the relational databases usually store data in tables, XML databases provide new possibilities to store the data, with their advantages and inconveniences. Some systems store data in text format, what eases the reusing of different information to XML. Other systems, use XML data format to store the information and then an external conversion of data is needed. This topic has been discussed before in this document, providing different points of view to the problem.

Even with the great advantages that would provide data regularization, a system with these features would not be successful if it is difficult to use. For this reason, an important effort is being made to improve user interfaces. This effort is more obvious in the XML databases, due to their recent emergence and that they are in the technological maturity process. In this aspect, these databases have gone already through a long development process, so that they can overtake the traditional relational databases. In the relational databases, the interfaces are not easy to use, but after many years in the market, there is a generation of users that is somehow comfortable using them.

5 Conclusions

Along this document, the most significant features of the XML native databases have been explored, in contrast to the relational databases to which the general public is more used to. This has allowed establishing a start point to appreciate the capabilities and debilities of these databases. Also, different tests were performed using examples and applications related to telemedicine.

During all these tests, the capabilities of these databases for data input and search presented a rich potential when compared with the relational databases, so much for the speed and for the reliability of the system. This has been checked by the medical staff asked in the tests performed.

In the same way, this improvement has to be preceded by a complete renovation of the interfaces of the existent applications. The lack of intuitive, easy and efficient interfaces has provoked a rise in the cost of medical staff training and has slowed down the implementation of these systems.

However, the capability of managing information from multiple sources, without the need of a previous format of that information, is what makes XML native databases a trustily solution for the future sanitary industry in the world.

Acknowledgments. This work has been partially supported by the *Department of Health, Government of Castile and Leon (Spain)* under the project GES39/VA05/10 and the *Spanish Ministry of Science and Innovation* under the project TIN2010-20529.

References

[1] Kroenke, D.M.: Database Processing Fundamentals, Design and Implementation. Prentice Hall International, New Jersey (2000)

[2] Cuong, N.V.: XML Native Database Systems. Czech Technical University, Prague (2006)

[3] Werner, G.: Seguridad en XML. Advantage Security, México DF (2003)
[4] Staken, K.: Introduction to Native XML Databases (2001), http://www.xml.com (retrieved October 15, 2009)
[5] Mabanza, N., Chadwick, J., Rao, G.S.V.R.: Performance Evaluation of Open Source Native XML databases. University of Fort Hare, Alice (2006)
[6] Meier, W.: Quick Start Guide eXist XML. Database (2003), http://exist.sourcefoge.net/quickstart.html (retrieved November 5, 2009)
[7] W3Schools Trainings, XML Tutorial. W3Schools Trainings (2006), http://www.w3schools.com/xml (retrieved October 20, 2009)
[8] Boyer, J., Dubinko, M., Klotz, L., Landwehr, D., Merrick, R., Raman, T.: XForms 1.0. IBM, Google, Novell & Cardiff (2007)
[9] Boag, S., Chamerlin, D., Fernandez, M., Florescu, D., Robie, J., Siméon, J.: XQuery 1.0: An XML Query Language. IBM & Oracle (2007)
[10] Berglund, A., Boag, S., Chamerlin, D., Fernandez, M., Kay, M., Robie, J., Simeon, J.: XML Path Language (XPath) W3C Recommendation 2.0. IBM & AT&T Labs (2007)

A Semantic Role-Based Approach for Ontology Learning from Spanish Texts

José Luis Ochoa, Maria Luisa Hernández-Alcaraz,
Rafael Valencia-García, and Rodrigo Martínez-Béjar

Abstract. The Semantic Web can be defined as an extension of the current Web in which information is provided with well-defined meaning, so that computers and people are able to work in a cooperative fashion. Ontologies are the de facto knowledge representation methodology in the Semantic Web. Ontology learning from Web documents is considered to be an important activity to promote the Semantic Web. In this paper, an automatic method for acquiring knowledge from Spanish texts is described. The approach presented here is based on semantic roles, which have been employed in our research for extracting semantic relations between concepts. The method makes it possible to represent multiple semantic relations. A set of experiments have been performed with this approach in the oncology domain that show promising results.

Keywords: Ontology learning, semantic role labeling, information extraction.

1 Introduction

Due to its size and the diversity of its textual information, the World Wide Web has become a precious resource for the acquisition of lexical information and for the compilation of corpora. Web sites contain information originally designed to be human-readable, so that manual process is required for making that information machine-readable. This process can be tedious, difficult, and time-consuming.

In [2] the Semantic Web was defined as an extension of the current Web in which information is provided with well-defined meaning, so that computers and people can work in a cooperative manner. In the Semantic Web, ontologies are used as a knowledge representation technology that is usually defined as a formal specification of a domain knowledge conceptualization. Ontologies have been

José Luis Ochoa · Maria Luisa Hernández-Alcaraz · Rafael Valencia-García
Rodrigo Martínez-Béjar
Facultad de Informática, Universidad de Murcia,
Campus de Espinardo 30100 Murcia Spain
e-mail: {joseluis.ochoa,mlhernandez,valencia,rodrigo}@um.es

A. Abraham et al. (Eds.): International Symposium on DCAI, AISC 91, pp. 273–280.
springerlink.com

applied in a number of different domains such as bioinformatics, finance, tourism, geographic systems, digital contents, digital libraries, and e-learning,.

Due to the outstanding importance of ontologies in these domains, different methodologies for designing and building ontologies have been proposed. In this respect, it can be said that the manual ontology construction process constitutes a major problem, since it involves a time-, resources consuming task [5].Hence, the generation and development of methods and software tools to support the construction of ontologies is a relevant research area, which is known as Ontology Learning. One of the most active subareas in Ontology Learning is the use of natural language texts to build ontologies.

As it could be expected, the vast majority of ontology learning methods have focused on the English language. In comparison with English language, Spanish has a much more complex syntax, and is nowadays the second most spoken language in the world[1]. These facts have led us to claim that the computerization of Internet domains in Spanish is of utmost importance.

In this paper, we propose a method for ontology learning from Spanish natural language texts based on the identification of semantic relations among concepts using semantic roles.

The rest of the paper is organized as follows. The proposed method is explained in Section 2. A validation of the ontology learning method in an oncology corpus is described in Section 3. Finally, conclusions and future work are put forward in Section 4.

2 Ontology Learning Process

The ontology learning process embraces three sequential processes, namely, *concept extraction, relations extraction* and *ontology construction* (see Figure 1). These stages are applied to each sentence in the text, with the subsequent extraction of the knowledge entities contained in them.

2.1 Concept Extraction Process

Through this process, terms representing concepts are identified. It is assumed that there exist both multiword and single word terms. By taking into account this assumption, two different methods have been implemented: the NC-Value algorithm [3], which allows to obtain the multiword terms candidates to represent concept, and RIDF[8], which has been employed to obtain terms formed by one word. Next, the stages of this process are described.

NLP stage

The main aim at this stage is the extraction of the morphosyntactic structure of each sentence. For this purpose, a set of NLP software tools including a sentence

[1] http://www.ethnologue.com/ethno_docs/distribution.asp?by=size

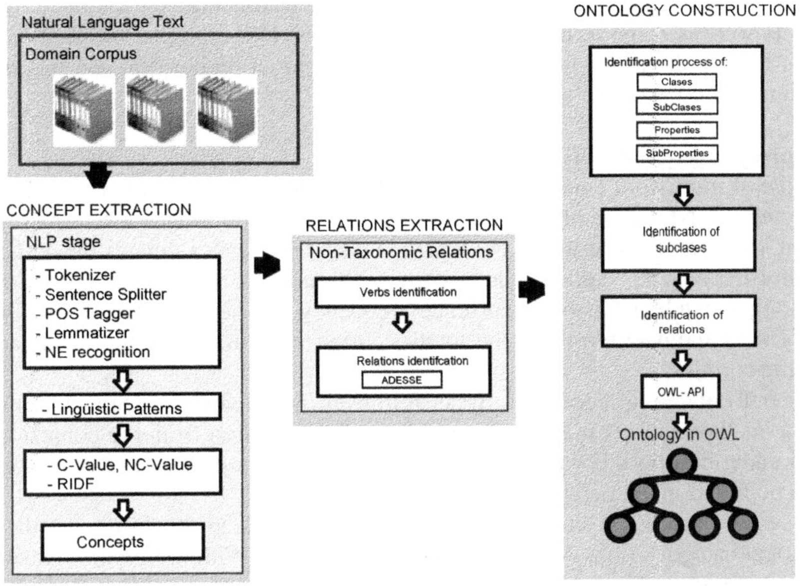

Fig. 1 Ontology Learning process

detection component, a tokenizer, a set of POS taggers, a set of lemmatizers and a set of syntactic parsers have been developed. For it, Freeling[2]2.2 has been employed.

Linguistic patterns stage

The candidate terms are identified by means of a hybrid method which uses a series of linguistic patterns in which the morphosyntactic structure of the terms is described. These patterns are domain-independent, and for their design it is possible to use a predetermined list of terms. These terms can be obtained from a number of sources such as defined ontologies, terminological databases, etc. In the work described here, the patterns have been defined manually from the corpus by a domain expert.

Table 1 Linguistic patterns

Linguistic Pattern	*Term*
NC + SPS+ NC	*cáncer de mama* (breast cancer)
NC + AQ	*Efecto secundario* (secondary effect)
NC + SP + DA + NC	*Parte de el cuerpo* (body part)
NC	*cáncer (cancer),*

[2] http://nlp.lsi.upc.edu/freeling/

Table 1 shows some of the morphosyntactic patterns obtained for the oncology domain as well as some terms matching the patterns. For example, the first row indicates that all the terms formed by a common noun, a preposition and a common noun will be candidate terms.

Multiword concept extraction stage

Once a list of multiword candidate terms has been obtained, this list is filtered out by applying the NC-value algorithm. For that, the system arranges the terms list according to the amount of words contained in each term and calculates the values for several parameters, namely, the occurrence frequency of the candidate term within longer candidates, the occurrence frequency of the candidate term, the length of the candidate term and the total occurrence frequency of the candidate term in the corpus.

In order to obtain an acceptable precision level in the candidate term list, the NC-value method [11; 3] uses the morphological information from the context of the term under question. For this, we consider that verbs, adjectives and nouns are likely to be found in the neighbourhood of a term [6].

The system processes context words and split them up according to their grammatical category (i.e., Adjectives, Verbs or Nouns). With the method developed by Grefenstette [6], a type of weight known as 'context weighting factor' is obtained. It calculates the probability of a context word appearing with a certain term. Table 2 shows the 5 first candidate multiword terms obtained for the oncology domain.

Table 2 First 5 candidate terms obtained

NC-Value	*Term*
339.001	*cáncer de mama* (breast cancer)
84.308	*efecto secundario* (secondary effect)
83.117	*tipo de cáncer* (cancer type)
72.8	*células cancerosas* (cancer cells)

Single word concept extraction stage

Residual IDF (RIDF) is defined as the difference between logarithms of actual document frequency and document frequency predicted by Poisson distribution [8].

$$RIDF(i) = Idf(i) + \log_2 \frac{1}{(1 - p(0;\lambda_i))})$$

where p is the Poisson distribution with parameter $\lambda_i = \frac{cf_i}{N}$ (the average number of occurrences of each word per document). $1 - p(0;\lambda_i)$ is the Poisson probability of a document with at least one occurrence of i.

2.2 Relation Extraction Process

Once the concepts have been identified in the corpus, the semantic relations of these concepts have to be obtained. In natural language, relations between concepts are usually associated with verbs. A number of systems for learning relationships are based on the extraction and identification of verbs [13, 15, 17]. In this work semantic roles and semantic class membership for Spanish verbs are used in order to extract and identify these relationships.

A semantic role is the relation between a syntactic constituent and a predicate. It defines the role of a verbal argument in the event expressed by the verb [10]. The semantic roles set developed in the Proposition Bank (PropBank) project [12] and in the FrameNet project [4] are the most widely used in the literature, and they are only for the English language. ADESSE [1] collects nearly 4,300 semantic roles of Spanish verbs in a syntactic database of nearly 160,000 clauses retrieved from a Spanish corpus of 1,5 million words.

```
<predicate lemma="disminuir">
<roleset id="modificación">
<roles>
  <role n="0" descr="decrease"/>
  <role n="1" descr="holder"/>
</roles>
</roleset>
```

La [$_0$**Terápia de Radiación**] disminuye el [$_1$**Tamaño de el tumor**]

The [$_0$**Radiation theraphy**] decreases the [$_1$**Tumor size**]

Fig. 2 An example of the *disminuir* frame in ADESSE.

Table 3 Some relations obtained from the corpus.

Sentence	**Semantic relation**
Un 20 por ciento de las personas infectadas con el virus de la hepatitis B desarrollarán cáncer de hígado. (About 20 percent of people infected with hepatitis B virus develop liver cancer)	hepatitis_B **desarrollar** cáncer_de_hígado (hepatitis_B **develop** liver_cancer)
...la terapia hormonal puede ser eficaz para controlar la metástasis de el cáncer de mama (...hormone therapy may be effective in controlling metastatic breast cancer)	terapia hormonal **controlar** cáncer de mama (Hormone_therapy **controls** breast_cancer)

The unfolding of this process is described next. The main verb of the current sentence is identified. Then, there is a search for the type of semantic relation associated with that verb in ADESSE. This search is conducted on the ADESSE relational knowledge base by means of the lemmatized word of the verbal expression. Once the type of relation associated with the main verbal expression in the current sentence is found, the system selects the concepts related to that verb. For this purpose, the system looks for concepts on both sides of the verb.

In order to detect ontological semantic relations among entities, a mapping between semantic relations and semantic roles has to be done. For instance, Figure 2 shows the ADESSE *disminuir (decrease)* frame. The example shows how the semantic role relates "Terapia de Radiación" (radiation therapy) and "Tamaño de el tumor" (tumor size). Some other examples of relations extracted from the corpus are shown in table 3.

2.3 Ontology Construction Process

In this process, the ontology is built from the elements previously extracted with the purpose of detecting classes, subclasses and properties of the ontology. In an OWL ontology, a property can be a datatype property or an object property. At this point, the system attempts to identify the subclasses of the concepts extracted at the first stage. The detected relations are then processed at the second stage.

Identification of subclasses: The *subclass_of* relations are detected by means of the name of the class. In case a class's name is made up of other classes names, then it would be a subclass of the first class. For instance, the Terapia_de_radiación (Radiation_therapy) is a subclass of the Terapia (Therapy) concept, since the token "Terapia" takes part of the token "Terapia_de_radiación".

Identification of relations: At this stage, concepts are related from the results obtained at the relation extraction stage. In order to identify the names of the properties, the lemmatized form of the verb is used.

3 Validation in the Oncology Domain

In this section, the experimental results obtained by our method are presented. As mentioned above, the experiment has been conducted in the oncology domain, where knowledge representation formalization has been claimed to be a relevant task [14].
Our experimental corpus has 96,458 words and comprises 19 documents. This corpus has been manually processed by domain experts, obtaining a total amount of 456 concepts or classes, 282 *subclass-of* relations, and 181 semantic relations (see Table 4).

In order to evaluate the performance of the method, recall and precision scores were calculated. These measures are the most commonly used for the assessment of statistical extraction systems, and trace their origins back to the Information Retrieval discipline.

As it can be seen in table 4, the results of the validation seem promising. It is worth noting, for instance, that the method obtains a precision value of 74% and a recall value of 70% in the detection of non-taxonomic relations.

Table 4 Evaluation results

	Classes	Subclass_of relations	Relations
Expert	456	282	181
System	512	150	233
Precision	69 %	59 %	74 %
Recall	75 %	64 %	70 %

4 Conclusions and Future Work

In this paper, an automatic method for acquiring knowledge from texts has been presented. This approach is based on the use of semantic roles from ADESSE in order to extract semantic relations between concepts.

Ontology building from free text is an important activity for the knowledge engineering community. One of the hottest trends in this research area is ontology learning from Web documents [9, 15], which is considered to be an important activity to promote the Semantic Web [13]. The approach presented in this work is totally automatic. Another key feature of this approach is that it works not only with taxonomies, but also with multiple semantic relations.

The authors in [15] present a methodology for the detection of non-taxonomic relations from Web texts. This methodology is based on the identification of relevant verbs, which are used as a knowledge basis for learning and tagging non-taxonomic relations automatically and without supervision. Both studies use linguistic patterns for obtaining taxonomic relations.

In [7] a methodology is introduced that has a similar common ground to ours. The aim of these authors is the development of a high-quality ontology, and for this purpose they use a combination of statistical and lexical-semantic methods.

Further validations of the system are planned by means of its application to texts from different medical domains and by using statistical methods for the analysis of the results obtained. Moreover, we intend to extend the system to cover axioms such as the work presented in [16]. The main forecast problem concerning axioms is, however, that the number of participants is a priori unknown. Notwithstanding this fact, the amount of axioms present in a text is irrelevant in comparison to the amount of other knowledge entities.

Acknowledgments. This work has been supported by the Spanish Ministry for Science and Innovation under projects TIN2010-18650 and TIN2006-14780, and by the Murcian Government under project BIO-TEC 06/01-05.

References

1. Albertuz-Carneiro, F.: Sintaxis, semántica y clases de verbos: clasificación verbal en el proyecto ADESSE. Actas del VI Congreso de Lingüística General, Las lenguas y su estructura (IIb) 2, 2015–2030 (2007)
2. Berners-Lee, T., Hendler, J., Lassila, O.: The semantic web. Scientific American 284, 34–43 (2001)

3. Barrón-Cedeño, A., Sierra, G., Drouin, P., Ananiadou, S.: An Improved Automatic Term Recognition Method for Spanish. In: Gelbukh, A. (ed.) CICLing 2009. LNCS, vol. 5449, pp. 126–136. Springer, Heidelberg (2009), doi:10.1007/978-3-642-00382-0_10
4. Filmore, C.J.: Framenet and the linking between semantic and syntactic relations. In: Proc. 19th International Conference on Computational Linguistics, COLING 2002 (2002)
5. Fortuna, B., Mladenič, D., Grobelnik, M.: Semi-automatic construction of topic ontologies. In: Ackermann, M., Berendt, B., Grobelnik, M., Hotho, A., Mladenič, D., Semeraro, G., Spiliopoulou, M., Stumme, G., Svátek, V., van Someren, M. (eds.) EWMF 2005 and KDO 2005. LNCS (LNAI), vol. 4289, pp. 121–131. Springer, Heidelberg (2006), doi:10.1007/11908678_8
6. Grefenstette, G.: Explorations in Automatic Thesaurus Discovery. Kluwer Academic Publishers, Norwell (1994)
7. Jiang, X., Tan, A.H.: CRCTOL: a semantic-based domain ontology learning system. Journal of the American Society for Information Science and Technology 61, 150–168 (2010), doi:10.1002/asi.21231
8. Manning, C.D., Schütze, H.: Foundations of Statistical Natural Language Processing. The MIT Press, Cambridge (1999)
9. Millan, M., Sánchez, D., Moreno, A.: Unsupervised Web-based Automatic Annotation. In: STAIRS 2008: Proceedings of the Fourth Starting AI Researchers' Symposium, The Netherlands (2008)
10. Moreda, P., Llorens, H., Saquete, E., Palomar, M.: Combining semantic information in question answering. Information Processing and Management (in Press, 2011), doi:10.1016/j.ipm.2010.03.008
11. Ochoa, J.L., Almela, A., Ruiz-Martínez, J.M., Valencia-García, R.: Efficient Multiword Term Extraction in Spanish. Application to the Financial Domain. In: Proceedings of ICIIT 2010, Lahore, Pakistan (2010)
12. Palmer, M., Gildea, D., Kingsbury, P.: The proposition bank: an annotated corpus of semantic roles. Computational Lingustics 31, 71–106 (2005), doi:10.1162/0891201053630264
13. Shamsfard, M., Barforoush, A.: Learning ontologies from natural language texts. International Journal of Human-Computer Studies 60, 17–63 (2004)
14. Sánchez, D., Moreno, A.: Learning Medical Ontologies from the Web. In: Riaño, D. (ed.) K4CARE 2007. LNCS (LNAI), vol. 4924, pp. 32–45. Springer, Heidelberg (2008), doi:10.1007/978-3-540-78624-5_3
15. Sánchez, D., Moreno, A.: Learning non-taxonomic relationships from web documents for domain ontology construction. Data & Knowledge Engineering 64, 600–623 (2008), doi:10.1016/j.datak.2007.10.001
16. Terrientes, L., Moreno, A., Sánchez, D.: Discovery of Relation Axioms from the Web. In: Bi, Y., Williams, M.-A. (eds.) KSEM 2010. LNCS, vol. 6291, pp. 222–233. Springer, Heidelberg (2010), doi:10.1007/978-3-642-15280-1_22
17. Valencia-García, R., Fernández-Breis, J.T., Ruiz-Martínez, J.M., García-Sánchez, F., Martínez-Béjar, R.: A knowledge acquisition methodology to ontology construction for information retrieval from medical documents. Expert Systems 25, 314–334 (2008), doi:10.1111/j.1468-0394.2008.00464.x

A Dynamical Characterization of Case-Based Reasoning Systems for Improving Its Performance in Highly Dynamic Environments

Luis F. Castillo, M.G. Bedia, M. Aguilera, and L. Uribe

Abstract. In this paper a mathematical technique is presented for modelling the generation of solutions in a standard-CBR methodology adapted to highly dynamic environments. In recent years, much research has focused on exploring how to improve CBR-systems to deal with dynamic environments where changes are difficult to model or predict and, consequently, the performance of the CBRs gets worse with time. High-level planning, reactive strategies or hybrid alternatives have been proposed to face this problem, but this contribution will not be related on particular techniques. We simply concentrate on formal aspects without establishing which should be the most adequate procedure in a subsequent implementation stage. The advantage of the presented scheme is that it does not depend on neither the problem nor the model of the environment. It consists in a formal approach that only requires, local information about the averaged-time spent by the system in obtaining a solution and an estimated measure about the dynamism of the environment.

1 Introduction

Case Based Reasoning (CBR) [Aamodt et al., 1994] is a software technique that tries to solve new problems by using or adapting solutions that we employed to solve old problems. It offers a reasoning paradigm that is similar to the way people routinely solve problems. Faced with a new problem, a human often relates

Luis F. Castillo
Industrial engineering at National University of Colombia
e-mail: lfcastilloos@unal.edu.co

M.G. Bedia · M. Aguilera
Computer Science at University of Zaragoza
e-mail: {mgbedia,maguilera}@unizar.es

L. Uribe
Computer Science at Autonomous University of Manizales
e-mail: auribe@autonoma.edu.co

A. Abraham et al. (Eds.): International Symposium on DCAI, AISC 91, pp. 281–289.
springerlink.com

the problem to one or more memory episodes and composes a solution from these episodes. As we have indicated, a CBR is often seen as a computational intelligence tool whose knowledge source is composed of a memory of past cases but, sometimes, it is more generally considered as a methodology to solve problems without establishing commitments in the implementation stages [Watson, 1999]. As a technological tool, CBR-based programs have been successfully applied to a variety of problems in several fields and disciplines. However, one of the main problems in deploying these systems is that have been designed to "static" domains (domains in which the system has unlimited amount of time to solve each problem and during this time, the state does not change) but most real-world (and most interesting) domains are dynamic [Mehta et al., 2010], [Mehta et al., 2010 (b)]. In the real-world, CBR-systems have time-constraints and must deal with dynamic conditions that changes over time. We can find in the literature several approaches in the CBR community to deal with this issue. Particularly interesting have been some proposals as; (i) "on-line case-based planning" [Ontanon et al., 2010], useful in real-time strategy games, and (ii) "dynamic case base reasoning", used on fault diagnosis and prognosis activities [Berenji et al., 2005]. Related to theses issues, we explored how stated the upper time-limit in the retrieval stage, in mathematical terms and in order to obtain an optimal solution in a dynamic environment. The aim focused on formal aspects without establishing which should be the most adequate procedure in a subsequent implementation stage. The advantage of the presented scheme is that it does not depend on neither the problem nor the model of the environment. The structure of the paper is as follows: in section 2, a brief overview about the CBR-concept, its cycle and its constitutive elements are illustrated; in section 3, the proposed mathematical model will be presented; in section 4, computational and formal aspects of our model will be stated and solved; section 5 is related to the implementation stage and empirical results and, finally, in section 6, conclusions and future work will be introduced.

2 A CBR Strategy: Strengths and Weaknesses

Case-based reasoning has been defined as a problem-solving technique based on the hypothesis that *reasoning is reminding*. CBR systems have proven useful in domains with weak models and a large body of unstructured and experiential knowledge. The generic CBR cycle consists of the following steps [Aamodt et al., 1994]: (1) retrieval of relevant (similar) cases from the repository based on cues derived from problem requirements, (2) reuse of applicable cases to suggest solutions to a new problem, knowledge-based revision of relevant cases, (3) testing-based verification and rule-based validation to ensure correctness, and (4) retention of past solutions and failures to enable learning. For simplicity, we do not consider stage 3 in our model. Moreover, stages 1 and 2 will be integrated together in a new stage called "generation of solution". Therefore, the cycle of a CBR will consist in: (1) a generation stage (that covers retrieving and adapting processes), and (2) an execution stage that ends in a retain process (see Figure 1).

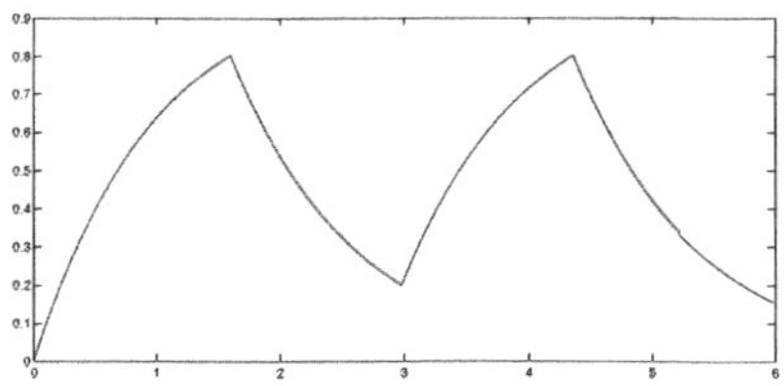

Fig. 1 Scheme of the adjustment function $a(t)$ for a CBR cycle in a dynamical environment: generation (retrieval and reuse phases) and execution (display of solutions) are intercalated

A variety of specific case retrieval techniques and case adaptation ones are been discussed. However, many of these proposals are not directly applicable in dynamics environments because real-time interactions are required [Urdiales et al., 2003]. We think there exists an important and transversal assumption that is not suited for real-time domains involving CBR systems, i.e. the CBR cycle produces a solution but the solution execution is delegated to some external module in a subsequent stage (that is, *execution* and *problem solving* are *decoupled.*) In connection with that assumption, it is also accepted that in a dynamic environment where the system does not know how changes on its surrounding are going to be, the system is built considering that the optimal solution to a problem should be composed by optimal intermediate solutions obtained in every cycle. In other words, it assumed that given a time period T, resulting from a temporal windows series $\sum_{i=1}^{n} \Delta t_i$, the best strategy is a greedy strategy, the one that holds $optimal(T) = \sum_{i=1}^{n} optimal(\Delta t_i)$. That statement is assumed but no demonstrated, and we feel that it embodies a *naive* and very extended error related to the design of artifacts in dynamical environments. In this work it is questioned if that assumption is right, i.e. if the optimal strategy, under conditions of uncertainty, must be a greedy strategy. In next sections it will be shown that this is not necessarily true.

3 A Dynamical Characterization of a CBR System

In this section we will try to give a satisfactory answer to this question: 'A CBR system in a dynamic environment, with the possibility of generating a good solution and without information about future states, does it makes sense to implement an inferior quality solution?'

3.1 Mathematical Preliminaries and Dynamical-Systems Framework

In this section a minimal model of CBR-methodology is introduced. The assumptions of the model will be quite simple in order to allow the conclusions of our model to be the most general:

1. **Base of solutions:** There exists a mechanism for generating solutions, that we represent as $\{x_1,.,x_i,x_{i+1},.,x_j,x_{j+1},...,x_m\}$, where $\{x_1,..,x_i\}$ are compatible with a temporal window Δt_1, $\{x_1,...,x_i,x_{i+1},..,x_j\}$ are compatible with Δt_2, etc. In other words, more adjusted solutions of the CBR will be more "expensive" (in time units) and simpler solutions will be easier to obtain. It will be called *adjustment of a solution in t*, and it is denoted by $a(t)$, a measure of similarity between the generated solution $x(t)$ and the ideal solution $x^*(t)$.

$$a(t) = \left\{ \begin{array}{ll} 1, & \text{if } x(t) = x^*(t) \\ 1 - \frac{|x(t)-x^*(t)|}{x(t)}, & \text{if } \frac{|x(t)-x^*(t)|}{x(t)} < 1 \\ 0, & \text{if } \frac{|x(t)-x^*(t)|}{x(t)} \geq 1 \end{array} \right\} \tag{1}$$

2. **Generation phase:** We propose a general functional relation between "the time to generate a solution" and "adjustment". It will be considered as a nonlinear function (the effort in obtaining more adjusted solutions grows in relative terms with time), so it can be assumed exponential, $a(t) = a_M(1 - e^{-t/\tau})$. This functional dependence, considering the interval $t \in \{t_0,t_1\}$ comes from a differential equation:
$$\frac{d}{dt}a(t) = \frac{1}{\tau}(a_M - a(t))$$
3. **Execution phase:** We also assume that it is known that the environment is going to change (without knowing the direction of the change). So it is also considered an exponential functional dependency between the adjustment of the obtained solution and time, that is, $a(t) = (e^{-t/\varepsilon})$. Considering $t \in \{t_1,t_2\}$, the solution comes from a equation as:
$$\tfrac{d}{dt}a(t) = -\tfrac{1}{\varepsilon}a(t)$$

Both cases can be combined, taking as values $\gamma(t) = \gamma_0$ (*generating a solution*), when $t \in \{t_0,t_1\}$, and $\gamma(t) = \gamma_1$ (*executing a solution*) if $t \in \{t_1,t_2\}$, obtaining the global behaviour equation:

$$\frac{d}{dt}a(t) = -\frac{1}{\varepsilon}a(t) + \frac{1}{\tau}a_M(1 - \gamma(t)) \tag{2}$$

The agent performance will be obtained only integrating the suitability of the system during the execution periods (the ones in which the agent is acting on the world). In order to solve equation (2), we will consider: (i) $\gamma_0 = 0$, and $\gamma_1 = 1$; (ii) the solution in a period T:

$$a(T) = \frac{1}{T}\int_0^T a(t)dt$$

will be defined by the average performance of the system, $\overline{p}(T)$, evaluated in $(0,T)$:

$$\overline{p}(T) = \frac{1}{T}\int_0^T \gamma(t)\cdot a(t)dt \tag{3}$$

The optimal solution of the system, $a_{opt}(t)$, is the one that maximizes $\overline{p}(T)$. If we consider $\gamma(t) = \{0,1\}$ then,

$$\begin{cases} \dot{a}(t) = \frac{1}{\tau}(1-a(t)) - \gamma(t)\cdot(\frac{1}{\tau}+a(t)\cdot(\frac{1}{\varepsilon}-\frac{1}{\tau}) \\ \dot{p}(t) = \gamma(t)\cdot a(t) \end{cases}$$

and it is wanted to find the set $\{\gamma_k(t_k)\}$ that maximizes $p(t)$. Discretizing,

$$\begin{cases} a_{k+1} - a_k = -h(\frac{1}{\tau}(1-a_k) - \gamma_k\cdot(\frac{1}{\tau}+a_k\cdot(\frac{1}{\varepsilon}-\frac{1}{\tau})) \\ \\ p_{k+1} - p_k = h(\gamma_k \cdot a_k) \end{cases}$$

where h is a temporal step, $k = 0,1,2,...,N$, so $a(0) = a_0$, $p(T) = p_N$, given $T = \{t_1,t_2,...,t_N\}$. For the sampled version the problem can be reformulated by the following statement (knowing that h is constant): "Find the set of decisions $\{\gamma_k(t_k)\}$ that maximize $\sum_{k=0}^{N} \gamma_k a_k$". That is, the $\{\gamma_k(t_k)\}$ values must be computed providing that,

$$p_N = \max_{\gamma_0,\gamma_1,\cdots\gamma_N} \sum_{k=0}^{N} \gamma_k a_k \tag{4}$$

which, since it starts in a_0 is denoted by $p_N^{MAX}(a_0)$. For solving the problem it is applied the Bellman Algorithm (*An optimal policy has the property that whatever the initial state and initial decision are, the remaining decisions must constitute an optimal policy with regard to the state resulting from the first decision.* [Bellman, 1957]). The algorithm computes the complete sequence $(\gamma_0,\gamma_1,\cdots\gamma_N)$, therefore,

$$p_N^{MAX}(a_0) = \max_{\gamma_0}[\gamma_0 a_0 + \max_{\gamma_1,\cdots\gamma_N} \sum_{k=1}^{N} \gamma_k a_k], \quad \text{where} \quad \mathrm{p}_{N-1}^{MAX}(a_1) \max_{\gamma_1,\cdots\gamma_N} \sum_{k=1}^{N} \gamma_k a_k,$$

Iterating, is obtained de sequence:

$$p_N^{MAX}(a_0) = \max_{\gamma_0}[\gamma_0 a_0 + \max_{\gamma_1}[\gamma_1 a_1 + \max_{\gamma_2}[\gamma_2 a_2 + ... + ... + \max_{\gamma_N}[\gamma_N a_N]]....]$$

For solving the system, it must be started from the last decision to the first. Since the last does not affect to the future, the maximization is local. In our case, it is:

$$\gamma_N(t_N) = \begin{cases} 1, & \text{if} \;\; a_N \in (a_M, 0), \; \dot{a}_N < 0 \\ 0, & \text{if} \;\; a_N \in (0, a_M), \; \dot{a}_N > 0 \end{cases}$$

$$p_0^{MAX}(a_N) = \begin{cases} a_N, & \text{if} \;\; a_N \in (a_M, 0), \; \dot{a}_N < 0 \\ 0, & \text{if} \;\; a_N \in (0, a_M), \; \dot{a}_N > 0 \end{cases}$$

Once we know what is the optimal decision in $\gamma_N(t_N)$, the previous instant $\gamma_{N-1}(t_{N-1})$ is computed, applying the following equation:

$$p_1^{MAX}(a_{N-1}) = \max_{\gamma_{N-1}}[\gamma_{N-1} a_{N-1} + p_0^{MAX}(a_N)], \quad \text{and it is known that}$$

$$a_N = a_{N-1} - \frac{h}{\tau}((1-a_{N-1}) - \gamma_{N-1}\cdot(1+a_{N-1}\cdot(\frac{\tau}{\varepsilon}-1)),$$

therefore given $\gamma_{N-1} = \{0,1\}$, there only have to be computed which one of the two cases is bigger:

$$a_{N-1} + p_0^{MAX}[(1-\tfrac{h}{\varepsilon})\cdot a_{N-1}] \gtrless p_0^{MAX}[(1-\tfrac{h}{\tau})\cdot a_{N-1} + \tfrac{h}{\tau}]$$

The equilibrium condition meets for a critical a_{N-1} value, denoted as a^*_{N-1}, that allows to rewrite the equation in the following way:

$$p_1^{MAX}(a_{N-1}) = \begin{cases} a_{N-1} + p_0^{MAX}[(1-\frac{h}{\varepsilon})\cdot a_{N-1}], & \text{if } a_{N-1} > a^*_{N-1} \\ p_0^{MAX}[(1-\frac{h}{\tau})\cdot a_{N-1} + \frac{h}{\tau}], & \text{if } a_{N-1} \leqslant a^*_{N-1} \end{cases}$$

The procedure can be repeated for $(k = 2,...,N)$ being obtained the $\{a^*_0, a^*_1, ..., a^*_{N-1}, a^*_N\}$ values by iteratively solving the equation:

$$p_{N-k}^{MAX}(a_{N-k}) =$$
$$\max_{\gamma_{N-k}}[\gamma_{N-k}a_{N-k} + p_{k-1}^{MAX}(a_{N-k} - \tfrac{h}{\tau}((1-a_{N-k}) - \gamma_{N-k}\cdot(1+a_{N-k}\cdot(\tfrac{\tau}{\varepsilon}-1)))\ \text{k=1,...,N.}$$

The classical model propels us to compute the better solution in the available time (every window of opportunity) and execute it until it stops being good enough (as it

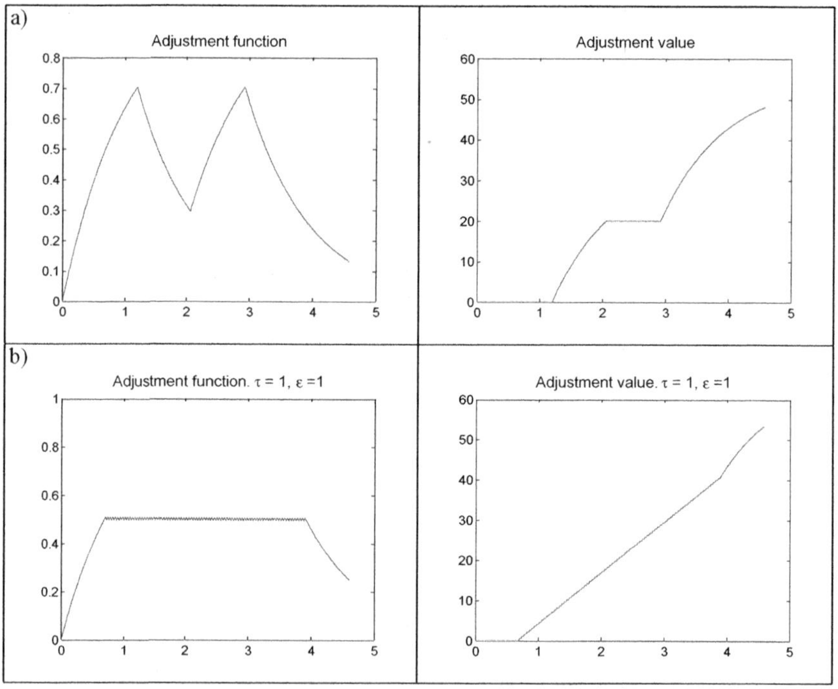

Fig. 2 Comparison of strategies and representation of the adjustment function $a(t)$ and the adjustment value $p(t)$ for: (a) the greedy strategy where every window of opportunity is optimized, and (b) the coupled solution CBR-environment (with $\tau = 1, \varepsilon = 1$) which offers the optimal solution.

is illustrated in Figure 2.a). Our results forces us to change that point of view (see figure 2.b): it is shown that the global result do not depend on the quality of the chosen solution, but in how that solution is coupled in the available time window. In a nutshell, the optimal solution can be captured under the following dictum: *"Using a CBR when the environment changes, the best behaviour is the one that maximizes the number of cycles with the world, being the optimal level fitness level determined by the dynamics of the environment characterized by the relation between τ and ε."* (See figure 2). In other words, the obtained result tells us the following: when the environment changes, quick and fast generation solutions are preferred than the ones that spent more time in being generated. Much to our surprise, there has been shown that in dynamical environments, solutions locally considered as "bad" give the optimal response with time. And there has also been shown that, when analyzing dynamical systems, does not matter how simple they are, the solution can be surprising when the global picture is took.

4 Implementation and Experimental Tests

Firstly, we state a mechanism of computational implementation based on a brute-force algorithm, in order to generate a solution space that represents all possible binary strings of length n, that is, 2^n . The solution space consists of strings of the form a_1-$a_2 - ... - a_n$, where $a_i = \{0,1\}$, and where $\{0\}$ means "generation of solutions' while that $\{1\}$ means "execution of solutions". The results have been illustrated in Table 1(case a) and were obtained using a computer Intel XEON Quad 2.0 Ghz with 32 Gb of RAM memory. The most relevant issue was the run into a java heap overflow for the case $n = 18$. In order to give a solution, it was proposed a new execution of the virtual machine of java with the configuration as follows: java -Xmx15360m -jar ModelSRecurrente.jar). It allows us to obtain new results for cases with $n = \{18,19,20,21,22,23,24,25\}$. The times for obtaining a solution in these cases are shown in the Table 1 (case a).

Secondly, we use an alternative technique base on genetic algorithms. In the proposed adaptive genetic algorithm, the design domain was initially discretized and initialized by a chromosome in the early stages of design. It was used the library JGAP to obtain the optimal solution. This seeding of the initial populations (600 unit per population) and an amount of 400 evolutions, provide the results illustrated in Table 1 (case b). The algorithm increases the performance of approaching higher levels of fitness, and allows the solution of problems with larger numbers of design variables. The solution is compared with the one-time brute-force approach, and it is demonstrated that finds a better solution with less computational cost.

As the temporal and spatial complexity of the algorithm is relevant, its implementation was proposed in the computational cluster service hold in the Autonomous University of Manizales (Colombia). It allowed the execution of services and applications in high speed networks. The initial test tried to parallelize the algorithm using the advantages of Condor software as job manager for distributing the computational weight between different nodes of the cluster. The results were not

Table 1 Table of time that the host needs to generate the whole space of solutions: (a) Brute-force algorithm; (b) genetic algorithm (population=600 units, evolutions=400).

		Case a		Case b
string	**n**	**time (ms)**	**n**	**time (ms)**
001010101010111111	18	17432	18	1596
0001010101010111111	19	37697554	19	1523
00101010101010111111	20	43082592	20	1441
001001010101010111111	21	39839192	21	1087
0010101010101010111111	22	52621296	22	1096
00101010101010101111111	23	39086117	23	1095
001010101010101010111111	24	58413732	24	1116
0001010101010101010111111	25	57040740	25	1131
00101010101001101010111111	–	—	26	1145
000101010101010101010111111	–	—	27	1154
0010011010101010100110111111	–	—	28	1169
00101010101010101010101111111	–	—	29	1189
001010101010101010101010111111	–	—	30	1209

completely satisfactory. As a consequence, it was defined a new technique that used a master server with Linux as operating system with 4 processors Quad Intel XEON 2.0 Ghz, 32 Gig RAM, 4 Drives 250 GB, with computer software for connection element cluster. The Host IP address was 200.21.104.84. It was installed java 1.6 using evolutionary computation. The results obtained and presented in table 1 (case b) demonstrated the genetic strategy does not need to be implemented in a parallel computing environment. It is emphasized the fact that in our model, and in the explored cases, a genetic algorithms obtains the optimal solution without errors (see first column of table 1) although the theoretical solution might be superior in some cases.

5 Overview and Conclusions

There has been criticised a traditional perspective about the use of CBR in dynamic environments characterized for:

- Decoupling the generation and execution phases with time.
- Considering the presence of uncertainty (about the future in the environment) only as 'noise', ignoring it as a source of opportunities.
- Since the system has not information about how the world is going to change, it is maximized the adjustment of particular solutions in the available time every window of opportunity.

Our mathematical and experimental study, reveal us:

- It is possible to develop a system where the generation and execution of solutions are coupled in time without explicitly knowing how the world is going to change.

- In dynamic environments, solutions that maximize the local adjustment do not necessarily find the global optimal strategy.
- The implementation processes demonstrated that a genetic strategy without errors in strings of dimension $n \in (1,30)$ does not need to be implemented in a parallel computing environment.

Acknowledgements. This work was supported in part by Colciencias (Learning Management System -LMS- Proyect, cod. 121948725660, num. 487, 2009).

References

[Aamodt et al., 1994] Aamodt, A., Plaza, E.: Case-Based Reasoning: Foundational Issues, Methodological Variations, and System Approaches. AI Commun. 7(1), 39–59 (1994)

[Bellman, 1957] Bellman, R.E.: Dynamic Programming. Princeton University Press, Princeton (1957) (republished 2003: Dover), ISBN 0486428095

[Berenji et al., 2005] Berenji, H., Wang, Y., Saxena, A.: Dynamic Case Based Reasoning in Fault Diagnostics and Prognostics. In: FUZZ-IEEE, Reno (May 2005)

[Mehta et al., 2010] Mehta, M., Ram, A.: Run-Time Behavior Adaptation for Real-Time Interactive Games. IEEE Transactions on Computational Intelligence and AI in Games 1(3) (September 2009)

[Mehta et al., 2010 (b)] Mehta, M., Ontañón, S., Ram, A.: Using Meta-reasoning to Improve the Performance of Case-Based Planning. In: McGinty, L., Wilson, D.C. (eds.) ICCBR 2009. LNCS, vol. 5650, pp. 210–224. Springer, Heidelberg (2009)

[Ontanon et al., 2010] Ontanon, S., Mishra, K., Sugandh, N., Ram, A.: On-Line Case-Based Planning. In: Computational Intelligence, pp. 84–119 (2010)

[Urdiales et al., 2003] Urdiales, C., Perez, J., Vázquez-Salceda, J., Sandoval, F.: A hybrid architecture for autonomous navigation using a CBR reactive layer. In: Proceedings of the 2003 IEEE/WIC International Conference on Intelligent Agent Technology (IAT 2003), Halifax, Canada, October 13-16, pp. 225–232. IEEE Computer Society, Los Alamitos (2003), ISBN 0-7695-1931-8

[Watson, 1999] Watson, I.: CBR is a methodology not a technology. Knowledge Based Systems Journal 12(5-6), 303–308 (1999)

Application of the Artificial Intelligence in Enterprise Quality Systems

Jose Amelio Medina, Carmen De Pablos, Lourdes Jimenez, and Jorge Peñas

Abstract. In a highly globalized market, a demand of a more specialized and flexible labour force is required. This situation often means a loss of knowledge in the organizations. Our paper offers some alternatives that can be applied to mitigate this threath.

Throughout this work, we will raise a proposal of improvement in the way firms manage quality in their processes according to the norm 9001:2008 [1], by means of the use of an Expert System. This proposal will allow a decrease of the problems that can appear as a consequence of the circumstances previously described.

Keywords: Artificial Intelligence, Expert Systems, Knowledge Management, Logic, Quality Systems, ISO 9001:2008.

1 Introduction

In a competitive market as the present one, the companies promote improvements that allow them to be different from their competitors. Amongst the different tools that firms often make use of, we can cite the following ones: methods and

Jose Amelio Medina
Computer Science Dept. University of Alcala, University Campus 28871 Alcalá de Henares
e-mail: `josea.medina@uah.es`

Carmen De Pablos
Business Economy Dept. Rey Juan Carlos University, Campus of Vicalvaro 28032 Madrid
e-mail: `carmen.depablos@urjc.es`

Lourdes Jimenez
Computer Science Dept. University of Alcala, University Campus 28871 Alcalá de Henares
e-mail: `lou.jimenez@uah.es`

Jorge Peñas
Computer Science Dept. University of Alcala, University Campus 28871 Alcalá de Henares
e-mail: `jorge.penasb@uah.es`

A. Abraham et al. (Eds.): International Symposium on DCAI, AISC 91, pp. 291–294.
springerlink.com

processes optimization, the implementation of quality systems and the development of information and communication technologies (ICT).

The implementation and development of these improvements, requires of qualified staff but, this option implies a dependency of the worker in the company.

In this situation, when the worker leaves or the company decides to let him/her go, a loss of knowledge takes place. Depending on the kind of job that the worker has been performing in the firm it can mean an important economic and image loss for the company.

Although the different quality norms such as ISO 9001:2008 [1] show this fact and advise to document the different firm's processes with their respective procedures within the company, however, this is not enough since, in many cases it is not done correctly or simply it is not realized. Therefore, an increase of the loss of knowledge takes place.

The prototype that we show in this article will allow us to reduce the loss of knowledge in the companies particularly from the analysis of the client's satisfaction according to ISO 9001:2008 [1], by means of the development of an expert system that is able to evaluate clients' answers and will help to make decisions with the purpose of learning from continuous improvement [3].

2 Description of the Problem

The loss of knowledge in the company has been studied by acclaimed authors as Kaplan and Norton (1996) [2], and this is one of the dimensions of the intellectual capital to be considered when evaluating the organization. There is a high level of consensus about the importance of this dimension in companies, and it has also been recognized by authors in this field [4], [5].

As far as companies need to be different from their competitors, they realize and experience that they can obtain it by means of greater quality products as well as by an optimization of processes. Companies trust and rely each time more in the norms as the ISO 9001:2008 and EFQM, since they will allow them to reach some high standards of quality.

Juran defines the quality as the product characteristics that satisfy clients' needs, enabling to the company to increase the satisfaction [6].

This assumption joint to the requirement of norm ISO 9001:2008 [1] in its point 8.2. *"Monitoring and measurement"* and more concretely in 8.2.1 section *"Customer satisfaction"*, indicates that a pursuit of client's satisfaction must be done with the purpose of analyzing the results and that it allows to obtain a feeding back of the input data.

With the purpose of keeping the requirement of this norm, the great majority of the companies evaluate client's satisfaction through two different ways: through shipment of a questionnaire to a group of clients that later they will analyze or through a pursuit of clients' orders and commercial's visits.

For this reason our prototype manages and evaluates client's satisfaction according to norm ISO 9001:2008 [1] and it allows to realize actions that are directed to improve the clients' perception on the company, product, etc.

2.1 Development of the System

The prototype has been developed as complement to the integral system that usually exists in any company, as for example the ERP and CRM. These systems usually are developed in SQL Server as data storage system independently of how it is the interface. From this process, data like internal resources, finance and human resources can be obtained, among others.

The development of the system has been realized from two points of view that we must not forget, one from the internal point of the own application by means of the data base in local way or in network, and from the external point the connection is realized via Web.

The knowledge acquisition that have been applied consisted in meetings with quality experts in enterprise and also in the analysis of the purpose of satisfying the evaluation requirement of client's satisfaction, described in section 8.2 of norm ISO 9001:2008 [9].

Nextly it is described how the questionnaire has been built: it is composed by 10 variables initially grouped in two categories. On one hand those that allow to evaluate products and services and, on the other hand, the degree of information and communication that we offer taking into account a question that compares it with the competition and other that realizes a global evaluation of the company, its shipment and later analysis of the obtained results. All together will the purpose of being able to apply improvements that guarantee clients' satisfaction.

The process of questionnaire shipment is realized using ICT as electronic mail answering system or a Web questionnaire or even using more traditional ways like the fax and telephone.This last option has the disadvantage that it cannot be documented.

With the purpose of optimizing the system four profiles have been settled down: user, intermediate controller, manager and an administrator. The profile establishes the level of access to the data in the system. From the collected data, the information is processed and stored in a data base, in our analysis we have made use of MySQL, next they are adapted and shown in the screen by means of the balance scorecard.

This balance scorecard (BS) [2] is based in the Olve at et., (2000) model [7] for the Heathrow airport and it serves as a reference in the development of our application.

For the accomplishment of the Expert system we have dealt with a human expert in the field in which we are working. This acquired knowledge has allowed us to define the variables,that later on have been turned into logic rules of the type If-then by means of the use of Karnaugh tables, and they have been implemented in CoCoA system (Computations in Commutative Algebra) [8] which has been previously used in other studies by our research group. This will allow us to analyze the data and establish corrective actions for each customer. This will also allow to detect nonseses in the data input.

3 Conclusions

In this paper we propose an Expert System that allows the improvement of the quality of the firm's integral systems. The system that we have built is based on the improvement and pursuit of the collected data in the section of the client's evaluation of the satisfaction according to the norm ISO 9001:2008.

Therefore, the system is centered on the optimization of the firm's resources as well as in the process automation of the evaluation of the client's satisfaction required by the norm ISO 9001:2008, enabling the information to the managerial positions and what it is more important, serving us as an aid in the decision making of the actions to realize from the obtained results in the answered questionnaires.

The application of this system to customer's satisfaction has allowed us to think about its potential extension to the rest of sections of the system according to norm ISO 9001:2008. Although it is not shown in this document, it allows to open important research lines in this field, by means of the implementation of intelligent systems that are able to analyze the results and make the decision making process easier. For instance, we mention aspects like the suppliers evaluation, verification and acceptance of orders, etc.

Therefore, and in view of the obtained results we can indicate that the use and management of this tool would reduce the response times in the SMEs and it would also allow to have information in a fast way and what it is more important, to get information about customer's opinions in real time . whenever firms use the system, problems can then be solved faster than before.

References

1. ISO-9001:2008, Quality management systems. Requirements, AENOR (2008)
2. Kaplan, R.S., Norton, D.P.: The Balanced Scorecard. Harvard Business Press, Boston (1996)
3. ISO-9004:2009, Managing for the sustained success of an organization. A quality management approach, AENOR (2009)
4. Edvinsson, L., Malone, M.S.: Intellectual capital. Harper Business (1997)
5. Euroforum, Medición del Capital Intelectual. Modelo Intelect, IUEE, Madrid (1998)
6. Juran, J.: Juran on leadership for Quality. Simon & Schuster, New York (2003)
7. Olve, N., Roy, J., Wetter, M.,, P.D.: A practical guide to using the balanced scorecard. Wiley, Chichester (1999)
8. CoCoA Team. CoCoA, a system for doing Computations in Commutative Algebra (2004), `http://cocoa.dina.unige.it`
9. Pajares, G.: Artifial Intelligence and Knowledge Management. RaMa (2005)

Adding Semantics to Research and Development Management

Carlos García-Moreno, Yolanda Hernández-González,
Maria Luisa Hernández-Alcaraz, Francisco García-Sánchez,
and Rafael Valencia-García

Abstract. Research and development (R&D) is one of the key drivers of innovation within businesses. It constitutes an important investment in a company's future. Accordingly, R&D management system should be effective and adequate. In this sense, the addition of semantics within this context could be crucial to the success of R&D initiatives. In this paper, a semantic based platform to assist in managing R&D projects is proposed. The platform takes advantage of the semantic information by classifying and annotating both the R&D projects and human resources by means of ontologies. The global behaviour of the system has been illustrated through a use case scenario in the software development domain.

Keywords: ontology, semantic web, research and development, knowledge management.

1 Introduction

Innovation means producing, assimilating and successfully exploiting a novelty in the economic and social fields, so bringing new solutions to problems and allowing to satisfy the needs of both individuals and society [10]. The need for innovation is even more pressing in the current economic situation, as a means to differentiate from competitors. In this production and marketing changing environment, customers are increasingly demanding better services at lower prices. The organization that is best suited for innovation has a clear competitive advantage over the rest. This work is concerned with obtaining and representing information by means of ontologies. Innovation in production processes can improve the quality of products and reduce its price, thereby improving productivity and competitiveness.

Carlos García-Moreno · Yolanda Hernández-González
Indra Software Labs, C\Acanto 11, 28045 Madrid

Maria Luisa Hernández-Alcaraz · Francisco García-Sánchez · Rafael Valencia-García
Departamento de Informática y Sistemas, Universidad de Murcia,
Campus de Espinardo 30100 Murcia

A. Abraham et al. (Eds.): International Symposium on DCAI, AISC 91, pp. 295–302.
springerlink.com

In addition, innovative products enable organizations to meet (and often anticipate) the customers' needs.

Research and development (R&D) management is defined as a general set of processes and procedures used to ensure that the organization makes all necessary tasks in order to achieve its objectives. It usually involves several phases, i.e., proposal submission, project selection, approved project administration and project deliverable dissemination. To support the critical decision making tasks in R&D project management, project related information needs to be shared among different parties, at various organizational levels, and with different knowledge backgrounds e.g. project applicants and reviewers with different research interests [6].

In the last few years some regulations for the R&D management have been defined, so enterprises can certify the "quality" of their innovation activities. The regulation provides guidelines beyond the requirements established by other management systems, considering both the effectiveness and efficiency of a R&D&I (Research, Development and Innovation) management system [11].

Due to the critical importance of research, development and innovation, different platforms for managing the development of new ideas that can go to market, have been recently proposed [6].

Semantic technologies [2] are currently achieving a certain degree of maturity. They provide a consistent and reliable basis to face the challenges for organization, manipulation and visualization of the data and knowledge. Consequently, the possibility of using knowledge-oriented query answering to exploit the benefits of semantics has become a top-class research challenge.

Ontologies are the paramount technology of the semantic web. An ontology can be defined as "a formal and explicit specification of a shared conceptualization"[8]. Ontologies provide a formal, structured knowledge representation, with the advantage of being reusable and shareable. Ontologies provide a common vocabulary for a domain and define -with different levels of formality- the meaning of the terms and the relations between them. Knowledge in ontologies is mainly formalized using five kinds of components: classes, relations, attributes, axioms and instances. Classes in the ontology are usually organized into taxonomies. Sometimes, the definition of ontologies has been diluted, in the sense that taxonomies are considered to be full ontologies [8].

In this paper, we present a semantic based platform to assist in managing R&D projects. The rest of the paper is organized as follows. The components that take part in the platform and its overall architecture are described in Section 2. In Section 3, a use case scenario in the software development domain is shown. Finally, conclusions and future work are put forward in Section 4.

2 Platform Architecture

The architecture of the proposed semantic repository is shown in Fig. 1. The repository is composed of 3 main ontologies: (1) R&D ontology, (2) the domain ontologies, and (3) the RDF (Resource Description Framework)-based semantic profiles. Next, these components are described in detail.

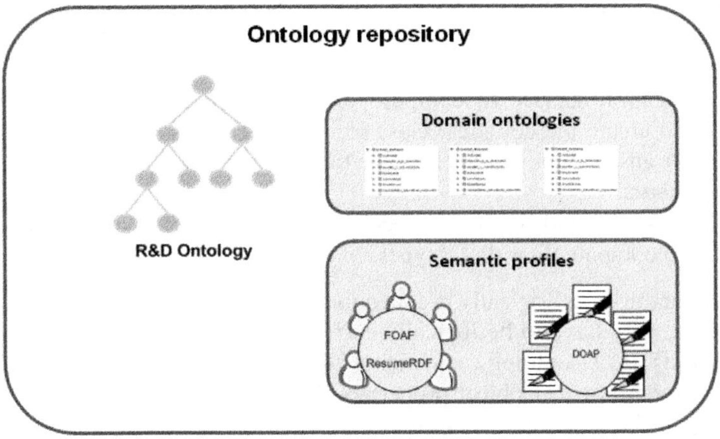

Fig. 1 System architecture.

2.1 R&D Ontology

For this work, a R&D ontology has been developed based on the Spanish innovation standards. The standards used to define this ontology are the following:

- "*UNE 166006:2006 - R&D&i Management: Technological Surveillance*": it describes the information related to the activities involved in the process of monitoring the technology [14].
- "*UNE 166001:2006 - R&D&i Management: Requirements of an R&D&I project*": it contains information that is relevant when implementing a R&D&i project [12].
- "*UNE-EN ISO 9001:2000 - Systems Quality Management: Requirements*": it emphasizes the possibilities concerning compatibility between the different management systems of an organization.
- "*INDRA Software Labs Quality Manual*": it was required to perform the use case.
- "*UNE 166005:2004 - UNE 166002:2002 application Guide*": it indicates how to apply the standard through the implementation of an innovation management system by experts [13].

The R&D ontology model has been designed from scratch using the Methontology methodology [3]. This ontology has been implemented using the second version of the "Web Ontology Language", OWL 2 [4]. In particular, OWL 2-DL, which is based on description logics (DL) and supports a number of important automatic DL inference services, has been used. These services can be provided by DL reasoners including HermiT, Pellet2, Fact++ or Racer [7] and are as follows:

- Consistency checking to ensure that an ontology does not contain any contradictory facts.

- Concept satisfiability to check whether it is possible for a class to have any instances. If a class is unsatisfiable, then defining an instance of the class will cause the whole ontology to be inconsistent.
- Classification service to compute subclass relations between every named class, to create the complete class hierarchy. The class hierarchy can then be used to answer queries such as getting all (or only) direct subclasses of a given class.
- Realization to find the most specific classes to which individuals belongs in order to compute their direct types.

An OWL ontology is logically a collection of domain axioms that must be satisfied; that is, they have to be logically correct for all kinds of domain parameters. The resulting R&D ontology contains a total of 359 classes, 91 individuals, 23 datatype properties, 294 object properties, 156 class axioms, 231 object property axioms, and 35 data property axioms.

2.2 Domain Ontologies

The "Domain ontologies" repository stores relevant knowledge about the application domain in which the platform is going to be applied. In the use case scenario described in section 3, an ontology in the software development domain is expounded.

2.3 Semantic Profiles

In this section, the semantic profiles used for representing both R&D projects and the organization's human resources are described. These semantic profiles are extracted from the information included in the databases of the organization. In order to do this, an application for extracting the semantics from these databases has been implemented. Next, these semantic profiles are described in detail.

2.3.1 Semantic Descriptions of R&D Projects

Description of a Project (DOAP[1]) is an RDF schema and XML vocabulary conceived to describe software projects. Particularly, it was created to convey semantically-enriched information associated with open-source software projects.

The vocabulary in DOAP is limited. For example, DOAP profiles do not take into account all the possible roles the participants can play in a R&D project. Therefore, a new DOAP-based RDF schema has been created by extending the vocabulary in order to improve the semantic description of a R&D project in an organization. Fig. 2 contains an example of a DOAP-based description of a project. There, the *doap_ext* prefix is used for referencing the name space of the DOAP extended vocabulary, including the *project_coordinator, project_participant,* and *project_manager* roles.

[1] http://trac.usefulinc.com/doap/

```
<rdf:RDF xmlns:rdf="http://www.w3.org/1999/02/22-rdf-syntax-ns#"
        xmlns:rdfs="http://www.w3.org/2000/01/rdf-schema#"
        xmlns:foaf="http://xmlns.com/foaf/0.1/"
        xmlns:doap="http://usefulinc.com/ns/doap#"
        xmlns:doap_ext="http://www.indra.es/ns/doap_ext#">
        <doap:Project>
           <doap:name>SONAR2</doap:name>
           <doap:shortdesc xml:lang="en">
                The SONAR2 Project focuses on the potential of Semantic Web
                technologies for representing and exploiting Financial Data.
           </doap:shortdesc>
           <doap_ext:project_coordinator>
                <foaf:Organization>
                        <foaf:name>Indra Software Labs S.L</foaf:name>
                </foaf:Organization>
           </doap_ext:project_coordinator>
           <doap_ext:project_participant>
                <foaf:Organization>
                        <foaf:name>Universidad de Murcia</foaf:name>
                </foaf:Organization>
           </doap_ext:project_participant>
           <doap_ext:project_manager>
                <foaf:Person>
                        <foaf:name>Carlos García Moreno</foaf:name>
                </foaf:Person>
           </doap_ext:project_manager>
        </doap:Project>
</rdf:RDF>
```

Fig. 2 An example of a DOAP description of a R&D project

2.3.2 Semantic Descriptions of Human Resources

Socio-semantic networks involve agents who produce, manipulate and exchange knowledge or information. Communities of scientists, free software developers, “wiki” contributors and webloggers are such instances, among others, of groups of distributed knowledge creation and processing — or knowledge communities [9].

In any organization, these agents can play many different roles in an R&D project in the ICT area: project manager, tester, programmers, etc. In particular, each actor can play different roles within the scope of a particular R&D project. For example, the project manager could also be part of the testing team.

The main objective of this ontology is to represent the roles and capabilities of the participants in a R&D project. The roles played by the human resources in the organization along with other semantic information are represented by means of FOAF (Friend-of-a-Friend), ResumeRDF and DOAP-based semantic profiles. As aforementioned, FOAF is an RDF representation of the people's personal information and their friends. ResumeRDF, on the other hand, is an ontology developed to express the information contained in a résumé, such as business and academic experience, skills, publications, certifications, etc. [1]. Finally, the extended DOAP-based profile defined in the previous section allows R&D projects to be semantically described.

The semantic profiles represent the experience of users in topics related to the domain and R&D ontologies. The human resources' semantic profile includes:

- The development projects in which they have been (or are) involved (DOAP).
- Their working experience in other companies (ResumeRDF).
- Other personal information (FOAF).

3 Use Case: Software Development Domain

Indra is a global company of technology, innovation, and talent, leader in high value-added solutions and services for sectors such as the transport and traffic, energy and industry, public administration and healthcare, finance, insurance, security and defence, telecom and media, etc. Indra operates in over 100 countries and has 30,000 employees worldwide who share their knowledge and experience in different sectors and countries to find innovative solutions to the challenges that clients face. Indra is the European company that most invests in R&D in its sector.

More concretely, Indra Software Labs is a subsidiary of Indra specialized in software development. It is characterized by the segmentation of the work in its various development centres and the creation and implementation of various methodologies, processes and tools for high productivity and quality products.

This platform is currently being implemented in Indra Software Lab. By taking into account the shortcomings of developing a new ontology from scratch, the ontologies developed under the scope of the DESWAP project [5] have been reused to represent the features and functional properties of the software projects.

Initially, representatives of the company are required to input and semantically describe some previous software projects of the company, which are stored in the ontology repository. Sesame RDF repository, backed up by a MySQL database, has been used to implement the ontology repository. Around 30 R&D software projects have been semantically described and inserted in the system. An average of 6 people per project participated in these projects, so a total of 90 different semantic descriptions of people were inserted into the system. The FOAF and ResumeRDF based profiles were inserted in the ontology repository through a web application. Once the participants of the projects have been inserted, the semantic description of each project has to be defined. In this second stage, the DOAP description of each project is manually defined. This description contains basically the name, a short description, the programming language, the platform of the project and the participants. Then, the software related topics are selected from the software ontology and included in the DOAP profile by means of the doap_ext:topic relation between the extended DOAP profile and the software ontology. For this, a web application for managing these profiles in the ontology repository has been developed.

The initial configuration of the ontology repository explained above was a difficult task and the people from the company involved in this task found it very tricky to describe the software projects and to annotate them in accordance with the software ontology. They found the tools implemented insufficient and with a lack of usability.

4 Discussion and Conclusion

Innovation is a key factor for success in today's business. This fact is even more accentuated in the current economic climate. In most cases, innovation can lead to an increase in profits by improving the quality of the outcome and decreasing production costs. Research and development (R&D) projects play a key role in the innovation process. They constitute the basis for companies to meet their business and strategic objectives. Traditional R&D management systems suffer from the problems derived from the necessity of sharing heterogeneous data among different organizational departments and levels.

Ontologies and semantic technologies have proven to be quite effective in capturing, defining, sharing and reusing the knowledge about a specific domain. In this work, we propose the utilization of ontologies to model R&D related data and the application of semantic technologies to build an enhanced R&D management system. The main benefits achieved by adding semantics to our R&D management system are the following:

1. Definition of a completely explicit information model: the existence of the ontologies serves as reference for communication (both among persons and computers) and helps to improve data quality and consistency.
2. Improved search capabilities: by describing R&D projects and participants profiles in terms of a well-defined and formal domain model.
3. Improved management of information: by exploiting the search and inference capabilities, more efficient management of information is achieved.

As further work, we plan to extend the scope of the application to cover the whole R&D&I management lifecycle. The ultimate goal is that R&D&I management in the organization is conducted in an integrated way. All the departments in the organization should be involved in this process, which must be cyclical with respect to use and generation of knowledge. Therefore, the entire lifecycle from the generation of ideas, to the analysis of the results of the executed projects, must be considered.

Acknowledgements. This work has been supported by the Spanish Ministry for Science and Innovation under project TIN2010-18650.

References

1. Bojārs, U., Breslin, J.G.: ResumeRDF: Expressing skill information on the Semantic Web. In: 1st International ExpertFinder Workshop, Berlin, Germany (January 2007)
2. Fensel, D.: Ontologies: A Silver Bullet for Knowledge Management and Electronic Commerce. Springer, Heidelberg (2002)
3. Fernández-López, M., Gómez-Pérez, A., Juristo, N.: METHONTOLOGY: From Ontological Art Towards Ontological Engineering. In: Spring Symposium on Ontological Engineering of AAAI, Stanford University, California, pp. 33–40 (1997)

4. Grau, B., Horrocks, I., Motik, B., Parsia, B., Patel-Schneider, P., Sattler, U.: OWL2 the next step for OWL. Web Semantics. Science, Services and Agents on the World Wide Web 6(4), 309–322 (2008)
5. Hartig, O., Kost, M., Freytag, J.-C.: Designing Component-Based Semantic Web Applications with DESWAP. In: Proceedings of the Poster and Demonstration Session at the 7th International Semantic Web Conference (ISWC), Karlsruhe, Germany (October 2008)
6. Liu, O., Ma, J.: A multilingual ontology framework for R&D project management systems. Expert. Syst. Appl. 37(6) (June 2010)
7. Sirin, E., Parsia, B.: Pellet: An OWL DL reasoner. In: Description Logic Workshop (DL 2004) (2004)
8. Studer, R., Benjamins, V.R., Fensel, D.: Knowledge engineering: Principles and methods. Data Knowl. Eng. 25, 161–197 (1998)
9. Rotha, C., Cointet, J.-P.: Social and semantic coevolution in knowledge networks. Social Networks 32, 16–29 (2010)
10. White paper: The Innovation Spanish System. COTEC Foundation for Research and Technology (2004)
11. UNE 166002:2006 – R&D&I Management: R&D&I Management System requirements
12. UNE 166001:2006 - Gestión de la I+D+i: Requisitos de un proyecto de I+D+i
13. UNE 166005:2004 - Guía de aplicación de la Norma UNE 166002:2002 EX al sector de bienes de equipo
14. UNE 166006:2006 - Gestión de la I+D+i: Sistema de Vigilancia Tecnológica

Abductive Reasoning for Semantic Matchmaking with Modular Ontologies

Viet-Hoang Vu and Nhan Le-Thanh

Abstract. Semantic matchmaking is defined as a process of finding possible matches between requests and supplies based on their logical relations. Recently, attempts have been done to formalize this process in Description Logics (DLs). We extend this formalization to Package-based Description Logics (P-DLs), the extensions of DLs for distributed and modular ontologies, for allowing the description of demands and offers to be represented in different terminologies. To support this task, we develop a distributed tableau-based method for computing concept abduction, a non-monotonic inference service specifically proposed for this kind of operation.

Keywords: abductive reasoning, tableau-based methods, distributed ontology, package-based description logics, semantic matchmaking.

1 Introduction

Recently, several efforts have been done in order to formalize matchmaking, the process of searching possible matches between requests and supplies, in a semantics-based framework such as Description Logics [1]. In this setting, demands and offers are represented as concepts with reference to a *common ontology* that specifies a conceptualization for the knowledge of the domain. Then, by using different inference services, the matching between the concepts can be semantically evaluated, based on their logical relationships, the result is thus more relevant than in other methods [6].

However, in reality, sometimes demands and offers may be specified not with reference to a single ontology, but to *different ontologies*, each of them

Viet-Hoang VU · Nhan LE-THANH
KEWI - Laboratory I3S - CNRS - Nice Sophia Antipolis University, France
e-mail: viethoang.vu@gmail.com, Nhan.LE-THANH@unice.fr

A. Abraham et al. (Eds.): International Symposium on DCAI, AISC 91, pp. 303–310.
springerlink.com

represents the knowledge about the domain from some point of view. To perform semantic matchmaking in such situations, we extend its formalization to the context of Packages-based Description Logics (P-DLs) [3], the extensions of DLs for distributed and modular ontologies.

To support this task, in this paper, we develop a distributed tableaux-based method for computing *concept abduction*, a non-monotonic inference service developed specifically for this kind of operation. The rest of the paper is organized as follows. Firstly, in the section 2, we recall some preliminaries on the P-DL $\mathcal{ALCP}^-$. Then, we introduce briefly the semantic matchmaking in the context of P-DLs in the section 3. Next, the distributed tableau-based method to compute abductive solutions is the subject of discussion in two sections 4 and 5. Finally, the section 6 is reserved for some conclusions.

2 Package-Based Description Logic $\mathcal{ALCP}^-$

Description logics (DLs) are a family of logic-based knowledge representation languages. Package-based description logics (P-DLs) are its extensions for distributed and modular ontologies [3]. In these formalisms, an (modular) ontology Σ is represented as a set of *packages* $\{P_i\}$. Each of them has its signature divided into disjoint subsets, a *local signature* $Loc()$ and an *external signature* $Ext()$. For any local term $t \in Loc(P_i)$, P_i is called the *home package* of t and t is called an *i-name*. If there is any j-name of P_j used in P_i, then we say P_i *directly imports* P_j and denote as $P_j \mapsto P_i$. P_i *indirectly imports* P_j , denoted by $P_j \mapsto^+ P_i$, if there is P_k, $k \neq i \neq j$, such that $P_j \mapsto P_k$ and either $P_k \mapsto P_i$ or $P_k \mapsto^+ P_i$. The *importing transitive closure* of P_i, denoted by P_i^+, is the smallest subset $\{P_j\} \subseteq \Sigma$ such that P_j is (directly or indirectly) imported by P_i. The importing relation is *acyclic* if, for all i, $P_i \notin P_i^+$, otherwise it's *cyclic*. Let $P_i^* = \{P_i\} \cup P_i^+$, a concept C is *understandable* by a package P_i if every term appeared in its description has a home package in P_i^*.

$\mathcal{ALCP}^-$ is the package extension of $\mathcal{ALC}$ with acyclic importations of concepts and roles. Its syntax and semantics are given in the figure 1 below.

$$
\begin{aligned}
(\top_j)^{\mathcal{I}_i} &= r_{ji}\left(\Delta^{\mathcal{I}_j}\right) \subseteq \Delta^{\mathcal{I}_i} \\
(\neg_j C)^{\mathcal{I}_i} &= r_{ji}\left(\Delta^{\mathcal{I}_j}\right) \setminus C^{\mathcal{I}_i} \\
(C \sqcap D)^{\mathcal{I}_i} &= C^{\mathcal{I}_i} \sqcap D^{\mathcal{I}_i} \\
(C \sqcup D)^{\mathcal{I}_i} &= C^{\mathcal{I}_i} \sqcup D^{\mathcal{I}_i} \\
(\forall RC)^{\mathcal{I}_i} &= \left\{a \in r_{ji}\left(\Delta^{\mathcal{I}_j}\right) \mid \forall b.\,(a,b) \in R^{\mathcal{I}_i} \rightarrow b \in C^{\mathcal{I}_i}\right\} \\
(\exists RC)^{\mathcal{I}_i} &= \left\{a \in r_{ji}\left(\Delta^{\mathcal{I}_j}\right) \mid \exists b.\,(a,b) \in R^{\mathcal{I}_i} \wedge b \in C^{\mathcal{I}_i}\right\}
\end{aligned}
$$

Fig. 1 Syntax and semantics of the P-DL $\mathcal{ALCP}^-$

Given an $\mathcal{ALCP}^-$ KB $\Sigma = \left\langle \{P_i\}, \{P_j \mapsto P_i\}_{j\neq i}\right\rangle$, a *distributed interpretation* is a tuple $\mathcal{DI} = \left\langle \{\mathcal{I}_i\}, \{r_{ji}\}_{P_j \in P_i^+}\right\rangle$, where each $\mathcal{I}_i = \left\{\Delta^{\mathcal{I}_i}, (.)^{\mathcal{I}_i}\right\}$ is a local interpretation of package P_i and $r_{ji} \subseteq \Delta^{\mathcal{I}_j} \times \Delta^{\mathcal{I}_i}$ is the *domain relation* that interprets the importing relation from P_j to P_i. Particularly, $r_{ii} = \mathrm{id}_{\Delta^{\mathcal{I}_i}} := \left\{(x,x) \mid x \in \Delta^{\mathcal{I}_i}\right\}$ is the identity mapping on the local domain $\Delta^{\mathcal{I}_i}$.

If $(a,b) \in r_{ji}$, b is called an "*image*" of a and denoted as $r_{ji}(a) = b$. Then,

- $r_{ji}(j:C) = \left\{b \in \Delta^{\mathcal{I}_i} \mid \exists a \in C^{\mathcal{I}_j}, (a,b) \in r_{ji}\right\}$.
- $r_{ji}(j:R) = \left\{(c,d) \in \Delta^{\mathcal{I}_i} \times \Delta^{\mathcal{I}_i} \mid \exists (a,b) \in R^{\mathcal{I}_j}, (a,c) \in r_{ji} \wedge (b,d) \in r_{ji}\right\}$.

Each $\mathcal{I}_i$ interprets normally a primitive concept A as subsets $A^{\mathcal{I}_i} \subseteq \Delta^{\mathcal{I}_i}$ and a role R as a subset $R^{\mathcal{I}_i} \subseteq \Delta^{\mathcal{I}_i} \times \Delta^{\mathcal{I}_i}$. The function is extended to provide the semantics for complex descriptions. The local interpretation $\mathcal{I}_i$ satisfies an axiom $C \sqsubseteq D$ ($C \equiv D$) if and only if $C^{\mathcal{I}_i} \subseteq D^{\mathcal{I}_i}$ ($C^{\mathcal{I}_i} = D^{\mathcal{I}_i}$) . It is a *model* of the package P_i, denoted by $\mathcal{I}_i \models P_i$, if it satisfies all of its axioms. The distributed interpretation $\mathcal{DI}$ is a *model* of a P-DLs KB $\Sigma = \left\langle \{P_i\}, \{P_j \mapsto P_i\}_{j\neq i}\right\rangle$, denoted as $\mathcal{DI} \models \Sigma$, if the followings hold:

1. $\bigcup_i \Delta^{\mathcal{I}_i} \neq \emptyset$.
2. (*One-to-one*) For all i, j, r_{ji} is an *injective partial* function, that is, for any $a \in \Delta^{\mathcal{I}_j}$ there is at most one image $b \in \Delta^{\mathcal{I}_i}$ of it.
3. (*Compositional consistent*) For all i, j, k, $i \neq j$, $P_j \in P_k^+$ and $P_k \in P_i^+$, $r_{ji} = r_{ki} \circ r_{jk} = \rho \cap (\Delta^{\mathcal{I}_j} \times \Delta^{\mathcal{I}_i})$[1].
4. For every j-concept C that is imported in P_i, $r_{ji}\left(C^{\mathcal{I}_j}\right) = C^{\mathcal{I}_i}$.
5. For every j-role R that appears in P_i, $r_{ji}\left(R^{\mathcal{I}_j}\right) = R^{\mathcal{I}_i}$.
6. For every j-role R that appears in P_i and every $(a,a') \in r_{ji}$, $(a,b) \in R^{\mathcal{I}_j}$ iff $(a', r_{ji}(b)) \in R^{\mathcal{I}_i}$.
7. $\mathcal{I}_i \models P_i$, for every i.

Two basic inference services in DLs systems are *concept subsumption* and *concept satisfiability*. The first one is to find out whether a concept is always more general than the other, and the second one is to to verify if a concept has a model. In P-DLs, they are contextualized, allowing different packages draw different conclusions depending on their own point of view.

Definition 1. *Given an* $\mathcal{ALCP}^-$ *KB* $\Sigma = \left\langle \{P_i\}, \{P_j \mapsto P_i\}_{j\neq i}\right\rangle$. Σ *is* consistent as witnessed by a package P_w *if* P_w^* *has a model* $\mathcal{DI} = \left\langle \{\mathcal{I}_i\}, \{r_{ji}\}_{P_j \in P_i^*}\right\rangle$ *, such that* $\Delta^{\mathcal{I}_w} \neq \emptyset$.

- *A concept* C *is* satisfiable as witnessed by P_w *if there exists a model of* Σ *such that* $C^{\mathcal{I}_w} \neq \emptyset$.

[1] '$\circ$' denotes the composition of relations and 'ρ' is the symmetric and transitive closure of $\bigcup_{P_j \in P_i^*} r_{ji}$.

- *A concept subsumption* $C \sqsubseteq D$ *is* valid as witnessed by P_w, *denoted by* $C \sqsubseteq_w D$, *if for every model* $\mathcal{I}$ *of* P_w^*, $C^{\mathcal{I}_w} \subseteq D^{\mathcal{I}_w}$.

3 Semantic Matchmaking with Distributed Modular Ontologies

Semantic Matchmaking is defined as a process of finding possible matches between demands and supplies, in which the evaluation of matching, based on logical relations, is computed in some semantics-based framework [6]. In such settings, five categories are distinguished:

1. *Exact match* : The supply fulfills completely the demand and vice versa.
2. *Full match* : The supply satisfies entirely the demand.
3. *Plug-in match* : The demand is more specific than the supply.
4. *Potential match* : The supply satisfies partially the demand.
5. *Partial match* : The demand conflicts with the supply.

Formalizing the semantic matchmaking in DLs, demands and supplies are represented as concepts D and S with reference to a common ontology $\mathcal{T}$. Their logical relations can thus be evaluated by using inference services provided by these formalisms. We extend this formalization to the context of P-DLs to allow the concepts to be specified in different ontologies. To do that, let Σ be some P-DLs knowledge base and $P_w \in \Sigma$ be some witness package such that D and S are understandable by P_w. The matching between D and C is evaluated with respect to (w.r.t) Σ as follows, if :

1. $D \equiv_w S$ then the match is exact w.r.t Σ as witnessed by P_w.
2. $S \sqsubseteq_w D$ then the match is full w.r.t Σ as witnessed by P_w.
3. $D \sqsubseteq_w S$ then the match is plug-in w.r.t Σ as witnessed by P_w.
4. $S \sqcap D \not\sqsubseteq_w \bot$ then the match is potential w.r.t Σ as witnessed by P_w.
5. $S \sqcap D \sqsubseteq_w \bot$ then the match is partial w.r.t Σ as witnessed by P_w.

The problem arises when there are possibly several potential matches for a given demand. Obviously, using only standard reasoning tasks like concept subsumption or satisfiability, we cannot differentiate them, making the highlighting of the most promising matches be impossible. To deal with this problem, a novel inference service, based on propositional abduction, named *concept abduction*, has been proposed to DLs in [5].

Definition 2. *Let* Γ *be some DLs KB and* S, D *are two concepts such that* $S \sqcap D \not\sqsubseteq_\Gamma \bot$. *A solution to the* concept abduction problem *(CAP)* $\mathcal{P}$ *for a given* $\langle S, D, \Gamma \rangle$ *is any concept* H *such that* $S \sqcap H \not\sqsubseteq_\Gamma \bot$ *and* $S \sqcap H \sqsubseteq_\Gamma D$.

Hence, the *maximal* solution H under subsumption relation ($\sqsubseteq_\Gamma$) can be used as a measurement of the *semantic distance* between concepts. Consequently, such abductive solutions can give us the indication of which are better among supplies potentially matching the demand. As in the context of P-DLs, demands and offers may be specified in different packages, the inference service must then be adapted as follows.

Definition 3. *Let $\Sigma = \{P_i\}$ be some P-DLs KB, $P_w \in \Sigma$ be some witness package and C, D are two concepts such that $C \sqcap D \not\sqsubseteq_w \bot$. A solution to the* distributed concept abduction problem *(DCAP) $\mathcal{DP}$ for a given $\langle C, D, \Sigma, P_w \rangle$ is any concept H such that $C \sqcap H \not\sqsubseteq_w \bot$ and $C \sqcap H \sqsubseteq_w D$.*

To devise an algorithm for computing solutions to a $\mathcal{DP}$, we combine the uniform tableaux method that was developed for or solving the (local) abduction problem in the context of the description logic $\mathcal{ALN}$[4], with the synchronous message mechanism of the distributed reasoning algorithm for P-DLs[2].

4 A Distributed Uniform Tableaux Method for $\mathcal{ALCP}^-$

To recall, the uniform method is based on tableau calculus for solving the concept abduction in $\mathcal{ALN}$. For this purpose, in *tableaux*, graph structures used to build consistent models for the knowledge base, two kinds of label are distinguished : one to denote a set of concepts and roles to which nodes and edges *belong* and the other to denote those they *don't belong to.* Following this separation, finding abductive solutions becomes more easily by analyzing the interaction of different labels of the same node. We extend this method to the distributed context of the P-DL $\mathcal{ALCP}^-$.

Definition 4. A (local) prefixed tableau τ *is a tree-like structure* $\tau = (V, E, v^r, T, F)$*, where:*
- *V is a finite set of nodes with $v^r \in V$ is the root of the tableau;*
- *E is a finite set of edges $e = \langle v, w \rangle$;*
- *$T()$ and $F()$ are labeling functions which assign for each node v sets of concepts and for each edge sets of roles .*

*For any $e = \langle v, w \rangle \in E$, if either $R \in T(\langle v, w \rangle)$ or $\neg R \in F(\langle v, w \rangle)$, then w is called a R-*successor *of v, and v is called the* predecessor *of w in τ.* Ancestor *is the transitive closure of predecessors and* descendant *is the transitive closure of successors.*

Let Σ be an $\mathcal{ALCP}^-$ KB, $P_w \in \Sigma$ be a witness package and C, D be two concepts that are understandable by P_w. For verifying if $C \sqsubseteq_w D$ w.r.t Σ, the distributed uniform algorithm works by starting with a tableau τ_0 that is locally initialized on the package P_w, where $V_0 = \{v^r\}$, $E_0 = \emptyset$, $T(v^r) = \{C \sqcap C_{\mathcal{T}_w}\}$ with $C_{\mathcal{T}_w} = \bigsqcap_{(C_i \sqsubseteq D_i) \in P_w} (\neg C_i \sqcup D_i)$ is the internalized concept of P_w, and $F(v^r) = \{D\}$. The tableau is then transformed using consistency preserving rules until no rule can be applied (see the figure 3). During this process, sometimes, knowledge need to be transferred from one local tableau to the others in order to reflect the importing relations. To do that, nodes can be copied across different tableaux. Each node v of a tableau τ is thus associated with $org(v)$ that represents the original node from which v is copied. If there is a node u in some tableau of P_j, such that $org(v) = org(u)$ and $P_i \in P_j^*$, then u is called an *image* of v, denoted by $u = v^{i \to j}$ and v a *pre-image* of u, denoted by $v = u^{i \leftarrow j}$. The synchronization of image and pre-images nodes is guaranteed by sending and receiving reporting messages (see the figure 2).

- $report^{(i \to j)}(v, C)$: *if* $v^{i \to j} \in V_j$ and $v^{i \to j}$ is not blocked *then* :
  -T) *if* $C \in T_i(v)$ *then* $T_j(v^{i \to j}) = T_j(v^{i \to j}) \cup \{C\}$.
  -F) *if* $C \in F_i(v)$ *then* $F_j(v^{i \to j}) = F_j(v^{i \to j}) \cup \{C\}$.
- $report^{(j \leftarrow i)}(v, C)$: create $v^{j \leftarrow i}$ if $v^{j \leftarrow i} \notin V_j$, and
  -T) *if* $C \neq \top_j$ and $C \in T_i(v)$ *then* $T_j(v^{j \leftarrow i}) = T_j(v^{j \leftarrow i}) \cup \{C\}$.
  -F) *if* $C \neq \bot_j$ and $C \in F_i(v)$ *then* $F_j(v^{j \leftarrow i}) = F_j(v^{j \leftarrow i}) \cup \{C\}$.
- $report^{(i \to j)}(\langle v, w \rangle, R)$: *if* $v^{i \to j} \in V_j$ *then* create $w^{i \to j}$ if $w^{i \to j} \notin V_j$; create $e = \langle v^{i \to j}, w^{i \to j} \rangle$ if $e \notin E_j$.
  -T) *if* $R \in T_i(\langle v, w \rangle)$*then* $T_j(e) = T_j(e) \cup \{R\}$.
  -F) *if* $\neg R \in F_i(\langle v, w \rangle)$ *then* $F_j(e) = F_j(e) \cup \{\neg R\}$.
- $report^{(j \leftarrow i)}(\langle v, w \rangle, R)$: *if* $v^{j \leftarrow i}$, $w^{j \leftarrow i} \in V_j$ *then* create $e = \langle v^{i \to j}, w^{i \to j} \rangle$ if $e \notin E_j$.
  -T) if $R \in T_i(\langle v, w \rangle)$, then $T_j(e) = T_j(e) \cup \{R\}$.
  -F) if $\neg R \in F_i(\langle v, w \rangle)$, then $F_j(e) = F_j(e) \cup \{\neg R\}$.

Fig. 2 Synchronization of local inference processes

1. $\sqcap$-**rules** :
   -T) *if* $(C_1 \sqcap C_2) \in T(v)$ and $\{C_1, C_2\} \nsubseteq T(v)$ *then* add both C_1, C_2 to $T(v)$.
   -F) *if* $(C_1 \sqcup C_2) \in F(v)$ and $\{C_1, C_2\} \nsubseteq F(v)$ *then* add both C_1, C_2 to $F(v)$.
2. $\sqcup$-**rules** :
   -T) *if* $(C_1 \sqcup C_2) \in T(v)$ and $\{C_1, C_2\} \cap T(v) = \emptyset$ *then* add either C_1 or C_2 to $T(v)$.
   -F) *if* $(C_1 \sqcap C_2) \in F(v)$ and $\{C_1, C_2\} \cap F(v) = \emptyset$ *then* add either C_1 or C_2 to $F(v)$.
3. $\exists$-**rules** :
   -T) *if* $(\exists R.C) \in T(v)$ and v has no R-successor w such that either $C \in T(w)$ or $\dot\neg C \in F(w)$ *then* add a new R-successor w of v with $T(w) = \{C\}$.
   -F) *if* $(\forall R.C) \in F(v)$ and v has no R-successor w such that either $C \in T(w)$ or $\dot\neg C \in F(w)$ *then* add a new R-successor w of v with $F(w) = \{C\}$.
4. $\forall$-**rules** :
   -T) *if* $(\forall R.C) \in T(v)$ and v has a R-successor w such that $C \notin T(w)$ *then* add C to $T(w)$.
   -F) *if* $(\exists R.C) \in F(v)$ and v has a R-successor w such that $C \notin F(w)$ *then* add C to $F(w)$.
5. CPush-**rules** :
   -T) *if* $C \in T(v)$ and $P_i \xrightarrow{C} P_j$ *then* send $report^{(i \to j)}(v, C)$.
   -F) *if* $\dot\neg C \in F(v)$ and $P_i \xrightarrow{C} P_j$ *then* send $report^{(i \to j)}(v, \dot\neg C)$.
6. CReport-**rules** :
   -T) *if* $C \in T(v)$, where C is $\top_j$ or $P_j \xrightarrow{C} P_i$ *then* send $report^{(j \leftarrow i)}(v, C)$.
   -F) *if* $\dot\neg C \in F(v)$, where either $\dot\neg C$ is $\bot_j$ or $P_j \xrightarrow{C} P_i$ *then* send $report^{(j \leftarrow i)}(v, \dot\neg C)$.
7. RPush-**rule** : *if* w is a R-successor of v and $P_i \xrightarrow{R} P_j$ *then* send $report^{(i \to j)}(\langle v, w \rangle, R)$.
8. RReport-**rule** : *if* w is a R-successor of v and $P_j \xrightarrow{R} P_i$ *then*
   * send $report^{(j \leftarrow i)}(v, \top_j)$; $report^{(j \leftarrow i)}(w, \top_j)$; $report^{(j \leftarrow i)}(\langle v, w \rangle, R)$.
9. r-**rule** : *if* $v^{(i \to j)} \in V_j$ and there is a k such that $P_i \in P_k^+$ and $P_k \in P_j^+$ but $\left(v^{(i \to j)}\right)^{(k \leftarrow j)} \notin V_k$ *then* send $report^{(k \leftarrow j)}\left(v^{(i \to j)}, \top_k\right)$.

Fig. 3 Uniform tableaux transformation rules

To ensure the termination of the transformation, a blocking strategy is required as normally: for any node $v \in V$, v is *blocked by* u if u comes before v in some enumeration and either $T(v) \subseteq T(u)$ or $F(v) \subseteq F(u)$.

Definition 5. *A prefixed tableau* $\tau = (V, E, v^r, T, F)$ *contains a* **homogeneous clash** *if either* $\{A, \neg A\} \subseteq T(v)$ *or* $\{A, \neg A\} \subseteq F(v)$. *It contains a* **heterogeneous clash** *if* $T(v) \cap F(v) \neq \emptyset$.

A local tableau is *complete* if no rule can be applied on it. Then, it's *locally closed* if it contains a clash, otherwise it's *locally open.* The set of local tableaux on packages $P_w^* \subseteq \Sigma$ connected by graph relations form a *distributed prefixed tableau* $D\mathcal{T}$. Then, $D\mathcal{T}$ is *closed* if some of its local tableaux is so, on the contrary, it's *open.*

Theorem 1. *Let* $\Sigma = \{P_i\}$ *be an* $\mathcal{ALCP}^-$ *knowledge base,* C, D *be two concepts and* $P_w \in \Sigma$ *be some witness package.* $C \sqsubseteq_w D$ *w.r.t* Σ *if and only if all distributed prefixed tableaux constructed by the algorithm starting with* $\tau_0 = (V, E, v^r, T, F)$ *initialized on* P_w *where* $V = \{v^r\}$, $E = \emptyset$, $T(v) = \{C \sqcap C_{T_w}\}$ *and* $F(v) = \{D\}$ *are closed. Particularly, if every local tableau contains a* T*-homogeneous clash then* $C \sqsubseteq_w \bot$ *w.r.t* Σ.

5 Solutions to the Distributed Concept Abduction

Following the distinction of *homogeneous clashes* and *heterogeneous clashes,* it can be seen that all we need to do is to find concept formulas which, when added to the initial tableau, will generate clashes in all of its derivations, among them at least one is heterogeneous. To do that, let $\theta_w = \{\tau_1, \ldots, \tau_n\}$ be the set of all local open tableaux for verifying $C \sqsubseteq_w D$ obtained on the package P_w. If θ_w is not empty, to find H, we start by building for each local tableau $\tau_i \in \theta_w$, two *closing sets* $cl^T(\tau_i)$ and $cl^H(\tau_i)$, of which the definition are given below :

Definition 6. *Let* $\tau = (V, E, v^r, T, F)$ *be a prefixed tableau. If* τ *is clash-free, we denote by* $cl^T(\tau)$ *and* $cl^H(\tau)$, *closing sets of* τ, *two disjoint sets of concept descriptions built from labels of* τ *such that, if we let* τ_T *and* τ_H *be respectively complete tableaux derived from* τ *by adding any element of* $ct^T(\tau)$ *and* $cl^H(\tau)$ *into* $T(v^r)$, *then* τ_T, τ_H *are closed, and:*
- *every clash in* τ_T *is homogeneous in some* $T()$*-label,*
- *every clash in* τ_H *is heterogeneous.*

Previously, from the definition of clashes, it can be seen that the only way to close a tableau is to generate *contradictions of concept names* in its nodes. Hence, let v be some node of the tableau,

- **if** A is a (maybe negated) concept name and $A \in T(v)$, **then** $\dot{\neg} A \in cl^T(v)$;
- **if** B is a (maybe negated) concept name, $B \in F(v)$ and $\dot{\neg} B \notin T(v)$, **then** $B \in cl^H(v)$.
- **if** v has a R-successor w and there is not another R-successor w' of v such that $w' \neq w$ in τ **then** :
  * **if** $R \in T(\langle v, w \rangle)$ **then** $\forall R.E \in cl^T(v)$ for each $E \in cl^T(w)$, **otherwise** $\forall R.F \in cl^H(v)$ for all $F \in cl^H(w)$ if $\forall R.F \notin cl^T(v)$.

It can be verified that any conjunctive expression H built from these sets such that, for each open tableau, at least one of its closing elements is appeared in H, will be the finding solution if $C \sqcap H \not\sqsubseteq_w \bot$. We characterize these conditions in the following theorem :

Theorem 2. *Let $\mathcal{DP} = \langle C, D, \Sigma, P_w \rangle$ be a distributed concept abduction problem in the P-DL $\mathcal{ALCP}^-$. Let $\theta_w = \{\tau_1, \ldots, \tau_n\}$ be the set of open and complete local prefixed tableaux for $C \sqsubseteq_w D$ w.r.t Σ obtained on the package P_w. If θ_w is not empty, for each $\tau_i \in \theta$, let $cl(\tau_i) = cl^T(\tau_i) \cup cl^H(\tau_i)$ and let* $\mathsf{choice}()$ *be some choice function.*

For any set $S = \langle H_1 = \mathsf{choice}(cl(\tau_i)), \ldots, H_n = \mathsf{choice}(cl(\tau_i)) \rangle$, let H be a conjunctive expression built from elements of S, $H ::= H_1 \sqcap \cdots \sqcap H_n$. If :

1) for every $\tau_i \in \theta$, $S \cap cl(\tau_i) \neq \emptyset$ and

2) there is $\tau_j \in \theta$ such that $S \cap cl^T(\tau_j) = \emptyset$

then H is a solution to the problem $\mathcal{P}$.

6 Conclusions

We have presented in the scope of this paper a distributed tableaux-based algorithm for solving the concept abduction problem in the context of $\mathcal{ALCP}^-$, a package extension of DLs for distributed and modular ontologies. This nonmonotonic inference service is essential to perform the semantic matchmaking in situations where demands and supplies are distributively specified with reference to different ontologies. In the future, we would like to extend this algorithm for more expressive P-DLs.

References

1. Baader, F., Calvanese, D., McGuinness, D.L., Nardi, D., Patel-Schneider, P.F. (eds.): The Description Logic Handbook: Theory, Implementation, and Applications. Cambridge University Press, Cambridge (2003)
2. Bao, J., Caragea, D., Honavar, V.: A tableau-based federated reasoning algorithm for modular ontologies. In: IEEE/WIC/ACM International Conference on Web Intelligence, pp. 404–410. IEEE Press, Los Alamitos (2006)
3. Bao, J., Voutsadakis, G., Slutzki, G., Honavar, V.: Package-based description logics. In: Ontology Modularization. Springer, Berlin (2008)
4. Colucci, S., Di Noia, T., Di Sciascio, E., Donini, F.M., Mongiello, M.: A uniform tableaux-based approach to concept abduction and contraction in aln. In: Contraction in ALN. Proc. of DL 2004 (2004). CEUR Workshop Proceedings, p. 104 (2004)
5. Colucci, S., Di Noia, T., Di Sciascio, E., Donini, F.M., Mongiello, M.: Concept abduction and contraction in description logics. In: Description Logics (2003)
6. Di Noia, T., Di Sciascio, E., Donini, F.M.: Semantic matchmaking as nonmonotonic reasoning: A description logic approach. Journal of Artificial Intelligence Research 29, 307 (2007)

Mitigation of the Ground Reflection Effect in Real-Time Locating Systems

Dante I. Tapia, Juan F. De Paz, Cristian I. Pinzón, and Javier Bajo

Abstract. Real-Time Locating Systems (RTLS) are one of the most promising applications based on Wireless Sensor Networks and represent a currently growing market. However, accuracy in indoor RTLS is still a problem requiring novel solutions. One of the main challenges is to deal with the problems that arise from the effects of the propagation of radio frequency waves, such as attenuation, diffraction, reflection and scattering. These effects can lead to other undesired problems, such as multipath and the ground reflection effect. This paper presents an innovative mathematical model for improving the accuracy of RTLS, focusing on the mitigation of the ground reflection effect by using Artificial Neural Networks.

Keywords: Wireless Sensor Networks, Real-Time Locating Systems, Ground Reflection Effect, Artificial Neural Networks.

1 Introduction

Wireless Sensor Networks (WSNs) allow us to obtain information about the environment and act on this, expanding users' capabilities and automating daily actions. One of the most interesting applications for WSNs is Real-Time Locating Systems (RTLS). The most important factors in the locating process are the kind of sensors used and the techniques applied for the calculation of the position based on the information recovered by these sensors. In addition, indoor locating needs still more development, especially with respect to accuracy and low-cost and efficient infrastructures [9] [14]. Therefore, it is necessary to develop Real-Time Locating Systems that allow performing efficient indoor locating in terms of precision and optimization of resources. This optimization of resources includes the reduction of the costs and size of the sensor infrastructure involved on the locating system. In this sense, the use of optimized locating techniques allows obtaining more accurate locations using even fewer sensors and with less computational requirements [9].

Dante I. Tapia · Juan F. De Paz · Cristian I. Pinzón · Javier Bajo
Computers and Automation Department, University of Salamanca.
Plaza de la Merced, s/n, 37008, Spain
e-mail: {dantetapia,fcofds,cristian_ivanp,jbajope}@usal.es, corchado@usal.es

A. Abraham et al. (Eds.): International Symposium on DCAI, AISC 91, pp. 311–318.
springerlink.com

There are several wireless technologies that may be used by indoor RTLS, such as RFID (Radio Frequency IDentification), Wi-Fi, UWB (Ultra-Wide Band), Bluetooth and ZigBee. However, independently of the technology used, it is necessary to establish mathematical models that allow determining the position of a person or object based on the signals recovered by the sensor infrastructure. Therefore, the position can be calculated by means of several locating techniques, such as signpost, fingerprinting, triangulation, trilateration and multilateration [4] [5]. However, all these techniques must deal with important problems when trying to develop a precise locating system that uses WSNs in its infrastructure, especially for indoor environments.

The electromagnetic waves transmitted and received by the wireless sensor infrastructure used by these systems are affected by some propagation effects, such as reflection, scattering, attenuation and diffraction [3]. Due to these effects, the energy of the transmitted electromagnetic waves is substantially modified between transmitter and receiver antennas in these systems. Thanks to the attenuation effect, it is possible to estimate the distance covered by a wave between a transmitter and a receiver antenna [1]. This is very useful to build RTLS based on these distances, as those based on trilateration [4]. However, reflection, diffraction and scattering effects lead to other problems such as the *ground reflection effect* [3], a kind of multipath propagation effect. Therefore, it is necessary to define new models and techniques that allow the improvement of accuracy in this kind of systems.

This paper proposes a new mathematical model aimed at improving the precision of RTLS based on wireless sensor networks, especially at indoor environments. This model uses Artificial Neural Networks (ANNs) as the main components to mitigate the ground reflection effect and calculate the position of the elements.

Next, Section 2 explains the problems that the ground reflection effect introduces in RTLS that are based on wireless sensor networks. Section 3 describes a new proposal for reducing the ground reflection effect by using ANNs. Section 4 depicts the experiments performed on a real scenario to validate the accuracy of the new model and also describes the obtained results. Finally, Section 5 presents the conclusions obtained so far and depicts the related future work intended to improve the proposed method, including new applications for it.

2 Background and Problem Description

Real-Time Locating Systems based on Wireless Sensor Networks can be seriously affected by some effects related to the electromagnetic waves propagation, especially indoors [9]. Some of these effects are reflection, scattering or attenuation, amongst others. Such effects can provoke which is known as multipath effect, and, more specifically to indoor RTLS based on WSNs, the *ground reflection effect* [2]. There are several related approaches focused on the study or the mitigation of the multipath or the ground reflection effect [16] [2] [13] [11] [12] [9] [4] [7]. However, these approaches just solve the problem partially and none of them are specifically intended to enhance the performance of Real-Time Locating Systems.

Real-Time Locating Systems calculate the position of each tag in the system from a set of measurements obtained from the electromagnetic waves transmitted amongst tags and readers (*e.g.*, RSSI or Received Signal Strength Indication), its quality (*e.g.*, LQI or Link Quality Indicator), its Signal to Noise Ratio (SNR) or the Angle of Arrival (AoA) to the reader, amongst many others. In an ideal environment, these measurements would be perfect, with no error or noise, and the calculation of tags positions would be exact. Nevertheless, in the real world, the electromagnetic waves are influenced by effects as reflection, scattering, attenuation and diffraction. Attenuation is, in fact, a desired effect for estimating distances from measurements such as the received power of signals (RSSI). RSSI can be used, indeed, in signpost, fingerprinting and trilateration techniques to estimate distances from signal received power. However, reflection, scattering and diffraction can make the readers to receive additional spurious signals that are undesired *copies* of the main signal. The reception of such spurious signals makes up the *multipath effect.* This effect is especially undesired when measuring parameters as the RSSI, the AoA or the TDOA (Time Difference Of Arrival). When the ground is the main responsible of waves reflections, multipath can be modeled as the ground reflection effect, which is described in the next subsection.

2.1 The Ground Reflection Effect

The effects that affect the propagation of the electromagnetic waves, such as reflection, scattering, attenuation and diffraction, can reduce or even increase the range of a radio transmission [3]. Specifically, these effects can be a major challenge when designing a RTLS based on WSNs, especially for indoor environments.

The detailed effects of phenomena as attenuation and reflection in the propagation of electromagnetic waves can be calculated by solving Maxwell's equations with some boundary conditions that model the physical characteristics of each object or medium involved [3]. As this calculation can be very complex or even the physical characteristics of each object can be even unknown, there are some approximations to model signal propagation and calculate range transmission. One of these approximations is the ray-tracing technique that simplifies electromagnetic wavefronts to simple particles. Physically, each wavefront is the locus of spatial points presenting the same phase for a certain electromagnetic wave. In the ray-tracing technique, each wavefront is considered to be a particle traveling from the transmitter to the receiver antennas. This is very useful to model reflection and refraction effects, although it ignores the scattering phenomenon [3]. An electromagnetic wave transmitted by a certain wireless source will be reflected, diffracted or even scattered by the multiple objects placed throughout the environment. This way, the antenna of the destination node will receive undesired *copies* of the transmitted signal. Even worse, these additional signals will be possibly delayed in time and shifted in frequency and phase. When a single ground reflection effect predominates in the multipath effect, a two-ray model, as shown in Figure 1, can be used.

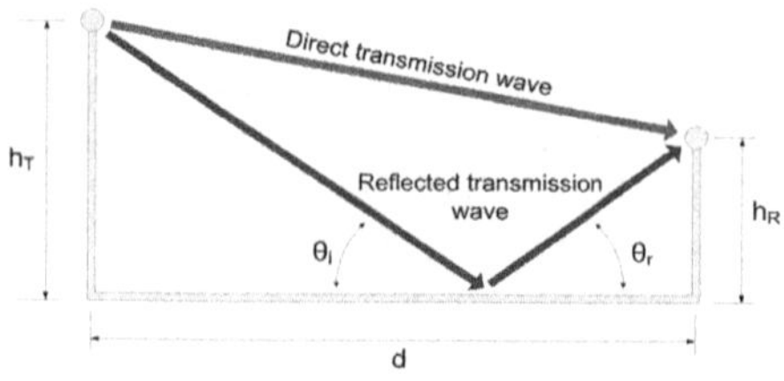

Fig. 1 Graphical representation of the ground reflection effect. Direct and reflected transmission waves travel from the transmitter to the receiver antennas, causing to be constructively or destructively added due to phases difference.

3 Mitigation of the Ground Reflection Effect

In ideal conditions, the modeling of the relationship between RSSI levels and distances between antennas has a decaying exponential shape. Nevertheless, when ground reflection effect is taken into account, the process of approximation of the relationship between the RSSI levels and the distances between antennas is complex and problematic. Therefore, it is necessary to use other models that allow considering the ground reflection effect in order to obtain a reliable estimation of the distances between tags and readers.

The model presented in this paper proposes the use of two Multi-Layer Perceptron [10] artificial neural networks to improve the precision of RTLS. On the one hand, the first MLP allows mitigating the ground reflection effect when estimating distances from power signal levels used to calculate the positions of users and objects by different locating techniques. On the other hand, the second MLP calculates the final positions of users and objects in the environment, using the output of the first MLP and acting, indeed, as a new locating technique that improves the precision of other compared techniques

For a certain range of RSSI values, there are fluctuations in the distance values regarding the RSSI levels. Thus, a certain RSSI value can mean distinct distances. In order to model the ground reflection effect we utilize time series applied to the first Multi-Layer Perceptron. Artificial Neural Networks allow forecasting a value according to the received historical values. Therefore, in this work the neural network is provided as inputs both the current detected RSSI value and the RSSI values detected in previous time instants. The neural network is made up of n input neurons, being n the time instants taken into account: $t, t - 1, ..., t - (n - 1)$. The intermediate layer of the neural network is configured following the Kolmogorov theorem [6] and choosing $2n + 1$ neurons.

In order to improve the forecast of the time series it was opted to incorporate the RSSI levels provided by other readers into the neural network. This way, the distances forecasting is done using a subset of the deployed readers in the system simultaneously. The neural network has k input groups with n neurons each of them. These n neurons correspond with the n values of the time series. Likewise, the k groups correspond with number of readers that are considered for the distance estimation. This number of readers is set in advance, thus selecting the readers with highest measured RSSI levels from the tag. The intermediate layer is made up of $2(k + n) + 1$ neurons, whereas output layer is formed by k neurons (*i.e.*, a neuron per each reader). The groups of input neurons are ordered according to the current RSSI level from highest to lowest. Therefore, the first output of the neurons is associated to the reader that received the highest RSSI level and so on.

Our proposed model captures data from the estimation of the positions by the trilateration algorithm. It stores these in a memory to subsequently use them to carry out the training of the second MLP. This way, this second MLP allows us to make the fastest estimations and is more responsive to variations in the distances resulting from the reflections of the waves emitted. Input data in the second MLP corresponds with the distances calculated by means of the first MLP from a pre-fixed number of readers and the position of the readers. These readers are selected according to the lowest distances they have to the tag. Output has two coordinates, one for each space coordinate. The number of neurons in the hidden layer is $2n + 1$, where n is the number of neurons in the input layer. Finally, there is one neuron in the output layer. The activation function selected for the different layers has been the *sigmoid*. Furthermore, the neurons exiting from the hidden layer of the neural network contain *sigmoidal* neurons. Network training is carried out through the *error backpropagation algorithm* [7].

4 Experiments and Results

In order to test the performance of this model into an indoor environment, we proceeded to deploy a WSN infrastructure made up of several ZigBee nodes (*i.e.*, readers and tags). These devices, called n-Core Sirius A for readers and Sirius B for tags, have both 2.4GHz and 868/915MHz versions and have several communication ports to connect to distinct devices, including a wide range of sensors and actuators. n-Core devices form part of the n-Core platform, which offers a complete API (Application Programming Interface) to access all its functionalities [8]. The ZigBee network was formed by 15 fixed Sirius A nodes acting as readers and distributed throughout three rooms. The total size of the monitored area was 19m per 19m. The distribution of the readers was done in this way so that each tag could be identified by several readers simultaneously. Therefore, the selected locating techniques (*i.e.*, signpost, fingerprinting and trilateration) could be applied using several simultaneous measurements. Firstly, as a previous step before the estimation of the tags positions, it was carried out the training of the neural network built to estimate the distances between nodes from the RSSI levels. A test tag was successively moved through different predefined location sequences (*i.e.*, zones inside the laboratory). This way, it was calculated the relationship of the measured RSSI levels with the real distances between the tag and the readers. For doing this, it was measured the detected RSSI levels between the tag and each of the 15 readers. Thus, the RSSI-distances measurements were used to make predictions in the time series. In total, 200 cases were generated for the training of the neural network. In addition, it was randomly chosen different positions throughout the zones to generate 100 new cases and estimate each position by means of both the neural network and other approximation methods to compare them. These other methods were SVR (Support Vector Regression) [15], a linear regression model and a logarithmic regression model. The calculation of the relationship between the RSSI levels and the distances in the training data set is necessary because the characteristics of the existing materials affect considerably to the detected distances.

As expected, the neural network obtained better results than SVR, the linear regression model and the logarithmic model because it presents a lower error for the distances estimation. The regression model obtained for the logarithmic regression fits in a very high grade the training data, as this model obtains an $R^2 = 0.9907$. Likewise, the linear regression model obtains an $R^2 = 0.8968$, which is also a high value. Basing on these R^2 values, both models can be considered as valid for the estimation. That is, the estimations made are significant and any other method that improves these results would also be valid. The errors for the MLP are lower than for the rest of the compared methods. Moreover, for the MLP the errors are concentrated in a certain range of RSSI levels. This allows creating reliable values outside some determined frequency ranges. Analyzing the dispersion of the error for each the compared model, shown in the Figure 2, it can be seen that the MLP offers the lowest dispersion and does not present so extreme values as SVR and linear regression do. Figure 2 (left) shows the box plot diagrams for the SVR, the regression models and the MLP. As can be seen, the MLP presents the lowest data variance and the minimum error.

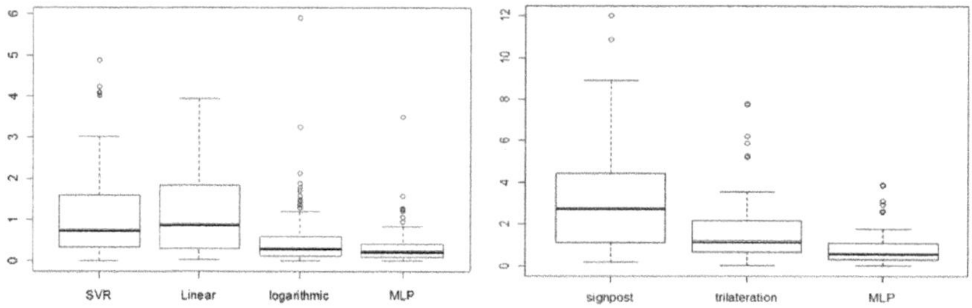

Fig. 2 Box plots representing the absolute error for the RSSI-distances relationship when using the different approximations (left) and the location errors for the different compared locating techniques (right).

The box plots representing the error information are presented in Figure 2 (right). As can be seen in the figure, the MLP provides lower estimation errors than the signpost and trilateration by themselves, that is, without modeling the RSSI and position behaviors.

5 Conclusions and Future Work

Amongst the wide range of Wireless Sensor Networks applications, Real-Time Locating Systems are emerging as one of the most exciting research areas. However, the operation of RTLS can be affected by undesired phenomena as the multipath effect, and more specifically, the ground reflection effect.

This paper proposes a new mathematical model aimed at improving the precision of WSN-based RTLS. The use of measurements from several readers as inputs of the MLP in the proposed model reduces even more the prediction error. This way, the ground reflection effect is mitigated and the approximations provided by other methods with high adjustment goodness, as the logarithmic regression model, were improved. This improvement in the distances forecasting is very relevant to estimate the positions of the tags, thus optimizing the overall calculations of locating techniques. In addition, the neural network responsible for calculating the final position reduces the error level of traditional methods from the information provided by them. The results obtained demonstrate that the use of ANNs allows improving the approximations provided by the locating techniques.

As future work it is planned the reduction of the readers necessary to perform the locating process, as well as the implementation in larger environments. Future work also includes the study of more detailed multipath models as Ricean and Rayleigh fading or shadowing [13].

Acknowledgments. This work has been supported by Spanish Ministry of Science and Innovation Project Ref. TRA2009_0096.

References

1. Barclay, L.W., I.O.E. Engineers.: Propagation of Radiowaves. Iet (2003)
2. Kim, E.S., Kim, J.I., Kang, I.-S., Park, C.G., Lee, J.G.: Simulation Results of Ranging Performance in Two-Ray Multipath Model. In: International Conference on Control, Automation and Systems, ICCAS 2008, pp. 734–737 (2008)
3. Goldsmith, A.: Wireless Communications. Cambridge University Press, Cambridge (2005)
4. Liu, H., Darabi, H., Banerjee, P., Liu, J.: Survey of Wireless Indoor Positioning Techniques and Systems. IEEE Transactions On Systems, Man, and Cybernetics, Part C: Applications and Reviews 37(6), 1067–1080 (2007)
5. Kaemarungsi, K., Krishnamurthy, P.: Modeling Of Indoor Positioning Systems Based On Location Fingerprinting. In: Twenty-Third Annual Joint Conference of the IEEE Computer and Communications Societies, INFOCOM 2004, vol. 2, pp. 1012–1022 (2004)
6. Katsuura, H., Sprecher, D.: Computational Aspects of Kolmogorov's Superposition Theorem. Neural Networks 7(3), 455–461 (1994)
7. Lecun, Y., Bottou, L., Orr, G.B., Müller, K.R.: Efficient Backprop. LNCS, pp. 5–50. Springer, Heidelberg (1998)
8. N-Core, N-Core: A Faster and Easier Way to Create Wireless Sensor Networks (2010), `http://Www.N-Core.Info` (retrieved October 27, 2010)
9. Nerguizian, C., Despins, C., Affès, S.: Indoor Geolocation with Received Signal Strength Fingerprinting Technique and Neural Networks. In: de Souza, J.N., Dini, P., Lorenz, P. (eds.) ICT 2004. LNCS, vol. 3124, pp. 866–875. Springer, Heidelberg (2004)
10. Nguyen, H., Chan, C.: Multiple Neural Networks for a Long Term Time Series Forecast. Neural Computing & Applications 13(1), 90–98 (2004)

11. Ray, J.K., Cannon, M.E., Fenton, P.C.: Mitigation Of Static Carrier-Phase Multipath Effects Using Multiple Closely Spaced Antennas. Navigation-Washington 46(3), 193–202 (1999)
12. Salcic, Z., Chan, E.: Mobile Station Positioning Using GSM Cellular Phone and Artificial Neural Networks. Wireless Personal Communications 14(3), 235–254 (2000)
13. Schmitz, A., Wenig, M.: The Effect of the Radio Wave Propagation Model in Mobile Ad Hoc Networks. In: Proceedings of the 9th ACM International Symposium on Modeling Analysis and Simulation of Wireless and Mobile Systems, Terromolinos, Spain, pp. 61–67 (2006)
14. Tapia, D.I., De Paz, J.F., Rodríguez, S., Bajo, J., Corchado, J.M.: Multi-Agent System For Security Control On Industrial Environments. International Transactions on System Science and Applications Journal 4(3), 222–226 (2008)
15. Vapnik, V.N.: Statistical Learning Theory. Wiley Interscience, Hoboken (1998)
16. Xie, J.J., Palmer, R., Wild, D.: Multipath Mitigation Technique in RF Ranging. In: Canadian Conference on Electrical and Computer Engineering, pp. 2139–2142 (2005)

Multiobjectivisation of the Antenna Positioning Problem

Carlos Segura, Eduardo Segredo, Yanira González, and Coromoto León

Abstract. *Multiobjectivisation* is a technique which transforms a mono-objective optimisation problem into a multi-objective one with the aim of avoiding stagnation. The transformation can be performed by the addition of artificial objectives or by the decomposition of the original objective function. Several well-known multiobjectivisation schemes, based on the addition of artificial objectives, are analysed. Also, some novel artificial objectives are suggested. The main advantages of these multiobjectivisation methods are their generality and ease of implementation. Different multiobjectivisation schemes have been applied to the mono-objective version of the Antenna Positioning Problem. Tests have been performed using NSGA-II, one of the most successful MOEAs. The experimental evaluation demonstrates that high quality results can be achieved by multiobjectivisation, when they are compared to the results obtained by the best mono-objective approaches.

Keywords: Multiobjectivisation, Diversity Maintenance, Antenna Positioning Problem, Multi-Objective Evolutionary Algorithms

1 Introduction

Optimisation may be defined as the process of finding the best solutions to a problem from the set of the candidate ones. In mono-objective optimisation only one objective is considered. Thus, a mono-objective optimisation problem can be described as an objective function f that maps a tuple of m parameters (decision variables) to a single objective y. Formally, $y = f(x)$ must be minimised or maximised, subject to $x = (x_1, x_2, ..., x_m) \in X$ where x is called the *decision vector*, X is the feasible region and y is the objective or fitness function. In multi-objective optimisation several

Carlos Segura · Eduardo Segredo · Yanira González · Coromoto León
Dpto. Estadística, I. O. y Computación, Universidad de La Laguna
La Laguna, 38271, Santa Cruz de Tenerife, Spain
e-mail: {csegura,esegredo,ygonzale,cleon}@ull.es

A. Abraham et al. (Eds.): International Symposium on DCAI, AISC 91, pp. 319–327.
springerlink.com

objectives are considered. These objectives are usually conflicting but must be simultaneously optimised [6]. In this kind of *Multi-Objective Optimisation Problems* (MOPs) a solution optimising every objective usually does not exist. In such cases, a non-dominated solution set, as close as possible to the optimal one, must be found. MOPs can be formally described as:

$$Optimize\ f(x) = (f_1(x), f_2(x), ..., f_k(x))\ \ subject\ to\ x \in X$$

where $f(x)$ is the objective vector, $f_i(x)$ is the i-th objective to be optimised, x is the decision vector and X is the feasible region in the decision space. While in single-objective optimisation the optimal solution is clearly defined, this does not hold for MOPs. Instead of a single optimum, there is a set or front of alternative trade-offs, known as a *Pareto-optimal* front, consisting of the non-dominated solutions y. Considering a maximisation problem, the Pareto-optimal front is constituted by the solutions $y \in X$ that:

$$\nexists z \in X / \ \forall i \in 1..k\ \ f_i(z) \geq f_i(y),\ \ \exists i \in 1..k\ \ f_i(z) > f_i(y)$$

Over the years, optimisation methods have evolved considerably. Several exact approaches have been designed for both, mono-objective problems and MOPs. However, exact approaches are unaffordable for many real world problems, so a wide variety of meta-heuristics has been developed with the aim of obtaining good quality solutions in a restricted time. Meta-heuristics can be classified into two groups: population-based and trajectory-based algorithms. In population-based algorithms, several solutions are taken into account simultaneously, while in trajectory-based ones only one solution is maintained. Among these techniques [17], Evolutionary Algorithms (EAs) and Multi-Objective Evolutionary Algorithms (MOEAs) are widely used in many fields. They are population-based algorithms inspired on the biological evolution. They have shown great promise for calculating solutions to large and difficult optimisation problems [7]. They can be easily adapted to afford different problems. However, depending on the characteristics of the decision space, premature convergence to local optima may occur. Maintaining a population is a great way to increase solutions diversity. However, even with the incorporation of a population, selection pressure may cause premature convergence. While in mono-objective optimisation, diversity is important to avoid stagnation, in MOPs it is necessary in order to cover the maximum possible regions of the decision space.

The term *multiobjectivisation* was introduced in [11] to refer to the reformulation of originally mono-objective problems as multi-objective ones. Since the fitness landscape is changed, multiobjectivisation can be useful to avoid local optima [9]. Although it can also produce a harder problem [2]. There are two different ways of multiobjectivising a problem. The first one is based on a decomposition of the original objective, while the second one is based on adding new objective functions. The addition of alternative functions can be performed by considering problem-dependent or problem-independent information.

APP (Antenna Positioning Problem) is an NP-Complete problem which arises in the engineering of mobile telecommunication networks [13]. Both mono-objective and multi-objective schemes have been used to deal with APP. A study of several mono-objective meta-heuristics applied to APP was presented in [12]. The problem-independent techniques achieved poorer quality solutions than those which incorporated problem-dependent information. The diversity maintenance mechanisms of the problem-independent techniques were not able to avoid stagnation. In order to escape from local optima, including problem-dependent information was necessary.

The main contributions of the current work are the following: advantages and drawbacks of different multiobjectivisation techniques based on the addition of new problem-independent objectives are analysed, and a performance study for the aforementioned multiobjectivisation techniques is carried out.

The remaining of the paper is structured as follows: the mathematical formulation for the APP is given in Sect. 2. In Sect. 3, a set of well known methods to multiobjectivise problems is presented. Also, the used optimisation method is detailed. Then, the experimental evaluation is described in Sect. 4. Finally, the conclusions and some lines of future work are given in Sect. 5.

2 APP: Mathematical Formulation

APP is defined as the problem of identifying the infrastructures required to establish a wireless network. It comprises the maximisation of the coverage of a given geographical area while minimising the *base stations* - BS - deployment. A BS is a radio signal transmitting device that irradiates any type of wave model. The region of the area covered by a BS is called a cell. In our definition of APP, BS can only be located in a set of potential locations. The APP mathematical formulation here used is presented in [1, 16]. In this formulation the fitness function is defined as:

$$f(solution) = \frac{Cover^{\alpha}}{Transmitters}$$

In the previous scheme a decision maker must select a value for α. It is tuned considering the importance given to the coverage, in relation with the number of deployed BS. As in [1, 16], the parameter α has been fixed to the value 2.

The geographical area G on which a network is deployed is discretised into a finite number of points or locations. Tam_x and Tam_y are the number of vertical and horizontal subdivisions, respectively. They are selected by communications experts, depending on several characteristics of the region and transmitters. U is the set of locations where BS can be deployed: $U = \{(x_1, y_1), (x_2, y_2), ..., (x_n, y_n)\}$. Location i is referred using the notation $U[i]$. The x and y coordinates of location i are named $U[i]_x$ and $U[i]_y$, respectively. When a BS is located in position i its corresponding cell is covered. The cell is named $C[i]$. In our definition we use the canonical APP problem formulation, i.e., an isotropic radiating model is considered for the cell. The set P determines the locations covered by a BS: $P = \{(\Delta x_1, \Delta y_1), (\Delta x_2, \Delta y_2), ..., (\Delta x_m, \Delta y_m)\}$. Thus, if BS i is deployed, the covered locations are given by the next set: $C[i] = \{(U[i]_x + \Delta x_1, U[i]_y + \Delta y_1), (U[i]_x +$

$\Delta x_2, U[i]_y + \Delta y_2), ..., (U[i]_x + \Delta x_m, U[i]_y + \Delta y_m)\}$. Being $B = [b_0, b_1, ..., b_n]$ the binary vector which determines the deployed BS, the next definitions holds for APP:

$$Transmitters = \sum_{i=0}^{n} b_i \qquad Cover = \frac{\sum_{i=0}^{tam_x} \sum_{j=0}^{tam_y} covered(i,j)}{tam_x \times tam_y} \times 100$$

where:

$$covered(x,y) = \begin{cases} 1 \; If \; \exists \, i/\{(b_i = 1) \wedge ((x,y) \in C[i])\} \\ 0 \; Otherwise \end{cases}$$

3 Optimisation Method

In the presented approach, an artificial objective function has been added to multiobjectivise the APP. The first objective has been selected as the fitness function of the APP, while for the second one, an artificial function which tries to maximise the diversity has been used. Since selection pressure is decreased, some low quality individuals could be maintained in the population. However, in the long term these individuals could help to avoid stagnation in local optima. One of the main challenges has been the selection of this artificial function. A comparison of a set of well-known schemes has been carried out. Moreover, two novel artificial objectives have also been tested.

Several options have been proposed to define the artificial objective [3]. The following ones have been taken into account:

- **Timestamp:** A timestamp represented by a counter is assigned as the artificial objective to be minimised.
- **Random:** A random value is assigned as the second objective to be minimised.
- **Inversion:** In this case, the optimisation direction of the original objective function is inverted and is used as the artificial objective.

Also, some schemes based on the Euclidean distance on the decision space has been defined. They have also been analysed:

- **DCN:** Distance to the closet neighbour of the population.
- **ADI:** Average distance to all individuals of the population.
- **DBI:** Distance to the best individual of the population, i.e., the one with highest APP fitness.

Two novel variants of the DBI scheme have also been considered. They are based on the addition of a threshold which penalises those solutions that may have a poor quality. In **DBI_TH1** a threshold is established over the APP objective function. Thus, individuals that are not capable to achieve the fixed threshold, are penalised by assigning a zero value to the second objective function. **DBI_TH2** is similar to the previous mechanism, but the threshold is established over the distance.

In order to test the aforementioned schemes, they were integrated in NSGA-II [4]. Tentative solutions are represented as binary strings with n elements, where n is the number of potential BS. Each gene determines whether the corresponding BS is deployed. The applied mutation operator has been the well-known *Bit Inversion Mutation* [14]. Each gene is inverted with a probability p_m. The used crossover operator has been the *Geographic Crossover* [15]. It exchanges the BS which are located within a given radius (r) around a randomly chosen BS.

4 Experimental Evaluation

In this section the experiments performed with different multiobjectivisation schemes are described. All multiobjectivisation schemes depicted in Section 3 have been tested. However, only the results for the four best alternatives are presented. Tests have been run on a Debian GNU/Linux computer with two Intel(R) Xeon(TM) at 2,66GHz and 1Gb RAM. The compiler which has been used is *gcc 4.3*. Two problem instances have been studied. The first one is a real world-sized problem instance [8]. It is defined by the geographical layout of the city of Malaga (Spain). This instance represents an urban area of 27.2 Km^2. The terrain has been modelled using a 450 x 300 grid, where each point represents a surface of approximately 15 x 15 m. The dataset contains 1000 candidate sites for the BS. The second instance is an artificial generated one. In this case, the terrain has been modelled using a 287 x 287 grid. The dataset contains 349 candidate sites for the BS.

Since we are dealing with stochastic algorithms, each execution was repeated 30 times. In order to provide the results with confidence, comparisons have been performed following the next statistical analysis [5]. First, a *Kolmogorov-Smirnov test* is performed in order to check whether the values of the results follow a normal (gaussian) distribution or not. If so, the *Levene test* checks for the homogeneity of the variances. If samples have equal variance, an ANOVA *test* is done. Otherwise, a *Welch test* is performed. For non-gaussian distributions, the non-parametric *Kruskal-Wallis* test is used to compare the medians of the algorithms. A confidence level of 95% is considered, which means that the differences are unlikely to have occurred by chance with a probability of 95%.

In the first experiment a comparison among the multiobjectivisation schemes is carried out. Each model has been executed with a stopping criterion of 4 hours. The analysis is performed in terms of the obtained fitness. Every model has been tested with the following parameterisation: $p_m = \frac{1}{n}$, $p_c = 1$, and $r = 30$.

The evolution of the average fitness values achieved by each considered model is shown in Fig. 1. Table 1 shows whether the row scheme is statistically better (↑), not different (↔), or worse(↓), than the corresponding column scheme. The best results were obtained by DBI. In fact, for the Malaga instance, DBI is statistically better than any of the other schemes. The multiobjectivisation schemes proposed as the best ones in [3] do not match with the multiobjectivisation schemes which obtain the best results in the current work. Therefore, the selection of the multiobjectivisation scheme is problem-dependent. Establishing the threshold over the fitness function has reported better solutions than establishing it over the alternative objective for the artificial instance. However, this has not held for the Malaga instance. Anyway,

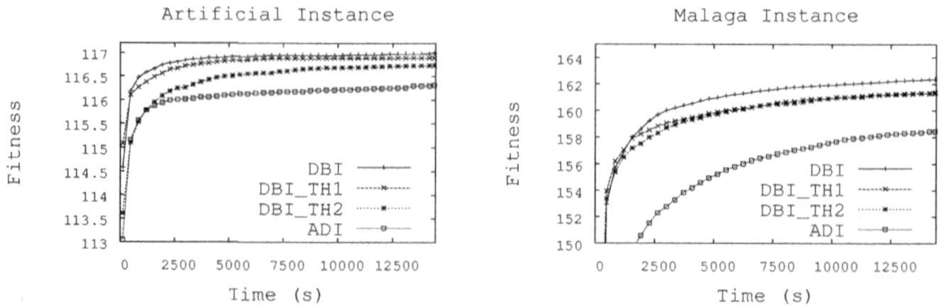

Fig. 1 Fitness evolution for the multiobjectivisation schemes

Table 1 Statistical comparison of multiobjectivisation schemes

	Artificial Instance				Malaga			
	DBI	DBI_TH1	DBI_TH2	ADI	DBI	DBI_TH1	DBI_TH2	ADI
DBI	↔	↔	↑	↑	↔	↑	↑	↑
DBI_TH1	↔	↔	↔	↑	↓	↔	↔	↑
DBI_TH2	↓	↔	↔	↑	↓	↔	↔	↑
ADI	↓	↓	↓	↔	↓	↓	↓	↔

in any of the cases, the fitness remains increasing after four hours of execution, showing the good capability of avoiding stagnation in local optima.

Results achieved by the multiobjectivisation schemes were able to improve the mono-objective models presented in [12] that did not include problem-dependent information. However, their quality was not as high as the ones obtained by the schemes which incorporated problem dependent information. The second experiment analyses whether, in the long term, the multiobjectivisation schemes are able to reach such high-quality solutions, and quantifies the impact over the performance when compared to the best approaches presented in the literature. The best multiobjectivisation model - DBI - is compared with the algorithm which achieved the best result in [12] (Iterated Local Search - ILS). In order to measure the improvement, the run-time behaviour is analysed using the ideas presented in [10]. Specifically, both schemes were executed using as finalisation condition the achievement of a certain quality level: the worst result achieved by ILS on 1 hour (L1), and the worst result achieved by ILS on 2 hours (L2). The success ratio is defined as the probability of achieving the required quality, considering a maximum amount of time. Figure 2 shows the run-length distribution of ILS and DBI for the quality level L2. Since ILS incorporates a high amount of problem-dependent information, it is able to obtain the high quality solutions very fast. DBI has a slower convergence, but it has also been able to obtain such high quality solutions. Considering a success ratio of 0.5 and the quality level L1, DBI requires 5.05 more times than ILS in the artificial instance. When considering L2, the ratio increases up to 13.61. In the case of the instance of Malaga, the ratio considering the quality level L1 is 10.57, while considering L2, it is 10.25.

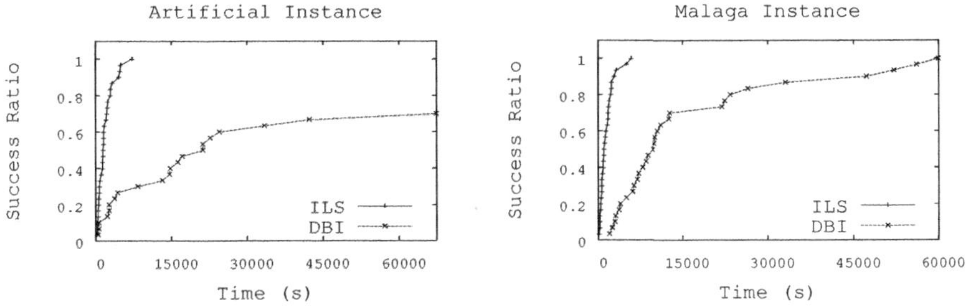

Fig. 2 Run-length distribution for the quality level L2

Although DBI has a slower convergence, it has been able to avoid local optima and obtain high quality solutions. In fact, by using executions of 24 hours DBI has been able to obtain the best currently found solutions for both analysed instances.

5 Conclusions and Future Work

In this paper we have tested the ability of multiobjectivisation to deal with a complex NP-Complete problem: APP. The used multiobjectivisation approach is based on adding new alternative objectives to the original fitness function of APP. The approach has been tested with NSGA-II, one of the most successful MOEA. Experimental evaluation has been carried out with several alternative objectives. The one which obtained the best results is based on using the Euclidean distance to the best individual in the population. However, it has been proved that the best-behaved multiobjectivization scheme depends on the problem over which it is applied. In the case of APP, the best multiobjectivisation method has been compared with the best mono-objective approach published in the literature. Although it has a slower convergence, in the long term it has been able to achieve solutions of similar quality. The main advantage of the multiobjectivisation method is its generality, and ease of implementation.

Future work will be focused in the analysis of other multiobjectivisation schemes. The usage of other problem-independent and problem-dependent alternative objectives should be analysed. Also, in order to reduce the time required to attain high quality solutions, a parallel MOEA should be tested. Since the appropriate alternative objective depends on the problem that is being solved, the hybridisation of multiobjectivisation and hyperheuristics seems a promising approach. Thus, the selection of which alternative objective must be used, could be performed in an automatic way.

Acknowledgements. This work was partially supported by the EC (FEDER) and the Spanish Ministry of Science and Innovation as part of the 'Plan Nacional de I+D+i', with contract number TIN2008-06491-C04-02 and by Canary Government project number PI2007/015. The work of Carlos Segura was funded by grant FPU-AP2008-03213. The work of Eduardo Segredo was funded by grant FPU-AP2009-0457. The work was also funded by the HPC-EUROPA2 project (project number: 228398) with the support of the European Commission - Capacities Area - Research Infrastructures.

References

1. Alba, E.: Evolutionary algorithms for optimal placement of antennae in radio network design. In: International Parallel and Distributed Processing Symposium, vol. 7, p. 168 (2004), http://doi.ieeecomputersociety.org/10.1109/IPDPS.2004.1303166
2. Brockhoff, D., Friedrich, T., Hebbinghaus, N., Klein, C., Neumann, F., Zitzler, E.: Do additional objectives make a problem harder? In: Proceedings of the 9th Annual Conference on Genetic and Evolutionary Computation, GECCO 2007, pp. 765–772. ACM, New York (2007), http://doi.acm.org/10.1145/1276958.1277114
3. Bui, L., Abbass, H., Branke, J.: Multiobjective optimization for dynamic environments. In: The 2005 IEEE Congress on Evolutionary Computation, vol. 3, pp. 2349–2356 (2005), doi:10.1109/CEC.2005.1554987
4. Deb, K., Pratap, A., Agarwal, S., Meyarivan, T.: A fast and elitist multiobjective genetic algorithm: NSGA-II. IEEE Transactions on Evolutionary Computation 6, 182–197 (2002)
5. Demšar, J.: Statistical comparison of classifiers over multiple data sets. Journal of Machine Learning Research 7, 1–30 (2006)
6. Ehrgott, M., Gandibleaux, X. (eds.): Multiple Criteria Optimization. State of the Art Annotated Bibliographic Surveys. International Series in Operations Research and Management Science, vol. 52. Kluwer Academic Publishers, Dordrecht (2002)
7. Eiben, A.E.: Handbook of Evolutionary Computation. IOP Publishing Ltd. and Oxford University Press (1998)
8. Gómez-Pulido, J.: Web site of net-centric optimization, http://oplink.unex.es/rnd
9. Handl, J., Lovell, S.C., Knowles, J.: Multiobjectivization by decomposition of scalar cost functions. In: Rudolph, G., Jansen, T., Lucas, S., Poloni, C., Beume, N. (eds.) PPSN 2008. LNCS, vol. 5199, pp. 31–40. Springer, Heidelberg (2008)
10. Hoos, H., Informatik, F., Hoos, H.H., Stutzle, T., Stutzle, T., Intellektik, F., Intellektik, F.: On the run-time behavior of stochastic local search algorithms for sat. In: Proceedings AAAI 1999, pp. 661–666 (1999)
11. Knowles, J.D., Watson, R.A., Corne, D.W.: Reducing local optima in single-objective problems by multi-objectivization. In: Zitzler, E., Deb, K., Thiele, L., Coello Coello, C.A., Corne, D.W. (eds.) EMO 2001. LNCS, vol. 1993, p. 269. Springer, Heidelberg (2001), http://portal.acm.org/citation.cfm?id=647889.736521
12. Mendes, S.P., Molina, G., Vega-Rodríguez, M.A., Gómez-Pulido, J.A., Sáez, Y., Miranda, G., Segura, C., Alba, E., Isasi, P., León, C., Sánchez-Pérez, J.M.: Benchmarking a Wide Spectrum of Meta-Heuristic Techniques for the Radio Network Design Problem. IEEE Transactions on Evolutionary Computation, 1133–1150 (2009)
13. Meunier, H., Talbi, E.G., Reininger, P.: A multiobjective genetic algorithm for radio network optimization. In: Proceedings of the 2000 Congress on Evolutionary Computation, pp. 317–324. IEEE Press, Los Alamitos (2000)
14. Segura, C., González, Y., Miranda, G., León, C.: A multi-objective evolutionary approach for the antenna positioning problem. In: Setchi, R., Jordanov, I., Howlett, R.J., Jain, L.C. (eds.) KES 2010. LNCS, vol. 6276, pp. 51–60. Springer, Heidelberg (2010)

15. Talbi, E.G., Meunier, H.: Hierarchical parallel approach for gsm mobile network design. J. Parallel Distrib. Comput. 66(2), 274–290 (2006), http://dx.doi.org/10.1016/j.jpdc.2005.09.006
16. Weicker, N., Szabo, G., Weicker, K., Widmayer, P.: Evolutionary multiobjective optimization for base station transmitter placement with frequency assignment. IEEE Transactions on Evolutionary Computation 7(2), 189–203 (2003), doi:10.1109/TEVC.2003.810760
17. Whitley, D.: An overview of evolutionary algorithms: Practical issues and common pitfalls. Information and Software Technology 43, 817–831 (2001)

Mobile Access System for the Management of Electronic Health Records of Patients with Mental Disability

M. Antón-Rodríguez, I. de la Torre-Díez, P. Gutiérrez-Díez*, F.J. Díaz-Pernas, M. Martínez-Zarzuela, D. González-Ortega, and J.F. Díez-Higuera

Abstract. This paper presents an electronic health records (EHR) management web system for patients with mental disability, allowing the access from mobile devices. The system gives priority to information security to guarantee the privacy and confidentiality of the data handle. Moreover, it provides features, which improves the typical use of common web systems, like accessing to the mobile camera from the web for taking images or videos and a complete interoperability with other systems using the medical standards HL7-CDA R2 and DICOM. The whole system is implemented using open technologies and free software.

Keywords: Web system; wireless access; electronic health records; medical standards HL7-CDA R2 and DICOM; communications security.

1 Introduction

Accessing to EHRs (Electronic Health Records) through mobile devices provides a number of advantages both for health centers and clinical staff, and for patients. Among these advantages are: accessing to patients' information in real time (from wherever and whenever), resource savings, improving the information management, and reducing the delay in health care. In the field of mental health, there are important epidemiological studies releasing relevant information about types and rates of more frequent disorders. However, a significant number of people with mental diseases remain unnoticed due to the incorrect identification of the symptomatology, the resistance to seek either help or information regarding these services, among others. Mobile technologies can offer a full potential for helping

M. Antón-Rodríguez · I. de la Torre-Díez · P. Gutiérrez-Díez · F.J. Díaz-Pernas
M. Martínez-Zarzuela · D. González-Ortega · J.F. Díez-Higuera
Department of Signal Theory, Communications and Telematics Engineering,
Telecommunications Engineering School, University of Valladolid, Valladolid, Spain
e-mail: {mirant,isator,pacper,marmar,davgon,josdie}@tel.uva.es, pgutdie@etsi.tel.uva.es
* Corresponding auhor.

A. Abraham et al. (Eds.): International Symposium on DCAI, AISC 91, pp. 329–336.
springerlink.com

people with cognitive problems and their supporting staff. The system presented in this paper, which we call EHRmobile, has as its main goal the remote access to the EHR of any patient with a cognitive disorder, through mobile devices (smartphone, PDA, tablet PC, etc.).

We performed a deep analysis of EHR systems of specialties like pediatrics [10], urgency [3], oncology [12], etc. Chew et al. [5] developed the OphthWeb application within a multidisciplinary project of EHRs in Singapore. Taddei et al. [16] accomplished a web-based system for EHRs of cardiology in an Italian health institute. There are other web applications, like CareWeb™, using the standard HL7 [11]. Becker & Sewell [4] presented an EHR system, InfoDOM, based on web technologies. Siika et al. [15] described the development and structure of an EHR system for patients with Human Immunodeficiency Virus (HIV) in Kenya. Cho & Park [6] developed an EHR system based on the Korean beta version of the International Classification for Nursing Practice (ICNP). The system was evaluated by 20 nurses and 57 patients, in 2 Korean hospitals. Karagiannis et al. [13] implemented a web-based EHR system (pEHR) that was proven by 22 physicians and 150 patients of 3 European hospitals. The system was developed to meet the needs of patients with a congenital heart disease, Parkinson or Diabetes type 2. The research team of the present work accomplished an web-system for ophthalmological EHR management, TeleOftalWeb [8] [9]. This system is in used in the the Institute of Applied Ophthalmobiology (*Instituto de Oftalmobiología Aplicada*, IOBA) of the University of Valladolid, Spain. TeleOftalWeb complies with the Health Level 7-Clinical Document Architecture Release 2.0 (HL7-CDA) standards for EHRs storage, and Digital Imaging and Communications in Medicine (DICOM 3.0) for medical images.

Mobile applications offer a chance to improve health services. However, nowadays there are few applications integrating mobile communications with EHRs. Velde & Brobbel [17] developed a mobile information system intended for cardiology field. Shyu et al. [14] performed a mobile EHR system for the family medicine department in the NTUH (National Taiwan University Hospital) from Taiwan. Countrywide, the Health Department (*Conselleria de Sanitat*) from the Community of Valencia allows the opportunity to access to the personal health record through its website, downloading the record updated and ciphered. Hence, patients can be better assisted out of the region [1] [2]. Nevertheless, some barriers still exist to achieve the extended use of this kind of systems: many EHR systems are still in a consolidation phase, information is not shared among different health systems, there is uncertainty in themes related to security and data protection…

Present paper shows the development of a web system for managing EHRs of patients with mental diseases intended for the socio-sanitary staff from Intras Foundation, accessing from a mobile or a desktop browser. Intras Foundation, Research and Treatment in Mental Health and Social Services (*Investigación y Tratamiento en Salud Mental y Servicios Sociales*), is a non-profit institution focusing its interest in improving the quality of living of people with sanitary, socio-sanitary and socio-labor integration requirements.

This paper is organized as follows. In section 2 the system presented is described differentiating among presentation, business and data layers. In section 3, conclusions are shown and, finally, the references used are listed.

2 EHRmobile System

EHRmobile is a web system for performing a complete management of EHRs of patients with mental disabilities. It includes the capability of interoperability with other systems thanks to the use of the standards HL7 and DICOM. The system provides two different versions depending on the access, from a PC and from a mobile device. Accessing from a PC allows the complete management of the system, while accessing from a mobile device provides functionalities for browsing and updating clinical data along with additional features related to the terminals used, and always prioritizing information security and privacy. Fig. 1 shows the schema of the proposed system.

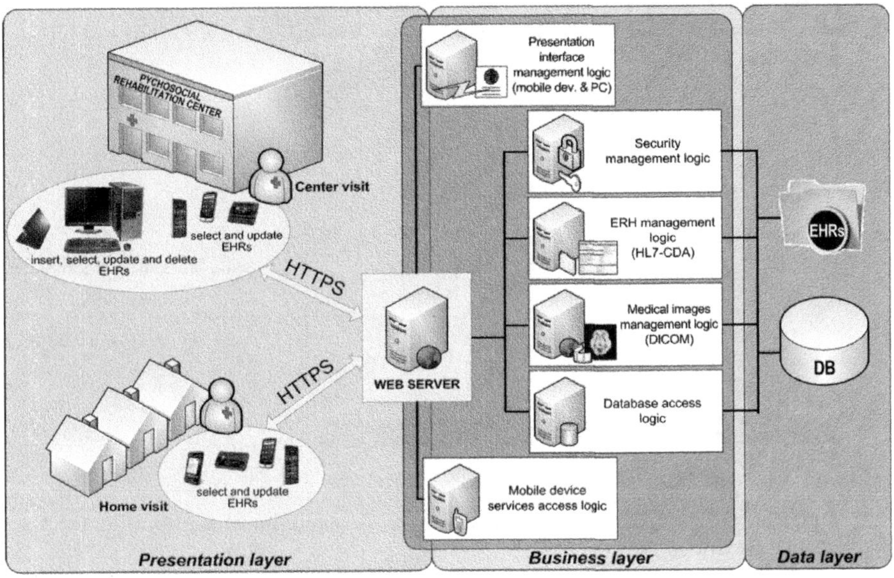

Fig. 1 Schema of the proposed system: EHRmobile.

2.1 *Presentation Layer*

This layer includes the graphic interface utilized by the user to access the system through the device browser. As previously said, the system provides two means of access, via PC or mobile device, so the type of device used is detected to provide it the appropriate design and features. Markup languages used to draw up both interfaces are different depending on the final user, XHTML-MP for mobile devices and HTML/XHTML for desktop web, and so are the designs due to the different specification of both ways of access, mainly the size of the display. W3C recommendations [18] have been followed.

Accessing from PC (see Fig. 2a), a complete functionality is achieve. It is possible to add new EHRs, search in them, update, and delete them. Also, it can be administered all about the users accessing the system, i.e. the socio-sanitary staff. Using a mobile device (see Fig. 2b), data can be consulted, updating those needed, always interacting with a friendly interface, considering the possibilities of the device. All functionalities are not allowed using a mobile device, but for an administrator user, since we considered nonviable creating an EHR from a 12-keypad. Time spent could be too high. Nevertheless, exploiting services provided by mobile devices, both pertinent medical images and videos can be taken and attached to the EHR.

(a) (b)

Fig. 2 (a) Home view from desktop version of the system, clinical staff profile. (b) View from mobile version of the system.

2.2 *Business Layer*

Business layer accomplishes the processing required for attending the user requests, hence, it includes all the system logic, completely developed under PHP5 using Apache as web server. This layer has been divided into different modules according to their functionalities:

- **Security management logic:** EHR information imply higher risk of breach of confidentiality and privacy, therefrom the importance of this module. Among the security measures taken are: users identification determining a privilege level within the system, securing communication established within the system, physical security… Personal identification is performed every time when trying to access the system, providing information according to the privilege level assigned to each user by Intras Foundation. In order to provide communications security we use OpenSSL, which implements the SSL (Secure Socket Layer) protocol. SSL support authentication and information privacy between ends using cryptography. Data exchange between client and server are cipher through a symmetric-key algorithm (DES, triple-DES, RC2, RC4 o IDEA), as well as the session key of the algorithm using a public-key algorithm. As a different key is generated for each transaction, even though it would be attacked, it could decipher future transactions. Hence, all the system communications use the HTTPS (Hypertext Transfer Protocol over Secure Socket Layer) protocol which includes a SSL based ciphering.

- **Presentation interface management logic:** It undertakes the task of detecting the device requesting access to the system in order to show the interface better fitting its features. We use a server PHP code including a list with the browsers more used, both for mobile and desktop. As it was previously said, detecting the device is not only important at design level providing a more compact and light web to the mobile terminal, but also at a functionality level, as from a desktop browser it is possible to access to the entire application, allowing inserting new records, delete them, and administering the system.
- **EHR management logic:** It is the module in charge of standardizing the EHRs observing the standard HL7-CDA, Health Level Seven-Clinical Document Architecture, in particular the version provided by the Department of Health from Castile and Leon (Spain) [7]. HL7 is the international standard more commonly-held for EHR storing and exchanging. It is based on XML. With the aim to not continuously work with the XML file of each patient, all data are stored in an encrypted MySQL database and, asynchronously, XML files are created or updated. By so doing, the temporal price when accessing to each EHR considerably diminishes as data reading or updating are directly performed against the database, and it is the parser who updates the XML data when it is necessary. In Fig. 3 it is shown the sequence diagram used by the parser for importing a new EHR. Fig. 3 b displays the sequence diagram when importing a user previously known by the system. Exporting EHRs are an easier action, generating the XML/HL7 file.
- **Medical images management logic:** It includes the code necessary for converting the medical images in DICOM (*Digital Imaging and Communication in Medicine*) format to classical digital images. DICOM is the internationally-recognized standard for medical images exchange.
- **Database access logic:** It is the module in charge of performing the database (MySQL) accessing. Database includes all the information required about patients (anamnesis, clinical observation, medication, and clinical appraisal), clinical staff, and system administration information. MySQL manager was chosen due to its easy handling from the server programming language used, PHP, and for being a popular open source database. As it has been previously said, the XML/HL7 parser uses the information stored into the relational database to

generate/update the XML files, and contrariwise, it gets through the XML file to insert its information into the database. This way, a quick access is provided to the final user, as the system interoperability remains.

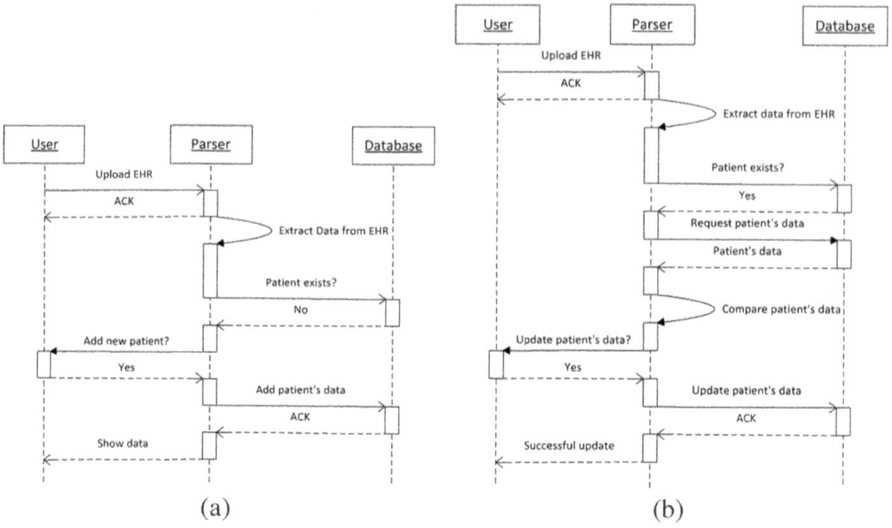

Fig. 3 Sequence diagram of the parser performing when importing a new patient from XML/HL7 file. (a) Importing a patient new to the system. (b) Importing a patient previously known, already registered in the system.

- **Mobile device services access logic:** As HTML5 standard is not yet finished, which it is supposed to provide a complete accessing to the mobile devices services. Meanwhile, to complete the access to the camera, we used Adobe® Flash®, JQuery, and PHP, allowing taking the images or videos from the corresponding web page and appending them to the patient record.

2.3 *Data Layer*

This layer is in charge of managing all the information in a persistent way, storing and supplying the data to the business layer. It includes information about the users (clinical staff), and the complete health records both in MySQL and XML files following the HL7-CDA R2 standard. This framework is also designed to provide a basis for storing medical images using DICOM standard, and digital images and videos, increasingly used in many medical specialties.

Consistency and speed in the access are critical factors for the smooth-running of the system. Hence, we opted for separating the access to the information from within the system and the interoperability to other systems.

3 Conclusion

Nowadays, applying mobile technologies to health assistant may open new possibilities: better access to relevant information, counseling and cooperation among health professionals, and patient care assistance at home. However, these new facilities must inexorably include security and privacy information issues. Furthermore, the use of EHR management systems, specifically framed in the mental health field which suffers from an ever-expanding perspective, can mean an important support for improving both the treatment quality of these patients and the work quality of the socio-sanitary staff.

Accordingly, this paper presents an electronic health records (EHR) information and management web system for patients with mental disability, allowing the access, from within the sanitary center and from the house call, which are routinely performed, via mobile device, prioritizing the secure and quickly communications and storing. The system also provides new functionalities, as they are enabling the image and video capture, and the interoperability to other systems by using the HL7-CDA R2 and DICOM standards.

At present, the system is on a trial basis with the final user (Intras Foundation), managing records from 87 patients with cognitive disabilities from Castile and Leon (Spain). The possibility of accessing to the updated information of the patients in the house calls means an important benefit for the Intras Foundation's staff and so, for its patients. Nevertheless, some issues have emerged, like the accessibility to the system from a mobile device, as a usual complaint is to have to type the URL accessing the system and, then, the patient's name. At this point, we pretend to incorporate the QR codes to give access to the patient's information. The future extension of the system will include the integration with Gradior, the rehabilitation system used by Intras Foundation and a great deal of associated centers, so as the data about the patient and his/her rehabilitation sessions.

Acknowledgement. This work has been partially supported by the *Department of Health, Government of Castile and Leon (Spain)* under the project GES39/VA05/10 and the *Spanish Ministry of Science and Innovation* under the project TIN2010-20529.

References

1. de Salud, A.V.: Nota de prensa: Puesta en marcha del proyecto 'Tu salud en el móvil (2009), `http://www.dep21.san.gva.es/deporihuela/?p=583` (last visited: September 2010)
2. de Salud, A.V.: (2010), `http://hclinica.serviciosmmc.com/index_es.php` (last visited: September 2010)
3. Amouh, T., et al.: Versatile Clinical Information System Design for Emergency Departments. IEEE Transactions on Information Technology in Biomedicine 9(2), 174–183 (2005)
4. Becker, M.Y., Sewell, P.: Cassandra: flexible trust management, applied to electronic health records. In: Proceedings of 17th IEEE Computer Security Foundations Workshop, pp. 139–154 (2004)

5. Chew, S.J., et al.: OphthWeb-cost-effective telemedicine for ophthalmology. Hong Kong Medical Journal 4, 300–304 (1998)
6. Cho, I., Park, H.: Development and evaluation of a terminology-based electronic nursing record system. Journal of Biomedical Informatics 36(4), 304–312 (2003)
7. de Sanidad, C., de Castilla y León, J.: Estándares de integración HL7 (2010), http://www.salud.jcyl.es/sanidad/cm/empresas/tkContent?idContent=500929&locale=es_ES&textOnly=false (last visited: November 2010)
8. De la Torre-Díez, I., et al.: Choosing the most efficient database for a Web-based system to store and exchange Ophthalmologic Health Records. Journal of Medical Systems (2010a) (in press), doi: 10.1007/s10916-009-9422-2
9. De la Torre-Díez, I., et al.: Performance evaluation of a Web-based system to exchange Electronic Health Records using queueing model (M/M/1). Journal of Medical Systems (2010b) (in press), doi: 10.1007/s10916-010-9555-3
10. Ginsburg, M.: Pediatric Electronic Health Record Interface Design: The PedOne System. In: Proceedings of the 40th Hawaii International Conference on System Sciences, pp. 1–10 (2007)
11. Halamka, J.D., et al.: CareWeb^{TM}, a web-based medical record for an integrated health care delivery system. International Journal of Medical Informatics 54, 1–8 (1999)
12. James, A., et al.: A Telematic System for Oncology Based on Electronic Health and Patient Records. IEEE Transactions on Information Technology in Biomedicine 2(1), 16–17 (2001)
13. Karagiannis, G.E., et al.: Web-based personal health records: the personal electronic health record (pEHR) multicentred trial. Journal of Telemedicine and Telecare 13, 32–34 (2007)
14. Shyu, F., et al.: Context-based Model for Mobile Electronic Medical Records. In: Proceeding of the Eighth International Conference on e-Health Networking, Application and Services, pp. 140–146 (2006)
15. Siika, A.M., et al.: An electronic medical record system for ambulatory care of HIV-infected patients in Kenya. International Journal of Medical Informatics 74(5), 345–355 (2005)
16. Taddei, A.: Development of an electronic medical record for patient care in cardiology. Computers in Cardiology 7(10), 641–644 (1997)
17. Velde, E.T., et al.: Application of Handheld Computers for Mobile Access to a Cardiology Information System. Computers in Cardiology 28, 157–160 (2001)
18. W3C, Basic Guidelines for Mobile Web Best Practices 1.0. En (2008), http://www.w3.org/TR/mobile-bp/ (last visited: September 2010)

Using Mobile Systems to Monitor an Ambulatory Patient

Ângelo Costa, Guilherme Barbosa, Tiago Melo, and Paulo Novais

Abstract. Medical diagnostics and vital signs monitoring demands more technological solutions to cope with new methods of treatment. Continuous monitoring and information processing tools are vital to a physician with several patients under his care. In this work, a system that relies on agents and mobile and wireless devices is presented. Its use with small scale sensors allows to collect and analyse vital data in real-time, triggering appropriate reactions in case of eminent danger. This includes real-time notifications to practitioners. In cases in which the physician is unable to divide his attention among all his patients, the system is able to drive his attention to one patient only and, when it is necessary, to another one, according to their medical state. This concept represents a breakthrough in terms of the physician's time and task management, being possible to apply it two major scenarios: patient recovery in a hospital environment or elderly living alone in a domestic environment. In that sense, we present a brief contextualiation of the problem as well as the architecture and technologies used to implement the proposed work.

1 Introduction

The increase in the quality of health care services, along with socio-economic growth and technological achievements are making life expectancy higher. Moreover, the latest United Nations (UN) reports on world population [1] show the current population ageing.

Due to the overall growth in the elderly population, constant monitoring of some specific groups of population has become increasingly necessary, namely in groups that suffer from chronic diseases and heart conditions. Although transport

Ângelo Costa · Guilherme Barbosa · Tiago Melo · Paulo Novais
CCTC, Departamento de Informática, Universidade do Minho, Braga - Portugal
e-mail: {acosta,pjon}@di.uminho.pt,
{gbarbosa,tiago.blackcode}@gmail.com

A. Abraham et al. (Eds.): International Symposium on DCAI, AISC 91, pp. 337–344.
springerlink.com

availability has increased over the years, frequent travel to health care facilities is not a viable solution, mainly in cases that need continuous care. Moreover, this can also have a negative impact on the patient's quality of life.

With the technological progress in the fields of electronics and computing, wearable and environmental sensors have become more compact and portable, while providing high precision measurements of the parameters they are sensing. This allows to develop applications that explore sensor's capabilities in mobile and portable devices that a patient can carry, without being intrusive or affecting the person's daily life.

Based on these ideas, we propose a portable monitoring system, implemented over a mobile device (e.g. PDA, smart-phone), capable of transmitting information collected by sensors worn by the patient to an external medical entity, which in turn will generate a medical diagnosis of the patient being monitored. In this case, the focus is on constant vital sign monitoring in order to anticipate heart failure. With these goals in mind, this system allows practitioners to provide instant and localized health-care services to those being monitored and avoid critical scenarios which can ultimately lead, in an unsupervised environment, to the patients' death.

1.1 Related Projects

Some projects exist that already have working devices and systems. Their development is important and allows to compare the features and, if proven useful, use the hardware and underlaying platforms to develop new applications with different purposes.

Vital Jacket [3] is a vital sign monitoring system integrated in the patient's clothing, combining both textile and micro-electronic components. VitalJacket is composed of a t-shirt that accommodates a small, lightweight and compact electronic device that constantly monitors the user's heartbeat. The electronic device has an integrated Bluetooth module for data transmission.

Plux [2] is a Portuguese company specialized in the design, development and production of systems for data acquisition and wireless sensors to study the electrophysiological activity. The data acquisition systems also communicate with the computing devices via Bluetooth.

2 iNumon- Independent User Monitoring

This project aims at being an answer to the previously presented problems. The question of on-body monitoring is a very complicated one, including challenges about reliability and choice of correct sources of data. It is also at steak the level of adaptation of the user to the entire system.

In the following pages, we will present our insights and the solutions that we propose in order to achieve the established goal. The project is organized into several main modules: the Sensor Network, the mobile devices and the server-side application. These modules are combined to develop a mobile monitoring system that can make the appropriate decisions. In this section we describe the architecture, the technologies, the developed modules and the results observed.

2.1 System Architecture

As mentioned before, the system proposed aims at a constant monitoring of relevant information about a patient, relying on the concept of Body Area Network (BAN). BANs will be operating over a mobile device that records the patient's condition through a sensor network and forwards the information collected to an external entity [12, 13]. A set of sensors must thus capture the specific condition of a patient in a discrete time interval and use such information for remote diagnose, to assess the patient health condition and detect eventual anomalies (Fig. 1).

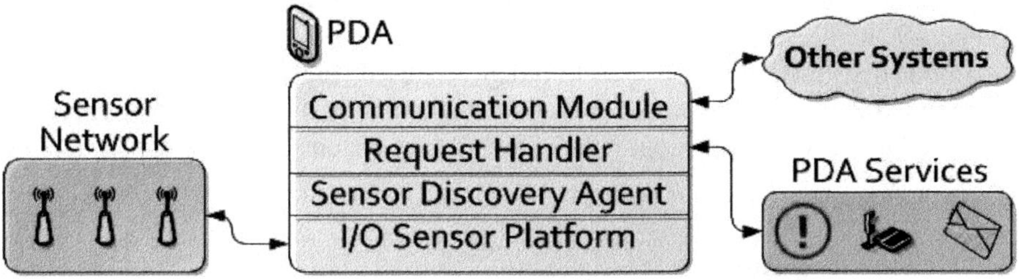

Fig. 1 The architecture of the proposed system.

This sensor network must be flexible enough to allow engaging or disengaging sensors without any kind of system reboot and without affecting the performance and efficiency of the system. Inside the BAN, a set of software agents will also reside, focused on providing interaction services to the remote entities [11]. These are able to process the result of the diagnose within the mobile device, making it possible to inform about the patient's condition after a successful diagnosis based on the information provided by his mobile device in the first place.

2.1.1 Agents' Characteristics

The agents that interpret the data provided by the sensors must only retain, manage and forward the information.

The forwarding of the information gathered by the sensors through a network to an external diagnosis entity is done by an agent called Sensor Discovery Agent (SDA) and has management-oriented characteristics.

These management abilities are associated with basic characteristics that this agent must comply with at the interaction level with the Sensor System Agents (SSA) and the I/O communication module in order to guarantee the existence of a data flow between the sensors and the respective SSA. Therefore, the SDA must receive notifications every time a sensor enters or exits the BAN through the I/O module. This allows the SDA to load or unload the sensor agents according to the system's needs.

Moreover, the SDA must gather information obtained by the SSA and forward it to an external entity so that a diagnosis of the patient's current state can be created.

To carry out information forwarding, a module called Communication Platform (CP) is being used. This module is responsible for the encapsulation of the available information and sending it along the network to the remote entity.

2.1.2 Communication Technologies

This project encompasses several methods and technologies of communication. In this section we describe the main technologies that were used to implement the proposed architecture.

In our first tests, the ZigBee technology was used (Figure 2) [5, 6, 7]. We have abandoned ZigBee in favour of Bluetooth. Specifically, the platform of reception was too large and did not support all of the mobile systems.

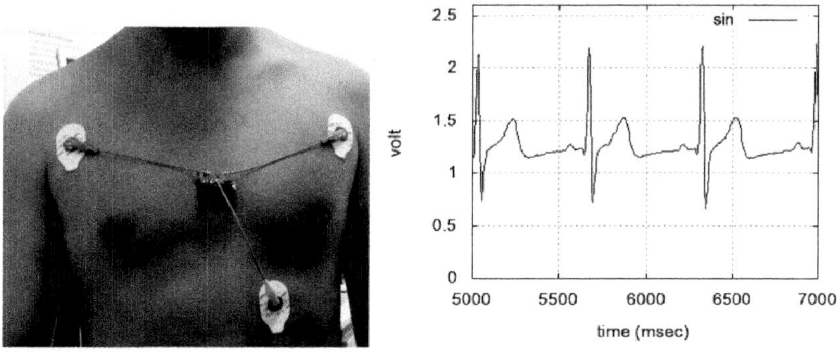

(a) EGC Capture with ZigBee communication.

(b) Result of the ECG Capture.

Fig. 2 Representation of the ECG device and the visual representation of the collected data

In terms of agent support and communication, the OSGi Platform is being used [8]. A Multi-agent System (MAS) [9, 10] supported by JADE Framework using FIPA-ACL standard and XML for the message content is our base supporting system to the developed agents.

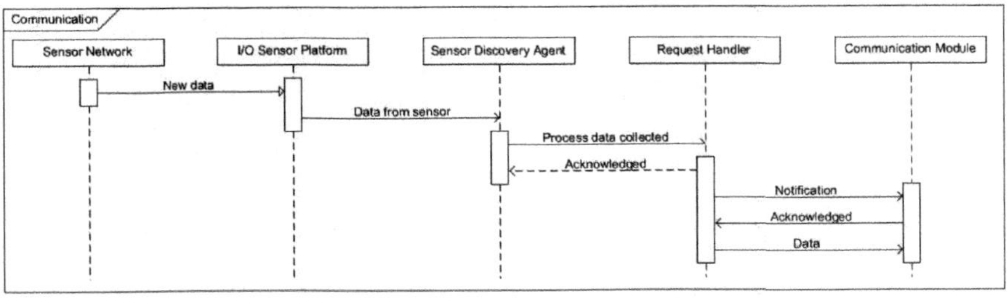

Fig. 3 Typical communication between the Agents and platforms

2.1.3 Sensor Monitoring

With the emergence of increasingly compact and precise sensors, the concept of Sensor Monitoring has become viable in the sense that a patient wearing a set of sensors can have a normal daily life without any discomfort or mobility difficulties. This technology is seen as very positive by the medical field as it allows to constantly control and monitor a patient's vital signs with no need for the him to be located inside any medical facility. Within this subject, the goal is to apply the concept of BAN as it is defined above, with some architecture specific variants that promote an intelligent, modular and flexible interaction between the different components.

The main challenge that arises from the use of sensors is the quality of the information received. One of the major problems here is the fact that sensors are not static and the user can, in fact, misuse them or damage them.

To monitor the sensors and measure the quality of the information provided by them, two different operating modules are considered: The first approach dictates that to communicate directly with the system's sensor hardware a low-level communication module is necessary. The module can be seen as a set of small software components that interact with a specific sensor and depend on the technology that the sensor uses to communicate with the monitoring device.

The second approach dictates that the agent-oriented platform must be able to read and filter the data coming through the sensors at a higher level. The agents that form this platform operate independently of the technology used by the sensors, which justifies the use of the first module mentioned before.

Every agent communicates directly with the respective software component that handles low-level communication to establish an information bridge and allow the agent to read and interpret the information.

The I/O Sensor Platform is thus responsible for encapsulating the different communication methods and protocols and will inform the system when a sensor enters or exits the BAN.

In Figure 4, a photo of the application running in a PDA is depicted. The information is from a live feed of the sensor platform and it is being processed in real-time. The data collected can be represented as the graph presented in the screen shows.

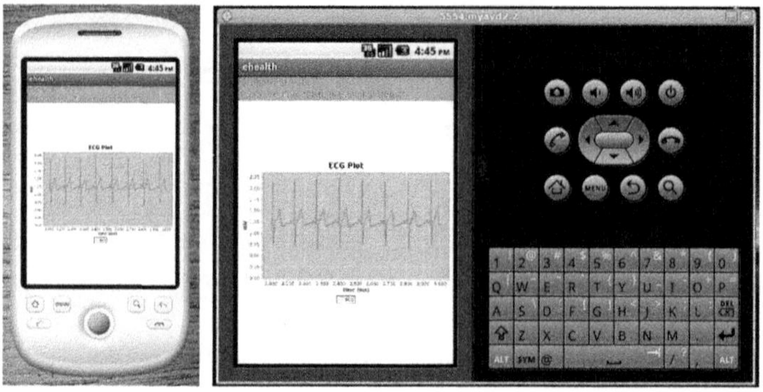

(a) Execution on Mobile Device. (b) Development interface.

Fig. 4 Two representations of the Advanced User Interface.

Currently we are testing the implementation of a portable ECG sensor platform in this system, which is reliable and more compact when compared with classical ECG sensors.

2.2 User Notification

Besides the monitoring component of the system, it is also convenient that the user of the device can be informed about his/her health condition. Moreover, location-based services can also be interesting in cases of emergency. This is implemented using several approaches, ranging from a notification to the user about his state using the interface of the application or audio alerts. With this level of flexibility, it is possible for a medical practitioner to fully support a patient regardless of his/her consciousness state.

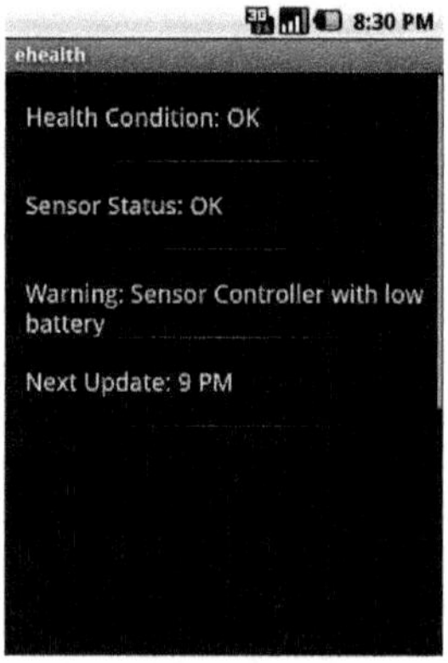

Fig. 5 User Mobile Interface

In the case of a user alert operation, the agent responsible for delivering this kind of information has a set of pre-determined definitions so that it can provide the user with very precise and clear information about his/her diagnosis.

This way, the user receives notifications if he/she needs some specific medical procedure concerning his/her clinical condition. On the other hand, if a text message or telephony operation is requested, the appropriate agents must be aware of the people directly involved in the patient's monitoring and therefore provide such configuration.

3 Results

The system presented in this paper essentially implements the collection of information on a ECG sensor, including the values of blood pressure and body temperature. The ECG data was collected by real sensors wore by a test subject. In addition, in order for more data to be available for testing purposes, a simulation platform was developed. This platform emulates data captured by an ECG sensor, following functions extracted from the real temporal series of values. The simulation can emulate both normal and abnormal health states and is running on an Android powered mobile device. The data collected about the user's body temperature and blood pressure value is used the same way as the one generated by the ECG simulation platform.

The development of the Android application and the Server has also the advantage of, by these being modular platforms, allow the services to be developed independently, without having to depend on the development of the sensor platform or any other module. Currently, the work focuses on the combined use of the Android platform and OSGi, with a special regard on the sensor fusion [15] theme.

From our point of view, the simulation platform is a first step that will support the rest of the development process. Its importance is even higher when we consider the risks of developing and testing in real environments, with real patients.

Another major feature is the fact that the project can communicate with other applications and projects in a transparent way. A concrete case is the integration with the iGenda project [12]. This agenda manager deals with the problems of automatic scheduling of events. In this case, the innovative factor is the inclusion of, for instance, the event of scheduling a visit to the physician or call the emergency system in case of an event involving critical readings of sensors or deterioration of the health condition.

4 Conclusion

The system presented in this paper covers different users (patients, physicians) and the use of a mobile device to capture the data and pre-process it, gives more mobility to the user, not interfering with his daily life. It is our conviction that the community has a lot to benefit from this project as it will improve the quality of life of people in need of constant health care.

Moreover, this project may indirectly relieve already congested medical institutions and services. Given the modular architecture of the system mostly relying on standard protocols of communication, the integration with other projects is possible, allowing to develop a wide range of services, focused on very different fields.

In future work we aim to explore a possible architecture for a complementary system to the mobile architecture presented here, which can achieve automatic and intelligent diagnose based on the information gathered. Moreover, another future goal is to provide the mobile device with learning skills This will focus on the interaction of the sensor agents and the data they gather from the sensors, and the user notification that will be taking place each time a diagnose is received by the mobile device.

Finally, a note about the use of a mobile device as a type of web-service with a set of properties and operations. This approach can be interesting in many other fields such as tourism, data synchronization or location-aware services.

References

1. Department of Economic and Social Affairs, Population Division: World Population Ageing. United Nations (2009)
2. Amini, S., Narasimhan, P.: Twitter Jacket, An automated activity and health monitoring solution for the elderly (2009)
3. Biodevices, S.A.: `http://www.biodevices.pt/VitalJacket`
4. Malan, D., Fulford-Jones, T., Welsh, M., Moulton, S.: CodeBlue: An Ad Hoc Sensor Network Infrastructure for Emergency Medical Care. In: MobiSys 2004 Workshop on Applications of Mobile Embedded Systems (2004)
5. Lehr, W., McKnight, L.W.: Wireless Internet access: 3G vs. WiFi. Telecommunications Policy 27(5-6) (2003)
6. Whitaker, R.M., Hodge, L., Chlamtac, I.: Bluetooth scatternet formation: A survey. Ad Hoc Networks 3(4) (2005)
7. Gislason, D.: Zigbee Wireless Networking. Newnes (2008)
8. Huang, H.-Y., Teng, W.-C., Chung, S.-L.: Smart Home at a Finger Tip: OSGi-based MyHome. In: Proceedings of the 2009 IEEE International Conference on Systems, Man, and Cybernetics (2009)
9. Sycara, K.: Multiagent Systems. Artificial Intelligence Magazine (1998)
10. Bagherzadeh, J., Arun-Kumar, S.: Flexible Communication of Agents based on FIPA-ACL. Electronic Notes in Theoretical Computer Science 159 (2006)
11. Latré, B., Braem, B., Blondia, C., Moerman, I., Demeester, P.: A Survey on Wireless Body Area Networks Wireless Networks. Springer, Heidelberg (2010)
12. Costa, Â., Novais, P., Costa, R., Corchado, J.M., Neves, J.: Multi-agent personal memory assistant. In: Demazeau, Y., Dignum, F., Corchado, J.M., Bajo, J., Corchuelo, R., Corchado, E., Fernández-Riverola, F., Julián, V.J., Pawlewski, P., Campbell, A. (eds.) Trends in PAAMS. AISC, vol. 71, pp. 97–104. Springer, Heidelberg (2010), ISBN 978-3-642-12432-7
13. Costa, R., Novais, P., Lima, L., Bulas-Cruz, J., Neves, J.: VirtualECare: Group Support in Collaborative Networks Organizations for Digital Homecare. In: Yogesan, K., Bos, L., Brett, P., Gibbons, M.C. (eds.) Handbook of Digital Homecare. Series in Biomedical Engineering (2009), ISBN: 978-3-642-01386-7
14. Gama, Ó., Carvalho, P., Afonso, J.A., Mendes, P.M.: QoS Deployment in Wireless e-Health and e-Emergency: Main Issues and a Case-study. In: Proc. 3th Symposium of Ubiquitous Computing and Ambient Intelligence 2008 (UCAmI 2008). AISC. Springer, Heidelberg (2008)
15. DeLong, P.R.: Interoperability & Sensor Fusion. Naval Engineers Journal 115 (2003)

Bluetooth-Based System for Tracking People Localization at Home

S. Orozco-Ochoa, X.A. Vila-Sobrino,
M. Rodríguez-Damián, and L. Rodríguez-Liñares

Abstract. Indoor location or positioning is not a mature technology but with the spread of wireless communication technologies applications are nowadays more feasible. In the telecare field, indoor location could be used for tracking people position at home, in order to know their daily routine and automatically detect abnormalities. This paper describes a Bluetooth-based method for indoor location, suitable for this kind of applications. The fingerprint algorithm uses the received signal strenght (RSS) parameter and a k-Nearest Neighbour classifier. For testing the algorithm an experiment was carried on a public educational building. The goal is to determine in which room a person is at any time. Results are satisfactory: 95% of correct classification.

1 Introduction

Recent years have witnessed a rapid commoditization of GPS hardware and related products, so that outdoor location is now accessible. Unfortunately, due to the fact that GPS coverage is good only for open spaces, indoor positioning must be based in other technologies [11]. The proliferation of mobile computing devices and local-area wireless networks has fostered a growing interest in using this infrastructure in location-aware systems and services.

One of the most promising areas of application of indoor location is telecare [2]. Telecare seeks to improve quality of life by assisting an independent living of elderly or disabled people at home. Moreover, in order to detect atypical situations, or even a progressive deterioration in physical or mental status [8], it would be useful to know the behavioral patterns of the person. Indoor location for telecare can be based on radio-frequency technologies (Bluetooth, RFID, Wi-Fi) or video cameras. This

S. Orozco-Ochoa · X.A. Vila-Sobrino · M. Rodríguez-Damián · L. Rodríguez-Liñares
Universidade de Vigo, Escola Superior de Enxeñería Informática
32004 Ourense, Spain
e-mail: {moroo,anton,mrdamian,leandro}@uvigo.es

A. Abraham et al. (Eds.): International Symposium on DCAI, AISC 91, pp. 345–352.
springerlink.com

paper describes a prototype of a Bluetooth-based indoor location system suitable for locating a person in his house; thus, the proposed location has a room level granularity. Our intention is to integrate this subsystem with other systems like vital signs monitoring or intelligent vision systems.

The paper is organized as follows: next section describes the state of the art of Indoor Location based on radio signal technology, section 3 presents the methodology used for training and validating the system, section 4 shows results of the experiment and finally, sections 5 and 6 have the discussion and conclusions.

2 State of the Art

GPS-based outdoor positioning systems are nowadays a mature technology. Receivers are integrated in car navigators, mobile phones, or even watches. However, developing robust indoor positioning systems is an open issue; as explained here with a brief overview of the state of the art of indoor positioning using radio signals. A deeper review can be found in [11].

Two strategies can be used to develop an indoor positioning system. The first one is based on especiallized hardware like RFID devices, ultrasound or ultra high frequency (UHF) band equipment; an example of this is the LANDMARC system [12]. The second approach uses existing wireless communication hardware like 802.11 WLAN or Bluetooth, and some distance dependent control parameter is chosen as a measurement unit. The RSS (Received Signal Strength) parameter, available both in WLAN and Bluetooth networks, has been widely used for this purpose. As signal strength decreases with distance, it can be used as an indirect measurement of the distance between the transmitter and the receiver. The 2 major drawbacks of this parameter are: noise and the non linear dependence on the distance.

Some indoor positioning systems are based on signal propagation models, e.g. RADAR [1]. But, the model definition is a challenge because the characterization of the radio channel is rather difficult; besides, the achieved accuracy is not very high. An alternative approach is fingerprinting. This method has two phases: an off-line phase, in which a map of observed RSS values (or radio map) is built, and an on-line phase, in which object location is obtained. The location process is based on data from the radio map and a classification technique. A well known classifier, which we chose for this work, is the k-Nearest Neighbour (kNN) algorithm [10]. The fingerprinting method is accurate and does not have the difficulties associated with propagation models; however, its deployment is generally more costly and its accuracy degrades easily due to the changing conditions of indoor environments.

The Bluetooth technology has attractive characteristics such as: low cost of devices, low power consumption and integration in a wide variety of portable devices. These characteristics make Bluetooth a good candidate for the system under developement. Work done in the field, like Forno's [5], presents a system in which transmitters vary the transmission power in order to achieve a more accurate localization. Chawathe [3] describes a methodology for optimizing the location of beacons. Genco [7] uses the *link quality* parameter in his learning system; it is based on genetic and neural network algorithms for training. All of these have a goal of accuracy improvement as much as possible. But, what we need is a system that, by working at room level granularity, shows a high percentage of correct classification.

3 Methodology

3.1 Experimental Setup

The experimental testbed is located within a public educational building. Its dimensions are 21 x 15 meters, and it includes offices for research students, laboratories and corridors. The area was divided in eight zones numbered 10, 20, 30, 40, 50 60, 70 and 80 (Fig. 1). There are 5 Bluetooth Access Points (APs) v2.1 USB 2.0 from Conceptronic. The APs are located to cover the whole area with overlapping zones. The mobile host is a Toshiba laptop M100-139 with Bluetooth 2.0 EDR.

The host is in Bluetooth 'inquiry mode' transmitting a discovery packet every 15 seconds, while APs are set to 'discoverable' responding to these packets with their identification (MAC Address) and RSS information. In order to get the RSS information from the Linux Bluetooth protocol stack (BlueZ), a python program is executed in the laptop. To cover the area, we marked 17 possible localizations for the laptop. These *inquiry points* are shown as x's in Fig. 1.

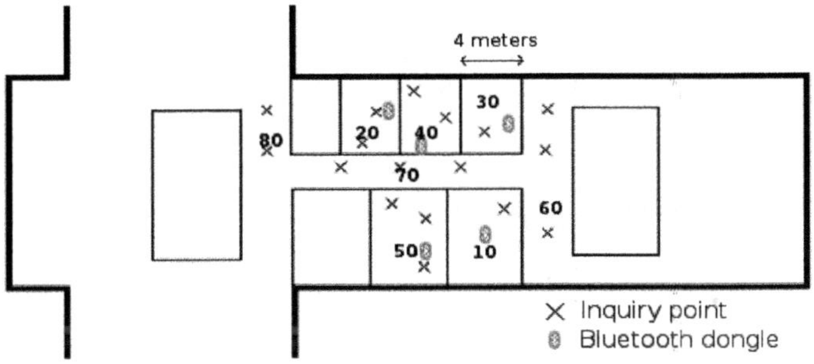

Fig. 1 Map of the experimental area with Bluetooth access points, zones and inquiry locations

3.2 Database

For each inquiry point m ($m = 1 \dots 17$), a total of 20 inquiries were performed (total 340 inquiries). Each inquiry i gives a set of vectors containing RSS values in dB magnitude, one vector per AP ($P = 1, \dots 5$).

$$\mathbf{O}^{p}_{mi} = \left\{ (o^{p}_{mi})^{(1)}, (o^{p}_{mi})^{(2)} \dots (o^{p}_{mi})^{(L_{mip})} \right\} \quad (1)$$

Lengths L_{mip} of vectors $\mathbf{O}^p_{mi}$ depend on the transmission conditions, and in our case they are between 0 (meaning out-of-range AP) and a maximum of 26. Instead of dealing with vectors of variable length, average values $\overline{o^p_{mi}}$ are calculated:

$$\overline{o^p_{mi}} = \begin{cases} \frac{1}{L_{mip}} \sum_{l=1}^{L_{mip}} (o^p_{mi})^{(l)} & L_{mip} > 0 \\ -128 & otherwise \end{cases} \tag{2}$$

where -128 is an arbitrary value and $-128 \ll \min\left((o^p_{mi})^{(l)}\right)$.

Then, our database is a matrix

$$\begin{pmatrix} \Omega_1 \\ \Omega_2 \\ \vdots \\ \Omega_{MI} \end{pmatrix} = \begin{pmatrix} \overline{o^1_{11}} & \overline{o^2_{11}} & \cdots & \overline{o^P_{11}} & C_1 \\ \overline{o^1_{21}} & \overline{o^2_{21}} & \cdots & \overline{o^P_{21}} & C_2 \\ \vdots & \vdots & & \vdots & \\ \overline{o^1_{MI}} & \overline{o^2_{MI}} & \cdots & \overline{o^P_{MI}} & C_{MI} \end{pmatrix} \tag{3}$$

with $P = 5$, $M = 17$, $I = 20$ and column C is the classification. This matrix was divided into a *training set* (inquiries $1 \ldots 10$) and a *validation set* (inquiries $11 \ldots 20$).

3.3 Classification Algorithm

For classification, the algorithm used is the kNN [9]. When an instance of the validation set is presented for evaluation, the algorithm computes its k closest neighbors in the training set. Then, a class is assigned to the instance by voting among those neighbors. In our case, the goal is to estimate the class (or location) of a given a vector from the evaluation set (Ω_i). The distance between vector Ω_i and each Ω_j from the training set is calculated as:

$$d(\Omega_i, \Omega_j) = \sqrt{\sum_{P=1}^{5} (\overline{o^P_i} - \overline{o^P_j})^2} \tag{4}$$

This Euclidean distance formula is used in the RSS-space to determine the k nearest neighbors.

Because the entire training data must be scanned to classify each test vector, instance-based learning is time-consuming for datasets of realistic size. Improved procedures such as *condense* and *reduce* [6] address this issue. *Condense* removes the data points which do not add more information, and show similarity with other training data. *Reduce* goes further by removing patterns which are not affecting the training result. These techniques make use of redundancy in data; besides, thet provide a minimal consistent subset of the training data. The subset gives the same performance in classification, as it would, if the full set was used. The outcomes from these methods are shown in section 4.1.

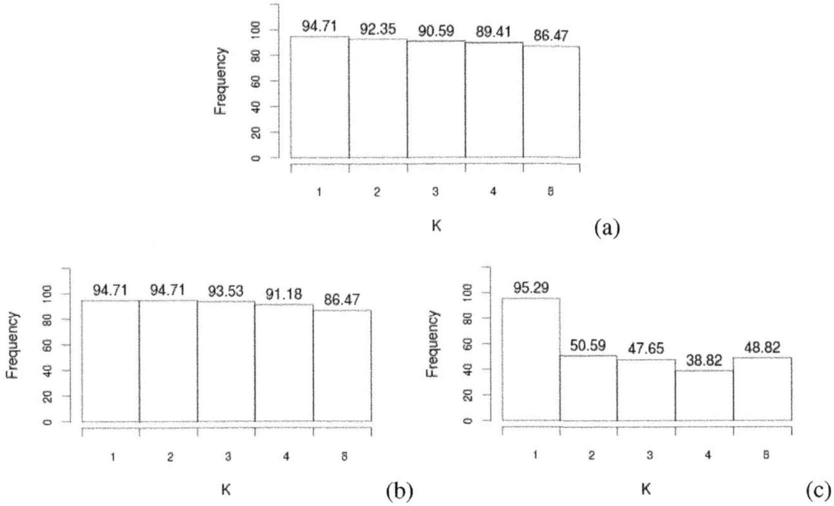

Fig. 2 Correct classification rates on: training set with cross-validation (a), validation set with full training set (b), validation set with the reduced set (c)

We used R[1], a well known software environment for statistical computing [4], for implementing the algorithm. R has many publicly available packages that extend core functionality, one of them is the package *class* [13] that includes classification functions.

4 Results

4.1 Evaluation of the Traning Set and Validation

Figure 2(a) shows results of cross-validation on the training set for k varying between 1 and 8. The best result is obtained when $k = 1$ (94.71% correct); then, the performance gradually decreases as k increases, 86.47% for $k = 8$.

Performance, in new situations and new data, must be tested once the capability of the training set for discriminating between different classes (locations) was evaluated. With this purpose, we devised a validation procedure with a new collection of data, the validation set. Fig. 2(b) shows correct classification rate for an increasing k. High correct classification rates are obtained, 94.71% (for $k = 1$, $k = 2$) down to 86.47% (for $k = 8$). In fact, we developed a 2-fold validation and results for the second fold were similar, 94.11% of correct classification for $k = 1$.

After applying *condense* and *reduce*, only 29 vectors are left in the training set. Fig. 2(c) shows correct classification rate for this reduced set. $K = 1$ yields good results (95.29% correct); but, for $k > 1$ performance drops to values around 50%.

[1] http://www.r-project.org/

4.2 Error Analysis

Given the small size of our training set, processing time is not a problem; therefore, we decided to use the full training set, and the value of parameter $k = 1$ for the kNN algorithm. The analisis of the errors shows which classes are most frequently confused and misclassified. Conclusions can be drawn from Table 1, such as:

- Class 10 shows 20% of misclassification, all assigned to adjacent class 50.
- Class 30 shows 10% of misclassification, all assigned to non-adjacent class 20.
- Class 40 shows 20% of misclassification, equally split between classes 20 and 30, both of them adjacent.
- Class 70 (corridor) with 6,6% of misclassification, equally split between adjacent classes 40 and 50.
- Classes 20, 50 and 60 have no misclassification error.

Table 1 Confusion matrix

	10	**20**	**30**	**40**	**50**	**60**	**70**	**80**
10	80	0	0	0	20	0	0	0
20	0	100	0	0	0	0	0	0
30	0	10	90	0	0	0	0	0
40	0	10	10	80	0	0	0	0
50	0	0	0	0	100	0	0	0
60	0	0	0	0	0	100	0	0
70	0	0	0	3.3	3.3	0	93.4	0
80	0	0	0	0	0	0	0	100

4.3 Accuracy

Bibliography about indoor location systems sometimes gives results in terms of *accuracy* or mean distance error. Although, it is not our goal to give the position in meters but in terms of *room number*; for comparison purposes the accuracy of our algorithm is calculated. As our classes correspond to rectangular areas of size between 20 and 45 square meters; we simulated our classes with circles of the same area. Estimation of the mean distance error depends on the results of the classification procedure as follows:

- If classification is correct, the mean distance error will be approximated by $r/2$. Being r the radius of the circle corresponding to the class.
- For points belonging to class A but assigned to an adjacent class B, the mean distance error will be approximated by $(r_A + r_B)$.
- For points belonging to class A but assigned to non-adjacent class B, the mean distance error will be approximated by $(r_A + r_B + \sum_{\langle c \rangle} 2r_C)$. Being $\langle c \rangle$ the set of classes between A and B.

This approximation along with the classification results from Table 1 give a mean distance error of 1.97 meters.

5 Discussion

In this paper we present our indoor location positioning system, designed for locating the room in which the person is. For $k = 1$, the system has a correct classification rate of 95%. By increasing the size of the training set, the error rate may be further reduced. Analysis of errors shows that, except in one case, all of them correspond to misclassification of adjacent classes. In a real situation most of these errors could be corrected using *trajectory* information. For instance, when going from room 30 to 40, 70 must be crossed; hence, a simple set of rules or *restrictions* describing the allowed *transitions* should improve performance.

Validation results also show that it is possible to use methods for reducing the training set while maintaining the performance. The reduced set has 29 samples, about 3 samples per class. From such a small sample, it can be understood why performance drops for values of $k > 1$.

We have an accuracy of 1.97 meters. In case of no errors, accuracy would be 1.45 meters. Therefore, it is possible to obtain even better accuracy by defining smaller regions, in that case each room should be subdivided in smaller zones. Of course, error rates would be higher since there is a trade-off between accuracy and precision. As a comparison, works like that of Forno [5] show an accuracy of 1.8 meter; RADAR's [1] accuracy is about 2-3 meters, and the comercial system Ekahau[2] has a room level accuracy such as ours.

6 Conclusions and Future Work

A system suitable for locating people at home using Bluetooth was presented in this paper. We showed that this Bluetooth-based system offers good precision for our purposes, 95% of correct classification within 2 meters accuracy. Moreover, features of Bluetooth such as: the chips come embedded in many devices and the low power consumption, make it an affordable technology. On the other hand, WLAN is a widely spread technology and the infrastructure is present in many public areas; therefore, we are exploring the possibility of merging information from Bluetooth and WLAN access points with fusion techniques. The system we proposed can be easily deployed in real world conditions, so we are developing a mobile phone application with our indoor positioning algorithm.

Acknowledgements. This work was supported by the Xunta de Galicia under the grant 08SIN002206PR.

References

1. Bahl, P., Padmanabhan, V.N.: Radar: an in-building rf-based user location and tracking system. In: Proc. 19th Annu. Joint Conf. on Computer and Communications Societies, vol. 2, pp. 775–784 (2000)

[2] http://www.ekahau.com/

2. Botsis, T., Hartvigsen, G.: Current status and future perspectives in telecare for elderly people suffering from chronic disease. Journal of Telemedicine and Telecare 14(4), 195–203 (2008)
3. Chawathe, S.S.: Beacon placement for indoor localization using bluetooth. In: 11th Int. IEEE Conf. on Intelligent Transportation Systems ITSC, pp. 980–985 (October 2008)
4. Dalgaard, P.: Introductory statistics with R, 2nd edn. Springer, New York (2008)
5. Forno, F., Malnati, G., Portelli, G.: Design and implementation of a bluetooth ad hoc network for indoor positioning. IEE Proc. Software 152(5), 223–228 (2005)
6. Gates, G.W.: The reduced nearest neighbor rule (corresp.). IEEE Transactions on Information Theory 18(3), 431–433 (1972)
7. Genco, A.: Three Step Bluetooth Positioning. In: Strang, T., Linnhoff-Popien, C. (eds.) LoCA 2005. LNCS, vol. 3479, pp. 52–62. Springer, Heidelberg (2005)
8. Hayes, T.L., Pavel, M., Kaye, J.A.: An unobtrusive in-home monitoring system for detection of key motor changes preceding cognitive decline. In: 26th Annu. Int. Conf. IEEE on Engineering in Medicine and Biology Society EMBS, vol. 4, pp. 2480–2483 (2004)
9. Izenman, A.J.: Modern Multivariate Statistical Techniques. Springer, New York (2008)
10. Lin, T., Lin, P.: Performance comparison of indoor positioning techniques based on location fingerprinting in wireless networks. In: Int. Conf. on Wireless Communications, Networking and Mobile Computing, vol. 2, pp. 1569–1574 (June 2005)
11. Liu, H., Darabi, H., Banerjee, P., Liu, J.: Survey of wireless indoor positioning techniques and systems. IEEE Transactions on Systems, Man and Cybernetics, Part C (Applications and Reviews) 37(6), 1067–1080 (2007), `http://dx.doi.org/10.1109/TSMCC.2007.905750`
12. Ni, L.M., Liu, Y., Lau, Y.C., Patil, A.P.: Landmarc: indoor location sensing using active rfid. In: Proc. 1st IEEE Int. Conf. on Pervasive Computing and Communications PERCOM, pp. 407–415 (March 2003)
13. Venables, W.N., Ripley, B.D.: Modern Applied Statistics with S, 4th edn., p. 495. Springer, New York (2002)

A Case Study on Agriture: Distributed HLA-Based Architecture for Agricultural Robotics

Patricio Nebot, Joaquín Torres-Sospedra, and Rafael Martínez

Abstract. In agricultural robotics, as in other robotic systems, one of the most important parts is the control architecture. This paper describes the definition of a new control architecture specially designed for groups of robots in charge of doing maintenance tasks in agricultural environments. This architecture has been developed having in mind principles as scalability, code reuse, abstraction hardware and data distribution. Moreover, it is important that the control architecture can allow coordination and cooperation among the different elements in the system. The architecture presented in this paper implements all these concepts by means of the integration of different systems, such as *Player*, *JADE* and *HLA*. The most important system is *HLA* because it not only allows the data distribution and implicit communication among the parts of the system, but also allows to operate with simulated and real entities at the same time, allowing the use of hybrid systems in the development of applications.

1 Introduction

Robotic applications have been widely spread to many different application fields in the last years. One of them could be the implementation of agricultural robotic systems in charge of doing a great variety of applications such as navigation into orange groves for maintenance tasks as spraying, detection of illnesses, detection and elimination of weeds, fertilize, or simply explore the grove to eliminate rests of trees by the pruning.

Patricio Nebot · Joaquín Torres-Sospedra
Departamento de Ingeniería y Ciencia de los Computadores, Universitat Jaume I,
Avda. Sos Baynat S/N, Zip Code 12071, Castellón, Spain
e-mail: `{pnebot,jtorres}@icc.uji.es`

Rafael J. Martínez
Robotics Institute, University of Valencia,
Apdo. Correos 2085, Zip Code 46071, Valencia, Spain
e-mail: `rafael.martinez@uv.es`

A. Abraham et al. (Eds.): International Symposium on DCAI, AISC 91, pp. 353–360.
springerlink.com

In the bibliography we can find some approaches of robotic systems used in the agriculture to perform different maintenance tasks, [1, 4, 3, 10, 9]. In general, most of the approaches are based on the use of a single complex robot which performs specific actions on the environment, but we have not found any cooperative system, with some different robots, for agricultural tasks.

In a cooperative robotic system, the control architecture is one of the most important parts to develop. This control architecture must give support for all the facilities of the system and forms the backbone of the robotic system. The right choice of the architecture can facilitate the specification, implementation and validation of the applications implemented for the system. In the literature, we can found some important architectures which are able to be used in our cooperative system.

- **Robot Operating System** (*ROS*): It is a framework for robot software development, providing operating system-like functionality on top of a heterogeneous computer cluster. This framework was originally developed in 2007 by the Stanford Artificial Intelligence Laboratory in support of the *Stanford AI Robot* project under the name *switchyard*. Although it is intended to be cross-platform, it is only fully supported by a few linux distributions.
- **Orca / Orocos:** *Orca* is another open-source framework, released under *LGPL* and *GPL* licenses, for developing component-based robotic systems. It was initially a part of an EU sponsored project, *OROCOS 2002*, but it was renamed to *ORCA*. This framework is based on *CORBA* and it provides the means for defining and developing the parts of complex robotic systems.
- **Umbra:** *Umbra* is a framework for modeling and simulation. This framework allows generating models and simulations for intelligent system development, analysis, experimentation, and control and supports the analysis of complex robotic systems. The models in Umbra include 3D geometry and physics of robots and environments. Model components can be built with varying levels of fidelity, so the models built with low fidelity for conceptual analysis can be gradually converted to high fidelity models for later phase detailed analysis. Within control environments, the models can be easily replaced with actual control elements.

All these alternatives are well known and widely used, and implement important concepts as abstraction of the hardware, code reuse, scalability, and some of them implement the capability to manage the possible coordination or cooperation among different entities which could be involved in the system to perform the different tasks. Moreover, some of them implement systems for allowing communication, implicit or explicit, between the elements in the system.

The research described in this paper is focused in the development of a cooperative system to control a team of mobile robots. The intention, is to use the system proposed in agricultural environments developing different maintenance tasks. As mentioned before, in any robotic system it is necessary to choose a control architecture giving support for the system. However, none of the architectures reviewed before fit well in our cooperative system requirements. In our case, including the capabilities of the other architectures, it is also desirable to have the architecture based on international standards and allowing simulations of the elements of the system in order to deploy or debug the applications without the necessity to move to the outdoor scenario. These concepts are not present in the cited architectures. For this reason we introduce a new control architecture called "*AGRIcultural architecTURE*" (*Agriture*), which integrates three different systems (*Player* [6], *JADE* [8] and *HLA* [7]) that interact among them to provide a solid control architecture for developing cooperative robotic system.

Following, a description of the proposed architecture, and the different levels which compose it, is presented in section 2. In section 3, some possible distributed applications are depicted. These applications show how the system makes use of some of the capabilities of the proposed system. Finally, the conclusions are presented, with some future research lines.

2 Agriture: Our Proposed Architecture

As mentioned, the purpose of this work is to develop a system able to control a team of mobile robots in agricultural environments. This system is based on the *Agriture* architecture which is the backbone for the system. *Agriture* as explained, is based in three subsystems: Player, JADE and HLA, each one giving different features to the architecture.

The most interesting system added to the architecture is the *HLA* (*High Level Architecture*) subsystem. *HLA* is an architecture to create complex simulations using simple components and it is defined under *IEEE Standard 1516-2010*. HLA simulations are divided into federates, "A *HLA* federate can be a computer simulation, a manned simulator, a supporting utility (such as a viewer or data collector), or even an interface to a live player or instrumented range" [5]. This property of the HLA systems opens a very big field of applications. It allows the use of both, real and simulated entities, at the same time, being these entities robots, environments, or any part of the system able to be simulated. This is specially useful when the system is optimized for a specific application, like the case of agricultural robotics. The relationship among the different parts is depicted in figure 1.

2.1 Architecture Design

Our new architecture, shown in figure 2, is composed of three layers. The physical layer is related to all the real or simulated devices (robots, cameras, ...) of our system. Each of them has its own specifications and communication protocols. They are the agents which interact with the real or simulated environment.

In the architecture middleware cooperation and coordination, communication between the elements and abstraction of the external devices (elements) are implemented. This middleware involves three parts, one of them is in charge of implicit communication and establishes the links and data exchange format between the

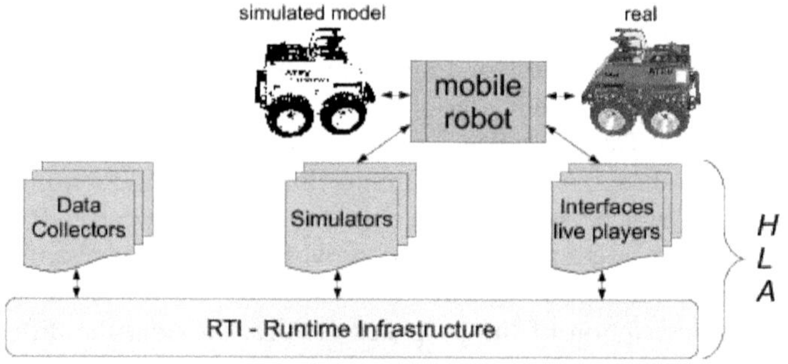

Fig. 1 HLA offers the possibility to employ at the same time real and simulated entities

different elements; another part is related to explicit communication, it is in charge of determining which information is transmitted and how the tasks are assigned to physical elements; the last one provides machine independence to the system because each robotic system is abstracted. The system does not work directly with the real robots, their abstraction allows to use basic commands to control the robots. In this way, the robot can be easily replaced by a different robot without having to recode the middleware. Only the abstraction module related to the robot has to be replaced.

Finally, the highest part of our architecture is composed by the distributed applications. These distributed applications are controlled by the middleware layer, which is in charge of assigning the different tasks to each specific element or device.

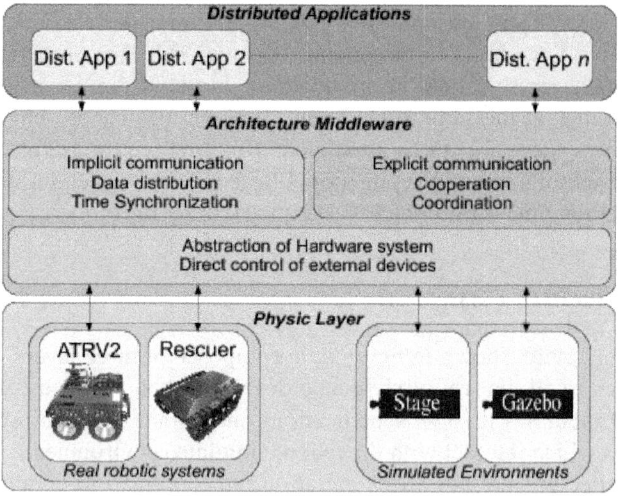

Fig. 2 The new distributed HLA-based architecture

2.2 *Architecture Implementation*

2.2.1 Physical Devices / Hardware

The physical layer of the architecture contains all the hardware elements of our system. They are directly managed by *Player*. In the experiments on orange groves, the following devices will be used:

- **ATRV-2:** The *ATRV-2* is a rugged four-wheel drive, differentially steered all-terrain robot vehicle for outdoor robotic research and application development. It is stable in wide varied terrains and it can traverse them easily.
- **RESCUER:** It is a mobile platform extremely solid and very appropriate for outdoor environments, even hazardous to access. It is appropriate for scientific research and some civil protection agencies use them.
- **GPS:** Each robot has is own *GPS* module, concretely a *GPS Pathfinder ProXT Receiver*. This device is composed by a sub-meter precision GPS receiver, an antenna, and an all-day battery. Moreover, it is totally cable-free, rugged and weatherproof, so it suitable for outdoor environments such as orange groves.
- **Vision system:** The vision system is composed by a *VGA FOculus* camera and an autoiris *Cosmicar* lens which can be used in outdoor environments.
- **LASER detector:** To perform the detection we have used a *HOKUYO's LASER* system. This detector provides a field-of-view around 240 degrees and its angular resolution is close 36 degrees with a scanning refresh rate of up to 10Hz.
- **WiFi Network:** This device is used to allow the wireless communication between the different robots.

2.2.2 The Middleware Layer

In this subsection, the three different systems which forms the middleware layer are described.

- **JADE:** JADE is a software framework designed to develop agent-based applications in compliance with the *FIPA* specifications for interoperable intelligent multiagent systems. It is a software framework fully implemented in Java language and simplifies the implementation of multiagent systems through a middleware and through a set of graphical tools that supports the debugging and deployment phases [2]. The *JADE* middleware implements an agent platform for execution and a development framework. Also, it provides some agent facilities such as life cycle management, naming service, message transport and parsing service, and a library of *FIPA* interaction protocols ready to be used [2].
- **The High Level Architecture:** *HLA* provides a general framework in which the researchers can structure and describe their final applications. In any case, *HLA* is neither a simulator nor a modelling tool. Furthermore, *HLA* does not generate data or simulations for you because it does not eliminate programming.
  The main objective of *HLA* is to generate systems (*Federations*) based on *reusable* components of different nature (*Federates*) which can *interact* among

them easily through a distributed, real-time operating system. To perform this task, there are three main components in the *HLA* architecture:

- A set of rules: *HLA* consists of a set of ten rules which must be obeyed in order to govern the overall system and to govern each participating component.
- *An Object Model Template:* The *Object Model Template*, *OMT*, provides a standard for defining and documenting the form, type and structure of the information shared within a simulation.
- *Interface Specifications:* The *Interface Specifications*, *IS*, describes the run-time services between each federate and the *RTI*. There are six classes of services. The *HLA Interface Specification* defines the way these services are accessed in an application programmer's interface (API).

- **Player/Stage:** One of the most widely used software nowadays for programming multirobot applications is the *Player/Stage* project [6]. *Player* is a network oriented device server that provides clients with network-oriented programming interfaces to access actuators and sensors of a robot. It employs a one-to-many client/server style architecture where one server serves all the clients of a robot's devices and it relies on a TCP-based protocol to handle communication between client and server. Accompanying *Player* is the robot simulator Stage, a lightweight, highly configurable robot simulator that supports large populations of robots and allows programmers to control these virtual robots navigating inside a virtual environment.

In the system, each one of the subsystems is in charge of a main task. In that way, the coordination and cooperation task, as well as the explicit communications, are relied to *JADE*, whereas the implicit communication, the data distribution and the control of the simulated entities are assigned to *HLA*. Finally, *Player* is in charge of the hardware abstraction and the control of the real entities in the system.

3 Possible Distributed Applications of Our Architecture

Our architecture can interact with real and simulated devices. Depending on the final application, a concrete device (real or simulated) can be used. For instance, an application can be run in a real orange grove in real time. After this running, some simulations can be performed off-line with the information related to this grove in order to optimize some parameters of the system.

Furthermore, we can consider the use of our architecture for hybrid applications in which real and simulated elements are used together.

In a first application, the real robots move in a free-obstacle environment (Fig. 3). The orange grove can be simulated but the interaction and navigation of the different robots can be real. For instance, the robots can be located into a warehouse but the information provided by the camera and GPS can be used to simulate that they are navigating into a real orange grove. In this application, the robot behavior in an orange grove can be tested without being there. In this case, the real robots will navigate into the simulated grove without the risk of damaging people. Moreover, the economical costs of optimizing the system (testing different algorithms for a task) are reduced because the outdoor field experiments are avoided.

Fig. 3 First application: HLA offers the possibility to virtualize an orange grove into a warehouse

The second hybrid application (fig. 4) consists in using both, real and virtual devices, in a real environment. In this system, a real robot could perform a virtual task in parallel with other real tasks. For instance, the orange threes can be virtually sprayed or the weeds located in the path can be virtually removed while the robot is navigating through the grove. In this way, we can test some different algorithms to perform a new task while the robot is running a real application. The real devices, such as cameras and other sensors can provide real information to the new virtual element, so the simulation can be more realistic. It can be also useful in the case in which the availability for simulations of the real devices is low, because we need the resources to perform a real application. In this case, the new virtual applications can be tested, in situ, while the required real task is being done.

Fig. 4 Application 2: HLA offers the possibility to virtualize an element/device of a real robot

4 Conclusions

In this paper, *Agriture*, a new architecture to implement an agricultural multirobotic system has been introduced. In *Agriture*, *Player/Stage* is in charge of the hardware (real or simulated) abstraction. The coordination and cooperation task are relied to *JADE* whereas the implicit communication is assigned to *HLA*. They have been selected to improve the *reusability*, *scalability* and *interoperability* of system.

Furthermore, we have introduced the use of *Agriture* either in real or simulated environments. It can be used in real applications or it can be applied to simulate experiments. In addition, the proposed architecture can be used in hybrid systems in which real and simulated elements (devices and environments) can interact.

Acknowledgments

This paper describes research carried out at the *Robotic Intelligence Laboratory* of *Universitat Jaume-I* and the *Institut de Robótica* of the *Universitat de Valencia*. Support is provided in part by the Generalitat Valenciana under project GV/2010/087, and by the Fundació Caixa Castelló - Bancaixa under project P1-1A2008-12.

References

1. Astrand, B., Baerveldt, A.J.: An agricultural mobile robot with vision-based perception for mechanical weed control. Auton. Robots 13, 21–35 (2002)
2. Bellifemine, F., Caire, G., Poggi, A., Rimassa, G.: Jade a white paper. EXP in Search of Innovation (Special Issue on JADE) 3(3), 6–19 (2003)
3. Blasco, J., Aleixos, N., Roger, J.M., et al.: Automation and emerging technologies: Robotic weed control using machine vision. Biosystems Engineering 83(2), 149–157 (2002)
4. Cho, S.I., Lee, D.S., Jeong, J.Y.: Automation and emerging technologies: Weed-plant discrimination by machine vision and artificial neural network. Biosystems Engineering 83(3), 275–280 (2002)
5. Dahmann, J.S., Calvin, J.O., Weatherly, R.M.: A reusable architecture for simulations. Commun. ACM 42(9), 79–84 (1999)
6. Gerkey, B., Vaughan, R., Howard, A.: The player/stage project: Tools for multi-robot and distributed sensor systems. In: Int. Conf. on Advanced Robotics, pp. 317–323 (2003)
7. IEEE: Peruse ieee p1516.2, draft standard for modeling and simulation (ms) high level architecture (hla)
8. Lab, T.I.: Jade - java agent development framework, `http://jade.tilab.com`(last Visited: 10/09/2007)
9. Li, Z., An, Q., Ji, C.: Classification of weed species using artificial neural networks based on color leaf texture feature. In: Computer and Computing Technologies in Agriculture II, vol. 2, pp. 1217–1225. Springer, Boston (2009)
10. Nejati, H., Azimifar, Z., Zamani, M.: Using fast fourier transform for weed detection in corn fields. In: IEEE Int. Conf. on Systems, Man and Cybernetics, pp. 1215–1219 (2008)

Tracking a Mobile Target Using Visual Servoing and Estimation Techniques

Carlos Alberto Díaz-Hernández,
José Luis Muñoz-Lozano, and Juan López-Coronado

Abstract. This paper is concerned with the visual servoing of a mobile robot in dynamic environments. We assume a target with maneuvering capabilities, and thus it can be hidden from the camera by the obstacles in the scene. These two problems must be taken into account in the control law to ensure correct servoing. The control law must consider the target movement to reduce tracking errors as small as possible. Moreover, the control law should consider visual loss management (reconstruction of the visual signal in case of occultation), and collision avoidance, estimating the obstacle motion. Here, we present a strategy to avoid the tracking error due to the movement of the target itself.

1 Introduction

Visual servoing provides very efficient solutions to control robot motions from a initial position to a precise goal [1]. It supplies high accuracy and good robustness to noise in image processing, camera calibration, and other setting parameters.

This work is concerned with the visual servoing of non-holonomic mobile robot in dynamic environments. A pan-camera, attached to the mobile robot, is used to perform the task of tracking a mobile target. Target dynamics contributes on visual error, so with the aim of reducing the tracking error, we must compensate the command, that is computed from visual information. The compensation is based on an estimate of the target movement.

In the literature there are several solutions to estimate target motion: in Ref. [2] a visual controller is presented, that is based on a classical method in automatic control to cancel tracking errors consists of compensating the target motion through an integral term in the control law. This scheme allows the controller to cancel the tracking errors only if the target has a constant velocity. Other approach, as in Ref. [3] that is based on feed forward control, estimates directly the velocity through the image measurements and the camera velocities.

Carlos Alberto Díaz-Hernández · José Luis Muñoz-Lozano · Juan López-Coronado
Universidad Politécnica de Cartagena, Cartagena 30202, España
e-mail: {ca.al.di.he, joselu.mlozano, jl.coronado}@gmail.com

A. Abraham et al. (Eds.): International Symposium on DCAI, AISC 91, pp. 361–368.
springerlink.com

This article is structured as follows. In section 2 we present the model of the serving robot. In section 3 we propose a first approach to control law that takes into account the target movement. Validation of the precedent control law is discussed in section 4. Finally, in section 5 we present the conclusions.

2 Modeling

We consider a robotic configuration composed of a non-holonomic robot equipped with a camera on a pan-platform (see Fig. 1). The configuration of the mobile base is described by a vector $Q = [x, y, \theta]^T$, where the pair (x, y) denotes the coordinates of the robot reference point M with respect to the scene frame $R_O\,(O, \overrightarrow{x}, \overrightarrow{y}, \overrightarrow{z})$ and the angle θ represents the orientation of the mobile base to the axis $(O, \overrightarrow{x})$. $R_M\,(M, \overrightarrow{x}_M, \overrightarrow{y}_M, \overrightarrow{z}_M)$ denotes the frame linked to the mobile base. P denotes the center of rotation of the pan-platform and D_x is the distance $\overline{MP}$. $R_P\,(P, \overrightarrow{x}_P, \overrightarrow{y}_P, \overrightarrow{z}_P)$ and $R_C\,(C, \overrightarrow{x}_C, \overrightarrow{y}_C, \overrightarrow{z}_C)$ represent respectively the frames attached to the pan-platform and the camera. θ_{Pl} defines the orientation of R_P in relation to R_M. The camera's optical center, denoted C, has the coordinates $[a, b, c]^T$ in R_P. Vector $\overrightarrow{z}_C$ matches up with the optical axis of the camera. Defining the vector $q = [l, \theta, \theta_{\text{Pl}}]^T$,

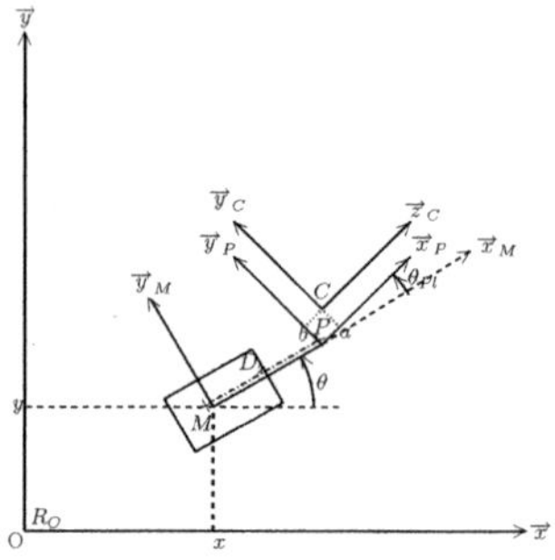

Fig. 1 The mobile robot with pan-platform.

where l denotes the curvilinear abscissa of M, the velocity screw of the camera with respect to R_O is expressed in R_C as:

$$T_C = J_q \dot{q}, \tag{1}$$

where

$$J_q = \begin{pmatrix} 0 & 0 & 0 \\ -\sin\theta_{\text{Pl}} & a + D_x \cos\theta_{\text{Pl}} & a \\ \cos\theta_{\text{Pl}} & -b + D_x \sin\theta_{\text{Pl}} & -b \\ 0 & -1 & -1 \\ 0 & 0 & 0 \\ 0 & 0 & 0 \end{pmatrix},$$

and

$$\dot{q} = [v, \omega, \omega_{\mathrm{Pl}}]^T .$$

The null rows of the Jacobian matrix J_q define the restricted movements of the camera. Thus, the velocity screw is reduced, by retaining only the controllable degrees of freedom (DOF) of the camera, as expressed in Eq. (2).

$$T_{\mathrm{r}} = J_{\mathrm{r}}\dot{q}. \tag{2}$$

where

$$J_{\mathrm{r}} = \begin{pmatrix} -\sin\theta_{\mathrm{Pl}} & a + D_x\cos\theta_{\mathrm{Pl}} & a \\ \cos\theta_{\mathrm{Pl}} & -b + D_x\sin\theta_{\mathrm{Pl}} & -b \\ 0 & -1 & -1 \end{pmatrix}.$$

In the pin-hole camera model, scene points are projected onto the image plane through a perspective projection. Thus, if $\underline{x}_p = [x, y, z]^T$ represents the coordinates of a point p of the scene in R_C, then, its projection P in the image plane in metric coordinates is $\underline{X}_P = [X, Y, f]^T$ computed by:

$$\underline{X}_P = \frac{f}{z}\underline{x}_p, \tag{3}$$

where f represents the focal length of the camera.

Let s the vector associated with the observed visual pattern. The velocity of the visual features, $\dot{s}$, in the image is related to T_C by:

$$\dot{s} = L_s T_c, \tag{4}$$

where L_s denotes the interaction matrix which is defined by Samson and Espiau [4]. This matrix depends on the nature of the visual characteristics considered (points, lines, etc.) and depth of the target from the camera.

The target is modeled by the four-corners of a square, thus s is defined by a vector containing the coordinates (X_i, Y_i) of these points. In the case of a single point, Chaumette [5] defines L_s as:

$$L_s = \begin{bmatrix} \frac{-1}{z} & 0 & \frac{X}{z} & XY & -(1+X^2) & Y \\ 0 & \frac{-1}{z} & \frac{Y}{z} & 1+Y^2 & -XY & -X \end{bmatrix}. \tag{5}$$

In order to match the dimensions of L_s and T_{r}, we remove the columns of L_s corresponding to the not allowed movement of the camera. Thus, the reduced interaction matrix is written as:

$$L_s = \begin{bmatrix} 0 & \frac{X}{z} & XY \\ \frac{-1}{z} & \frac{Y}{z} & 1+Y^2 \end{bmatrix}. \tag{6}$$

The final interaction matrix for the chosen pattern is:

$$L_s = \left[L_{s1}^T, L_{s2}^T, L_{s3}^T, L_{s4}^T\right]^T . \tag{7}$$

3 Control

The task function e is computed from the visual features [4, 6]:

$$e(q,t) = C(s(q,t) - s^*), \tag{8}$$

where s is the current value of the visual features for task e and s^* its desired value; C is the combination matrix and allows to enclose more information to compute the command than the number of camera DOF. A practical solution is to define $C = \widehat{L_s}^+$, where $\widehat{L_s}^+$ is the left pseudo-inverse of an estimated of L_s.

Usually, the control law is obtained from the following equation that constrains to set an exponential decoupled decreasing of the task function:

$$\dot{e} = -\lambda e, \tag{9}$$

where $\lambda > 0$. From Eq. (8), we obtain:

$$\dot{e} = \frac{\partial e}{\partial q}\dot{q} + \frac{\partial e}{\partial t} = CL_s J_r \dot{q} + \frac{\partial e}{\partial t}, \tag{10}$$

where $\frac{\partial e}{\partial t}$ represents the variation of e due to target movement.

Taking into account the Eqs. (9) and (10), Chaumette [5] deduces the next control law:

$$\dot{q} = J_r^{-1}\left(-\lambda e - \widehat{\frac{\partial e}{\partial t}}\right). \tag{11}$$

Chaumette [5] estimates $\frac{\partial e}{\partial t}$ by the recurrence:

$$\left(\widehat{\frac{\partial e}{\partial t}}\right)_{k+1} = \left(\widehat{\frac{\partial e}{\partial t}}\right)_k + \mu e_k, \tag{12}$$

where $\left(\widehat{\frac{\partial e}{\partial t}}\right)_0 = 0$ and μ is a predefined gain. This estimator, type integrator, is based solely on successive measurement values of e, which is also based on visual characteristics.

Bensalah [3] has proposed a measure of the apparent movement of the target in the image, this measure is based on the difference between total variation of e and the change of e due to camera motion.

$$\left(\widehat{\frac{\partial e}{\partial t}}\right) = \hat{\dot{e}} - \left(\widehat{\frac{\partial e}{\partial q}}\right) T_C. \tag{13}$$

In the discrete domain (period Δt), it is expressed as:

$$\left(\widehat{\frac{\partial e}{\partial t}}\right)_k = \frac{e_k - e_{k|k-1}}{\Delta t} - \left(\widehat{\frac{\partial e}{\partial q}}\right)(T_C)_{k-1}, \tag{14}$$

where $e_{k|k-1}$ is the predicted value for e_k computed from e_{k-1} and the applied velocity screw at $k-1$ instant. Equation (14) is the measure equation for the evolution models of $\frac{\partial e}{\partial q}$, which are the basis for a Kalman filter.

4 Results

We present in this section the control law evaluation for tracking task. The mobile camera must track a visual target (a rectangle of four points easily detectable). The target robot with the visual pattern follows a straight line path with a constant velocity.

The reference pattern is defined by the vector $s^* = [\,-0.1,\ 0.1,\ -0.1,\ -0.1,\ 0.1,\ -0.1,\ 0.1,\ 0.1]^T$. This pattern will be observed by the camera only when 1) the axis $(C, \vec{z})$ pass through the square pattern center, 2) the image plane is parallel to pattern plane and 3) the camera's optical center is located to desired distance $d^* = 2m$ from the pattern plane.

The visual servoing has been evaluated, taking into account that $\partial e/\partial t = 0$ and estimating $\partial e/\partial t$ through the Eqs. (12) and (14).

Figures 2, 4, and 7, show the evolution of s. In all these graphs, we make comments on the left superior point of s. s^* is drawn as *reference point*. *Starting point* and *ending point* correspond to the observed points at initial and final instant of simulation, respectively. The line, connecting the starting and ending points, represents the evolution of the point. We encircle the ending point and reference point to highlight the error.

Figures 3, 5, and 8 display the error $\varepsilon = s - s^*$ for $\underline{X}_1$. The vertical and horizontal error components are plot in dash and solid lines, respectively. A single point has been represented, because the absolute value of the vertical and horizontal components of each point has the same behavior and it makes easier the visualization.

Figures 6 and 9 show the estimated value of the target velocity in the image plane, denoted as $\frac{\partial s}{\partial t}$. In a similar way as in Fig. 3, the components corresponding with $\underline{X}_1$ are visualized, in dash and solid lines for the vertical and horizontal components, respectively.

4.1 Visual Tracking without Target Movement Estimation

The control law has been evaluated taking into account that the target movement is not meaningful. Under this configuration, the task is not correctly realized, because not all the error components converge to zero, specifically those components related with the horizontal error observed in the image plane, which converge in magnitude

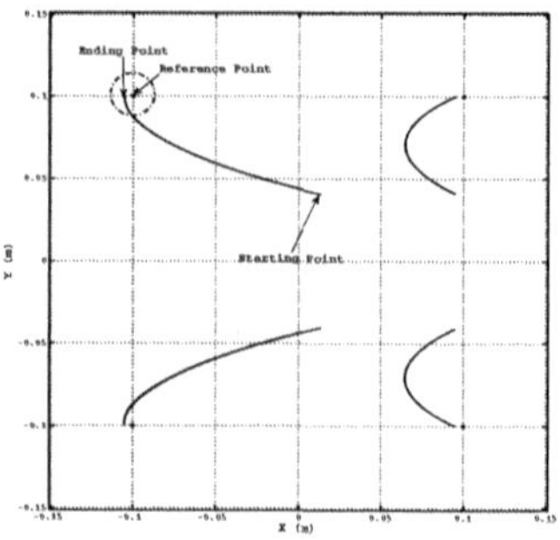

Fig. 2 Evolution of visual index s considering $\partial e/\partial t = 0$. Observed error between the *reference point* and the *ending point* is in the horizontal components of s (encircled).

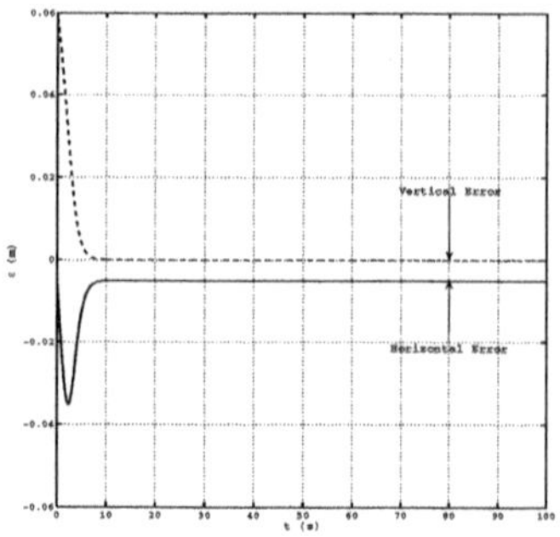

Fig. 3 Evolution of visual error ε considering $\partial e/\partial t = 0$ (only one point is plotted). At steady state, the vertical component is null, however, its horizontal component converges to a constant value proportional to target velocity in R_C.

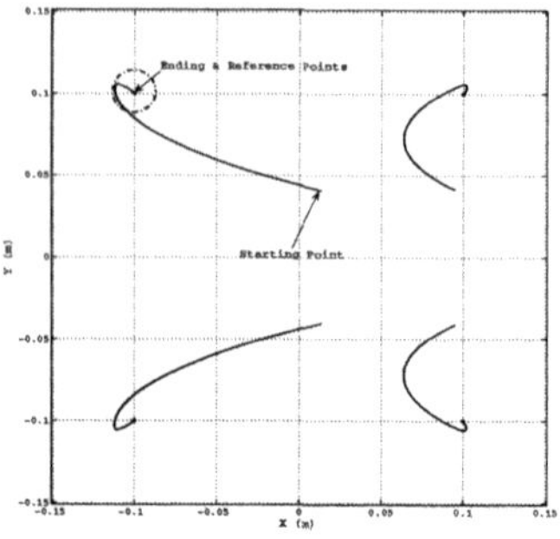

Fig. 4 Evolution of visual index s using Chaumette estimator. Differently to Fig. 2, the *ending point* coincides with the *reference point*, we can see that s exceeds s^* before to attain s^*.

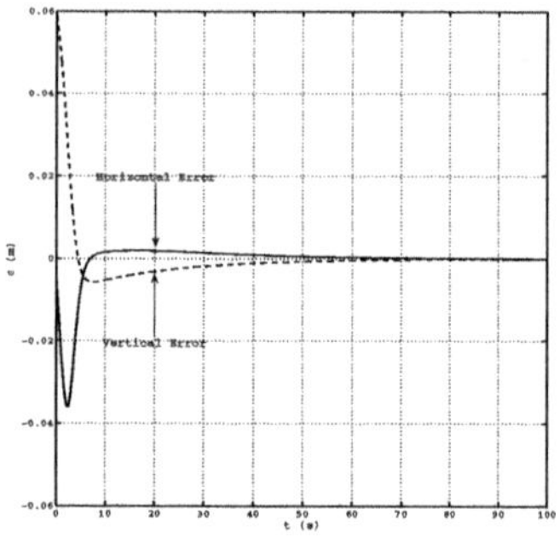

Fig. 5 Evolution of the error ε using Chaumette estimator (only one point is plotted). Differently to Fig. 3, in this case, the horizontal component of error converges to zero.

to a same value. It was observed that the magnitude of this error is directly related with the lineal velocity of the target.

As it was mentioned previously, Fig. 2 shows that the error is presented only in the abscissas of the image coordinates; correspondingly, Fig. 3 shows that the vertical component of the error tends to zero, and it implies that the desired distance d^* has been reaching, however the error in the horizontal component of ε indicates that the line that passes through the point C and the center of the square pattern does not coincide with the axis $(C, \overrightarrow{z})$. The error of the vertical components tends to zero, however, the magnitude of the horizontal error tends to a constant value, it indicates that the image plane is parallel to the pattern plane.

4.2 Visual Tracking with Target Movement Estimation

The control law expressed by Eq. (11) has been evaluated using estimators for the own velocity of the target proposed by Chaumette and Bensalah, expressed by Eqs. (12) and (14) respectively. In both cases the error observed in the image, ε, tends to zero and, thus, the task function expressed by Eq. (8) is correctly executed. This result can be observed in Figs. 4-5 and Figs. 7-8.

Although in both cases the task function is correctly executed, the dynamic obtained is different; Bensalah estimator allows the error, ε, tends to the final value

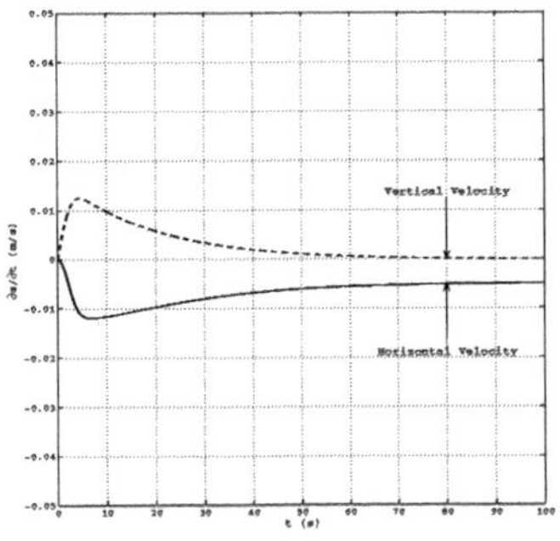

Fig. 6 Evolution of estimated $\frac{\partial s}{\partial t}$ using Chaumette estimator. Vertical components are null when the camera is positioned at distance d^* from the target, horizontal components converge to a value that is proportional to the target's linear velocity.

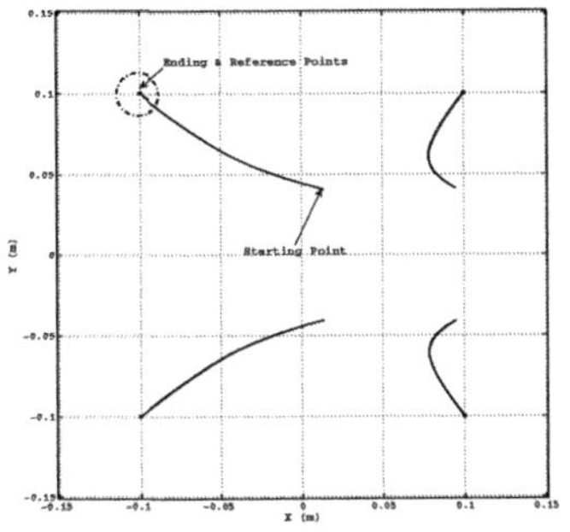

Fig. 7 Evolution of visual index s using Bensalah estimator. Note, similar to Fig. 4, the *ending point* coincides with the *reference point*, but, in this case, s do not exceeds s^*.

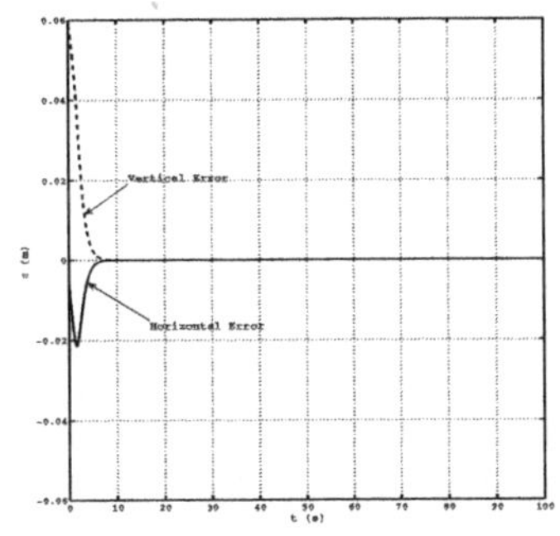

Fig. 8 Evolution of the error ε using Bensalah estimator (only one point is plotted). Note, differently to Fig. 5, the error in both components converge to zero, however, in this case its convergence is more rapid than that of Chaumette's estimator.

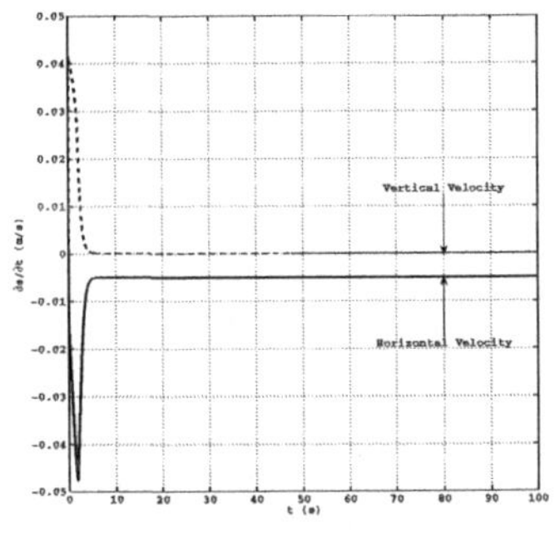

Fig. 9 Evolution of $\widehat{\frac{\partial s}{\partial t}}$, using Bensalah estimator. Similar to Fig. 6, at steady state the vertical component is null, and the horizontal components converges to the target own image plane velocity. However the convergence time in this case is faster than Chaumette's estimator.

faster than Chaumette estimator. Figures 5 and 8 show the dynamic of the error in both cases. This dynamic is directly related with the estimated value of $\frac{\partial e}{\partial t}$, which is shown in Figs. 6 and 9. The effect of $\widehat{\frac{\partial e}{\partial t}}$ calculated with Chaumette estimator arise that s exceeds s^*; in Fig. 4 this effect is shown by the curve generated by s in the proximity of s^* and in Fig. 5 when ε overcomes the t axis in the graph.

The fast convergence of Bensalah estimator is mainly due to it takes into account only information from the instant t and $t-1$, avoiding the effect of the error observed at the beginning of the task, which is generally larger in magnitude.

5 Conclusion

In this paper we has presented the controller for a robotic system, composed by a non-holonomic mobile base and a camera mounted on a pan-platform, from inverse kinematics. The controller exploits the properties of visual servoing method. The robotic system allows the driver to perform tasks of visual tracking of a moving target by a non-holonomic robot in dynamic environments.

In order to increase the accuracy of the tracking system, in this article, we have focused on the compensation of the command using an estimate of the movement of the target itself. The application of this compensator allows an efficient execution of the robotic tasks; the results show that the estimator proposed by Bensalah allows fast convergence and an accurate estimate of the target velocity and therefore the convergence of the task function. The simulated results show the feasibility, effectiveness and proper functioning of the controller. In the future, we will extend this work to make the target occultation management and obstacle avoidance method.

References

1. Hutchinson, S., Hager, G., Corke, P.: A tutorial on visual servo control. IEEE Transactions on Robotics and Automation 12(5), 651–670 (1996)
2. Chaumette, F., Santos, A.: Tracking a moving object by visual servoing. In: Proc. of 12th World Congress IFAC, July 1993, vol. 9, pp. 409–414 (1993)
3. Bensalah, F.: Estimation du mouvement par vision active. Ph.D. dissertation, Université de Rennes 1, France (1996)
4. Espiau, B., Chaumette, F., Rives, P.: A new approach to visual servoing in robotics. IEEE Transactions on Robotics and Automation 8(3), 313–326 (1992)
5. Chaumette, F., Hutchinson, S.: Handbook of Robotics: Visual servoing and visual tracking, pp. 563–583. Springer, Heidelberg (2008)
6. Samson, C., Espiau, B., Le Borgne, M.: Robot control: the task function approach. Oxford University Press, Oxford (1991)

CoolBOT: An Open Source Distributed Component Based Programming Framework for Robotics*

A.C. Domínguez-Brito**, F.J. Santana-Jorge, S. Santana-de-la-Fe, J.M. Martínez-García, J. Cabrera-Gámez, J.D. Hernández-Sosa, J. Isern-González, and E. Fernández-Perdomo

Abstract. Programming robotic systems is not an easy task, even developing software for simple systems may be difficult, or at least cumbersome and error prone. In this document it is presented a C++ distributed component based programming framework for robotics socalled CoolBOT, around which we have started an open source initiative which is freely available via **www.coolbotproject.org** termed *The CoolBOT Project*. This framework, started initally as a framework for easy software integration in robotics, has been improved in order to allow, not only easy integration using the CBSE paradigm, but also transparent distributed system computation.

1 Introduction

Despite there is no established standard methodology for robotic software development, in the last ten years many approaches based on the CBSE (Component Based Software Engineering) paradigm [1] have blossomed in the robotics community, some significant ones are: G^{en}oM/BIP [2][3], Smartsoft [4], OROCOS [5], ORCA [6], OpenRTM-aist [7] and ROS [8]. CoolBOT [9] is a CBSE C++ framework, designed and developed in our laboratory,

A.C. Domínguez-Brito · F.J. Santana-Jorge · S. Santana-de-la-Fe
J.M. Martínez-García · J. Cabrera-Gámez · J.D. Hernández-Sosa
J. Isern-González · E. Fernández-Perdomo
Instituto Universitario SIANI and the Departamento de Informática y Sistemas, Universidad de Las Palmas de Gran Canaria, Spain
e-mail: adominguez@iusiani.ulpgc.es

* This work has been partially supported by the Research Project *TIN2008-06068* funded by the Ministerio de Ciencia y Educación, Gobierno de España, Spain, and by the Research Project *ProID20100062* funded by the Agencia Canaria de Investigación, Innovación y Sociedad de la Información (ACIISI), Gobierno de Canarias, Spain, and by the European Union's FEDER funds.

** Corresponding author.

A. Abraham et al. (Eds.): International Symposium on DCAI, AISC 91, pp. 369–376.
springerlink.com

also aimed at easing software development for robotic systems. In its last operating version we have endowed it with facilities for distributed computation. In this paper we will introduce briefly the main features of CoolBOT as a distributed CBSE framework for robotics.

2 CoolBOT: Overview

CoolBOT [9] is a distributed C++ component oriented programming framework aimed to robotics. This is a programming framework that follows the CBSE paradigm for software development. The key concept in the CBSE paradigm is the concept of *software component* which is a unit of integration, composition and software reuse [10][11]. Thus, complex systems might be composed of several ready-to-use components.

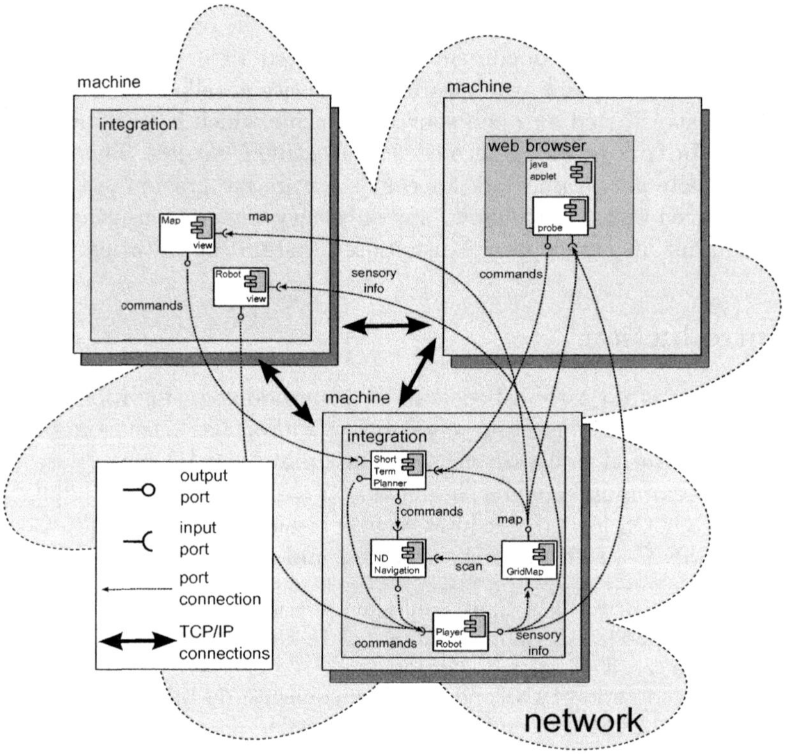

Fig. 1 CoolBOT. Secure Navigation System.

Fig. 1 gives a global view of a real example of software developed using CoolBOT. The example shows a secure navigation system for an indoor mobile robot based on the ND+ algorithm for obstacle avoidance [12]. It has been implemented attending to [13]. In CoolBOT we can find three types of distributed software components: *components*, *views* and *probes*. The system in the figure is organized using two CoolBOT integrations (processes), one only formed by CoolBOT component instances, and the other one containing two CoolBOT view instances. The former one consists of four component instances, namely: `PlayerRobot` (this is a wrapper component for hardware abstraction using the Player/Stage project framework [14]), `GridMap` (this component maintains a grid map of the surroundings of the robot built using robot range laser scans, it also generates periodically a 360° virtual scan for the ND+ algorithm), `NDNavigation` (implements the ND+ algorithm) and `ShortTermPlanner` (a planner which uses the grid map for planning paths in the robot surroundings). On other machine another integration is shown hosting two views through which we can control and monitor the system remotely. In addition, in another machine, there is a web browser hosting a Java applet using a CoolBOT probe to connect to some of the components of the system in order, also, to control and monitor the system remotely through a web interface.

3 CoolBOT Software Components: Components, Views and Probes

CoolBOT *components* are *active objects* [15][11]. They are modeled as *port automata* [16]. Fig. 2 provides a view of the structure of a CoolBOT component. Components can be seen as "data-flow machines". In fact, the model of computation in CoolBOT follows the *Flow Based Programming* (FBP) paradigm [17]. Components intercommunicate among them only through its external interfaces which are formed by *input and output ports*. When connected, they form *port connections*, as depicted on Fig. 1 (in the figure for the sake of clarification they have been simplified). Through them, components interchange discrete units of information termed *port packets*. Thence, a component's external interface comprises all its input and output ports, its types, and the port packets it accepts through them. Port connections are unidirectional (from output to input port), and follow a publish/subscribe paradigm of communication [11]. Output and input ports may be defined out of a set of available typologies. The typologies of the input and output ports involved in a port connection determine the pattern and semantics of the communication through it. There are specific typologies of connections to indicate events to other components; to send port packets which get stored in a fifo in the receiver end (the input port); to publish a *master copy* of port packets where its subscribers can access it, etc (more details in [9]). In addition as we can see on Fig. 2 there are two special ports in any component: the *control port* and the *monitoring port*, the rest of ports are user defined. Those two ports allow an external supervisor (i.e. another component, a view or a probe) to control and monitor a given component. As active objects, CoolBOT components can organize its rut-time using multiple threads of execution as depicted on Fig. 2. The synchronization among them is guaranteed by the underlying framework infrastructure.

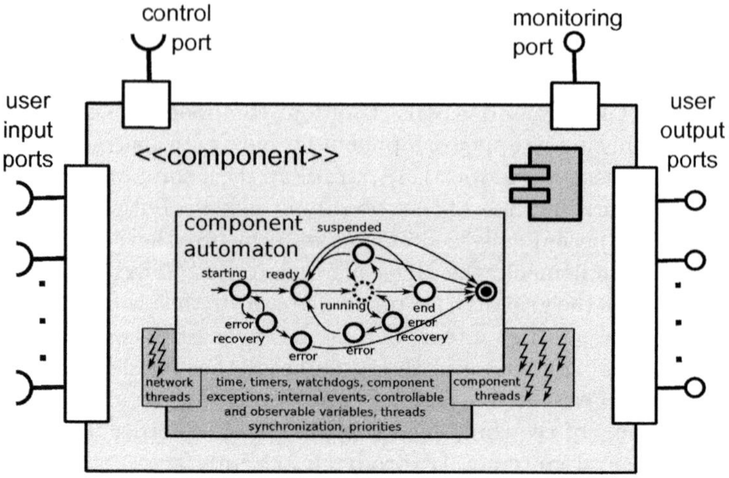

Fig. 2 CoolBOT. Component structure.

Inspired by one of the "good practices" proposed by the authors of CARMEN [18]: "one important design principle ... is the separation of robot control from graphical displays", we have introduced in CoolBOT the concept of *view* as an integrable, composite and reusable graphical interface available for CoolBOT system integrators and developers. Thus, CoolBOT *views* are graphical interfaces which, as software components, may be interconnected with any other component, view or probe in a CoolBOT system. In Fig. 3 is depicted the structure of a view in CoolBOT. As shown, CoolBOT views are also endowed with an external interface of input and output ports. Internally, a view is a graphical interface, in fact, the current views already developed and operational which are available in CoolBOT have been implemented using the GTK graphical library [19].

As depicted in Fig. 3 a CoolBOT *probe* is provided with an external interface of input and output ports, and likewise component and views, as software components, this allows them to be interconnected with other components, views or probes. *Probes* are devised as interfaces for interoperability with non

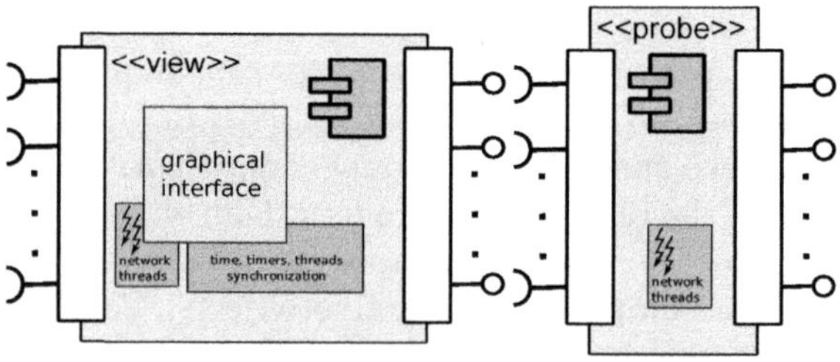

Fig. 3 CoolBOT. View and probe structures.

CoolBOT software. As of now we have used CoolBOT probes to interoperate with Java applets inserted in a web browser (shown in Fig. 1).

4 Distributed Computation

As depicted in Fig. 1, CoolBOT provides means for distributed computation. A given system can be mapped on a set formed by different machines sharing a computer network. Each machine can host one or several CoolBOT *integrations*. A CoolBOT *integration* is an application (a process) which integrates instances of CoolBOT components, views and probes. Port connections among components, views and probes are transparently multiplexed using TCP/IP connections between software components in different integrations. In the same integration port connections are supported using thread synchronization resources (locks, conditional variables, mutexes and shared memory). Integrations can be instantiated in any machine and port connections can be established dynamically at any moment, wheter local or remote.

Transparent distributed computation was one of the main concerns we had when designing CoolBOT network support for system integration. The main idea was to make network communications as transparent as possible to developers (and software components). Thus, at system level, to connect two component instances instantiated in different CoolBOT integrations should be as easy as connecting them when instantiated locally in the same integration. In particular, we follow three main principles related to transparent distributed computation facilities: transparent network inter component communications, network decoupling of component's functional logic, and incremental design.

Transparent Network Inter Component Communications. In order to make network communications transparent to components, views and probes, we have developed a protocol termed *Distributed CoolBOT Component Communication Protocol (DC3P)* to multiplex port connections over TCP connections established among the software components involved. In the current version of CoolBOT, only the TCP protocol is supported for network connections. The integration of the UDP protocol is under development, and it is expected for next CoolBOT version. DC3P has been implemented using the TCP/IP socket wrappers and the marshalling facilities provided by the ACE library [20]. DC3P is transparent to CoolBOT users in the sense that is only used internally by the framework itself, as illustrated in Fig. 4. In its current version, the DC3P protocol consists of the following packets:

- *Port Type Info* (request & response): For asking type information about input and output ports through network connections. This allows port connection compatibility verification when establishing a port connection through the network.

- *Connect* (request & response): For establishing a port connection over TCP/IP.
- *Disconnect* (request & response): To disconnect a previous established port connection.
- *Data Sending*: Once a port connection is established over TCP/IP, port packets are sent through it using this DC3P packet.
- *Remote Connect* (request & response): For establishing port connections between two remote component instances. Permits to connect component instances remotely.
- *Remote Disconnect* (request & response): To disconnect port connections previously established between two remote component instances.
- *Echo Request & Response*: Those DC3P packets are useful to verify that the other end in a network communication is active and responding.

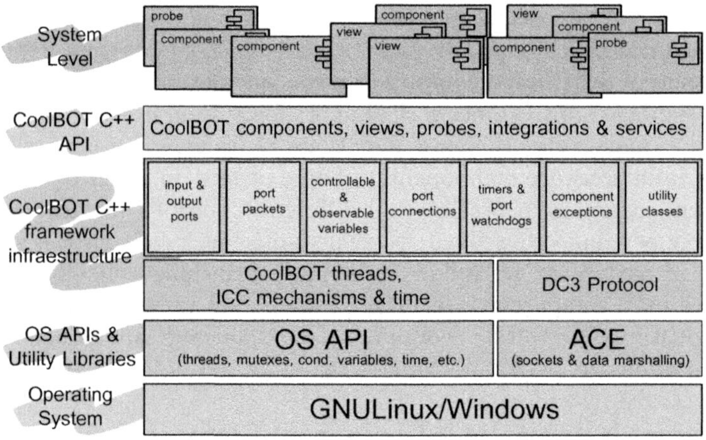

Fig. 4 Abstraction layers in CoolBOT.

All DC3P packets and port packets sent through port connections are marshalled and unmarshalled in order to be correctly sent through the network. We have used the facilities ACE provides for marshalling/demarshalling based on the OMG Common Data Representation (CDR) [21]. In general, port packets are user defined. In order to make their marshalling/demarshalling as easy and transparent as possible for developers, a description language describing port packets can be used, much like CORBA IDL [22]. A description language compiler generates a C++ skeleton class for each port packet where the code for marshalling/demarshalling is part of the code generated automatically. In addition, we have endowed also CoolBOT with a rich set of C++ templates and classes to support marshalling and demarshalling of port packets (or any other arbitrary C++ class).

Network Decoupling of Component's Functional Logic. Another important aspect for network communication transparency is the decoupling of network communication logic from the functional logic of a software component. Thus, each component, view or probe is endowed with a pair of network threads, and *output network thread*, and an *input network thread*, which are responsible for network communications using DC3P. CoolBOT guarantees transparently thread synchronization between them and the functional threads of the component. The network threads are mainly idle, waiting to have port packets to send through open network port connections, or to receive incoming port packets that should be redirected to the corresponding component's input ports. At instantiation time, it is possible to deactivate the network support for a component instance (and also for views and probes instances). In this manner, the component is not reachable from outside the integration where it has been instantiated, and evidently network threads and the resources they have associated are not allocated.

Incremental Design. In future versions of CoolBOT, it is very possible that the set of DC3P protocol packets grow with new ones. In order to allow an easy integration of new DC3P packets in CoolBOT, we have applied the *composite* and *prototype* patterns [23] to their design. These design patterns, jointly with the C++ templates and classes to support marshalling and demarshalling provide a systematic and easy manner of integrating new DC3P packets in future versions of the framework.

5 Conclusions

In this document we have presented a C++ distributed component based programming framework for robotics socalled CoolBOT, around which we have started an open source initiative which is freely available via **www.coolbotproject.org** termed *The CoolBOT Project*. This framework, started initally as a framework for easy software integration in robotics, has been improved in order to allow, not only easy integration using the CBSE paradigm, but also trasparent distributed system computation.

References

1. Heineman, G.T., Councill, W.T.: Component-Based Software Engineering. Addison-Wesley, Reading (2001)
2. Mallet, A., Pasteur, C., Herrb, M., Lemaignan, S., Ingrand, F.: GenoM3: Building middleware-independent robotic components. In: IEEE International Conference on Robotics and Automation (2010)
3. Bensalem, S., Gallien, M., Ingrand, F., Kahloul, I., Thanh-Hung, N.: Designing autonomous robots. IEEE Robotics and Automation Magazine 16(1), 67–77 (2009)
4. Schlegel, C., Haßler, T., Lotz, A., Steck, A.: Robotic Software Systems: From Code-Driven to Model-Driven Designs. In: Proc. 14th Int. Conf. on Advanced Robotics (ICAR), Munich (2009)
5. The Orocos Project (2010), `http://www.orocos.org`

6. Brooks, A., Kaupp, T., Makarenko, A., Williams, S., Oreback, A.: Towards component-based robotics. In: IEEE International Conference on Intelligent Robots and Systems, Tsukuba, Japan, pp. 163–168 (2005)
7. Ando, N., Suehiro, T., Kotoku, T.: A Software Platform for Component Based RT-System Development: OpenRTM-Aist. In: Carpin, S., Noda, I., Pagello, E., Reggiani, M., von Stryk, O. (eds.) SIMPAR 2008. LNCS (LNAI), vol. 5325, pp. 87–98. Springer, Heidelberg (2008)
8. ROS: Robot Operating System (2010), `http://www.ros.org`
9. Domínguez-Brito, A.C., Hernández-Sosa, D., Isern-González, J., Cabrera-Gámez, J.: CoolBOT: a Component Model and Software Infrastructure for Robotics. In: Software Engineering for Experimental Robotics, April 2007. Springer Tracts in Advanced Robotics Series, vol. 30, pp. 143–168. Springer, Heidelberg (2007)
10. Brugali, D., Scandurra, P.: Component-based robotic engineering (part i) [tutorial]. IEEE Robotics Automation Magazine 16(4), 84–96 (2009)
11. Brugali, D., Shakhimardanov, A.: Component-based robotic engineering (part ii). IEEE Robotics Automation Magazine 17(1), 100–112 (2010)
12. Minguez, J., Osuna, J., Montano, L.: A "Divide and Conquer" Strategy based on Situations to achieve Reactive Collision Avoidance in Troublesome Scenarios. In: IEEE International Conference on Robotics and Automation, New Orleans, USA (2004)
13. Montesano, L., Minguez, J., Montano, L.: Lessons Learned in Integration for Sensor-Based Robot Navigation Systems. International Journal of Advanced Robotic Systems 3(1), 85–91 (2006)
14. Vaughan, R.T., Gerkey, B., Howard, A.: On Device Abstractions For Portable, Reusable Robot Code. In: IEEE/RSJ International Conference on Intelligent Robot Systems (IROS 2003), Las Vegas, USA, October 2003, pp. 2121–2427 (2003)
15. Ellis, C., Gibbs, S.: Active Objects: Realities and Possibilities. In: Object-Oriented Concepts, Databases, and Applications. ACM Press, Addison-Wesley (1989)
16. Stewart, D.B., Volpe, R.A., Khosla, P.: Design of Dynamically Reconfigurable Real-Time Software Using Port-Based Objects. IEEE Transactions on Software Engineering 23(12), 759–776 (1997)
17. Paul Morrison, J.: Flow-Based Programming, 2nd Edition: A New Approach to Application Development. CreateSpace (2010)
18. Montemerlo, M., Roy, N., Thrun, S.: Perspectives on standardization in mobile robot programming: the carnegie mellon navigation (carmen) toolkit. In: Proceedings of IEEE/RSJ International Conference on Intelligent Robots and Systems (IROS 2003), October 2003, vol. 3, pp. 2436–2441 (2003)
19. The GTK+ Project (2010), `http://www.gtk.org`
20. Schmidt, D.C.: The Adaptive Communication Environment, ACE (2010), `http://www.cs.wustl.edu/~schmidt/ACE.html`
21. Object Management Group, The Common Object Request Broker: Architecture and Specification, ch. 15, Sec. 1-3 (2002), `http://www.omg.org/cgi-bin/doc?formal/02-06-01`
22. Object Management Group, OMG IDL: Details, `http://www.omg.org/gettingstarted/omg_idl.htm`
23. Gamma, E., Helm, R., Johnson, R., Vlissides, J.: Design Patterns: Elements of Reusable Object-Oriented Software. Addison-Wesley Professional Computing Series. Addison-Wesley, Reading (1995)

An ICT Solution with Real-Time Tracking Capacities for Improving the Incidence Management Timing in the Transportation of Industrial Equipments

Asier San Nicolás, Ignacio Angulo, Asier Perallos, and Nekane Sainz

Abstract. This paper explores the use of ICT technologies in good traceability. In fact, it presents a solution for real-time tracking and fleet management which can be applied in the transportation of manufacture equipment. The system can manage routes and operations and provides onboard support to the carriers, enabling an agile response to incidences happened during transport. We describe both through its technical implementation as a test case used to validate the basic functionality of a first developed prototype. Furthermore, some future works are presented in order to improve some capacities of the solution.

Keywords: goods traceability, transportation of industrial equipment, fleet management, incidence recovery.

1 Introduction

Nowadays there are numerous technological advances which can contribute to a higher quality and productivity in the supply chain management activities. The traceability of goods is a key activity in the supply chain which has to be performed in an effective way in order to achieve a higher competiveness of the companies. At this moment, with regard to the traceability of goods, not relying on new technologies is unthinkable [1].

In fact there are some industries in which the traceability is a very critical requirement. For example, concerning to agricultural industry, Regulation 178/2002 requires the traceability of foods in the European Union in order to know the source of each food and the route followed to reach its destination [2]. Other critical industry is the pharmaceutical one, where lately a lot of trends are arising. In

---

Asier San Nicolás · Ignacio Angulo · Asier Perallos · Nekane Sainz
Deusto Institute of Technology - DeustoTech, University of Deusto, Avenida de las Universidades 24, 48007 - Bilbao, Spain
e-mail: {asier.sannicolas,ignacio.angulo,perallos}@deusto.es, nekane.sainz@deusto.es

A. Abraham et al. (Eds.): International Symposium on DCAI, AISC 91, pp. 377–384.
springerlink.com

this industry, solutions for medicines tracking have to be deployed as far as possible in order to prevent counterfeit of drugs, assure their source, so as to guarantee that they arrive where they are expected to.

An unexplored sector is the manufacturing one. Specifically, the distribution and transportation of equipments, where the main goal is that all components of an equipment arrive together to their destination as soon as possible, in order to assemble the equipment and begin to produce manufactured goods. The traceability of this kind of components, as well as the capacity to get an efficient fleet management, route planning and incidences recovery, will increase the productivity of the manufacture companies. It is a consequence of the improvement of the way in which the distribution of the components that compound each equipment is done and the reduction of the time needed to resolve problems during their distribution.

Information and Communications Technologies (ICT), due to the latest advances in positioning, wireless communications and radio frequency systems, have become the new hope for improving such activities. In fact, the result of the work described in this paper is an ICT solution for real-time tracking and fleet management which can be applied in the transportation and distribution of manufacture equipment. Furthermore, this innovative solution is able to manage routes and operations, as well as onboard support to the carriers, enabling an agile response to incidences happened during transport. In the second section of this paper, the related work is analyzed. In the third section the context of our work is presented, as well as its functional requirements, challenges and innovative contributions. Then, in the fourth section we provide a detailed description of the technical solutions. In the fifth section a first prototype is presented and some results obtained in a test scenario. Finally, the future work and conclusions are presented.

2 Related Work

It is a common requirement in different kind of industries to track the route of each product being shipped from supplier to customer. The importance of tracking goods has reached such a point that is being regulated by law in some countries. For example in Europe, Regulation 178/2002 requires the traceability of all food from farms to the end of the supply chain. This is one of the reasons because most of the projects about tracking are focused on the food industry, as for example the Trace FP6 project (http://www.trace.eu.org) or those ones based on RFID technology and described in [3] [4] [5]. There are other critical sectors as the pharmaceutical one. Thus, the PharmaX initiative is proposed to shed light on the pharmaceutical traceability and overall-process regulation. This system ensures that all pharmaceutical supply chain participants can integrate with each other, resulting in information sharing, consistency checking and anti-counterfeit [6].

Other related challenge is the use of ICT to improve logistic processes in intermodal transportation, as is being done in TIMI project (www.proyecto-timi.es).

When performing the above projects and other similar ones, actually they are seeking the following benefits: processes automation, turnover increase, and stock management improvement [7]. In fact, there are many other sectors where new technologies are applied to improve goods tracking and achieve these goals. For example, Galeria Kaufhod in retail clothing market is using RFID to automate logistics processes of the store. Thus, it is possible to improve the inventory or count the entry of goods, among other benefits [8]. Other example is found in Dell Company. Changing from barcodes to RFID they have reached a big improvement in their logistic: read-accuracy improved from 95.8% to 99.8% and Mean-Time-To-Repair was reduced by 38%; the Return-on-Investment was 122% [9].

The use of this kind of ICT tracking solutions in manufacturing industry is a great innovation because there are not a lot of previous references in this sector. Furthermore, the transportation of industrial equipment has some additional technological challenges to be faced, such as interoperability with manufacturing information systems, real-time fleet and goods traceability, management and monitoring of the route and job actions, providing of onboard information to carriers, and real-time incidence management and recovery. The result of our work is an ICT solution with real-time tracking capacities for improving the incidence recovery timing in the transportation of industrial equipments. That is, an innovative characteristic is that it provides an ICT-based support for all the above processes.

3 Context of Work and Functional Requirements

The scope of the project covers all the stages included in the distribution process of industrial equipment. Functionalities of the project are described below.

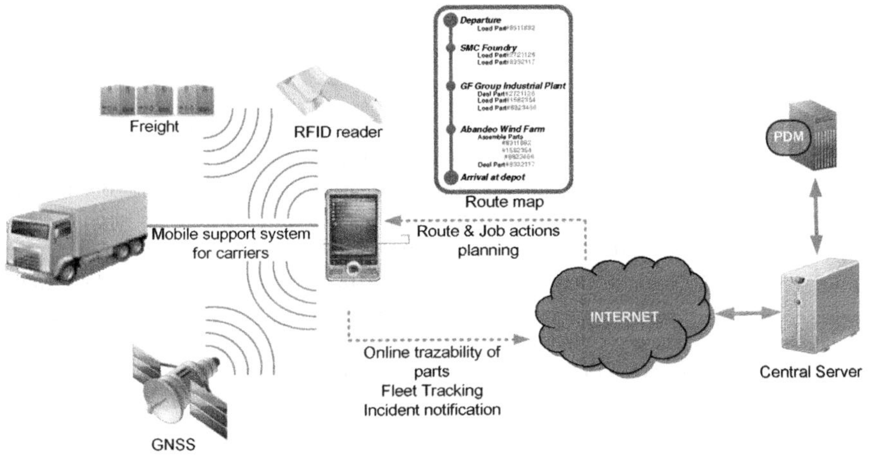

Fig. 1 Arquitecture of the proposed solution

a) Planning, route monitoring and job actions management: The planning server receives the list of actions to be carried out, it connects to the Product Data Management system (PDM) to get the set of required materials, their physical characteristics and realize its storage points or potential dealers in order to generate truck routes for every transport service. These routes determine actions to be performed by the staff of the company indicating the stops for each vehicle, the order they must be carried out and detailed list of every job action to be executed at every point on the route. The most common actions included are the collection or loading of materials, unloading and subsequent assembly and installation of different components that make up each end equipment, so as maintenance activities required by them. Once established, the routes are sent to the staff by the "carriers supporting device" (whose function is detailed below) to be executed in the shortest possible time. It should be mentioned that after delivery of the route of each vehicle it can be dynamically altered during the course if it is necessary.

b) Fleet online tracking and traceability of parts: Each vehicle has a specific route that gives the list of actions to carry out. In order to validate the correct execution of each job action, each vehicle is permanently monitored by the planning server, storing and displaying the position of each vehicle in real time. Moreover, all parts manipulated by the system are labeled with an RFID tag that allows tracking and ensures proper selection of different parts involved in each job action. Any difference occurred in each route is notified immediately to reduce its impact.

c) Carriers supporting device: Each employee responsible for a vehicle has a mobile device that will guide the development of assigned tasks. This system, through a user friendly graphic interface, includes driving support to reach each stop on the route established, shows the list of job actions to be carried out at each point of the route and allows validating the implementation of these actions through the traceability system. In addition, internally the device is permanently connected to the server to reveal the positioning of the vehicle and the moments in which takes place every planned action. Finally the user can notify all relevant incidences that have happened such as an accident or breakdown, filling out a simple form that is immediately sent to the server.

d) Incidences management: Information sent by the previously described device allows detecting significant deviations from the planned time and rescheduling the routes and job actions assignments to different involved vehicles in order to minimize the time needed to resolve the problem. The system distinguishes between manual incidence (required by a worker) and automatic (directly sent by the mobile device) and from emergency incidence, which must be notified immediately to the planning server, or informative because it does not significantly affect how the transport service is done. This system creates a repository that stores all the incidences generated by each employee or company car.

4 Technical Solution

All features explained in the previous section have been developed thanks to the use of different technologies. At this point, we are going to describe each function of the system from a technical point of view:

a) Planning, route monitoring and job actions management: The fundamental component of the project, also called Planning Server, has granted access to certain tables managed by the database management system used by the business software suite installed in the company. Through such access Planning Server reads tasks list and manages status assigned to each task in terms of demanded, under development and performed. By SOAP messages to a web service developed for the Product Data Management system (PDM), system takes all the information about the different parts needed for each task by storing features like size, weight, stock, place of collection and delivery, and all actions needed for final assembly of parts. Using a simple algorithm, the server divides the list of actions between the fleet of the company, setting the routes for each vehicle in accordance with their own physical limitations and geographical criteria. This route is stored in what has been called a route document, which is an XML formatted document that includes hierarchically the stops on the route, the actions to be carried out at each stop and parts involved in every action. For each piece, in addition to its relevant data such as the name of the piece, description, etc. a unique identifier is stored coincident with the ID code stored in the RFID tag assigned. Both route points as actions are displayed in the order they should be performed. The following diagram shows the general structure of the route document:

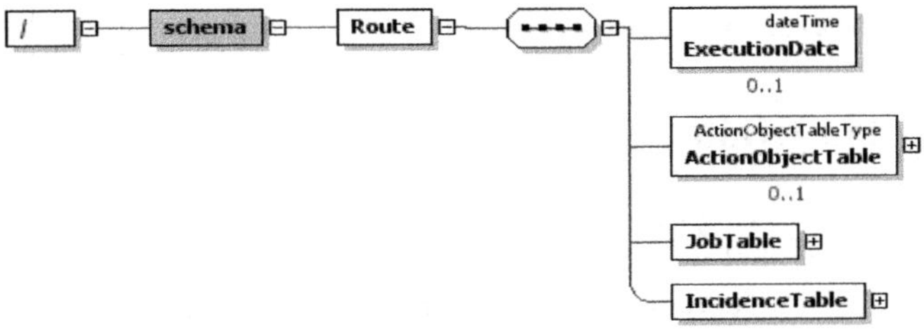

Fig. 2 Route document XML schema

Any transfer of information between Planning Server and each onboard mobile device on vehicle (using wireless technologies such as GPRS or HSPA) will be carried out by sharing this document, in which the mobile device is storing all the advances in the development of the activity imposed on each route. Thus, the importance of the document is as relevant that it allows comparing the planned and real times in the implementation of the tasks and carrying out the traceability of shipped parts. When a vehicle suffers an emergency incidence or if the deviation between the scheduled time and the execution is critical, the route of one or more vehicles can be replanned by simply updating XML file for each mobile device.

b) Fleet online tracking and traceability of parts: The onboard mobile device, implemented on a smartphone with HSPA and GPRS connectivity and buit-in GPS receiver, is responsible for periodically sending the position of each vehicle, so all vehicles can be monitored through a basic real-time fleet tracking system as well as storing routes performed by each vehicle for further analysis. Traceability of the parts is performed by the carrier with the help of a portable RFID reader that connects mobile device through Bluetooth connection allowing validate all parts involved in each action and notify immediately every error in parts selection to the Planning Server. Subsequently, the information stored by the tracking system can locate the source of each piece installed in an equipment and the current location of each piece supplied.

c) Support system for carriers: The system developed for carriers is also hosted as a resident application in an onboard mobile device. The interface displayed to each carrier is a representation of the route to be done, which is stored in the XML document. Using this application, driver can see which stops should be done, what action has to perform, an estimated planning and what parts or material are involved in a particular action. In addition, through GPS, system notices the driver when is arriving to a point of the route, helping him find the place.

d) Incidences management: The incidences that have been explained above are recorded and if necessary sent through the manager of the mobile application. When materials are being recorded by RFID and a product that should not be read is registered (part is not included in the route document) or when concluding an activity without validating document specifications, an automatic incidence is generated and written in the XML file, indicating the details and where it has occurred. Apart from this, the application allows writing an incidence manually by the driver himself filling out a simple form. If in this form the incidence is marked as urgent, it is sent to the scheduler at the time. The scheduler can replan the route of one or more trucks, shorten the route of the truck concerned or just do nothing.

5 Testing and Evaluation

In order to validate the system developed at this moment, a test route has been created for a truck with 3 stops: AgerBide, Radimer and BaraCons, which must initially be accessed sequentially. At each stop, a number of actions must be made: in AgerBide, to load propeller blades; in Radimer, to charge propeller body and propeller base and discharge propeller blades; finally, in BaraCons, to assemble propeller body and propeller base. See the route details in Table 1.

Table 1 Places and actions

Place	Longitude	Latitude	Actions	Parts
AgerBide	-3,031389	43,330278	Charge	P. Blades.
Radimer	-3,036667	43,30611	Charge Discharge	Body, Base. P. Blades
BaraCons	-2,991667	43,297222	Assemble	Body, Base

This route will test the basic functionality of the current system: communication platform, fleet and goods tracking, on board support information, and incidence detection. Other capacities, such as intelligent planning of routes and job actions, will be developed and tested in the future.

Once the carrier is authenticated by the mobile solution, the route of Table 1 is downloaded from the Planning Server. Once route is shown by the mobile application, transport service starts driving to the first destination. Figure 3 shows at the left, how the Planning Server is receiving positioning from each vehicle.

When vehicle approaches the desired place (AgerBide), the location icon in the mobile application changes of color and warns us. Once in destination, the driver chooses the option "Actions TODO" in the application menu. After choosing the action "Charge", mobile device establishes Bluetooth connection with the RFID reader and records appropriate part ID (propeller blades) to complete the action.

Once all actions assigned in the first stop are correctly performed, carrier drives to the second one, Radimer, where two actions are planned. The first one, which is unload propeller blades, is performed validating the RFID code, and in the second action, which should be loading propeller body and propeller base, RFID reader only registers propeller body, so when employee ends this action, the application will alert him indicating that some materials are missing. Accepting this issue, an automatic incidence is automatically sent to the Planning Server. The route will be changed by adding an additional stop where retrieve lost (see Figure 3).

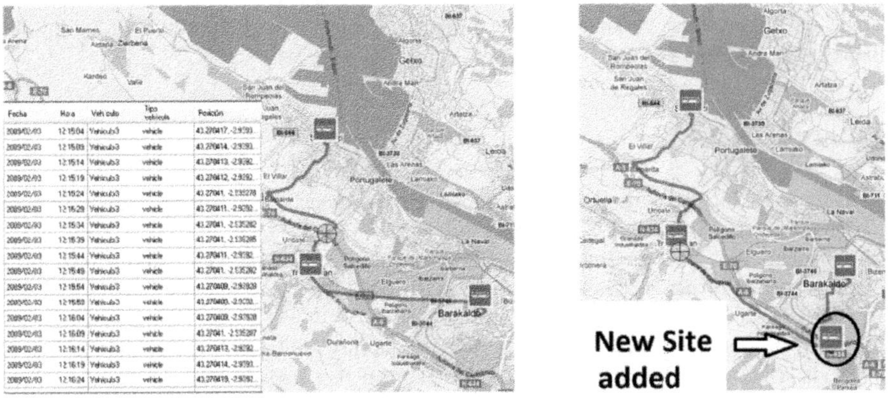

Fig. 3 Monitoring and replanning

This route is sent to our mobile device and with the updated route we can continue the job. Now we have a new stop (Machinery Retuerto) at which we need to collect the material we needed (propeller base). Thus, arriving at BaraCons, we can assemble all the components we have.

After finishing the test we can conclude that we obtain the expected results, detecting incidences in real time and minimizing the time needed to manage them.

6 Conclusion and Future Work

The result of our current work is an ICT solution with real-time tracking capacities for improving the incidence recovery timing in the transportation of industrial equipments. It is considered an innovative solution because it faces technological challenges concerning to this transport, such as interoperability with manufacturing information systems, real-time fleet and goods traceability, management of the route and job operations, onboard information to carriers, and real-time incidence management and recovery. Now, a first prototype has been successfully validated.

The current implementation of the system uses passive RFID tags that should be verified by a portable HF RFID reader which connects to the mobile device by simply sending ID codes of the RFID tags. These passive tags can be replaced by active ones which can store more information and improve the reach of the reader. In that way, the employee does not have to validate each individual piece with the portable RFID reader, but can perform the tasks of loading and unloading of materials being automatically validated. Consequently the traceability of goods can be carried out in a non-intrusive way (without modifying the behavior of transport staff). This will made the system portable to other areas such as rail or pharmaceutical. Other important aspect of future work must be to improve the way in which the planning of the routes and job actions is done, because a non-automatic process is used in this moment. A higher level of automation in the task of rescheduling, based-on the use of Artificial Intelligence techniques, is a desirable issue.

Acknowledgments. This work has been funded by the Ministry of Industry, Tourism and Trade of Spain under Avanza funding program (Grant TSI-020100-2008-582). Special thanks to Avangroup Business Solutions, S.L. for their support.

References

1. Cleland-Huang, J., Settimi, R., Romanova, E., Berenbach, B., Clark, S.: Best Practices for Automated Traceability. Computer 40(6), 27–35 (2007)
2. Schwägele, F.: Traceability from a European perspective. In: 51st International Congress of Meat Science and Technology (ICoMST), Meat Science, vol. 71(1), pp. 164–173 (2005)
3. Abad, E., et al.: RFID smart tag for traceability and cold chain monitoring of foods: Demonstration in an intercontinental fresh fish logistic chain. Journal of Food Engineering 93(4), 394–399 (2009)
4. Kelepouris, T., Pramatari, K., Doukidis, G.: RFID-enabled traceability in the food supply chain. Industrial Management & Data Systems 107(2), 183–200 (2007)
5. Manikas, I., Manos, B.: Design of an integrated supply chain model for supporting traceability of dairy products. Int. Journal of Dairy Technology 62(1), 126–138 (2009)
6. Huang, G.Q., Qin, Z., Qu, T., Dai, Q.: RFID-enabled pharmaceutical regulatory traceability system. In: 2010 IEEE International Conference on RFID-Technology and Applications (RFID-TA), Guangzhou, China, pp. 211–216 (2010)
7. Bertolini, M., Bottani, E., Rizzi, A., Volpi, A.: The Benefits of RFID and EPC in the Supply Chain: Lessons from an Italian pilot study. The Internet of Things, Part 4, 293–302 (2010)
8. Al-Kassab, J., Blome, P., Wolfram, G., Thiesse, F., Fleisch, E.: RFID in the Apparel Retail Industry: A Case Study from Galeria Kaufhof. In: Unique Radio Innovation for the 21st Century, Part 4, pp. 281–308 (2010)
9. Crowl, S., Mares, V., Moore, M.: Radio Frequency Identification (RFID) application at dell computer. In: IEEE Engineering Management Conference, Austin, TX, USA, pp. 28–30 (2006)

Levels of Adaptation and Control

Sebastian Bader and René Leistikow

Abstract. To realise the vision of ubiquitous computing, we need to control heterogeneous and dynamically changing device ensembles, that is software components distributed over different devices encapsulating hardware functions or services. The resulting control systems must be real-time capable and able to describe their actions. Here we propose a layered architecture, which is currently under development. It is based on a probabilistic model providing the real-time assistance functions, and on symbolic descriptions. These description can be used (a) to synthesise the probabilistic model, and (b) to explain the resulting actions taken by the controller.

1 Introduction and Motivation

Controlling dynamic and heterogeneous ensembles is a major problem and will be of central importance for the field of ubiquitous computing / distributed intelligent systems in the future. With the advent of ubiquitous computing techniques their intelligent control becomes more and more important. Already now smart environments are equipped with numerous sensors and actuators. Some are still visible but some are already hidden in the environment as envisioned by Marc Weiser in the seminal paper [1]. This makes the interaction with the embedded technique quite hard for the non-experienced user. To complicate things further, the device ensembles are dynamic. That is, devices enter and leave the environment – for example mobile computing devices carried by users. Nonetheless, we would like to have a smart control system, able to cope with changes and able to support the user while achieving his goals. Therefore, we propose a layered architecture based on probabilistic models, adaption systems, and a high-level logic description.

Sebastian Bader · René Leistikow
MMIS, Computer Science, University of Rostock
e-mail: `firstname.lastname@uni-rostock.de`

A. Abraham et al. (Eds.): International Symposium on DCAI, AISC 91, pp. 385–388.
springerlink.com

2 Requirements and Goals

We are aiming at a system able to control environments equipped with a number of different devices, that change over time. Such ensembles can for example be found in smart meeting rooms consisting of hard wired projection equipment, computers and other infrastructure as well as mobile devices like laptops, smart phones and the like. But also modern homes contain many intelligent and remotely controllable devices, like entertainment systems, home automation equipment, security systems and others, as well as mobile devices carried by the inhabitants. Therefore our system must be able to control such heterogeneous and dynamic infrastructure.

Another major requirement of a control system is the real-time capability. That is, the possibility to assist the user instantaneously. Therefore, the system can neither rely on logic based planning systems nor on other non-time-bounded inference procedures. Furthermore, the system needs to be able to explain the decision taken in a human-understandable manner – that is in some symbolic way. To address both issues, we propose to use a probabilistic controller, which is constructed from and adapted by a symbolic system. The generation of probabilistic systems from formal description has been described in [2]. But so far we have not been able to adapt the resulting systems dynamically to changes in the environment.

To summarise: a controller must be able to support users in dynamically changing heterogeneous environments in a real-time manner and must be able to provide explanations for actions taken.

3 A Layered Architecture for Adaptive Control

The design of our architecture is based on the following insights: (a) different information is valid for different time frames only, (b) assistance should be provided in real-time and (c) we need a symbolic description to provide explanations. All three points are detailed below. The time of validity of information varies. There are

I1. *Immutable truths:* the laws of mathematics and physics
I2. *Slowly changing information:* the number and functionality of devices
I3. *Faster changing information:* time of the day and the tiredness of the user
I4. *Instantaneous information:* position of the user

As mentioned above, we are aiming at a controller which is able to react in real-time. Therefore, most assistance provided by the system should be as provided in a reflex-like manner. But there are also slower adaptations within such a system. Namely those that depend on the slower changing information (I2 and I3). For example there might be different assistance required during day- and night-time. We can distinguish the speed of reaction as follows:

R1. *Adaptation layer:* adaptation to slow, structural changes of the environment (I2), that is to new devices or changing functionality.

R2. *Optimisation layer:* optimisation of the underlying system with respect to faster changing information (I3), that is the adaptation of the internal probability distributions and attached action sequences.
R3. *Reflexes:* instantaneous reactions to changes of the environment (I4)

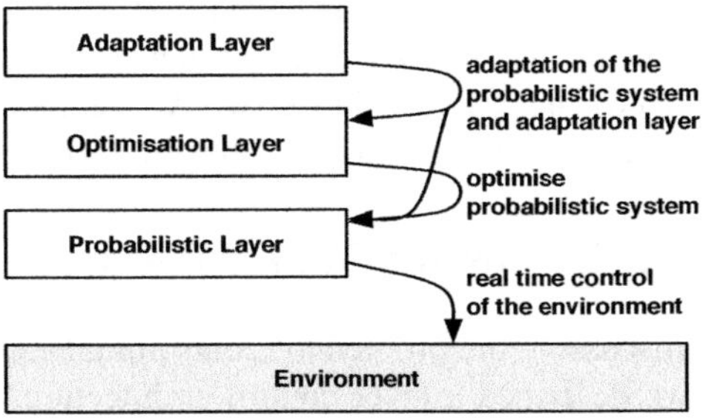

Fig. 1 The layers of the proposed architecture. The probabilistic layer at the bottom provides the actual assistance by controlling the environment. The adaptation layer optimises the probabilistic one by fine-tuning the parameters. The inference layer on top controls both, the probabilistic layer as well as the adaptation layer.

Based on these layers of adaptation, we propose the architecture depicted in Figure 1. In this system, the bottom layer, that is the reactive layer is implemented as a probabilistic state based system as for example a hidden Markov model [3]. Actions are attached to the internal states of the system, and those actions are executed whenever the system enters the state (meaning the state becomes the most likely one). For example, whenever the system enters the state corresponding to a situation requiring more light, a lamp could be switched on.

Depending on the faster changing information (I3), this system is optimised to different contexts. For example if the outside lighting conditions change, the concrete actions attached to the states are modified appropriately. But also the overall structure of the probabilistic model can be altered by adding new states, or removing unnecessary ones. Even though simple solutions for this exist, the optimal strategy for these adaptation is subject to further research.

The adaptation layer is based on a symbolic description of device capabilities as well as user behaviours and preferences. While the former are usually expressed in PDDL [4], the later can for example be captured in CTT, or CTML-models [5]. Both descriptions can be compiled into a probabilistic model with action sequences attached to states as follows: Based on the description of the human behaviour and the current context, we can compile a hidden Markov model, basically capturing all possible changes of the environment. We can furthermore infer which goals the user tries to accomplish, a process known as intention recognition. Based on these goals, the current state of the environment and the formal description of the device actions, we can construct action sequences to support the user in the specific situation. Those action sequence can now simply be attached to the states and be executed whenever the system enters the state.

Due to the fact that the whole probabilistic model is synthesised from a formal description, the system is able to explain its actions. As described above, a state of the probabilistic model corresponds to a certain situation with respect to the state of the world and the user's intention. Therefore, the system actually knows what it believes, namely that the user is trying to achieve a certain goal. On this basis and using the formal descriptions of the device actions, the system can provide an explanation of the taken actions, and thus address one of the major issues in the area of ubiquitous computing, namely the explainability [6].

For this to work, we need formal descriptions of both, the user behaviour and his preferences and of the devices of the environment. Those descriptions can be provided by means of user-independent task models generated by application experts, a user-dependent preference description provided by the user, and PDDL-fragments provided by the devices themselves. All fragments have to be collected by the controller and afterwards used to synthesise the probabilistic model.

4 Conclusions and Future Work

After discussing the problem of controlling heterogeneous dynamic device ensembles, we described a layer control architecture. The resulting system is based on a probabilistic state-based model with actions attached to states and a higher level controller, consisting of two layers: the first modifying the probability distributions and action sequences of the underlying model and the second to adapt the overall structure of the system based on formal descriptions.

We are currently building a system based on the proposed architecture and planning to use it for the control of a smart meeting room equipped with numerous sensors and actuators. Further steps will include the investigation of the formal properties of the system. In particular the correctness of the internal transformation routines and the consistency while modifying the underlying probabilistic model.

The control architecture described above has so far been designed as a central component controlling distributed devices. An interesting and challenging alternative to this central system is a distributed implementation.

References

1. Weiser, M.: SIGMOBILE Mobile Computing and Communications Review 3(3), 3 (1999), doi:10.1145/329124.329126, `http://dx.doi.org/10.1145/329124.329126`
2. Bader, S., Burghardt, C., Kirste, T.: From Symbolic to Probabilistic Models. In: Mileo, A., Delgrande, J.P., Merico, D. (eds.) Proceedings of the First International Workshop on Logic-Based Interpretation of Context: Modelling and Applications. CEUR Workshop Proceedings, vol. 550, pp. 7–8 (2009)
3. Rabiner, L.R.: Proceedings of the IEEE 77(2), 257 (1989)
4. Ghallab, M., Isi, C.K., Penberthy, S., Smith, D.E., Sun, Y., Weld, D.: PDDL - The Planning Domain Definition Language. Tech. rep., CVC TR-98-003/DCS TR-1165, Yale Center for Computational Vision and Control (1998)
5. Paterno, F., Santoro, C.: The concurtasktrees notation for task modelling. Tech. rep (2001)
6. Kyng, M.: Making dreams come true: or how to avoid a living nightmare. In: Proceedings of the 12th ACM International Conference on Ubiquitous Computing, SEP 201, pp. 1–2. ACM, New York (2010)

Application of Artificial Neural Networks for Inflow Estimation of Yuvacık Dam Catchment Area

Bahattin Yanık, Melih Inal, and Erhan Butun

Abstract. Inflow data for longer length at a reservoir site is necessary for proper planning and operation of the reservoir. However presently for most of the reservoirs, the measured length of inflow data is insufficient for use in planning and operation. Artificial neural networks (ANNs) have been applied within the field of hydrological modeling for over a decade but relatively little attention has been paid to the use of these tools for flood estimation in catchments. Modeling of non-linearity and uncertainty associated with rainfall-runoff process has received a lot of attention in the past years. We analyzed the potential of neural network models for the estimation of inflow for Yuvacik Dam Catchment. Multilayer feed-forward neural networks were developed to model the relationships between known rain, snow depth and temperature data. Results suggest that artificial neural network model can be simple, robust, reliable and a cost-efficient tool for environmental inflow determination at the catchment area.

Keywords: Artificial Neural Networks, inflow estimation, dam catchment area.

1 Introduction

Development of physically-based models requires an understanding of all the physical processes which impact a natural process and the interactions among them. Since identification of the relationships among these physical processes is very difficult, data-driven approaches have recently been utilized in hydrological

Bahattin Yanık
ISU, Kocaeli Water and Seewage Administration
e-mail: ybahattin@hotmail.com

Melih Inal
Technical Education Faculty, Kocaeli University
e-mail: minal@kocaeli.edu.tr

Erhan Butun
Civil Aviation College, Kocaeli University
e-mail: ebutun@kocaeli.edu.tr

A. Abraham et al. (Eds.): International Symposium on DCAI, AISC 91, pp. 389–397.
springerlink.com

modeling. Artificial neural networks are one of the widely used data-driven approaches for modeling hydrological processes. Rainfall-Runoff modeling has a significant role in operational flood management procedures like flood forecasting, flood warning and design of hydraulic systems. Though many conceptual or physically based models are popular, black box models are considered as very useful tools for operational hydrologists particularly in hydrological data scarcity scenarios. A vast variety of black-box-rainfall runoff models have been proposed and demonstrated in recent years. Alcazar and at al, analyzed the potential of neural network models for the estimation of environmental flow values in gauging sections and reaches under a natural flow regime in the watershed of the Ebro River, Spain, with a view to a future application in both ungauged and/or regime-altered sections [1]. Remesan and at al, explores the ability of two artificial intelligence techniques, namely Neural Network Auto Regressive with exogenous input (NNARX) and adaptive neuro-fuzzy inference system, to model the rainfall runoff phenomenon effectively from antecedent rainfall and runoff information. Specifically, to illustrate applicability of these techniques, two year (1994-1995) rainfall-runoff data from Brue catchment of The United Kingdom were used [2]. Partal and Cigizoglu aims to estimate and predict the suspended sediment load in rivers by a combined wavelet–ANN method. Measured data were decomposed into wavelet components via discrete wavelet transform, and the new wavelet series, consisting of the sum of selected wavelet components, was used as input for the ANN model. The wavelet–ANN model provides a good fit to observed data for the testing period [3]. Estimation of future monthly river flows for Guvenc River, Ankara is conducted using various artificial neural network models. Success of artificial neural network models relies on the availability of adequate data sets. A direct mapping from inputs to outputs without consideration of the complex relationships among the dependent and independent variables of the hydrological process is identified. In this study, past precipitation, river flow data, and the associated month are used to predict future river flows for Guvenc River [4]. Dawson and at al, uses data from the Centre for Ecology and Hydrology's Flood Estimation Handbook (FEH) to predict T-year flood events and the index flood (the median of the annual maximum series) for 850 catchments across the UK. When compared with multiple regression models, ANNs provide improved flood estimates that can be used by engineers and hydrologists [5]. Blazkova and Beven applied a continuous simulation approach to the estimation of flood frequency for a dam site in a large catchment (1186 km2) in the Czech Republic. The models used allow for the simulation of both high intensity and low intensity rainfall events, and snowmelt events, over subcatchments in contributing to the flood frequency distribution [6]. Based on the mid-to-long-term hydrological phenomena is a typical fuzzy system of the basic thinking, Guohua and at al, combined cause-andeffect and statistical analysis with fuzzy analysis and chooses predictors such as rainfall and atmospheric circulation in earlier stage which effect the annual maximum peak discharge. Then sets up the fuzzy model of pattern recognition that used to the Xiangtan station of the Xiangjiang river [7]. Applicability of non linear model based on ANN with a random component embedded is explored for Pawana reservoir in Upper Bhima River Basin,Maharashtra, India. Suitability of time lagged recurrent networks (TLRNs) with time delay, gamma and laguarre memory structures is investigated for predicting seasonal (June to October) reservoir inflow with a monthly time step [8].

2 Study Area

TheYuvacık Dam Cachment Area, chosen as the study area, it is located at the northwest of the Turkey, in Marmara Region (Fig. 1). It has an approximate watershed area of 25786 km2 and all cachment area is divided four different sub catchment area: Kirazdere, Kazandere, Serindere and Contribute. Reprensative coefficients for four sub catchment areas were listed in Table 1. There are 12 gauging stations within the watershed.

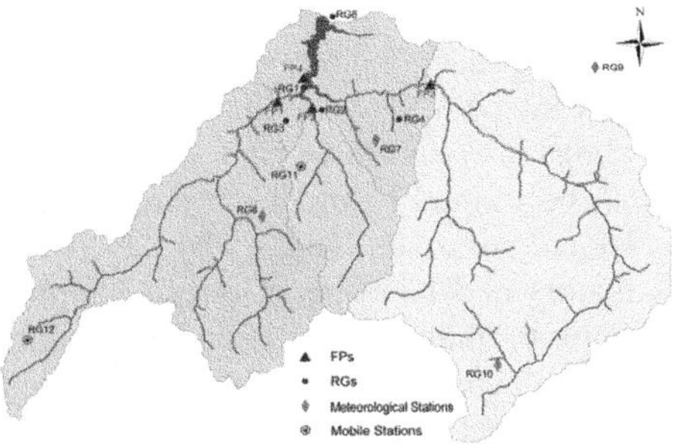

Fig. 1 Yuvacık Dam Cachment Area

Table 1 Reprensative coefficients for four sub catchment areas

	Kirazdere Sub-Catchment (FP1)	Kazandere Sub-Catchment (FP2)	Serindere Sub-Catchment (FP3)	Contribute Sub-Catchment (Fcont)
	79,54 km^2	23,10 km^2	120,53 km^2	34,69 km^2
RG1				0,247
RG2		0,11		0,094
RG3	0,058			
RG4			0,08	0,247
RG6				0,299
RG7		0,21	0,07	0,113
RG8	0,92	0,38		
RG9			0,35	
RG10			0,5	
RG11	0,022	0,3		

3 ANN Model

A neural network is a powerful data modeling tool that is able to capture and represent complex input/output relationships. The motivation for the development of neural network technology stemmed from the desire to develop an artificial system that could perform "intelligent" tasks similar to those performed by the human brain. Neural networks resemble the human brain in the following two ways:

1. A neural network acquires knowledge through learning.
2. A neural network's knowledge is stored within inter-neuron connection strengths known as synaptic weights.

The true power and advantage of neural networks lies in their ability to represent both linear and non-linear relationships and in their ability to learn these relationships directly from the data being modeled. Traditional linear models are simply inadequate when it comes to modeling data that contains non-linear characteristics. The most common neural network model is the multilayer perceptron (MLP). This type of neural network is known as a supervised network because it requires a desired output in order to learn. The goal of this type of network is to create a model that correctly maps the input to the output using historical data so that the model can then be used to produce the output when the desired output is unknown. Since that time research into ANNs has expanded and a number of different network types, training algorithms and tools have evolved. Given sufficient data and complexity, ANNs can be trained to model any relationship between a series of independent and dependent variables. That's why ANNs are considered to be a set of universal approximators and have been usefully applied to a wide variety of problems that are difficult to understand, define, and quantify—for example, in finance, medicine, engineering, etc. In the context of this paper, ANNs are trained to represent the relationship between a range of catchment descriptors and associated flood event magnitudes. There is no need for the modeler in this case to fully define the intermediate relationships (physical processes) between catchment descriptors and flood event magnitudes—the ANN identifies these during the learning process.

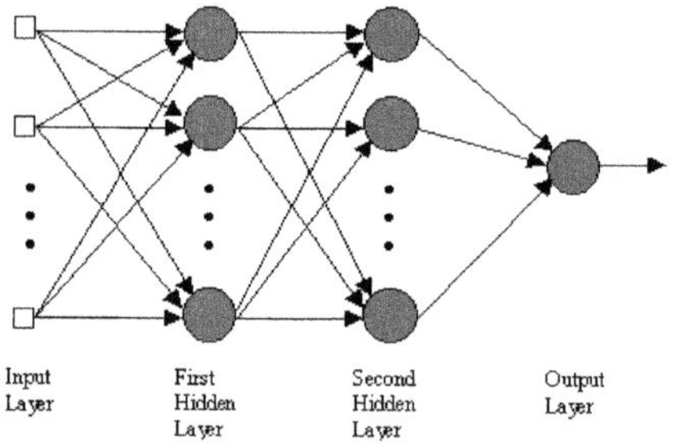

Fig. 2 A graphical representation of an MLP

Although there are now a significant number of network types and training algorithms, this paper will focus on the Multi-Layer Perceptron (MLP). In this case, the ANN has three layers of neurons (nodes) - an input layer, at least one hidden layer and an output layer. Each neuron has a number of inputs (from outside the network or the previous layer) and a number of outputs (leading to the subsequent layer or out of the network). A neuron computes its output response based on the weighted sum of all its inputs according to an activation function (in this case the logistic sigmoid). The network is trained by adjusting the weights that connect the neurons using a procedure called error backpropagation. With backpropagation, the input data is repeatedly presented to the neural network. With each presentation the output of the neural network is compared to the desired output and an error is computed. This error is then fed back or backpropagated to the neural network layer by layer and used to adjust the weights such that the error decreases with each iteration and the neural model gets closer and closer to producing the desired output. This process is known as training. The basic feedforward network performs a non-linear transformation of input data in order to approximate the output data. The number of input and output nodes is determined by the nature of the modeling problem being tackled, the input data representation and the form of the network output required. The number of hidden layer nodes is related to the complexity of the system being modeled. The interconnections within the network are such that every neuron in each layer is connected to every neuron in the adjacent layers. Each interconnection has associated with it a scalar weight which is adjusted during the training phase. The hidden layer nodes typically have sigmoidal transfer functions.

4 Learning Algorithms

In our study we used several different variations of the backpropagation training algorithm, each of them having a variety of different computation and storage requirements. They are summarized the training algorithms used in the searching procedure of the model with the highest level of accuracy.

Table 2 Training algorithms used in the searching procedure

Algorithm	Description
Gradient Descent with Adaptive Learning Rate (GDX)	Faster training than Gradient Descent, but can only be used in Batch training mode
Levenberg-Marquardt (LM)	Faster training algorithm for networks with moderate size, with ability of memory reduction for use when the training data set is large

Used technique requires the data set to be divided into two subsets: training, test and validation set. The training set consists of 1334 days data, between 1999 and 2009 years and is used for computing the gradient and updating the network weights and biases and the training procedure monitors the Mean Squared Error (MSE) of the validation set for 30000 epoch and as soon as the error starts to increase the training stops and returns the weights at the phase where the MSE is 10^{-6}. In catchment areas, wetness is most important parameter for estimating inflow rate. For this reason, first 10 and then 2 days data were used to train models shown in Fig. 3 and 4 respectively.

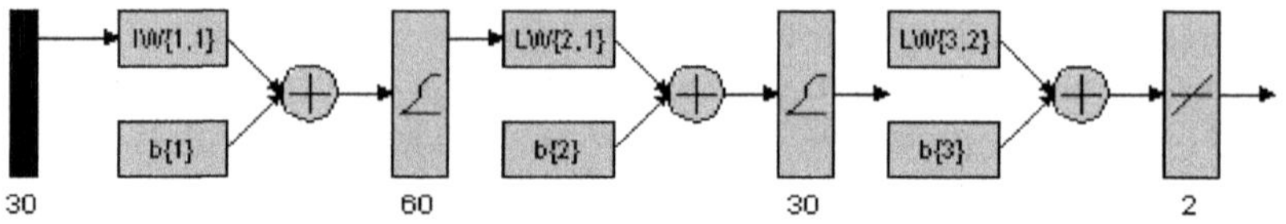

Fig. 3 Training model for 10 days data

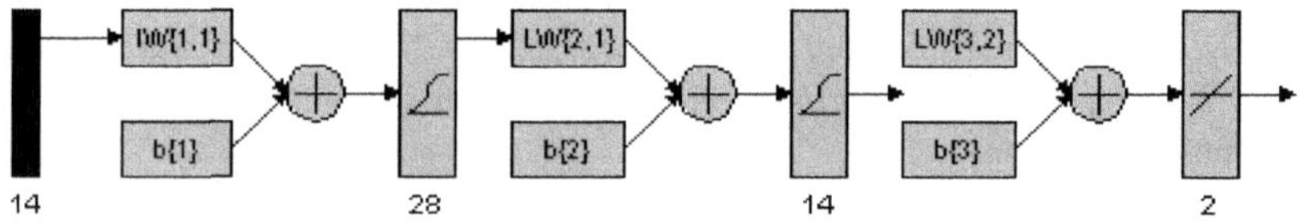

Fig. 4 Training model for 2 days data

Results after training for 10 and 2 days are listed in Table 3 and 4 respectively.

Table 3 Training model results for 10 days data

	Model	MSError (Fcont Ftotal)	MSError_test (Fcont Ftotal)	Test Min. Error Value and Index	Test Max. Error Value and Index
Levenberg-Marquardt and TanSigmoid	fcont net_ftotal_onceki_ft_ve_cont_trainlm_tansig_norm_karsiz ftotal	0.1027 10^-3 0.0972 10^-3	470.7057 514.3509	-143.9269 32 -137.7739 32	47.3614 30 81.0542 30
	fcont net_ftotal_onceki_ft_ve_cont_trainlm_logsig_norm_karsiz ftotal	0.0965 10^-3 0.1029 10^-3	277.7908 198.6201	-89.7367 14 -62.1639 14	55.9895 38 39.8763 38

Table 3 *(continued)*

Gradient Descent with Adaptive Learning Rate And	**fcont net_ftotal_onceki_ft_ve_cont_traingdx_tansig_norm_karsiz ftotal**	***0.0468*** ***0.0697***	***1.0136*** ***0.8691***	***-4.7527 39*** ***-2.7440 37***	***2.7652 14*** ***4.3125 36***
	fcont net_ftotal_onceki_ft_ve_cont_traingdx_logsig_norm_karsiz ftotal	***0.0571*** ***0.0842***	***0.7391*** ***3.2633***	***-3.0132 35*** ***-3.3831 32***	***5.1533 30*** ***14.9845 30***

Table 4 Training model results for 2 days data

	Model	**MSError (Fcont Ftotal)**	**MSError_test (Fcont Ftotal)**	**Test Min. Error Value and Index**	**Test Max. Error Value and Index**
Levenberg-Marquardt and TanSigmoid	fcont net_ftotal_onceki_ft_ve_cont_trainlm_tansig_norm_karsiz_2delay ftotal	0.0108 0.0110	2.9823 10^3 3.9625 10^3	-235.2003 51 -273.9539 51	283.3554 37 327.7100 37
	fcont net_ftotal_onceki_ft_ve_cont_trainlm_logsig_norm_karsiz_2delay ftotal	0.0104 0.0097	7.0734 10^3 9.0979 10^3	-222.6024 21 -257.3828 21	438.4752 39 490.4559 39
Gradient Descent with Adaptive Learning Rate And LogSigmoid	**fcont net_ftotal_onceki_ft_ve_cont_traingdx_tansig_norm_karsiz_2delay ftotal**	**0.0753** **0.1418**	**0.3788** **2.3798**	**-4.3932 40** **-9.8993 40**	**1.7595 41** **8.5992 38**
	fcont net_ftotal_onceki_ft_ve_cont_traingdx_logsig_norm_karsiz_2delay ftotal	**0.0789** **0.1498**	**0.6700** **2.9510**	**-1.0902 22** **-3.1601 22**	**6.7246 38** **14.9323 38**

According to Table 3 and 4, Model 1 and Model 2 are chosen according to their performances. They are marked with bold characters. They are trained with 10 years data. Performance of these two models are given in Figure 5 and 6 respectively.

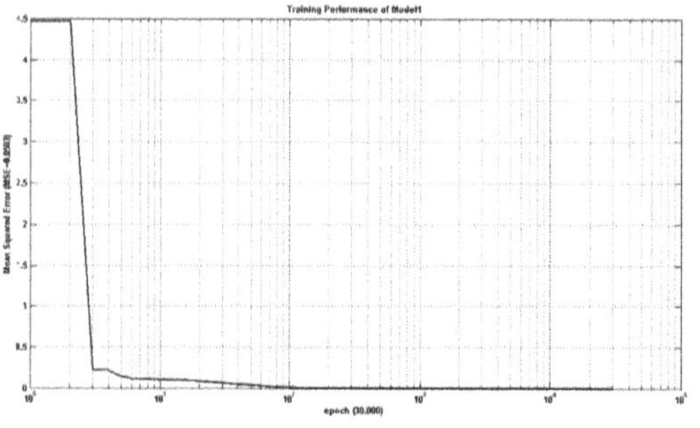

Fig. 5 Training performance of Model 1.

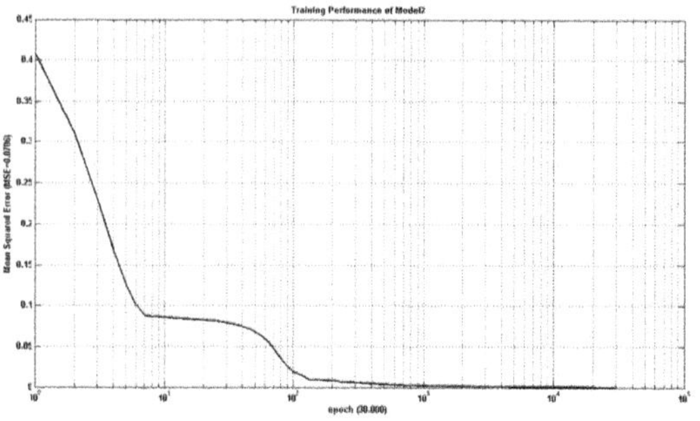

Fig. 6 Training performance of Model 2.

After training, both of the models are tested by using year 2010's data and results are shown in Figure 7.

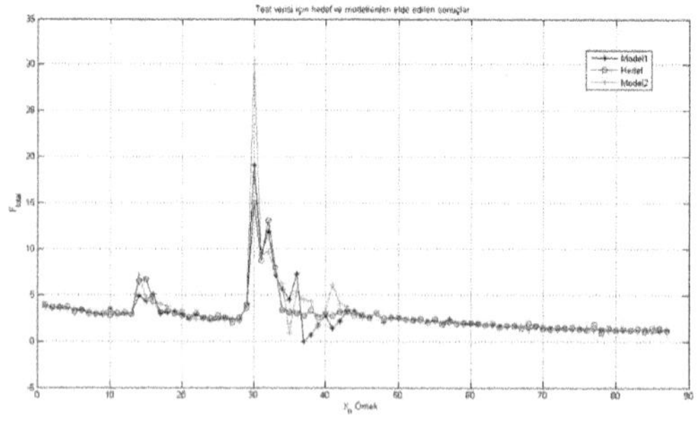

Fig. 7 Training performance of Model 1.

5 Conclusions

The results of this study supported artificial neural network models as simple, robust, reliable and cost-efficient tools for inflow determination of Yuvacik Dam Catchment Area. We found two models capable of good estimations of inflow with a probability of approximately 90%, for 12 gauging stations under natural flow regime in the Catchment Area. Overall results are satisfactory, considering the extreme conditions are slightly over-estimated. One of the major weaknesses of ANN models is that they may fail to generate good estimates for extreme events. However, since the ANN model has the chance to be trained well for regular events it has higher chances of producing reliable results Thus, it is very important to be able to identify the extreme events. Classification of an input vector as a regular or an extreme event may provide important information for the planning and management of water dam.

References

1. Alcázar, J., Palau, A., Vega-Garcí, C.: A neural net model for environmental flow estimation at the Ebro River Basin, Spain. Journal of Hydrology 349, 44–55 (2008)
2. Remesan, R., Shamim, M.A., Han, D., Mathew, J.: ANFIS and NNARX based Rainfall-Runoff Modeling. In: IEEE International Conference on Systems, Man and Cybernetics, pp. 1454–1459 (2008)
3. Partal, T., Kerem Cigizoglu, H.: Estimation and forecasting of daily suspended sediment data using wavelet–neural networks. Journal of Hydrology 358, 317–331 (2008)
4. Kentel, E.: Estimation of river flow by artificial neural networks and identification of input vectors susceptible to producing unreliable flow estimates. Journal of Hydrology 375, 481–488 (2009)
5. Dawson, C.W., Abrahart, R.J., Shamseldin, A.Y., Wilby, R.L.: Flood estimation at ungauged sites using artificial neural networks. Journal of Hydrology 319, 391–409 (2006)
6. Blazkova, S., Beven, K.: Flood frequency estimation by continuous simulation of subcatchment rainfalls and discharges with the aim of improving dam safety assessment in a large basin in the Czech Republic. Journal of Hydrology 292, 153–172 (2004)
7. Guohua, H., Xing, Y., Hehua, S.: Medium-Long Term Forecast of the Annual Maximum Peak Discharge at the Xiangjiang River Basin on Fuzzy Method. In: International Forum on Information Technology and Applications, pp. 200–204 (2009)
8. Kote, A., Jothiprakash, V.: Reservoir Inflow Prediction Using Time Lagged Recurrent Neural Networks. In: First International Conference on Emerging Trends in Engineering and Technology, pp. 618–623. IEEE, Los Alamitos (2008)

Integrating 3D Animated Characters with Adaptative Tests

Carina S. Gonzalez

Abstract. In this paper we present a 3D animated character developed with Blender that evaluates the knowledge of a user in a particular subject using an adaptive test. The avatar was built with the Game Engine of Blender and Python. Our adaptive test is based on XML as data manager and CLIPS as inference engine. Both are connected with a new library in Python that we had to develop for this purpose. Some characteristic of our project include: a) mesh deformation in real time (dynamic creation and edition of IpoCurves in Game Engine), b) Interaction with external applications to Blender (such as CLIPS) using pipes, c) use of threads for concurrent execution of tasks, d) use of Text to Speech.

Keywords: 3D animation character, Adaptive tests.

1 Introduction

The cognitive-emotional processing of information is vital for learning and avatars can be used to transmit non-verbal information improving the communication [1]. In this sense, the objective of this work is to present in a friendly way adaptative tests to the users. So, in this paper we will focus on the developed 3D character and the structure of the adaptative test presented to the user.

The developed 3D animated avatar has a humanoide shape which function is formulating the questions to the user using adaptative tests. The 3D avatar developed in Blender is not limited to the modeled and animation, but there is a complex development in Python [2].

The system is formed by three fundamental modules: a) Question Manager (QM): responsible of select and evaluate questions for the adaptive test. It has been developed in Python, b) User Interface (UI): compose by the avatar developed in Blender, and c) Inference Engine (IE): responsible of select the level of knowledge of user through the process of rules in CLIPS [3].

Carina S. Gonzalez
Departamento de Ingeniería de Sistemas y Automática
Universidad de La Laguna
e-mail: `cjgonza@ull.es`

A. Abraham et al. (Eds.): International Symposium on DCAI, AISC 91, pp. 399–406.
springerlink.com

The IE and the QM determine the results of the test, that are stored in XML files. These data are process by the UI using a module developed in Python. The interaction among these modules is carry out by the GameEngine of Blender, so our animated character has the total control of the system leading the interaction with the user.

The following sections shows the differents parts of our system. First of all, we present the how we have modeled and animated the 3D avatar, then we show how we have iluminated the scenes, after the TextToSpeech (TTS) module is explained and finally how is produced the interation between the avatar and the adaptative test is described.

2 Modeled and Animation

The modeled and animation of the 3D avatar was completely made in Blender [4]. The 3D character is formed by a head and all those elements related to it. At the time of approaching the eyes, we decided to create an eye Pixar style. This type of eye is characterized to give much personality to the character by means of the depth. This is easy to simulate, for it is created the eye from 4 parts: ocular globe, rainbow, cheat on and pupil. Of course, the eyes have to be complemented with eyelashes. For its implementation it's has been used a plane with alpha channel. This technique consist in to insert a plane, of the approximated thickness of an eyelash. An image with black lines is created (the eyelashes) and the rest is transparent (using a channel alpha). This image is mapped in the rectangle created previously and it is curved to obtain the effect of curly of the eyelashes.

The hair is one of the elements that more realism gives to an animated character, but, simultaneously, it is one of the most complicated part to approach. The hair has been modeled from a mesh to which the wished form has occurred it. Next, this mesh has been divided by heights and made slow parent of each one of the heights carried out weak. Finally an offset has been assigned to these heights obtaining a realistic movement.

The 3D avatar is located in a stage formed by several elements (Figure 1). There is a table that serves as support for the head, which is within a room with metallic walls. All this structure has been illuminated with centers of type lamp. The illumination of the room has not been created with a predefined scheme, it has been proved until finding the wished effect.

The hole left in one of the walls has filled up with a window. Through this hole and the ceiling, can be observed the sky, that has been created using two planes. In the sky it is possible to observe the way a cloud movement takes place. This was obtained by means of a plane with a applied texture. Through the window, an incessant traffic of air vehicles can be observed. This effect helps to give the impression that we are in a great futurist city. The user can interact and control different elemens of the stage through buttons.

Fig. 1 Stage with the 3D avatar

In order to represent the answers given by the users, a picture has been designed to show the chosen option. After formulating the question, the user must introduce his election by means of keyboard. After this, the picture will reflect the answer until the moment at which a new answer is asked for to him (that is, when finishing of formulating the following question).

2.1 Gestures

In order to give her a little more life and naturalness to the character, we decided to create a script that execute a set of gestures ramdomly. Within these gestures we found the movement of head, eyes, mouth, etc. Basically the algorithm that has been implemented is explained below.

In some especific times (regulated by timer) it is generated a number random that will determine the gesture. When the gesture finishes the character remains in delay until the following gesture is generated.

During the execution of the previous algorithm different types of gestures take place. The gestures can be developed through animation curves (in Blender called "IPO curves"):

a) IPOs from GameEngine: in order to activate animations supported by the module of games; in our case, animations of rotation and position;

b) IPOs from code: we invoked from code the IPOs that uses relative vertex keys. We do not have left more alternative than to call to our method for the execution of this type of IPO from Python.

c) Motion: in some occasions we used this type of animation due to its simplicity.

2.2 Animation

Due to the versatility that allows we used Relative Vertex Keys (RVK). The different relative keys used (related or not to the movement of the mouth) are the following:

a) Keys used for Key blinking: closed straight eye and closed left eye.

b) Keys used for Key speech: inferior lip slightly lowered; movement of right comisura; movement of left comisura; superior lip slightly raised; tongue down; tongue arrives; key inferior lip raised; lowered superior lip; smile.

Through these keys, it is possible that the mouth adopts the wished form. We have been used a method in Python to control the animations. The values of the keys are stored in a configuration file that reads the 3D avatar when being sent. In this file the buccal groups and the letters belonging to each of them are identified themselves, with the optimal values for their representation. This allows an extense range of action: from being able to sharp the simulation of the movement of each buccal group, to the generalization of these movements creating the corresponding file to different languages.

2.3 Illumination

The illumination was created with standard lamps of Blender distributed in an arbitrary form. There are many elements without illumination, because has not been considered necessary. These elements are: a) the hair: when having a very dark colour, the illumination would not be appreciated; b) the eyelashes and the eyebrows: by the same reason that the hair; c) the bellboys and the ships. In the case of the eyes, two lamps were located to give something of light. The stage was illuminated adding lamps as the surroundings were created. To create the illumination of the most important part, the head, we used a predefined scheme.

For the illumination of the head a classic scheme in the cinema was used. This system uses three light sources: a) key light: it is the light located in front of the head, spot type (directed light) and is strongest of the scene; b) light of stuffed: it is the central light that has circular form, lamp type, to be able to fill up better the hollows lazy by the key light; c) back light: it is the light located behind the head. In the Blender it is a light of spot type of smaller force than the other two.

3 TextToSpeech

The TTS is a tool that allows to transform text written in spoken text. In our case, the TTS was implemented by means of Speech API of Microsoft. This API allows to make speech functions from the code in Python.

The first problem when trying to communicate Blender with the TTS is produced when it calls to the TTS, because the execution in Blender is left blocked until the execution of the sintetyzer of voice ends. This is totally unacceptable that the voice and the movement of the mouth would be totally out of phase. In order to resolve this problem it has been necessary to use a thread-demon which allows that the independet execution of the TTS and of Blender. Once established this independence we were with another problem: the possibility that a new thread is executed, or rather demon, without has finished the previous one. In order to solve this, we places a semaphore that it indicates if the execution of the thread has finished, so that it allows only a new call to the TTS when the previous one has finished.

4 Control System

The GameEngine is the module of game of Blender that allow us to carry out very complex operations supported in scripts in Python. Next, we explained the basic structure of the logic of games used in our project.

The control system, developed in Python, is the kernel of the logic of games. The control kernel makes the following functions: a) to make the mouse appears; b) to initialize the CLIPS; c) to load the profile of the user; d) to load all the necessary files for the execution; e) to pass to the following question and f) to send the blinking.

The speech in the GameEngine, is implemented in two functions: a) *hablar.py:* to control the Speech API, that is as well in charge to initiate the sound of the voice. This bookstore has many parameters, which it allows to make changes on the voice of the character in a quite simple way; b) *moverboca.py:* its function is to cause that the mouth of the character moves while speaking. The way in which the mouth appears in the configuration file.

Another script that is executed on continuous way is gestos.py. This file causes that the character makes random gestures eventually and its main function is to modify the value of the variable "k". This value is obtained from random way and establishes that type of gesture is due to execute, explained in the section 2.1. Each gesture is activated when the variable "k" is in certain interval.

The motor of 3D avatar inference is implemented in CLIPS and it was needed to communicate with it from Blender. So, we have implemented a tunnel between the process of Blender and the process of CLIPS, using the bookstores of management of processes of Python. The working of this tunnel is quite simple, but is quite complicated o implement; the idea is to have a process of CLIPS sent in background, and that 3D avatar is sending to it the entrances and gathering the exits; that is to say, that 3D avatar is the user who is introducing the adecuated commands in the CLIPS and reading the corresponding exits. The user can select the choices of the answer through letters of the alphabet, pressing it in the keyboard.

5 Adaptive Tests

For our project we used the Bayesian adaptative test as solution [5,6,7]. With this kind of adaptive tests we can make a dinamic assessment of the kwnoledge of student, where depending of his answers the difficulty is incremented or decremented.

The structure of the adaptative test is stored in XML format. A test is divided in syllabyus (A), a syllabus in topics (T), the topic in concepts (C), the concepts in questions (P), and each question has associated a variable number of answers (Figure 2).

In the Bayesian net we have two phases: a) *evaluation phase*: use the results of the previous phase in the propagation of probabilities, an a punctuation system to determine the final punctuation obtained for the student and b) *assessment phase*: use the part of the net with concepts, questions and their relations. In this phase, the concepts that the student knows is determined trough his answers.

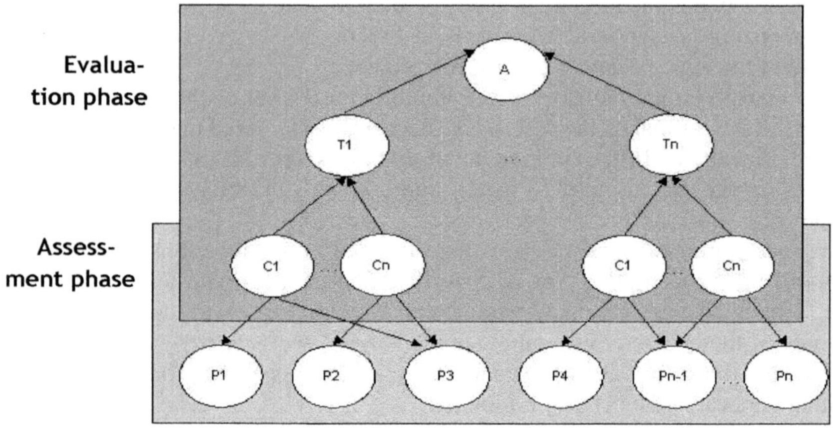

Fig. 2 Structure of the adaptative test

The concepts has four difficulty levels: simple, intermediate, complex and advanced. However, each questions has associated two probabilities: a) the probability of to know the answer and b) the probability of guess the answer (or aleatory probability). So, the probability to answer correctly the questions is the sum of a) plus b). Nevertheless, we distinguish three difficulty levels in questions: low, media, high.

On the other hand, regarding to the user, we distinghish different bayesian sets according to the knwoledge: novel, intermediate, advanced and expert. The level of knwoledge varies related with the test execution. The level asigned a priori is novel. In summary, the probabilistics sets implemented in our test are:

-*Concepts*: simple, intermediate, advanced and complex.
-*Questions*: low, media, high.
-*Users*: novel, intermediate, advanced y expert.

The evaluation process has been implemented in a function in Python (clipsxml.py) and by rules in CLIPS (Figure 3).

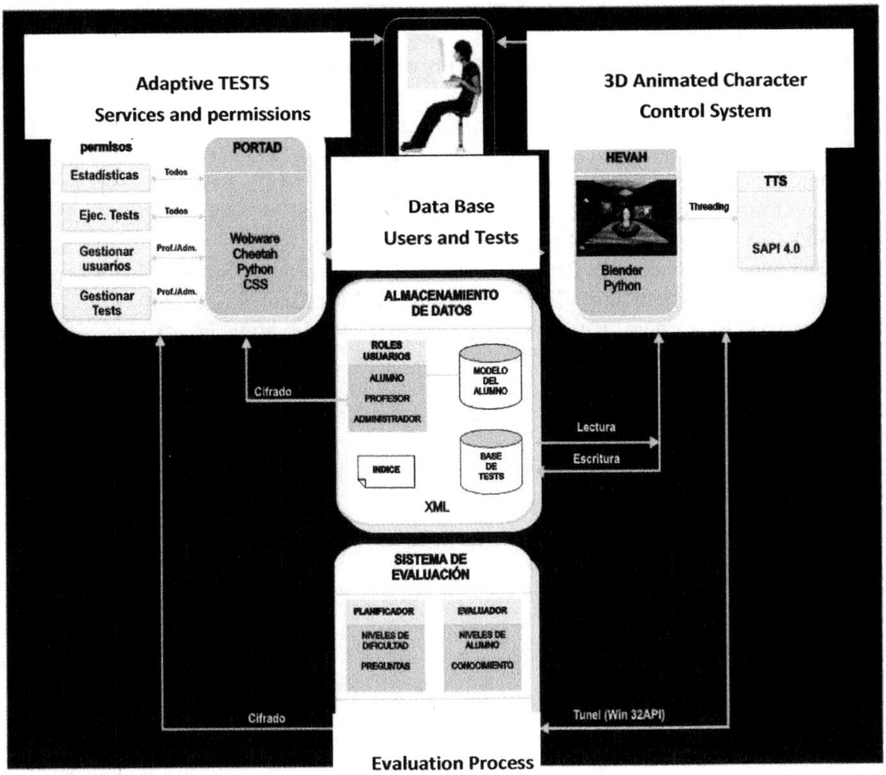

Fig. 3 Architecture of the system

6 Conclusions

In this paper we have presented an intelligent system composed by a 3D animated avatar developed in Blender and Python, and a module of Bayesian adaptative tests based in XML, CLIPS and Python. Several phases of the developing process has been presented as well as the found problems and the implemented solutions. We use this system to teach a topic of Memory Hierarchy in the Computer Architecture's syllabus in the School of Computer Science Engineering at University of La Laguna. We are currently working about Bayesian adaptative tests mixed with conceptual maps and implemented collaborative learning functions in the system.

References

[1] Dontschewa, M., Künz, A., Kovanci, S.: Development of 3D avatars for professional education. In: Shumaker, R. (ed.) VMR 2009. LNCS, vol. 5622, pp. 154–158. Springer, Heidelberg (2009)

[2] González, C.S., González, D., Santos, F., Ramos, Z.: Hevah Project. In: Blender Conference 2004 (2004)
[3] González, D., Santos, F., Ramos, Z.: Proyecto Final de Carrera. Escuela Superior de Ingeniería Informática (2004), http://hevah.cyc.ull.es
[4] Blender Organization, http://www.blender.org/
[5] Moreno, L., González, E., González, C.S., Piñeyro, J.D.: Towards a support for autonomus learning process. In: Omatu, S., Rocha, M.P., Bravo, J., Fernández, F., Corchado, E., Bustillo, A., Corchado, J.M. (eds.) IWANN 2009. LNCS, vol. 5518, pp. 582–585. Springer, Heidelberg (2009)
[6] Guzmán, E., Conejo, R., Pérez-de-la-Cruz, J.-L.: Improving Student Performance Using Self-Assessment Tests. IEEE Intelligent Systems 22(4), 46–52 (2007)
[7] Guzmán, E., Conejo, R.: Towards efficient item calibration in adaptive testing. In: Ardissono, L., Brna, P., Mitrović, A. (eds.) UM 2005. LNCS (LNAI), vol. 3538, pp. 402–406. Springer, Heidelberg (2005)

Depth-Wise Multi-layered 3D Modeling

S.S. Mirkamali and P. Nagabhushan

Abstract. 3D modeling is an emerging trend both in the areas of machine vision and computer graphics. With the current 3D modeling systems the user can virtually travel around a scene and see the foreground objects. A 3D model would be more realistic if the user could also go into the depth of a scene from a specific view and see the obscured objects as well as foreground objects. The aim of this paper is to present a new scheme of 3D modeling which capable of segmenting a specific view of a scene into depth-wise multiple layers followed by layer peeling. The previous works in the area of depth-wise segmentation is reviewed in detail. Later the issue of layer peeling which is causing holes in the residual image is discussed and the solutions to the problem are reviewed exhaustively.

Keywords: Multi-layered segmentation, 3D Modeling, Layer peeling, Inpainting.

1 Introduction

Three-dimensional (3D) modeling is an emerging trend both in the areas of machine vision and computer graphics as we presently witness the appearance of 3D-TVs, 3D video games, and3D virtualtours. Thediverseapplications of 3D modeling have necessitatedtheirimprovement whichhas been partly achieved through the introduction and use of somecomplex and realistic models. One such improvement can be achieved by segmenting a 3D model into multiple layers. Most of the layer segmentation methods separate layers of an image into the foreground (moving objects) and the background layers (static objects). However, there are some methods of layer segmentation which separate an image into a multiple layers regardless of the depth order of the objects. The first layer comprises all those objects that are simultaneously visible irrespective of their depth. The subsequent layers will include the occluded objects.

S.S. Mirkamali
Department of Studies in Computer Science, University of Mysore, Mysore, India
e-mail: s.s.mirkamali@ieee.org

P. Nagabhushan
Currently on leave from University of Mysore, BESTAR: Bangalore Educational Society For Technology Advancement and Research, Bangalore, India
e-mail: pnagabhushan@hotmail.com

A. Abraham et al. (Eds.): International Symposium on DCAI, AISC 91, pp. 407–414.
springerlink.com

The proposal is to introduce a novel scheme of layer segmentation called *depth-wisemulti-layered 3D modeling*. This technique would be capable of segmenting a 3D model into multiple layers on the basis of the depth and shape of objects perceivedin the layers.The process will start with an available sequence of images (or a video) captured from different views of a static scene which may contain a clutter of objects positioned in a particular location and continueswith the *depth-wise segmentation*of the 3D model from a specific view. In this paperdepth-wise segmentation is theprocess of segmenting an image into multiple layers based on depth of objects followed by*peeling off*the layers. Figure 1 illustrates an example of an expected layered view of a scene along with its depth-wise segmented layers and the anticipated residual images after peeling off the layers.

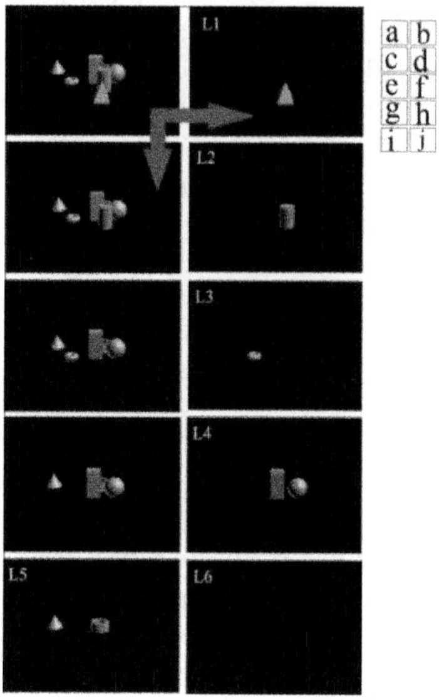

Fig. 1 a) A view from of a static scene. b) The first layer of the view. c) Expected result of peeling off the first layer (L1) from the selected view. d) The second layer (L2) e) Remaining objects of the scene after peeling off the L1 and L2. f) The third layer of the selected view. g) Expected objects of the scene behind the first three layers. h) The fourth layer of the selected view. i) Expected result of peeling off the first four layers (L1, L2, L3 and L4) which is the anticipated to be the fifth layer. j) Background layer.

Theoretically, peeling off the upper layers of a 3D image should reveal the obscured portion of objects present in the hidden layers. In reality, the act of layer peeling will caused holes in the residual image (Fig 2.b). To complete the holes (Fig 2.c) many methods have been introduced in the literature which are called *inpainting* algorithms. Most of the inpainting algorithms can either fill up the small holes (like cracks) using known data on the boundary of the holes or complete them from available textures of an alternative image. A novel inpainting algorithm is required which is capable of filling up of the holes in the residual images using information in the other views.

(a) (b) (c)

Fig. 2 a) Selected view from a sequence of images of a static scene. b) Residual image showing a hole caused by peeling off the first layer of image c) Expected image after inpainting.

2 Literature Review

Since the proposed scheme is expected to contain such components as depth-wise segmentation and inpainting, a detailed review of the related works for each component is given in a separate sub-section.

2.1 Depth-Wise Segmentation

Depth-wise segmentation is the process of segmenting a 3D image into multiple layers based on depth of objects followed by peeling offthe layers. Based on the given definition, this section is divided into two parts of depth reconstruction and layer segmentation.

2.1.1 Depth Reconstruction

There are mature algorithms for depth estimation from stereo images and exhaustive surveys of stereo methods available in the literature[1].Most of the dense two-frame stereo correspondence algorithms[2] first enumerate all possible matches at all possible disparities, then select the best set of matches in some way. This is a useful approach when a large amount of ambiguity may exist in the computed disparities. An alternative approach is to use methods inspired by classic (infinitesimal) optic flow computation. Here, images are successively warped and motion estimates incrementally updated until a satisfactory registration is achieved. These techniques are most often implemented within a coarse-to-fine hierarchical refinement framework[3-6].

A survey of Multiview stereo (MVS) methods canbe found in[7]. Many of these methods (e.g., voxel-based approaches[8-9]) aim to build a 3D model for a single object and are usually not applicable to large scale sceneries due to the high computational complexity and memory space requirement. The approaches based on multiple depth maps[10] are more flexible, requiring fusing view-dependent depth maps into a 3D model. Merrell et al.,[11] described a quick depth map fusion method to construct a consistent surface among multiple frames. Recently, a 3D modeling system with a hand-held camera is introduced by Pollefeys et al.,[12].In their method,depth maps are obtained by combining measurements of multiple pairstracked or matched between consecutive views and multi-view relations.

2.1.2 Layer Segmentation

Many of the problems in machine vision and computer graphics are formulated using the concept of layer segmentation. These problems can be categorized into two main classes of video object segmentation and image-based rendering.

For multi-object segmentation from the multi-view video of a dynamic scene Reid et al.,[13] proposed an algorithm uses Maximum A Posterior (MAP) estimation to compute the parameters of a layered representation of the scene, where each layer is modeled by its motion, appearance and occupancy. Some other methods formulate the problem of multi-view object segmentation based on layer separation using Epipolar Plane Image (EPI)[14], a volume constructed by collecting multi-view images taken from equidistant locations along a line.

Another popular application of the layer separation is in image-based rendering approaches. In[15], Shade et al., proposed the sprite with depth and the layered depth image (LDI). LDI is a view of the scene from a single input camera view, but with multiple pixels along each line of sight. Correspondingly, the depth map is also multi-valued for each pixel. One major problem in this rendering method is that holes may occur in the rendered view due to undersampling or disocclusion. The LDI tree[16] is a modified LDI approach which combines multiple reference views into a single hierarchical representation, which maintains the resolution of each reference view in the data structure. On the other hand, even if holes do happen, they may be removed through algorithms such as splatting or meshing.

Zhu et al.,[17] extract set of epipolar plane images (EPIs) from a long image sequence, and a panoramic depth map which is generated by analyzing the EPIs. They use layering to represent occluded regions and different spatial resolutions of objects with different depth ranges in different layers. Recently, Lee et al.,[18] proposed a method to layer a scene, in a single pass, to compute DOF blur. To create layers, they use layered rendering method. However, this method does not protect the shape of the objects of the scene and it may decompose them into different layers.

2.2 Inpainting

Inpainting refers to the specific image restoration task of reconstructing an image with a missing or damaged region. Digital inpainting was first proposed by Bertalmio et al.,[19] and subsequently many approaches followed. Most relevant and related works in this area are classified into two categories, image and video inpainting, in the following sub-sections.

2.2.1 Image Inpainting

Image inpainting algorithms can roughly be grouped into 3 types: Variational and partial differential equation (PDE), statistical, and exemplar based. The variational and PDE based methods solves optimization problems constrained on the known data on the boundary of the hole. An inpainting solution is obtained by finding the restored image which minimizes the functional. This optimization problem can be solved by using calculus of variation which will lead to the problem specific Euler-Lagrange PDE which has to be evolved numerically. A few examples of this type of methods can be found in the work by Ballester et al.,[20].

At the second corner of the triangle we have the statistically based methods[21], in which statistical models of image content are constructed and used for sampling new image content in the hole. These methods are mainly based on techniques for texture synthesis and as such works well for regions with stochastic and irregular textures.

Exemplar based methods[22-23]search for similar image patches, usually in the image surrounding the hole, and paste a filling of the hole using the found patches. Criminisi et al.,[23] choose the filling order based on a gradient measure on the boundary to the hole, which leads to an order where patches at high gradient edges going into the hole are visited first.

2.2.2 Video Inpainting

Most of video inpainting algorithms fundamentally evolved from image inpainting approaches. A Markov Random Field (MRF) is used to model the local distribution of the pixel and new texture is synthesized by querying the existing texture using a patch-based direct sampling process. This forms the basis of the space–time video completion scheme in which the inpainting is formulated as a global optimization problem. The hole of the video is filled by using spatio-temporal patches sampled from the existing video. Cheung et al. introduced a space–time patch model based on probabilistic learning with applications to inpainting[24]. Inpainting is treated as a reconstruction problem and the epitomes in this case are learned from the observed pixels. Inferring the missing pixels from the condensed epitomes leads to severe over smoothing of the reconstructed pixels. A video completion scheme based on motion layer estimation followed by motion compensation and texture completion has been proposed by Zhang et al.,[25]. After removing a particular motion layer, motion compensation is used to complete moving objects and non-parametric texture synthesis is used to complete the static background regions. The inpainted layers are then warped into every video frame to complete the holes. An object-based video inpainting algorithm based on a modular approach for a scene consist of stationary background and moving foreground regions is proposed by M. Vijay Venkatesh et al.,[26] using replacement. A novel sliding-window based dissimilarity measure in a dynamic programming framework is introduced to inpaint repetitive moving objects. unlike other inpainting algorithms, which is described in this survey, this technique can handle large regions of occlusions, inpaint objects that are completely missing for several frames, and has minimal blurring and motion artifacts.

3 Discussion

So far we have discussed a variety of successful methods of depth reconstruction, depth-wise segmentation and inpainting in details. However, there are many issues in each domain still remain unsolved.

Depth of a scene is estimated in many different ways using three different sources: i) monocular images ii) stereo images iii) multi-view stereo. The first group of algorithms, which use monocular images, employs global structure of the image as well as prior knowledge of that. Employing the prior knowledge of known specific classes of objects of a scene provides a good depth cues. However, preparing a knowledge-base of all the objects is almost impossible. The second group of works tries to recover the depth of a scene employing stereo images. Using a pair of calibrated cameras to capture the stereo images is the difficulty of these methods. Recently, many of researchers used a sequence of images or multi-view images instead of stereo images with the advantage of easy capturing of the images and the accessibility of more details of objects of a scene to reconstruct the depth map.

Progress in the field of layer segmentation can be traced through at least two different scientific disciplines. In object segmentation, a view is layered into two main layers, foreground (moving objects) and background (static objects). In image-based reconstruction the target image is separated into some constant number of layers, where the first layer contains pixels to be displayed in the scene and whose remaining layers represent occluded objects of the scene. Unfortunately, in these methods the ordering of layers does not match the ordering of objects of the scene based on their depth. Layer separation of still images is adapted to solve the problem of Depth-of-Field recovery. An image is separated into layers based on a constant distance value with respect tothe pinhole camera lens. However, there is no method in the literature to address the problem of segmenting an image into layers while preserving both the depth order and shape of the objects of a scene in the respective layers.

As image inpainting needs to be visually pleasant and clear, many algorithms have been specifically tailored for completing only small holes or cracks. Although there are methods that have dealt with large holes most of these methods use either a knowledge-base or repetitive moving objects of the scene to complete these types of holes. There are no reports in the literature showing that there are methods to address the problem of completing a hole without using a database of objects or without repeating the same texture of the scene.

4 Conclusion

A novel scheme of 3D modeling called depth-wise multi-layered 3D modeling is introduced in this paper. The major thrust isapplying a new method of depth-wise segmentation of a view of a scene and filling up the holes in the residual images caused by layer peeling. In the course, depth estimation is adapted as a part of the depth-wise segmentation technique. If the objectives of this paper achieved, the field of 3D modeling will have the advantage of making a 3D model capable of separating and manipulating the layers of a scene captured from multiple views.

References

[1] Scharstein, D., Szeliski, R.: A Taxonomy and Evaluation of Dense Two-Frame Stereo Correspondence Algorithms. International Journal of Computer Vision 47, 7–42 (2002)

[2] Hannah, M.J.: Computer Matching of Areas in Stereo Images, Ph.D. Thesis, Stanford University (1974)

[3] Barron, J.L., et al.: Performance of optical flow techniques. International Journal of Computer Vision 12, 43–77 (1994)

[4] Bergen, J., et al.: Hierarchical model-based motion estimation. In: Sandini, G. (ed.) ECCV 1992. LNCS, vol. 588, pp. 237–252. Springer, Heidelberg (1992)

[5] Quam, L.: Hierarchical warp stereo. In: Readings in Computer Vision: Issues, Problems, Principles, and Paradigms, pp. 80–86 (1984)

[6] Szeliski, R., Coughlan, J.: Spline-Based Image Registration. International Journal of Computer Vision 22, 199–218 (1997)

[7] Seitz, S.M., et al.: A Comparison and Evaluation of Multi-View Stereo Reconstruction Algorithms. In: CVPR, pp. 519–528 (2006)

[8] Seitz, S.M., Dyer, C.R.: Photorealistic Scene Reconstruction by Voxel Coloring. International Journal of Computer Vision 35, 151–173 (1999)

[9] Vogiatzis, G., et al.: Multi-view stereo via volumetric graph-cuts. In: CVPR, vol. 2, pp. 391–398 (2005)

[10] Bradley, D., et al.: Accurate multi-view reconstruction using robust binocular stereo and surface meshing. In: CVPR, pp. 1–8 (2008)

[11] Merrell, P., et al.: Real-Time Visibility-Based Fusion of Depth Maps. In: IEEE 11th International Conference on Computer Vision, ICCV 2007, pp. 1–8 (2007)

[12] Pollefeys, M., et al.: Visual Modeling with a Hand-Held Camera. International Journal of Computer Vision 59, 207–232 (2004)

[13] Reid, I., Connor, K.: Multiview segmentation and tracking of dynamic occluding layers. Image and Vision Computing 28, 1022–1030 (2010)

[14] Oo, T., et al.: The separation of reflected and transparent layers from real-world image sequence. Machine Vision and Applications 18, 17–24 (2007)

[15] Shade, J., Gortler, S.J., He, L.-w., Szeliski, R.: Layered depth images. Computer Graphics (SIGGRAPH 1998), pp. 231–242 (1998)

[16] Chang, C.-F., Bishop, G., Lastra, A.: LDI tree: a hierarchical representation for image-based rendering. In: Computer Graphics (SIGGRAPH 1999), pp. 291–298 (1999)

[17] Zhigang, Z.: 3D LAMP: a New Layered Panoramic Representation, pp. 723–723 (2001)

[18] Lee, S., et al.: Depth-of-field rendering with multiview synthesis. In: SIGGRAPH Asia 2009, pp. 1–6 (2009)

[19] Bertalmio, M., Sapiro, G., Caselles, V., Ballester, C.: Image inpainting. In: SIGGRAPH 2000, pp. 417–424 (2000)

[20] Ballester, C., et al.: Filling-in by joint interpolation of vector fields and gray levels. IEEE Transactions on Image Processing 10, 1200–1211 (2001)

[21] Zhu, S.C., et al.: Filters, Random Fields and Maximum Entropy (FRAME): Towards a Unified Theory for Texture Modeling. International Journal of Computer Vision 27, 107–126 (1998)

[22] Criminisi, A.: Object Removal by Exemplar-Based Inpainting, pp. 721–721 (2003)

[23] Criminisi, A., et al.: Region filling and object removal by exemplar-based image inpainting. IEEE Transactions on Image Processing 13, 1200–1212 (2004)
[24] Cheung, V., et al.: Video epitomes. In: CVPR, vol. 1, pp. 42–49 (2005)
[25] Zhang, Y., et al.: Motion Layer Based Object Removal in Videos. In: WACV/MOTIONS 2005, pp. 516–521 (2005)
[26] Vijay Venkatesh, M., et al.: Efficient object-based video inpainting. Pattern Recognition Letters 30, 168–179 (2009)

Semi-supervised Learning for Unknown Malware Detection

Igor Santos, Javier Nieves, and Pablo G. Bringas

Abstract. Malware is any kind of computer software potentially harmful to both computers and networks. The amount of malware is increasing every year and poses a serious global security threat. Signature-based detection is the most widely used commercial antivirus method, however, it consistently fails to detect new malware. Supervised machine-learning models have been used to solve this issue, but the usefulness of supervised learning is far to be perfect because it requires that a significant amount of malicious code and benign software to be identified and labelled beforehand. In this paper, we propose a new method of malware protection that adopts a semi-supervised learning approach to detect unknown malware. This method is designed to build a machine-learning classifier using a set of labelled (malware and legitimate software) and unlabelled instances. We performed an empirical validation demonstrating that the labelling efforts are lower than when supervised learning is used, while maintaining high accuracy rates.

1 Introduction

Malware is any computer software intentionally designed to damage computers. Although 'fame and glory' were the main goals of malware writers in the past, more recently, their reasons have evolved into economic considerations [9].

Commercial anti-malware solutions generally base their main detection systems on signature databases [7]. A signature is a unique sequence of bytes that is always present within malicious executables and in the files already infected by that malware. The main problem of such an approach is that specialists have to wait until new malware has damaged several computers to generate a signature file and they can provide a suitable solution. Suspect files that are later subjected to analysis are

Igor Santos · Javier Nieves · Pablo G. Bringas
Laboratory for Smartness, Semantics and Security (S^3Lab), DeustoTech - Computing, University of Deusto, Avenida de las Universidades 24, 48007 Bilbao, Spain
e-mail: {isantos,jnieves,pablo.garcia.bringas}@deusto.es

A. Abraham et al. (Eds.): International Symposium on DCAI, AISC 91, pp. 415–422.
springerlink.com

compared with this list of signatures. When the signatures match, the file being tested is flagged as malware. Although this approach has been demonstrated to be effective when threats are known in advance, signature methods cannot cope with code obfuscation, with previously unseen malware, or with large amounts of new malware [11, 12].

Machine-learning-based approaches train classification algorithms that detect new malware (i.e., 'in the wild'), relying on datasets composed of several characteristic features of both malicious and benign software. Schultz et al. [14] were the first to introduce the concept of applying machine-learning models to the detection of malware based on their respective binary codes. Specifically, they applied several classifiers to three different feature sets: (i) program headers, (ii) strings and (iii) byte sequence. Later, Kolter et al. [4] improved Schulz's results [14], by applying n-grams (i.e., overlapping byte sequences) instead of non-overlapping sequences. This approach employed several algorithms, achieving the best results with a boosted[1] decision tree. In a similar vein, substantial research has focused on n-gram distributions of byte sequences and data-mining [8, 19, 12].

Machine-learning classifiers require a high number of labelled executables for each of the classes (i.e., malware and benign datasets). Nevertheless, it is quite difficult to obtain this amount of labelled data for a real-world problem such as malicious code analysis. To generate these data, a time-consuming process of analysis is mandatory, and in the process, some malicious executables can avoid detection. Within the full scope of machine-learning, several approaches have been proposed to deal with this issue.

Semi-supervised learning is a type of machine-learning techniques that is specially useful when a limited amount of labelled data exists for each class. These techniques create a supervised classifier based on labelled data and predict the label for all unlabelled instances. The instances whose classes have been predicted with a certain threshold of confidence are added to the labelled dataset. The process is repeated until certain conditions are satisfied (a commonly used criterion is the maximum likelihood found by the expectation-maximisation technique). These approaches improve the accuracy of fully unsupervised methods (i.e., no labels within the dataset) [1].

In light of this background, we propose here the first approach that employs a semi-supervised learning technique for the detection of unknown malware. In particular, we utilise the method *Learning with Local and Global Consistency* (LLGC) [18] able to learn from both labelled and unlabelled data and capable of providing a *smooth* solution with respect to the intrinsic structure displayed by both labelled and unlabelled instances. For the representation of executables, we choose the byte n-gram distribution, a well-known technique that have achieved significant results with supervised machine learning (e.g., [4, 8, 12]). However, the presented semi-supervised methodology is scalable to any representation susceptible to be represented as a feature vector. Summarising, our main findings in this paper are: (i) we describe how to adopt LLGC for unknown malware detection, (ii) we determine

[1] Boosting is a machine-learning technique that builds a strong classifier composed by weak classifiers [13].

the optimal number of labelled instances and we evaluated how this parameter affects the final accuracy of the models and (iii) we show that labelling efforts can be reduced in the industry, while still maintaining a high rate of accuracy.

The remainder of this paper is organised as follows. Section 2 provides the background regarding the representation of executables based on byte n-gram frequencies. Section 3 describes the LLCG method and how it can be adopted for unknown malware detection. Section 4 describes the experiments and presents results. Finally, Section 5 concludes the paper and outlines avenues for future work.

2 Byte n-Gram Representation

Byte n-grams frequencies distribution is a well-known approach for training machine-learning classifiers to detect unknown malicious code [14, 4, 8, 15, 19, 12]. To obtain a representation of the executables by the use of byte n-grams, we need to extract every possible sequence of bytes and their appearance frequency. Specifically, a binary program $\mathcal{P}$ can be represented as a sequence of ℓ bytes b as $\mathcal{P} = \{b_1, b_2, b_3, ..., b_{\ell-1}, b_\ell\}$. A byte n-gram sequence g is defined as a subset of consecutive bytes within an executable file where $g \subseteq \mathcal{P}$ and it is made up of bytes b, such as $g = (b_1, b_2, b_3, ..., b_{n-1}, b_n)$ where n is the length of the byte n-gram g. Therefore, a program $\mathcal{P}$ is composed of byte n-grams such as $\mathcal{P} = (g_1, g_2, ..., g_{\ell-1}, g_\ell)$ where ℓ is the total number of possible n-grams of a fixed length n.

```
4D 5A 90 00
03 00 00 00
04 00 00 00
FF FF 00 00
```

Fig. 1 Machine code example.

Consider an example based on the machine code snippet shown in Fig. 1; the following byte bi-grams can be generated: $g_1 = (\texttt{4D, 5A})$, $g_2 = (\texttt{5A, 90})$, $g_3 = (\texttt{90, 00})$, $g_4 = (\texttt{00, 03})$, $g_5 = (\texttt{03, 00})$, $g_6 = (\texttt{00, 00})$, $g_7 = (\texttt{00, 00})$, $g_8 = (\texttt{00, 04})$, $g_9 = (\texttt{04, 00})$, $g_{10} = (\texttt{00, 00})$, $g_{11} = (\texttt{00, 00})$, $g_{12} = (\texttt{00, FF})$, $g_{13} = (\texttt{FF, FF})$, $g_{14} = (\texttt{FF, 00})$, and $g_{15} = (\texttt{00, 00})$.

We use 'term frequency - inverse document frequency' $(tf - idf)$ [6] to obtain the weight of each byte n-grams, whereas the weight of the ith n-gram in the jth executable, denoted by $weight(i, j)$, is defined by: $weight(i, j) = tf_{i,j} \cdot idf_i$, where the term frequency $tf_{i,j}$ [6] is defined as: $tf_{i,j} = \frac{m_{i,j}}{\sum_k m_{k,j}}$ where $m_{i,j}$ is the number of times the n-gram $t_{i,j}$ appears in an executable e, and $\sum_k m_{k,j}$ is the total number of n-grams in the executable e.

On the other hand, the inverse document frequency idf_i is defined as: $idf_i = \frac{|\mathcal{E}|}{|\mathcal{E} : t_i \in e|}$ where $|\mathcal{E}|$ is the total number of executables and $|\mathcal{E} : t_i \in e|$ is the number of documents containing the n-gram t_i.

Finally, we can obtain a vector $\mathbf{v}$ composed of byte n-gram frequencies, $\mathbf{v} = ((g_1, weight_1), ..., (g_{m-1}, weight_{m-1}), (g_m, weight_m))$, where g_i is the byte n-gram and $weight_i$ is the value of $tf - idf$ for that particular n-gram.

3 Overview of LLCG

Learning with Local and Global Consistency (LLGC) [18] is a semi-supervised algorithm that provides *smooth* classification with respect to the intrinsic structure revealed by known labelled and unlabelled points. The method is a simple iteration algorithm that constructs a smooth function coherent to the next assumptions: (i) nearby points are likely to have the same label and (ii) points on the same structure are likely to have the same label [18].

Formally, the algorithm is stated as follows. Let $\mathscr{X} = \{x_1, x_2, ..., x_{\ell-1}, x_\ell\} \subset \mathbb{R}^m$ be the set composed of the data instances and $\mathscr{L} = \{1, ..., c\}$ the set of labels (in our case, this set comprises two classes: malware and legitimate software) and $x_u(\ell + 1 \leq u \leq n)$ the unlabelled instances. The goal of LLGC (and every semi-supervised algorithm) is to predict the class of the unlabelled instances. $\mathscr{F}$ is the set of $n \times c$ matrices with non-negative entries, composed of matrices $F = [F_1^T, ..., F_n^T]^T$ that match to the classification on the dataset $\mathscr{X}$ of each instance x_i. with the label assigned by $y_i = \text{argmax}_{j \leq c} F_{i,j}$. F can be defined as a vectorial function such as $F : \mathscr{X} \rightarrow \mathbb{R}^c$ to assign a vector F_i to the instances x_i. Y is an $n \times c$ matrix such as $Y \in F$ with $Y_{i,j} = 1$ when x_i is labelled as $y_i = j$ and $Y_{i,j} = 0$ otherwise Considering this, the LLGC algorithm performs as follows:

if $i \neq j$ *and* $W_{i,i} = 0$ **then**
  Form the affinity matrix W defined by $W_{i,j} = \exp\left(\frac{-||x_i - x_j||^2}{2 \cdot \sigma^2}\right)$;
Generate the matrix $S = D^{-1/2} \cdot W \cdot D^{-1/2}$ where D is the diagonal matrix with its (i,i) element equal to the sum of the i-th row of W;
while $\neg$ *Convergence* **do**
  $F(t+1) = \alpha \cdot S \cdot F(t) + (1 - \alpha) \cdot Y$ where α is in the range $(0,1)$;
F^* is the limit of the sequence $\{F(t)\}$;
Label each point x_i as $\text{argmax}_{j \leq c} F^*_{i,j}$;

Fig. 2 LLGC algorithm.

The algorithm first defines a pairwise relationship W on the dataset $\mathscr{X}$ setting the diagonal elements to zero. Suppose that a graph $G = (V, E)$ is defined within $\mathscr{X}$, where the vertex set V is equal to $\mathscr{X}$ and the edge set $\mathscr{E}$ is weighted by the values in W. Next, the algorithm normalises symmetrically the matrix W of G. This step is mandatory to assure the convergence of the iteration. During each iteration each instance receives the information from its nearby instances while it keeps its initial information. The parameter α denotes the relative amount of the information from the nearest instances and the initial class information of each instance. The information is spread symmetrically because S is a symmetric matrix. Finally, the algorithm sets the class of each unlabelled specimen to the class of which it has received most information during the iteration process.

4 Empirical Validation

The research question we seek to answer through this empirical validation is the following one: *What is the minimum number of labelled instances required to assure a suitable performance using LLGC?* To this end, we collected a dataset comprising 1,000 malicious executables and 1,000 benign ones. For the malware, we gathered random samples from the website VxHeavens[2]. Although they had already been labelled according to their family and variant names, we analysed them using Eset Antivirus[3] to confirm this labelling. For the benign dataset, we collected legitimate executables from our own computers. We also performed an analysis of the benign files using Eset Antivirus to confirm their legitimacy.

Hereafter, we extracted the byte n-gram representation for each file in the dataset for $n = 2$. This specific length was chosen because it is the number of bytes a operation represented by an operational code needs in machine code and it is a widely-used n-gram length in the literature (e.g., [14, 4]) Because the total number of features we obtained was high, we applied a feature selection step based on a Document Frequency (DF) measure, which counts the number of documents in which a specific n-gram appears, selecting the 1,000 top ranked byte n-grams. This concrete number of features was chosen because it provides a balance between efficiency and accuracy and it has been proven to be effective [8].

Next, we split the dataset into different percentages of training and tested instances. In other words, we changed the number of labelled instances from 10% to 90% to measure the effect of the number of labelled instances on the final performance of LLGC in detecting unknown malware. We did not use cross-validation because in the validation we do not want to test the performance of the classifier when a fixed size of training instances is used repeatability. Otherwise, we employ a variable number of training instances and try to predict the class of the remaining ones using LLGC in order to determine which is the best training set size. In this case, the training instances are the labelled ones whereas the unlabelled ones are the ones in the test dataset. In particular, we used the LLGC implementation provided by the *Semi-Supervised Learning and Collective Classification* package[4] for the well-known machine-learning tool WEKA [2]. Specifically, we configured it with a transductive stochastic matrix W [18] and we employed the Euclidean distance.

To test the approach, we measured the *True Positive Ratio* (TPR), i.e., the number of malware instances correctly detected divided by the total number of malware files: $TPR = TP/(TP+FN)$ where TP is the number of malware cases correctly classified (true positives) and FN is the number of malware cases misclassified as

[2] http://vx.netlux.org/

[3] http://www.eset.com/

[4] Available at:
`http://www.scms.waikato.ac.nz/$\sim$fracpete/projects/collective-classification/downloads.html`

legitimate software (false negatives). We also measured the *False Positive Ratio* (FPR), i.e., the number of benign executables misclassified as malware divided by the total number of benign files: $FPR = FP/(FP+TN)$ where FP is the number of benign software cases incorrectly detected as malware and TN is the number of legitimate executables correctly classified. Furthermore, we measured *accuracy*, i.e., the total number of the hits of the classifiers divided by the number of instances in the whole dataset: $Accuracy(\%) = (TP+TN)/(TP+FP+TP+TN)$ Besides, we measured the *Area Under the ROC Curve* (AUC) that establishes the relation between false negatives and false positives [17]. The ROC curve is obtained by plotting the TPR against the FPR. All the these measures refer to the test instances.

Fig. 3 and Fig. 4 show the obtained results. The best results in terms of AUC were obtained with a training set containing 50% of labelled instances. These results indicate that we can reduce the efforts of labelling software in a 50% while maintaining a AUC higher than 88%. In terms of accuracy, the best results were achieved with a training size of 65%.

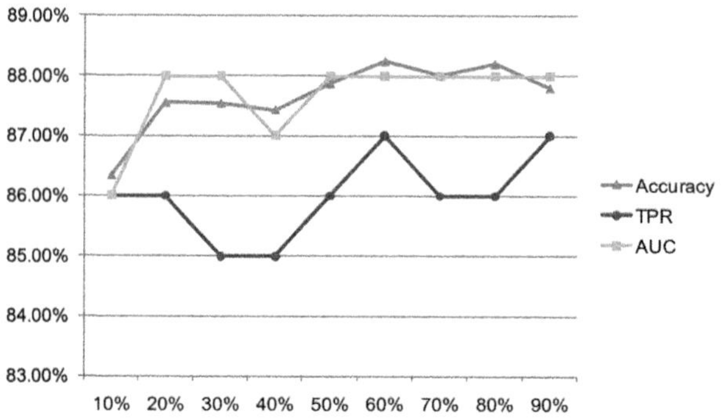

Fig. 3 Accuracy, TPR and AUC results. The X axis represent the percentage of labelled instances. The best AUC results with a 50% size for the labelled dataset.

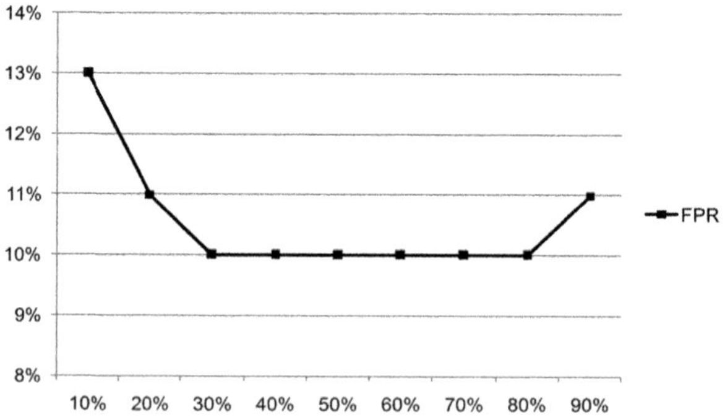

Fig. 4 FRP results. The X axis represent the percentage of labelled instances. In particular, the best results were obtained with a size greater than 30%.

Previous supervised learning obtains better results (above 90% of accuracy [14, 4, 8]) than this semi-supervised approach. However, the main contribution of this paper is the reduction in the number of required labelled instances while maintaining a relative high precision. We consider that these results are significant for the anti-malware industry. The reduction of the efforts required for unknown malware can help to deal with the increasing amount of new malware.

However, because of the static nature of the features we used with LLGC, it cannot counter *packed* malware. Packed malware is produced by cyphering the payload of the executable and having it deciphered when finally loaded into memory. Indeed, broadly-used static detection methods can deal with packed malware only by using the signatures of the packers. Accordingly, dynamic analysis seems to be a more promising solution to this problem [3]. One solution for this obvious limitation of our malware detection method is the use of a generic dynamic unpacking schema such as PolyUnpack [10], Renovo [3], OmniUnpack [5] and Eureka [16].

5 Concluding Remarks

Unknown malware detection has become an important topic of research and concern owing to the growth of malicious code in recent years. Moreover, it is well known that the classic signature methods employed by antivirus vendors are no longer completely effective in facing the large volumes of new malware. Therefore, signature methods must be complemented with more complex approaches that provide the detection of unknown malware families. While machine-learning methods are a suitable approach for unknown malware, they require a high number of labelled executables for each classes (i.e., malware and benign datasets). Since it is difficult to obtain such amounts of labelled data in a real-word environment, a time-consuming process of analysis is mandatory.

In this paper, we propose for the fist time the use of a semi-supervised learning approach for unknown malware detection. This learning technique does not need a large amount of labelled data; it only needs several instances to be labelled. Therefore, this methodology can reduce efforts in unknown malware detection. By labelling 50% of the software, we can achieve results with more than 86% of accuracy.

Future work will be focused on three main directions. First, we plan to extend our study of semi-supervised learning approaches by applying more algorithms to this issue. Second, we will use different features for training these kinds of models. Finally, we will focus on facing packed executables with a hybrid dynamic-static approach.

References

1. Chapelle, O., Schölkopf, B., Zien, A.: Semi-supervised learning. MIT Press, Cambridge (2006)
2. Garner, S.: Weka: The Waikato environment for knowledge analysis. In: Proceedings of the New Zealand Computer Science Research Students Conference, pp. 57–64 (1995)

3. Kang, M., Poosankam, P., Yin, H.: Renovo: A hidden code extractor for packed executables. In: Proceedings of the 2007 ACM Workshop on Recurring Malcode, pp. 46–53 (2007)
4. Kolter, J., Maloof, M.: Learning to detect malicious executables in the wild. In: Proceedings of the 10th ACM SIGKDD International Conference on Knowledge Discovery and Data Mining, pp. 470–478. ACM, New York (2004)
5. Martignoni, L., Christodorescu, M., Jha, S.: Omniunpack: Fast, generic, and safe unpacking of malware. In: Proceedings of the 23rd Annual Computer Security Applications Conference (ACSAC), pp. 431–441 (2007)
6. McGill, M., Salton, G.: Introduction to modern information retrieval. McGraw-Hill, New York (1983)
7. Morley, P.: Processing virus collections. In: Proceedings of the 2001 Virus Bulletin Conference (VB 2001), pp. 129–134. Virus Bulletin (2001)
8. Moskovitch, R., Stopel, D., Feher, C., Nissim, N., Elovici, Y.: Unknown malcode detection via text categorization and the imbalance problem. In: Proceedings of the 6th IEEE International Conference on Intelligence and Security Informatics (ISI), pp. 156–161 (2008)
9. Ollmann, G.: The evolution of commercial malware development kits and colour-by-numbers custom malware. Computer Fraud & Security 2008(9), 4–7 (2008)
10. Royal, P., Halpin, M., Dagon, D., Edmonds, R., Lee, W.: Polyunpack: Automating the hidden-code extraction of unpack-executing malware. In: Proceedings of the 22nd Annual Computer Security Applications Conference (ACSAC), pp. 289–300 (2006)
11. Santos, I., Brezo, F., Nieves, J., Penya, Y.K., Sanz, B., Laorden, C., Bringas, P.G.: Idea: Opcode-sequence-based malware detection. In: Massacci, F., Wallach, D., Zannone, N. (eds.) ESSoS 2010. LNCS, vol. 5965, pp. 35–43. Springer, Heidelberg (2010)
12. Santos, I., Penya, Y., Devesa, J., Bringas, P.: N-Grams-based file signatures for malware detection. In: Proceedings of the 11th International Conference on Enterprise Information Systems (ICEIS). AIDSS, pp. 317–320 (2009)
13. Schapire, R.: The boosting approach to machine learning: An overview. Lecture Notes in Statistics pp. 149–172 (2003)
14. Schultz, M., Eskin, E., Zadok, F., Stolfo, S.: Data mining methods for detection of new malicious executables. In: Proceedings of the 22nd IEEE Symposium on Security and Privacy, pp. 38–49 (2001)
15. Zubair Shafiq, M., Khayam, S.A., Farooq, M.: Embedded malware detection using markov *n*-grams. In: Zamboni, D. (ed.) DIMVA 2008. LNCS, vol. 5137, pp. 88–107. Springer, Heidelberg (2008)
16. Sharif, M., Yegneswaran, V., Saidi, H., Porras, P.A., Lee, W.: Eureka: A framework for enabling static malware analysis. In: Jajodia, S., Lopez, J. (eds.) ESORICS 2008. LNCS, vol. 5283, pp. 481–500. Springer, Heidelberg (2008)
17. Singh, Y., Kaur, A., Malhotra, R.: Comparative analysis of regression and machine learning methods for predicting fault proneness models. International Journal of Computer Applications in Technology 35(2), 183–193 (2009)
18. Zhou, D., Bousquet, O., Lal, T., Weston, J., Schölkopf, B.: Learning with local and global consistency. In: Advances in Neural Information Processing Systems 16: Proceedings of the 2003 Conference, pp. 595–602 (2004)
19. Zhou, Y., Inge, W.: Malware detection using adaptive data compression. In: Proceedings of the 1st ACM Workshop on AISec, pp. 53–60. ACM, New York (2008)

Self-organized Clustering and Classification: A Unified Approach via Distributed Chaotic Computing

Elena N. Benderskaya and Sofya V. Zhukova

Abstract. The paper describes a unified approach to solve clustering and classification problems by means of oscillatory neural networks with chaotic dynamics. It is discovered that self-synchronized clusters once formed can be applied to classify objects. The advantages of distributed clusters formation in comparison to centers of clusters estimation are demonstrated. New approach to clustering on-the-fly is proposed.

Keywords: clustering, classification, distributed clusters, chaotic neural network.

1 Introduction

Clustering is formulated like the search process of n-input objects division into previously unknown number of m groups. The division should be the best from the point of maximum similarity of objects within each cluster (intra-cluster similarity is high) and at the same time maximum dissimilarity of objects that belong to different clusters [6, 10, 11] (inter-class similarity is low). Clustering techniques are developed as heuristics capable to recognize input topological peculiarities at the extent of similarity measure universality. The similarity measure depends greatly on mutual disposition of elements in the input dataset. If we have no a priori information about groups topology (ellipsoidal, ball-shaped, compact, scattered due to some distribution, etc.) then the risk to choose a wrong metric, or grid size or vitality parameter or density distribution (depending on joining criteria of a method) is very high [11].

Elena N. Benderskaya
St. Petersburg State Polytechnical University, Faculty of Computer Science,
Russia, 194021, St.Petersburg, Politechnicheskaya 21
e-mail: helen.bend@gmail.com

Sofya V. Zhukova
St. Petersburg State University, Graduate School of Management,
Russia, 199004, St.Petersburg Volkhovsky Per. 3
e-mail: sophya.zhukova@gmail.com

A. Abraham et al. (Eds.): International Symposium on DCAI, AISC 91, pp. 423–431.
springerlink.com

Clustering as one of fundamental problems in pattern recognition and data processing covers in its general statement also classification, forecasting and segmentation problems. Though there are hundreds of clustering techniques [6, 11] organized in groups of partitional [20] (e.g. k-means, h-means, canopy clustering, PAM clustering), hierarchical [12] (e.g. UPGMA, UPGMC, WPGMC, WPGMA), density-based (e.g. DBSCAN, OPTICS, DENCLUE), grid-based [7] (e.g. STING, WaveCluster, CLIQUE), probability-based [10] (e.g. EM algorithm, COBWEB, Gaussian mixture model, Bayesian clustering), bio-inspired methods [9, 13, 15, 19] (SOM, ART, neural gas, ACCM, DSC algorithm) it is hard to choose an adequate clustering method when there is no a priori information about peculiarities of mutual disposition of objects in input dataset. To cope with large set of algorithms voting principal and visual-aided analytics are applied. Visual-aided approach [11] consists in expert control of computational process by setting different input parameters, interpretation of the results and directing the algorithms towards the solution that better describes the underlying phenomena. The key issue of the first approach is the expert participation in computational process that makes pattern recognition system automated, but not automatic. Voting principal [21] is based on simultaneous application of different clustering techniques and choice of the most frequent clustering answer as the final one due to the majority rule. This approach is enormously resource consuming as many of clustering algorithms are NP-complete. This paper aims to make next step in the development of automatic high quality clustering system with reduced computing complexity that demonstrate capabilities to clustering on-the-fly and classification.

2 An Alternative Approach to Clustering

Under high quality clustering technique is understood right clustering solutions obtained in respond to each of complex clustering problem in world known FCPS dataset [18] without any special expert tuning the method. So the developed clustering system is set in the hardest conditions with no prior information about topology and number of clusters. In comparison partitioning methods ask the user to provide number of clusters; density-based methods require maximum radius of the neighborhood and minimum number of points in a neighborhood; probability-based methods need data on probability density models of clusters (distribution hypothesis); genetic clustering algorithms need prior statement of fitness function; neural network clustering techniques rely on prior knowledge about valid distance metric or minimum number of nearest neighbors, or vigilance parameter, or number of clusters; hierarchical clustering require information about clusters topology, otherwise there is a great risk to obtain erroneous results.

Bio-inspired clustering techniques (genetic algorithms, neural networks, swarm-based networks, ant colonies) reduce computational complexity of traditional clustering techniques by means of their parallel interpretation and implementation [9, 17]. Unfortunately when no information about clustering data is available solution quality should be somehow improved on the next stages of data processing and this is a serious obstacle to the development of automatic clustering systems.

Oscillator paradigm

In recent years promising results obtained in nonlinear dynamics, and neurophysiology predetermined formation of oscillator paradigm that combines knowledge from neurobiology, molecular physics, electronics, chaos and synchronization theories [5, 8, 14, 16]. One of most perspective techniques that apply internal synchronization of coupled chaotic oscillators to clustering is described in [1]. It was improved greatly in [2, 3], in terms of automatically estimated local scale (neighborhood parameter), improved processing of output dynamics and matrix interpretation of system's overall evolution. After implementation of preceding modifications proposed technique does not require any more a priori information about topology and number of clusters to generate high quality results. Comparative analysis of OCNN clustering results with other 43 clustering techniques is given in [4]. In this paper this clustering technique based on oscillatory chaotic neural network (OCNN) [2, 3] capable to cluster objects in p-dimensional feature space is applied to accomplish also on-the-fly clustering, classification and multiple classification procedures in one unified way. To demonstrate the results in a vivid form we use data from FCSP dataset (2D, 3D). Chaotic neural networks can be classified as oscillatory neural networks with chaotic transfer function. Highly unstable dynamics and distributed data processing were combined in OCNN model.

Phenomenology of chaotic neural network

OCNN is a recurrent neural network with one layer of n neurons. Each neuron corresponds to one point in the input dataset which in general case consists of n objects, each described by p features (p-dimensional image). OCNN is a dynamic neural network, where each processing unit changes its state depending on the dynamics of all other neurons:

$$\begin{cases} y_1(t+1) = \dfrac{1}{C_1}\sum\limits_{i=1}^{N} w_{1i} f(y_i(t)), t = \overline{1, T_n} \\ y_2(t+1) = \dfrac{1}{C_2}\sum\limits_{i=1}^{N} w_{2i} f(y_i(t)), t = \overline{1, T_n} \\ \ldots \\ y_N(t+1) = \dfrac{1}{C_N}\sum\limits_{i=1}^{N} w_{Ni} f(y_i(t)), t = \overline{1, T_n} \end{cases}, \tag{1}$$

$$C_i = \sum_{i \neq j}^{N} w_{ij}, \quad i, j = \overline{1, N}, \tag{2}$$

$$W = \{w_{ij}\} = \exp\left(-d_{ij}^2 / 2a\right), \quad i, j = \overline{1, N}, \tag{3}$$

$$f(y(t)) = 1 - 2y^2(t), \tag{4}$$

where N – number of neurons, w_{ij} - strength of linkage between elements i and j, d_{ij} - euclidean distance between neurons i and j (each neuron represents a point from input dataset described by p coordinates), a – local scale calculated by means of triangulation metric [2], T_n – evolution period. The initial state of neural network is described by random values in the range [-1, 1]. Information about input dataset is given to OCNN by means of linkage coefficients calculated via (3). Number of neurons N correspond to n objects in the input dataset. Transfer function of each neuron is calculated via (4). Overall evolution of coupled chaotic oscillators is described by (1), where C_i value allows to limit output values in the range of [-1; 1].

The key point of the OCNN functioning is the emergence of cooperative dynamics between neuron's outputs via time. After some transition period they start to change states synchronously. To explore the chaotic neural dynamics (oscillatory clusters) visualization of OCNN output evolution is provided. The deep analysis of the output dynamics helps to discover a new type of synchronization – fragmentary synchronization [3].

3 Methodology of OCNN Clustering

In case of system described by (1)-(4) at present formal mathematics gives only partial but not general solutions. In this paper high dimensional OCNN is investigated by means of computer modeling. Clustering technique consists of 4 stages. On the first stage OCNN is initialized due to (3) and random assignment of neuron outputs from the range of [-1; 1]. Second stage (learning) is aimed to form oscillatory clusters by means of self-organizing OCNN dynamics during transition period via (1). Third stage (preparatory) consists in gathering dynamics about neurons fluctuations during some observation period via (1). Fourth stage (clustering) aims to translate oscillatory clusters into object clusters.

Detection of distributed clusters

Exponential growth of dynamics dependence from initial conditions is one of the main chaos indicators [14]. In OCNN each of neurons evolves in chaotic manner due to logistic map (4). If clustering results depend on initial conditions then applicability of OCNN to pattern recognition stands to be questionable. Nevertheless it is possible if OCNN parameters (linkage coefficients w_{ij}) are set in a proper way. All trajectories of outputs evolution remain to be chaotic. Thus, a trajectory in the phase space of OCNN undoubtedly depends on initial conditions. Yet, mutual trajectories disposition in n-dimensional phase space stands to be invariable to start point. That means that mutual synchronization is insensitive to initial conditions and as a result the obtained solution of clustering problem is insensitive either. To be particularly stressed is the coincidence on the third stage clusters membership, number and size though the initial states of neurons can be absolutely different.

In other words OCNN stay to be chaotic all the time but manages to produce stable solution if the mean field is appropriately calculated [2]. There appear very interesting logics in mutual relations of neurons expressed in cooperative oscillation dynamics. These logics can hardly be described by standard mathematical techniques, but after all can be successfully applied to solve various clustering problems.

On Fig. 1 are demonstrated the results on both simple and complex image clustering produced by OCNN. The more complicated input dataset the more intricate becomes the structure of oscillatory clusters. Correct clustering of input image from Fig. 1.*d* becomes possible due to the phenomenology of fragmentary synchronization when each oscillatory cluster is characterized by its unique melody. Within trajectories that comprise one cluster there can be no instant coincidence between neurons phases and amplitudes and yet the clusters integrated dynamics differs from the dynamics of trajectories from other clusters.

For each of test clustering problems from fundamental clustering problems suite [18] was fixed the fact of cluster formation insensitiveness to initial conditions and correct clustering results were obtained for each of the test datasets.

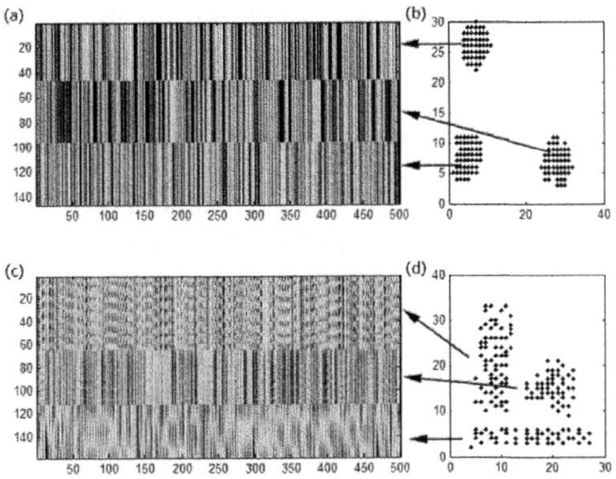

Fig. 1 Fragmentary synchronization of OCNN outputs: (a) – OCNN dynamics statistics comprised by completely synchronized neurons within each of clusters in respond to simple input image; (b) – simple input dataset comprised by 46, 50, 50 points in three clusters (clusters are compact); (c) – complex input image comprised by 65, 47, 46 points in three clusters (clusters are loose); (d) - OCNN dynamics statistics comprised by fragmentary synchronized neurons in respond to complex input image.

Thus the main accent in the proposed approach is made on temporal dimension. The neural network sway helps to overcome the geometrical uncertainty by means of distributed character of decentered oscillations, no centers of clusters are evaluated. The fourth stage is realized by means of algorithm proposed in [3] that allow to detect complete, phase and fragmentary synchronization.

4 Combining Clustering and Classification

Solutions of complex pattern recognition problems deal with clustering and classification processes. When there is no information about typical representatives of classes (or group labels assigned to objects) clustering is preliminary accomplished. There are two main approaches to use clustering results. First considers clustering as the mechanism to get group labels and clustered image becomes training data for classification algorithms that either constructs classifier (discriminant rule) in the form of surfaces or class centers. (In case of unknown number of clusters there is the need to combine centers of clusters to reflect more closely real number of groups in the input dataset). This approach in fact doubles classification time. Second approach generalize clustering results in the form of computing centers of clusters with further comparison of new object with centers of clusters as their typical representatives in order to classify new object. Thus classification process can be realized in two different ways: classification with fixed classes and classification with changing classes (dynamic clustering). If a new object belongs to a class that previously was not recognized wrong classification take place, as pattern recognition system can't generate the answer "I don't know" without fuzzification [19]. Thus modern pattern recognition system somehow should combine both classification and clustering abilities to reduce the computational complexity and to increase clustering and classification quality.

Classification by means of OCNN

In this paper it was discovered that OCNN is capable not only to cluster input images of different complexity [18], but also to classify new objects. Clustering answer in the form of synchronized chaotic sequences is stored in the memory of OCNN. When new object is added to the input dataset for further classification a neuron joins structure of OCNN and its individual chaotic trajectory is calculated on the base of OCNN dynamics stored in OCNN memory.

It is important to stress that scaling constant used in (3) is not recalculated for the modified dataset but corresponds to the one estimated for the clustered image. If the classified point belongs to the neighborhood of one of the clusters in the image on the language of OCNN it means that individual dynamics of neuron k is seized by one of the closest synchronized clusters and thus classification takes place.

Classification process accomplished by OCNN is demonstrated on Fig. 2.c. In respond to 2D image (input dataset) comprise by146 points (Fig. 2.*b*) OCNN generates chaotic oscillatory clusters (Fig. 2.*a*), which are considered to be memory (146 trajectories correspond to 146 points in the input dataset). Let us classify point one by one 7 points added to the input dataset (Fig. 2.*d*). Classification process consists in calculation trajectories of 7 neurons on the base of previous OCNN dynamics (OCNN memory), with further comparison of this 7 trajectories with oscillatory clusters stored in OCNN memory. As one can see 7 trajectories form 3 different groups. And if one compares each of these groups with clusters on Fig. 2.*a* the corresponding coincidence can be noticed.

To check ability of OCNN to generate "I don't know" answer 8 points (Fig. 2.*f*) with far location from three clusters are added to the input dataset. These 8 points are classified independently one by one. As a result on Fig. 2.*e*. OCNN generates 8 trajectories that evolve asynchronously to each of three oscillatory clusters demonstrated on Fig. 2.*a*. On the language of oscillations this means that no decision on this 8 points membership in any of the three clusters can be made. Thus 8 "I don't know" answers are produced.

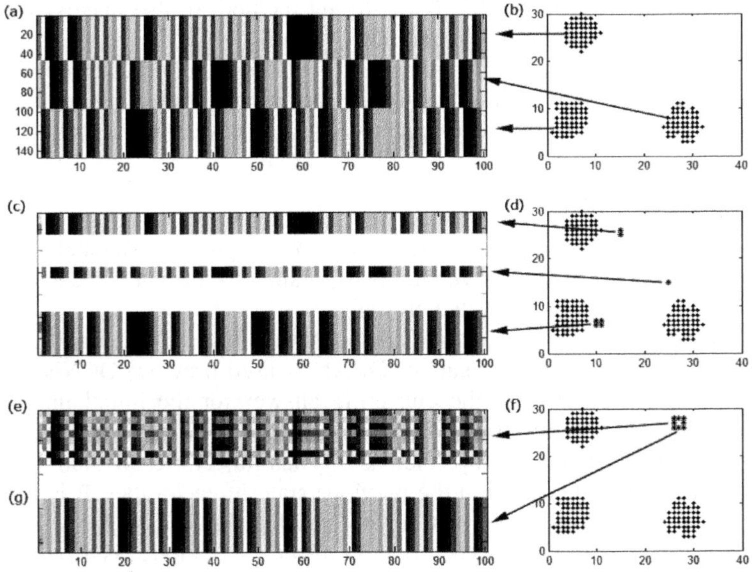

Fig. 2 Classification and on-the-fly clustering by means of OCNN: (a) – input dataset to be clustered, comprised of 146 points; (b) – 3 oscillatory clusters (OCNN memory) comprised by 46, 50, 50 neurons trajectories; (d) – 7 points marked with asterisks to be classified by OCNN; (c) – 2,1,4 neurons trajectories are generated in respond to 7 points (the three 2, 1, 4 groups of trajectories fully coincide with those stored in OCNN; (f) – 8 points to be classified, clustered on-the-fly; (e) –8 trajectories are asynchronous to each of 3 oscillatory clusters stored in OCNN, 8 "I do not know answers" are generated; (g) – 8 synchronous trajectories form a new oscillatory cluster different from the three stored in OCNN memory.

Multiple classifications by means of OCNN

Computational complexity of clustering method based on OCNN is predetermined by global linkage between elements. The analysis of OCNN mathematical model comprised by n difference equations (1) reveals the way how to cut down clustering time.

Each of OCNN neurons state depends on the states of all other neurons. And this extends greatly clustering time. To overcome this it is proposed to use matrix form of OCNN evolution.

$$\begin{aligned} &Y(t+1) = [W * f(Y(t))]]./C, \\ &Y(t) = (y_1(t), y_2(t), \ldots y_N(t))^T, \\ &C = (C_1, C_2, \ldots, C_N), \\ &W = \{w_{ij}\}, i = \overline{1,N}, j = \overline{1,N}. \end{aligned} \tag{5}$$

The identity of (1) and (5) comes from mathematical apparatus of linear algebra that says that left matrix and vector multiplication ends with consequent sum of element wise multiplications. This interpretation opens the opportunity for future hardware implementation of OCNN clustering method. It also makes possible to conduct multiple classifications in parallel.

Independent calculation of 7 trajectories (Fig. 2.*b*) via (5) thus can be provided in parallel and this makes classification system more efficient in comparison to Kohonen's network, where objects can be classified only consequently.

Clustering on-the-fly

A lot of existing clustering techniques do not support incremental clustering. Hereon it is proposed to check whether it is possible to form new clusters without recalculation of previously revealed clusters.

The on-the-fly clustering process is demonstrated on Fig. 2.*f*, Fig. 2.*g*. To provide on-the-fly clustering by means of OCNN is used previous OCNN dynamics (Fig. 2.*a*) that corresponds to the clustering answer for the initial input dataset (Fig. 2.*b*). Linkage coefficients are calculated for each of 8 added points (Fig. 2.*f*) and the scaling constant a is not recalculated. Then trajectories of each 8 new neurons are calculated depending on the previous state of each other. It is noted that the states of all other 146 neurons are not recalculated. On Fig. 2.*g* is demonstrated the formation of a new synchronous cluster, its "melody" is different to any of the three clusters generated by OCNN previously (Fig. 2.*a*). Thus clustering on-the-fly allows to reduce computational complexity twice, because there is no need to recluster the whole input dataset.

5 Conclusions

It was discovered that OCNN is capable not only to cluster but to classify objects on the basis of previous oscillatory dynamics without reclustering of the whole image. What is more crucial OCNN manages to reveal whether an object belongs to any of clusters at all – if classified object is located far from each of clusters OCNN produces "I don't know" answer. Due to the introduced interpretation of mathematical apparatus OCNN can be applied to accomplish multiple simultaneous classifications. Clustering on-the-fly abilities of OCNN were investigated. Without any recalculation of overall system's dynamics OCNN is applicable to incremental clustering. To summarize clustering, classification, on-the-fly clustering, multiple classifications can be processed within one unified approach based on oscillatory chaotic neural network.

References

1. Angelini, L., Carlo, F., Marangi, C., Pellicoro, M., Nardullia, M., Stramaglia, S.: Clustering by inhomogeneous chaotic maps in landmine detection. Phys. Rev. Lett. 86, 89–132 (2001)
2. Benderskaya, E.N., Zhukova, S.V.: Clustering by chaotic neural networks with mean field calculated via delaunay triangulation. In: Corchado, E., Abraham, A., Pedrycz, W. (eds.) HAIS 2008. LNCS (LNAI), vol. 5271, pp. 408–416. Springer, Heidelberg (2008)

3. Benderskaya, E.N., Zhukova, S.V.: Fragmentary synchronization in chaotic neural network and data mining. In: Corchado, E., Wu, X., Oja, E., Herrero, Á., Baruque, B. (eds.) HAIS 2009. LNCS (LNAI), vol. 5572, pp. 319–326. Springer, Heidelberg (2009)
4. Benderskaya, E.N., Zhukova, S.V.: Dynamic Data Mining: Synergy of Bio-Inspired Clustering Methods in Data Mining. Book 2, INTECH (2011) ISBN: 978-953-307-1417-4
5. Borisyuk, R.M., Borisyuk, G.N.: Information coding on the basis of synchronization of neuronal activity. Bio Systems 40(1), 3–10 (1997)
6. Han, J., Kamber, M.: Data Mining. In: Concepts and Techniques. The Morgan Kaufmann Series in Data Management Systems, Morgan Kaufmann, San Francisco (2005)
7. Ilango, M.V.: A Survey of Grid Based Clustering Algorithms. International Journal of Engineering Science and Technology 2(8), 3441–3446 (2010)
8. Kaneko, K.: Phenomenology of spatio-temporal chaos. In: Directions in Chaos, pp. 272–353. World Scientific Publishing Co., Singapore (1987)
9. Kaski, S.: Data exploration using self-organizing maps. Mathematics, Computing and Management in Engineering Series, vol. 82, p. 57 (1997)
10. Kumar, B.V., Mahalanobis, A., Juday, R.D.: Correlation Pattern Recognition. Cambridge University Press, Cambridge (2006)
11. Maimon, O., Rokach, L. (eds.): Data Mining and Knowledge Discovery Handbook, 2nd edn. Springer, Heidelberg (2010)
12. Murtagh, F.: A Survey of Recent Advances in Hierarchical Clustering Algorithms. The Computer Journal 26(4), 354–359 (1984)
13. Pedrycz, W., Weber, R.: Special issue on soft computing for dynamic data mining. Applied Soft Computing (8), 1281–1282 (2008)
14. Pikovsky, A., Rosenblum, M., Kurths, J.: Synchronization: A Universal Concept in Nonlinear Sciences. Cambridge University Press, Cambridge (2003)
15. Pöllä, M., Honkela, T., Gao, X.-Z.: Biologically Inspired Clustering: Comparing the Neural and Immune Paradigms. Studies in Computational Intelligence, vol. 129, pp. 179–188 (2008)
16. Schweitzer, F.: Self-Organization of Complex Structures: From Individual to Collective Dynamics. CRC Press, Boca Raton (1997)
17. Liu, S., Dou, Z.-T., Li, F., Huang, Y.-L.: A new ant colony clustering algorithm based on DBSCAN. Machine Learning and Cybernetics 3, 1491–1496 (2004)
18. Ultsch, A.: Clustering with SOM: U*C. In: Proc. Workshop on Self-Organizing Maps, Paris, France, pp. 75–82 (2005)
19. Valente de Oliveira, J., Pedrycz, W.: Advances in Fuzzy Clustering and its Applications. Wiley, Chichester (2007)
20. Velmurugan, T., Santhanam, T.: A survey of partition based clustering algorithms in data mining: An experimental approach. Information Technology Journal (10), 478–484 (2011)
21. Zhuravlev, Y.I., Ryazanov, V.V., Senko, O.V., et al.: The program system for intellectual data analysis, recognition and forecasting. WSEAS Transactions on Information Science and Applications 2(1), 55–58 (2005)

Cognition and Digital Ecosystems

Cecilia Ciocan and Ioan Ciocan

Abstract. The paper aims to establish a framework for analysing the role of cognitive science in the development of digital ecosystems. It is well known that cognitive science has plenty borrowed from artificial intelligence, therefore building models based on the co-evolution of cognitive human and artificial agents, which could provide answers in a considerably shorter time than people usually need for fulfilling the same tasks. Organizations, as complex social and human systems, regardless of the information technologies they use, were and still are subjected to knowledge. This happens both at the level of members' individual behaviour, as well as at a large scale with regard to groups and organizations behaviour: relations between goals, means, results, and the relationship between the organization and its environment. The development of knowledge ecosystems is analyzed (regarded) as a modelling activity, which involves developing an informatic model. Developing knowledge ecosystems emphasizes the importance of cognitive processes in all their aspects, involving the simulation of new and flexible ways of cooperation and work in the network by the means of dynamic association and self-organizing developing structures through open source.

Keywords: knowledge management, cognitive processes, digital ecosystems.

1 Introduction

As a result of the ecologic unbalance caused by the industrialized countries both on their territory as well as on the territory of other states it could be found that the nature does not always have available the means for restoring the ecologic balance, in many cases being required the restoring intervention of the human being.

The sustainable economic development assumes a new attitude of the human being toward the environment, modifying the relations between the human being and the nature, ensuring the balance between the human activities and the natural environment where they are carried out. That is why; the sustainable economic development is aimed at all the segments of the social life.

Cecilia Ciocan · Ioan Ciocan
Department of Economic Cybernetics,
Academy of Economic Studies, Bucharest, Romania
e-mail: {cecilia.simion,ionutciocan}@yahoo.com

A. Abraham et al. (Eds.): International Symposium on DCAI, AISC 91, pp. 433–442.
springerlink.com

In the last years, a new concept have appeared and have developped at the border between biology, ecology, economics and business world: the digital ecosystem. In fact this is an adaptive complex system. In biology, an ecosystem is "a system of organisms which live in a habitat, together with that aspects regarding the physical environment with which interact". (The New Shorter Oxford English Dictionary, 1993).

The notion of "ecosystem" within an economic context was used for the first time by Zeleny (Zeleny et al, 1990) for justifying the necessity of some businesses which should be aimed to human requirements and in the same time should comply with rules of behaviour of nature (Business Ecosystem). Thus, nowadays this concept is not defined very clearly. Some specialists use the term of "Business Ecosystem" for describing a systemic model of a company involved in a tight relationship with business environment and distribution networks. Ecosystem is a dynamic complex of microorganisms and their life environment which interact in a functional unit.

The digital ecosystems have a self-organization capacity appearing whenever the interactions between agents reach a complex linear behaviour level. The achievement of self-organization within knowledge-based ecosystems assumes the necessity of a superior capacity of reproduction of the components with a minimum level of intervention from the part of human agents.

Regarding the organization, the Business Digital Ecosystem (DBE) might be defined as „a self-organizer evolutive system that aimes the creation of a digital software for small organizations" (Holland, 1995), which support the local and regional development using open technologies and promote evolutionist business models for developing small organizations.

The agents within a digital ecosystem resemble the biological individuals, meaning that they interact, move and die. In conclusion, the digital ecosystem consists of agents interacting between one another, but also with the external environment just like within the real ecosystems in the nature.

The organizations co-evoluated within a social ecosystem, because co-evolution cannot take place isolated. Therefore, a social ecosystem consist of organizations, and not of individuals, and from here the qualificative "social" derives. Mitleton-Kelly proves that any social ecosystem is an evolutive complex system, so a more evoluated type of system than adaptive complex system.

The knowledge-based digital ecosystems are aimed at exceeding the existent barriers and promoting some innovative forms of software creation, sharing of knowledge and composing of communities, thus allowing the long-term progress and the competitiveness of companies acting in the field of healthcare. The purpose of the initiative regarding digital business ecosystems consists of simulating new and flexible ways of cooperation and work in a network by dynamic association and self-organization evolution through the intermediary of an open source structure. The open source model refers to a decentralized initiative, opened to a diverse range of participants situated in different localities, which makes control more difficult.

There is a tight analogy between ecologic communities and business communities, that explains why the companies, as biologic microorganisms, function within a dense networks of interactions, from domestic economy to global economy. Biologic ecosystems and economic ecosystems are adaptive complex system and follow the same profound rules. There is an essential difference between these systems: the human ability of taking conscious decisions, while biologic organisms have no conscience of the same type. In the digital ecosystems, the agents are intelligents and able to plan the future with certain accuracy. On the other hand, business ecosystems compete to acquire new members.

Biological ecosystems and knowledge-based ecosystems have a series of common properties like interaction, interdependence, emergent behaviour, self-organization, feedback, non-linearity etc.

Both types of systems operate as organisms and not as a machine. In addition, neither biological ecosystems nor digital business ecosystems are capable of optimizing their own behaviour. In business ecosystems, the company represents the equivalent of organisms within the biological ecosystem.

The digital ecosystem strongly imposed itself after the European Union launched a series of projects and a program cancelling the disparity between the performances of small and medium enterprises of Europe and of the United States. These ecosystems consisted of "digital species" occupying a "digital environment". The digital species can be software programs and components, applications, services, knowledge, business models, learning modules, conceptual framework, architectures and legislation.

The digital ecosystem is a methodological proposal of economical and technological innovation containing a program that addresses a large number of economic users and services found in interaction. Economic users interact between one another, they develop and adapt to a dynamic digital environment thus serving the economic requirements in a continuous change imposed by economy.

2 Biology of Digital Ecosystems: Profiles for Adapting

Companies are developed within a business ecosystem, which assumes the necessity of appearance of a platform for administering such ecosystem.

The ecosystem consists both of an environment and of a set interacting agents or entities, which reproduce in that environment. The agents' needs modify from one place and from one moment to another, and from this point of view the profile for adapting a digital ecosystem is complex and dynamic, resembling the genetic algorithm of a biological ecosystem.

The success of a digital business ecosystem is determined by three factors, which are productivity, as fundamental factor; robustness, for the purpose of accumulation of competitive advantages from various sources and the ability to transform when the environment changes; possibility to create new niches and opportunities for the new companies.

The companies have multiple relations with their partners, which can be direct or indirect, formal or informal. Relations can be:

- vertical: direct complementary relationship, cooperation, subcontracting;
- horizontal: possibility of substitution, competition, but also cooperation;

- transversal: common goal, agreements among sectors;
- intangibility: common culture, behaviour standards, etc.

An important role within digital business ecosystems is occupied by the relations of trust, as any kind of economic transactions requires a certain level of trust between the parties involved. By trust it is understood a certain degree of safety, a fundamental aspect for small and medium enterprises operating within a complex regulation environment.

The relations of trust have a central role for digital business activities as any kind of economic transactions requires a certain level of trust between the parties involved in the relevant transaction. An important condition of the achievement of trust is a certain degree of "safety", fundamental aspect for small and medium enterprises operating in a complex regulation environment.

The rules of digital ecosystems are:

- leadership: One or more companies have the leading role. Thus the leader of the company have to elaborate a common vision of all members of business ecosystems and the leader guides the competences. The leadership could be developped and influence the evolution of business ecosystem;
- keystone and platform;
- common central competences;
- biodiversity: The actors of a business ecosystem are eterogen (companies, institutions, labour unions, groups with special interests). The actors of a business ecosystem come from different industrial branches. There is a convergence of industries. The same company can belong to more business ecosystems;
- competition: so at the intra-ecosystem (accesion to the role of leader), as at the inter-ecosystem (competition among business ecosystems), there is a strong dynamics of competition.

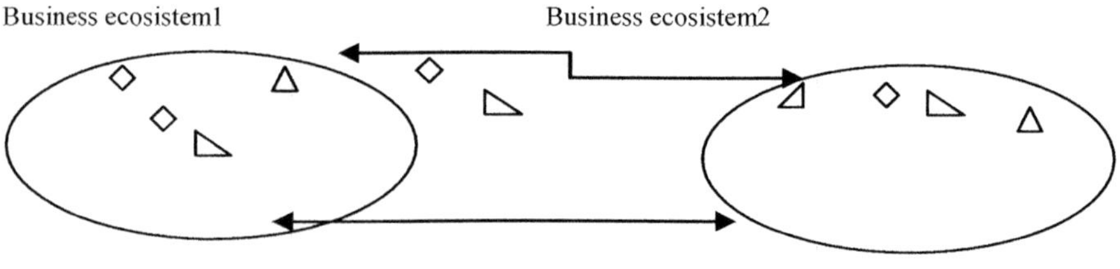

Fig. 1 Relationship between ecosystems

The achievement of self-organization within digital business ecosystems assumes the necessity of a superior capacity of reproduction of the components with a minimum level of intervention from the part of human agents. The objective of achievement of self-organization within digital business ecosystems suggests the necessity of a superior capacity of reproduction of the components with a minimum level of intervention from the part of human agents, thus increasing the number of difficulties for quality control.

The application of digital ecosystems can solve complex and dynamic problems. The medical business world is a dynamic and complex environment generating local, general, human and social interactions. The information technology has a significant role in the elaboration of systems that should solve such dynamic and complex problems.

The major advantage that the digital ecosystems present compared to other methods is represented by the capacity of dynamic adaptive self-organization. For the solutions generated by digital ecosystems to be useful they should be efficient from the IT point of view and should solve the relevant problems. Thus, the digital ecosystem is useful in the business environment if it succeeds to achieve a balance between the freedom of self-organization of the system and the capacity to find useful solutions, taking into account its internal dynamics and it should satisfy the adaptation criteria imposed by users.

In designing a useful digital ecosystem they shall also take into calculation the error possibilities, they shall make attempts to adapt the possibilities of stability and diversity. In order to avoid the appearance of problems in the system they shall also take into calculation the introduction of negative feedback mechanisms.

The performance indicator of a digital ecosystem is given by users' satisfaction by the achievement of measurable, adaptive solutions of complex dynamic problems. At digital ecosystems, the diversity should be put in balance with the efficiency of adaptation. The system should react efficiently at environment changes adapting rapidly so that it would avoid sudden modifications that can lead to losing control on the situation occurred.

Regarding digital ecosystems, diversity have to be balanced with the eficiency of adaptation. The system have to react eficiently to environment changes, adapting rapidly so that to avoid the sudden changes that could lead to the loss of control on the appeared situation.

The ecosystem addresses a large number of economic users and services found in interaction and operates in two steps:

- a decentralized fundamental network of services provided, with relations from peer to peer.
- structure operating locally (habitat) and being aimed at identifying solutions that would satisfy the relevant local constraints.

Through the intermediary of this double process they obtain optimum levels of resolution of problems and constraints locally. This is a digital ecosystem where the autonomous mobile agents represent various services provided by the participating economic entities which carry out services in local habitats.

The diversification of activities and of services required forces agents to grow on a permanent basis in order to cope with such requirements. In order to cope with the permanent procedures, the pressures in the dynamics of economy, of social pressures and changes from the business environment many small and medium enterprises were established in economic entities and networks reflecting the structure of economic sectors within the base of users of the digital ecosystem.

3 Cognitive Architecture within Digital Ecosystems

The cognitive processes are determined by the preoccupation for generating new knowledge by individual knowledge labour, but especially collaborative labour, by using knowledge-based IT instruments, in a computerized intellectual environment and, sometimes, even an intelligent one, being populated with natural and artificial intelligent agents and systems. Under these conditions it becomes possible to elaborate a *global cognitive model,* which can integrate in a unitary explanatory structure the knowledge, methods and projects regarding the knowledge labour, the actual cognitive environment, just like the work groups and practice communities established in the intellectual healthcare environment.

Cognitive processes are distributed between the members of a social group in time so that the final result of a previous event can change the outcome of future events involving coordination between internal and external structures. The cognitive processes can be automated or controlled. The automated cognitive processes are represented by those unconscious, rapid processes carried out with no effort, while the controlled cognitive processes are represented by those conscious, slow processes involving effort and attention and being easily modifiable.

The development of the human society is achieved by knowledge and learning. Knowledge is dependent of the learning process. Information, knowledge and communications are at the base of scientific, economical, technological, social, cultural, medical processes/events etc. The limitative factor in development shall be more and more linked to knowledge and learning, to the individual's capacity to assimilate and develop new technologies, to use them in new fields of activity.

Cognitive architectures are frequently created to model human performance in multimodal multiple task situation. Cognitive systems are evaluation from many points of view: adaptive behaviour, dynamic behaviour, flexible behaviour, development, evolution, learning, knowledge integration, vast knowledge base, natural language, real-time performance, and brain realization. Two key design properties that underline the development of any cognitive architecture are memory and learning. Together, learning and memory form the rudimentary aspects of cognition on which higher-order functions and intelligent capabilities, such as deliberative reasoning, planning and self-regulation are built. Organization of memory depends on the knowledge representation schemes.

The notion of knowledge ecosystem could be presented from the definition of digital business ecosystem in which digital environment is replaced by an environment consisting of knowledge. For Magnan, F. s.a., Knowledge Ecosystem term is used for „describe a practice comunity what use collaborative aplications to build knowledge in a bottom-up way".

Table 1 Taxonomy of cognitive architecture

Cognitive architectures		
Symbolic	**Emergent**	**Hybrid**
Memory - Rules based memory - Grapf based memory	Memory - globalist - localist	Memory - localist distributed - symbolic connectionist
Learning - inductive - analytical	Learning - associative - competitive	Learning - bottom-up - top-down

The knowledge ecosystems represents complex adaptive systems having the shape of virtual networks interconnected by people, knowledge and technical means whereby the knowledge is created, organized, selected, sinthetised and distributed to the other entities/ systems of environment which need knowledge. Among knowledge ecosystems and these entities / systems a symbiothic relationship is created whereby the knowledge is created and recreated uninterrupting, in a feedback process of transformation from implicit knowledge in explicit knowledge and inverse, through mechanisms and flows with a dynamic character and being in a continous evolution and change. The networks within ecosystems and the other systems intermit and intercondition each other, in an intime process of permanent transfer and of recreating knowledge depending on new appeared data and information, on new appeared situation and on ways of solving them. Knowledge ecosystem aime to create, organize and operate with knowledge in an unified and dynamic way. Using the newest methods and technologies of knowledge management, sustained by methods and models based on agents, Knowledge Ecosystems come into prominence in more and more scientific and applied fields, especially in those which generate and distribute knowledge (scientific research, education, technologic innovation etc).

4 Knowledge Management

The knowledge management intends to explain key subjects like: organizational adaptation, survival and increase of competences under the conditions of a continuous evolution of the environment. Essentially, the knowledge management includes organizational processes, which would profit from the synergetic combinations between the avalanche of the data generated by the information technology and creativity and the innovative potential of human resources.

Knowledge management is accompanied today by project management and invention management, then by change management, risk management or strategic management.

The approach of knowledge management as a process leads to optimizing the creation, to the cooperation of the creators of knowledge and to the balance of the knowledge assets market. This process should be understood as a spiral of transformation of the tacit knowledge into an explicit one. Thus, the tacit knowledge represents a cognitive scheme of mental models, personalized and formal, formed of two components: technique/know-how and cognitive/beliefs. Representative for the tacit knowledge are the ideals, the intuition, the premonition and the inner feeling. The explicit knowledge is given by encrypted elements easily transmissible.

Knowledge management should be concentrated on the instruments of cooperation of the program on the change of knowledge, on team work and on portals of knowledge, so that it would organize in an efficient manner a large volume of information, it would filter the essential contents and it would gain access to the corporative knowledge.

In knowledge economy, the success of companies is more and more a relational success. The companies aime more and more to intensify cooperation with research laboratories, universities and inclusive other private companies. The recognition of an organization or of a country will be more and more caused by the ability and eficiency with which will know to access, to assimilate and to use information.

Many corporations implemented softwares of distributing knowledge and information among subsidiaries, stimulation of the innovation process being realised through training courses and continous education. The transfer of knowledge among the subsidiaries of the same companies records a major weight in the international transfer of knowledge.

The fight for the control of standards lead to create partnerships among companies. The complexity of technical specifications of nowadays, a lot of applications cause divergent evolution in innovation process. The control and enforcement of the own technical specifications is an important factor in the success of economic activity.

5 Intelligent Technologies

The cognitive models are based on individual cognitive processes moulded through neuronal networks, fuzzy systems, agent and multi-agent systems based on knowledge.

Such a methodology frequently used in the development of multi-agent systems is the knowledge-based methodology. A large part of the methods of intelligent multi-agent systems was developed starting from their approach in the methodology of knowledge.

A business model is created for describing the roles and relationships among a company, his customers, partners and suppliers. Thus in a business model are specified the roles of each entities, details of structures of taking decisions, also strategies of specific business.

At almost all levels of decision process, interaction among players is significant.

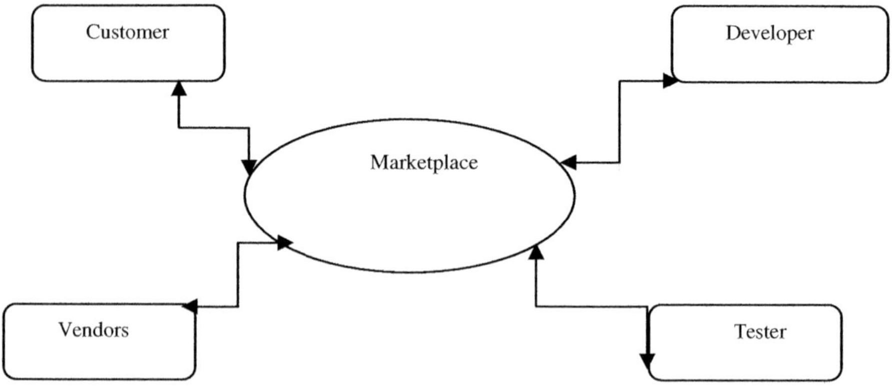

Fig. 2 Service of ecosystem

It is aimed:

- To create new global business environment, based on digitalization, virtual organization in business network;.
- To secure technologic and business information in the systems of economic cooperation, in the systems of public administration and systems of services;
- To build some innovation pole, to generate and to capitalize the knowledge, to build self-organized and self-multiplication mechanisms.

6 Trends and Strategy

Within actual business environment there are known some key elements that will influence the future of mankind. These are: the rise of population and demographic changes, the climatic changes, the mirage of energetic crise, globalisation, accelerate and exponential development of technology, etc.

We could say that the trends within business environment are determined by macroeconomic evolution, either microeconomic conjuncture. The most important macroeconomic trends of evolutions are demographic, social, technologic and business. One of the recent trends in business is collaboration of big companies with small and medium sized enterprises for realising some products or performe some services.

Small and medium sized enterprises meet a series of dificulties in participating or creating global digital markets. The key problem in creating some digital markets is to choose some specific strategies as strategic partnerships, agreements regarding sharing market segments like, for example, franchising and schemes similar with franchising.

As individual participants to digital markets, small sized enterprises can meet big dificulties because, in an initial phase of some kind of digital markets, the price transparency will raise more than transparency of product characteristics generally (quality, delivery term, services level and business knowledge).

7 Conclusions

Strategic involvement and managerial ingeniousness is necessary regarding the combination of computing facilities of intelligent assistance with consolidated organizational practices referring the innovation, learning, and cooperating interactivity. Being a developing field, the application of information and communication technology are away from the maxim level.

Using information and communication technology give chance to business of small and medium sized dimensions to compete at domestic or global level with big corporations.

Information and communication technology presents an important role in adopting the best practices regarding intelligent assistance of activities of learning, of innovating and of knowledge management within digital ecosystems.

References

[1] Bray, D.: Knowledge Ecosystems: Technology, Motivations, Processes, and Performance (online), Dissertation, Emory University (2008), http://papers.ssm.com/sol3/papers.cfm?abstract_id=1016486#PaperDownload

[2] DEBI Institute, Digital Ecosystems Simulation Prototype, [online], Curtin University of Technology, Australia (2007), http://www.debi.curtin.edu.au

[3] Dutta, S.: Strategies for implementing knowledge-based systems. IEEE Transactions in Engineering Management (1997)

[4] Eliasmith, C., Anderson, C.H.: Neural Engineering: Computation, Representation and Dynamics in Neurobiological Systems. MIT Press, Cambridge (2003)

[5] Holland, J.H.: Hidden Order. How Adaptation Builds Complexity. Perseus Books, Cambridge (1995)

[6] Mitleton-Kelly, E.: Organisation as Co-Evolving Complex Adaptive Systems. In: British Academy of Management Conference, London (1997)

[7] Nachira, F.: Towards a Network of Digital Business Ecosystems Fostering the LocalDevelopment (2002), http://www.europa.eu.int/information_society/topics/ebusiness/godigital/sme_research/doc/dbe_discussionpaper.pdf

[8] Nadel, L. (ed.): Encyclopedia of Cognitive Science. Nature Publishing Group, London (2003)

[9] Sun, R., Zhang, X., Mathews, R.: Modelling meta-cognition in a cognitive architecture. Cognitive Systems Research (2006)

[10] Duch, W., Oentaryo, R.J., Pasquier, M.: Cognitive Architectures: Where do we go from here? (2007)

[11] Wooldridge, M.: An Introduction to Multiagent System. Wiley, Chichester (2002)

Author Index

GPSR Compliance

The European Union's (EU) General Product Safety Regulation (GPSR) is a set of rules that requires consumer products to be safe and our obligations to ensure this.

If you have any concerns about our products, you can contact us on ProductSafety@springernature.com

In case Publisher is established outside the EU, the EU authorized representative is:

Springer Nature Customer Service Center GmbH
Europaplatz 3
69115 Heidelberg, Germany

Batch number: 10755047

Printed by Printforce, the Netherlands